AAPG TREATISE OF PETROLEUM GEOLOGY
REPRINT SERIES

The American Association of Petroleum Geologists
gratefully acknowledges and appreciates the leadership and support
of the AAPG Foundation in the development of the
Treatise of Petroleum Geology.

RESERVOIRS I

PROPERTIES

COMPILED BY
EDWARD A. BEAUMONT
AND
NORMAN H. FOSTER

TREATISE OF PETROLEUM GEOLOGY
REPRINT SERIES, NO. 3

PUBLISHED BY
THE AMERICAN ASSOCIATION OF PETROLEUM GEOLOGISTS
TULSA, OKLAHOMA 74101, U.S.A.

Library of Congress Cataloging-in-Publication Data

Reservoirs.

    (Treatise of petroleum geology reprint series ; no. 3)
    Bibliography: v. 1, p.
    Contents: 1. Properties.
    1. Petroleum--Geology.   2. Gas, Natural--Geology.
I. Foster, Norman H.   II. Beaumont, E. A. (Edward A.)
III. American Association of Petroleum Geologists.
TN870.5.R43   1987   553.2 '82   87-19599
ISBN 0-89181-402-7 (v. 1)

Treatise of Petroleum Geology
Advisory Board

TABLE OF CONTENTS

RESERVOIRS I
PROPERTIES

GENERAL

PORE GEOMETRY, PERMEABILITY AND FLUID INTERACTION

PRESSURES

FRACTURES

PRODUCTION GEOLOGY

TABLE OF CONTENTS

RESERVOIRS II
SANDSTONES

TABLE OF CONTENTS

RESERVOIRS III
CARBONATES

INTRODUCTION

The reprint volumes (I) Reservoirs: Properties, (II) Reservoirs: Sandstones, and (III) Reservoirs: Carbonates, belong to a series that is part of the *Treatise of Petroleum Geology*. The *Treatise of Petroleum Geology* was born during a discussion we had at the Annual AAPG Meeting in San Antonio in 1984. When our discussions ended, we had decided to write a state-of-the-art textbook in petroleum geology, directed not at the student, but at the practicing petroleum geologist. The project to put together one textbook gradually evolved into a series of three different publications: the Reprint Series, the Atlas of Oil and Gas Fields, and the Handbook of Petroleum Geology; collectively these publications are known as the *Treatise of Petroleum Geology*. Using input from the Advisory Board of the Treatise, we designed this entire effort so that the set of publications will represent the state-of-the-art in petroleum exploration knowledge and application. The Reprint Series collects the most up-to-date, previously published literature; the Atlas is a collection of detailed field studies to illustrate all the various ways oil and gas are trapped; and the Handbook is a professional explorationist's guide to the latest knowledge in the various areas of petroleum geology and related fields.

Papers in the various volumes of the Reprint Series are meant to complement the chapters of the Handbook. Papers were selected mainly on the basis of their usefulness today in petroleum exploration and development. Many "classic papers" that led to our present state of knowledge are not included because of space limitations.

We divided the general topic of reservoirs into three separate volumes: (I) properties, (II) sandstones, and (III) carbonates. We grouped papers that discuss properties common to all reservoirs in the volume entitled *Reservoirs: Properties*. The volumes, *Reservoirs: Carbonates* and *Reservoirs: Sandstones* contain papers dealing specifically with those two general types of reservoirs. Originally, the subject of reservoirs was intended to be one volume, but the number of papers suggested for it became so large that it had to be divided into three separate volumes. Therefore, if you examine the content of the papers in all three volumes you will notice that there is much overlap in content.

Volume I, *Reservoirs: Properties*, comprises five parts: (1) General (papers dealing with general reservoir properties), (2) Pore geometry, permeability, and fluid interactions, (3) Pressures, (4) Fractures, and (5) Production geology. Papers in this volume generally describe properties or production techniques that are common to all reservoirs.

Volumes II and III, *Reservoirs: Sandstones* and *Reservoirs: Carbonates*, are each divided into three sections: (1) a section of general papers, (2) depositional environments, and (3) diagenesis.

Edward A. Beaumont
Tulsa, Oklahoma

Norman H. Foster
Denver, Colorado

GENERAL

BULLETIN OF THE AMERICAN ASSOCIATION OF PETROLEUM GEOLOGISTS
VOL. 34, NO. 5 (MAY, 1950), PP. 943-961, 13 FIGS.

INTRODUCTION TO PETROPHYSICS OF RESERVOIR ROCKS[1]

G. E. ARCHIE[2]
Houston, Texas

ABSTRACT

There is need for a term to express the physics of rocks. It should be related to petrology much as geophysics is related to geology. "Petrophysics" is suggested as the term pertaining to the physics of particular rock types, whereas geophysics pertains to the physics of larger rock systems composing the earth.

The petrophysics of reservoir rocks is discussed here. This subject is a study of the physical properties of rock which are related to the pore and fluid distrubution. Over the past few years considerable study has been made of rock properties, such as porosity, permeability, capillary pressure, hydrocarbon saturation, fluid properties, electrical resistivity, self- or natural-potential, and radioactivity of different types of rocks. These properties have been investigated separately and in relation, one to another, particularly as they pertain to the detection and evaluation of hydrocarbon-bearing layers.

GENERAL

This paper is concerned with rocks and their fluids *in situ*, particularly for the detection and evaluation of hydrocarbon deposits penetrated by a bore hole. Fundamentally, therefore, the study is one of pore size distribution and fluid distribution of each phase (oil, gas, water) within the pores of the rock.

The subject must not be limited to permeable rocks containing hydrocarbons, but should include the impermeable layers and permeable layers containing water as well. This must be done in order to distinguish between them.

DISCUSSION

Rocks are heterogeneous. Therefore, the pore-size distribution as well as the fluid distribution within the pores may be complicated, particularly from the microscopic point of view. We are dealing with a heterogeneous material, together with a great many varying conditions within this heterogeneous material. This, no doubt, is the reason for early belief that a quantitative approach to the problem might never be attained.

When rocks are studied from a macroscopic viewpoint, however, a definite continuity is found. A correlation of rock properties has resulted in the discovery that definite relations or trends exist between rock characteristics. If pieces of rock representing each transition phase of a formation are studied, definite trends are noted. No matter how thoroughly a single piece of rock is studied, even on a microscopic scale, it is not possible to predict the properties of a formation as a whole. This should not be taken to mean that fundamental research on a microscopic scale is not of great importance in the study of rock porosity.

Though permeable rocks are, by nature, heterogeneous, their characteristics

---

[1] Presented before Houston Geological Society, September 12, 1949. Manuscript received, October 15, 1949.

[2] Shell Oil Company. The writer expresses his appreciation to the Shell Oil Company for permission to publish this paper and to the many authors having published information on the subjects covered here. An extensive bibliography would be needed to refer the reader to all pertinent articles.

follow definite trends when considering a formation as a whole. Relations between the basic rock pore properties may be indicated somewhat as follows.

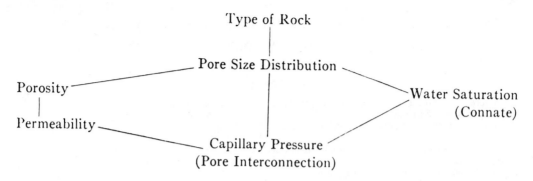

Type of rock, as here referred to, is a formation whose parts have been deposited under similar conditions and have undergone similar processes of later weathering, cementation or re-solution, as, for example, the upper Eocene Wilcox, lower Frio, Woodbine, or Bartlesville in a particular area.

The connecting lines are meant to portray the fact that a specific formation or rock type will have certain effective pore-size distributions which will produce a particular family of capillary pressure curves. The pore-size distribution controls the porosity and is related to the permeability and water saturation. Further, a certain rock will exhibit a relation between porosity and permeability.

ROCK TYPE-POROSITY-PERMEABILITY RELATION

A broad relationship exists between porosity and permeability of a formation. Figure 1 shows a plot of the measured values of these properties of cores from the upper Eocene Wilcox sandstone at Mercy, Texas, and the Nacatoch sandstone at Bellevue, Louisiana. The scattering is great, but it must be remembered that the only reason a trend exists at all is that the formation as a whole was deposited under a similar environment; individual parts (local environment) may differ from the whole. The trends shown in Figure 1 may be represented by a line. The average trend for different formations is shown in Figure 2. Note the paralleling of trends of different formations. (Only limited data are available in the high permeability range; therefore the lines have not been extended beyond the values shown.) Rocks indicated on the left, those having low porosity for a certain permeability, have relatively large pores, for example, the oölitic limestone. Those on the right have a high porosity for the permeability, indicating a smaller pore size, for example, the poorly sorted shaly sands of the Gulf Coast and the poorly sorted, shaly, calcareous Nacatoch sandstone.

It is interesting to find that the increase in permeability with increase in porosity is of the same general order for many of the formations. An increase in porosity of about 3 per cent produces a ten-fold increase in permeability. This is striking in view of the fact that the formations are widely different. For example,

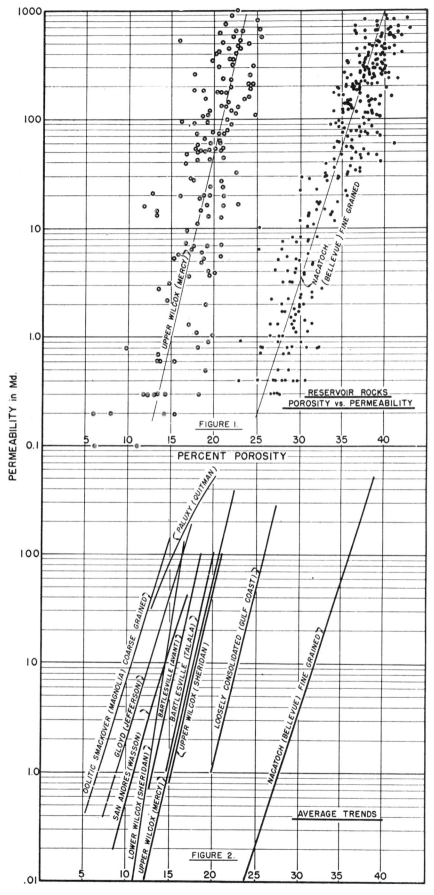

FIGS. 1 and 2.—Average relation between porosity and permeability for different formations.

3

the Smackover limestone, on the average, has a permeability of 1 millidarcy when the total porosity is only 7 per cent; the Nacatoch sandstone for the same permeability has an average total porosity of about 28 per cent.

It will be noted that some of the trends are not essentially parallel with the others: (1) Paluxy sandstone is believed to be of dune origin and has grains of uniform size; as a result, the average pore size is larger than for water-deposited sandstones and the permeabilities are higher, particularly in the lower range of porosity; (2) lower Wilcox has probably undergone considerable change since deposition (cementation of the fine pores and solution causing larger pores) and has a steeper trend than the other formations.

It appears that as diagenesis continues, the porosity-permeability trend moves toward the left. Compare the loose Gulf Coast sand trend with the harder sandstones where the smaller pores are absent and the larger pores are possibly enlarged.

Figure 2 indicates that an approximation of the probable permeability can be made when the porosity of a formation is known. It must be remembered, however, that the relation is an average of a large number of data from a formation and, when applying the relation to the same formation at a different locality, the assumption is made that the rock structure is similar.

If the type of void structure throughout the formation is similar, the relation is, of course, more significant. However, even at the same location, certain parts of a formation may deviate considerably from the average trend because the pore structure or type of rock changes appreciably. An example of this is shown in Table I. This small interval is purposely chosen from several hundred feet of Eocene Wilcox formation.

TABLE I

Depth	Permeability	Porosity
10,450–51	1.0	14.1
10,451–52	1.0	17.1
10,452–53	0.5	15.1
10,453–54	0.8	15.0
10,454–55	2.3	16.3
10,455–56	1.1	16.1
10,456–57	3.4	16.0
10,457–58	0.1	11.2
10.458–60	72.6	14.6
10,460–61	82.0	14.9
10,461–62	45.2	12.9
10,462–64	140.0	14.6
10,464–67	11.4	10.4

The interval 10,458–64 feet has a porosity of about 14.5 per cent and a permeability of about 100 millidarcys, whereas the interval immediately above, from 10,450–58 feet, has a porosity of about 16 per cent and only 1 millidarcy permeability. The former interval is clean and well sorted and therefore not typical of the lower Wilcox. Therefore, this relation of porosity versus permeability is only valid in a general way for a formation as a whole, and may not apply to a small integral part. The average data shown in Figure 2 does, however, help one visualize relative pore size of different types of formations.

ROCK TYPE-CAPILLARY PRESSURE-CONNATE WATER-PERMEABILITY RELATION

Air permeability of a dry rock sample is a measure of the average contributing effect of pores of all sizes. This average is not sufficient for complete analysis because the pore size must be considered in order to obtain permeability *"in situ"* (relative permeability) and fluid distribution. Capillary pressures of rocks help to analyze this average figure by obtaining what may be called effective pore-size distribution.

Capillary pressures of rocks have been discussed in the literature regarding connate water, and recently the relation between permeability and capillary pressure has been presented.[3] In order to tie the captioned relation more closely to rock type, capillary-pressure curves of different types of permeable formations are presented; the pore structures are due to a wide variety of geologic processes, that is, (a) sedimentation with little alteration, (b) alteration by solution, (c) redeposition or cementation. A comparison of capillary-pressure curves for different rock types and for the same type with varying permeabilities may then be made.

The following types of permeable formations are presented.

1. Friable sandstone, having high permeability and well sorted grains (Pennsylvanian sand, Healdton, Oklahoma)
2. Friable sandstone, poorly sorted grains, grading to shaly sandstone in the low-permeability range; partly cemented (upper Eocene Wilcox formation, Mercy, Texas)
3. Friable sandstone, poorly sorted grains, shaly and calcareous; comparatively high porosity for permeability (Nacatoch formation, Bellevue, Louisiana)
4. Hard sandstone, heavily cemented and considerable resolution; comparatively low porosity for permeability (lower Eocene Wilcox formation, Sheridan, Texas)
5. Limestone, crystalline texture, rock material of original deposition, porosity being of secondary nature, consisting of small interconnected vugs due to solution (San Andres limestone, West Texas)
6. Limestone, finely granular to earthy texture, siliceous; comparatively high porosity for permeabilities (Devonian cherty limestone, West Texas)

In types 1, 2, and 3, the pore size, shape of pore, and interconnection of the pores are controlled mainly by the original deposition, altered little by later cementation or solution. Therefore, pore structure is due mainly to the manner in which the fragments were deposited, that is, the sorting action and packing of the grains due to wave action and later compaction. The pore structure in type 4 is due mainly to later cementation and solution; in other words, the pore structure originally due to sedimentation has been altered considerably. The effective pore structure in type 5 is due almost entirely to solution with some redeposition. The origin of 6 is controversial.

Families of capillary pressure curves for each of these types are shown in

---

[3] W. R. Purcell, "Capillary Pressures: Their Measurement Using Mercury and the Calculation of Permeability Therefrom," *Petrol. Tech.*, 1, 39 (1949). *T. P. 2544.*

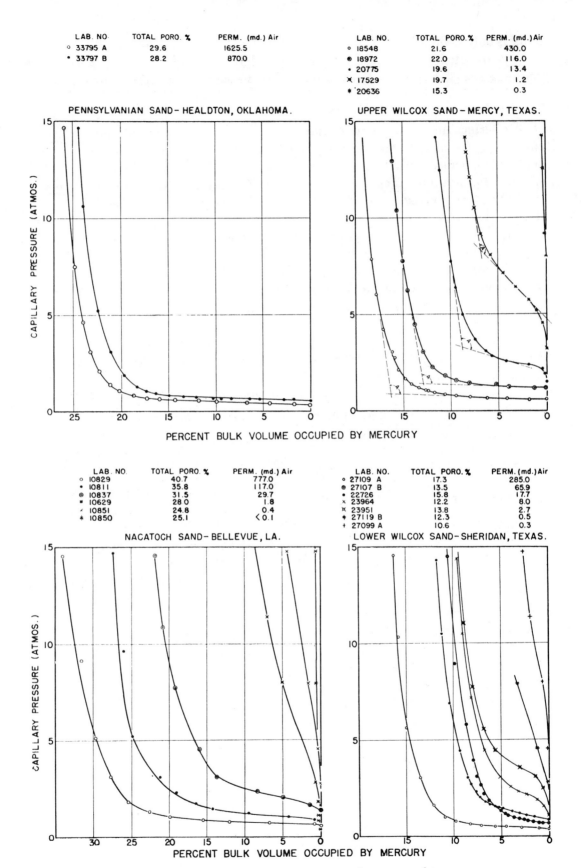

LAB. NO.	TOTAL PORO. %	PERM. (md.) Air
○ 33795 A	29.6	1625.5
• 33797 B	28.2	870.0

LAB. NO.	TOTAL PORO. %	PERM. (md.) Air
○ 18548	21.6	430.0
◉ 18972	22.0	116.0
• 20775	19.6	13.4
✕ 17529	19.7	1.2
✴ 20636	15.3	0.3

PENNSYLVANIAN SAND – HEALDTON, OKLAHOMA.

UPPER WILCOX SAND – MERCY, TEXAS.

PERCENT BULK VOLUME OCCUPIED BY MERCURY

LAB. NO.	TOTAL PORO. %	PERM. (md.) Air
○ 10829	40.7	777.0
• 10811	35.8	117.0
◉ 10837	31.5	29.7
✳ 10629	28.0	1.8
⟋ 10851	24.8	0.4
✳ 10850	25.1	< 0.1

LAB. NO.	TOTAL PORO. %	PERM. (md.) Air
○ 27109 A	17.3	285.0
◉ 27107 B	13.5	65.9
• 22726	15.8	17.7
✕ 23964	12.2	8.0
⋈ 23951	13.8	2.7
✦ 27119 B	12.3	0.5
+ 27099 A	10.6	0.3

NACATOCH SAND – BELLEVUE, LA.

LOWER WILCOX SAND – SHERIDAN, TEXAS.

PERCENT BULK VOLUME OCCUPIED BY MERCURY

FIG. 3.—Families of capillary pressure curves for some sandstones.

6

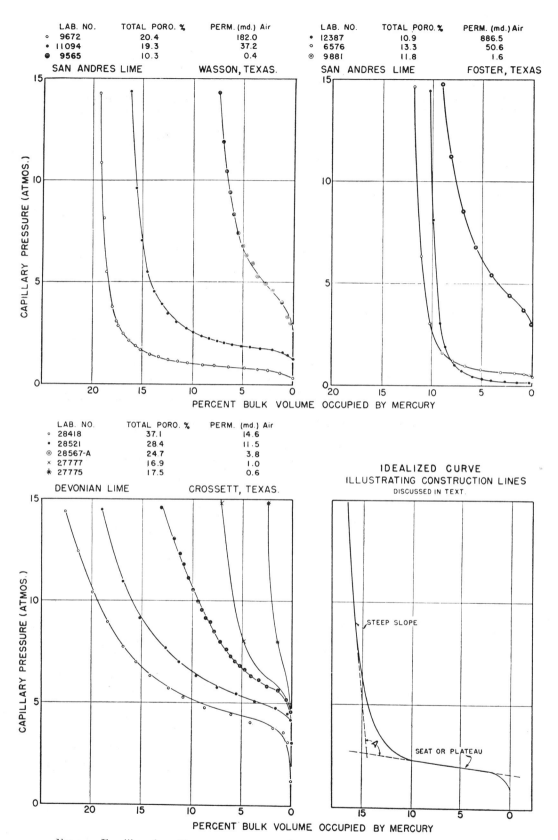

LAB. NO.	TOTAL PORO. %	PERM. (md.) Air
○ 9672	20.4	182.0
• 11094	19.3	37.2
◉ 9565	10.3	0.4

SAN ANDRES LIME   WASSON, TEXAS.

LAB. NO.	TOTAL PORO. %	PERM. (md.) Air
• 12387	10.9	886.5
○ 6576	13.3	50.6
◎ 9881	11.8	1.6

SAN ANDRES LIME   FOSTER, TEXAS

LAB. NO.	TOTAL PORO. %	PERM. (md.) Air
○ 28418	37.1	14.6
• 28521	28.4	11.5
◎ 28567-A	24.7	3.8
× 27777	16.9	1.0
✳ 27775	17.5	0.6

DEVONIAN LIME   CROSSETT, TEXAS.

IDEALIZED CURVE
ILLUSTRATING CONSTRUCTION LINES
DISCUSSED IN TEXT.

FIG. 4.—Families of capillary pressure curves for some limestones; also idealized curve.

Figures 3 and 4. These curves depict results obtained by injecting mercury into the cores. In order to show porosity in the graphs, the abscissae are "per cent bulk volume occupied by mercury." For example, it may be noted in Figure 3 that for sample No. 33795A, permeability 1625 millidarcys, the curve, with increasing pressure, approaches the abscissa 27.7, its effective porosity. These charts representing data from suites of cores show how the capillary-pressure curve, permeability, and porosity are related.

Certain general conclusions can be drawn by comparing the graphs. It can be seen that all of the samples having appreciable permeability exhibit a plateau, or seat, and a steep slope (idealized curve, Fig. 4). Examination of the capillary pressure curves will reveal that two straight lines (representing plateau and steep slope) can be drawn in and define the curves fairly well. The angle formed by extending these characteristic lines is useful. This angle "A" increases as the permeability of each type of rock decreases until one line appears and the plateau disappears.

The families of curves exhibit a striking similarity for the same permeabilities, regardless of type of formation or origin of porosity. Of course there are differences; for example, the rocks with high permeability for porosity exhibit much steeper steep slope (for example, San Andres limestone) because of a less amount of small pore space. Formations with comparatively low permeability for porosity, however, exhibit a more gentle steep slope because of the many small pores (Nacatoch and Devonian limestone). The curves of the Devonian limestone, which approaches chalk in texture, differ most from the others. It has a proportionally large amount of fine to very fine pores.

Photomicrographs of specimens of each type of rock investigated are presented in Plates I and II in order to show the visible difference in rock texture and pore structure. Notice the large pores in one of the San Andres limestone specimens, the medium pores in the sandstones and the very fine pores in the Devonian limestone and shaly sandstones. The rock textures include crystalline limestone, medium to very fine granular sandstones and very fine granular, almost chalky, Devonian limestone.

OTHER ROCK PROPERTIES RELATED TO PORE-SIZE DISTRIBUTION

If it were possible to measure the fundamental properties (exact pore size and fluid distribution) *in situ* of formations penetrated by the bore hole, the volume of the hydrocarbon in place and the productivity of the layer could be calculated. However, it is practically impossible as yet to get a direct measurement of the factors, porosity, permeability, hydrocarbon saturation, and thickness of the layer, in place, by coring or other physical measurements.

Complete recovery of cores can not be assured and all permeable cores recovered are invariably contaminated with the drilling fluid, or fluid conditions have changed because of pressure and temperature changes on bringing the core to the surface. Therefore, we must resort to indirect measurements, such as elec-

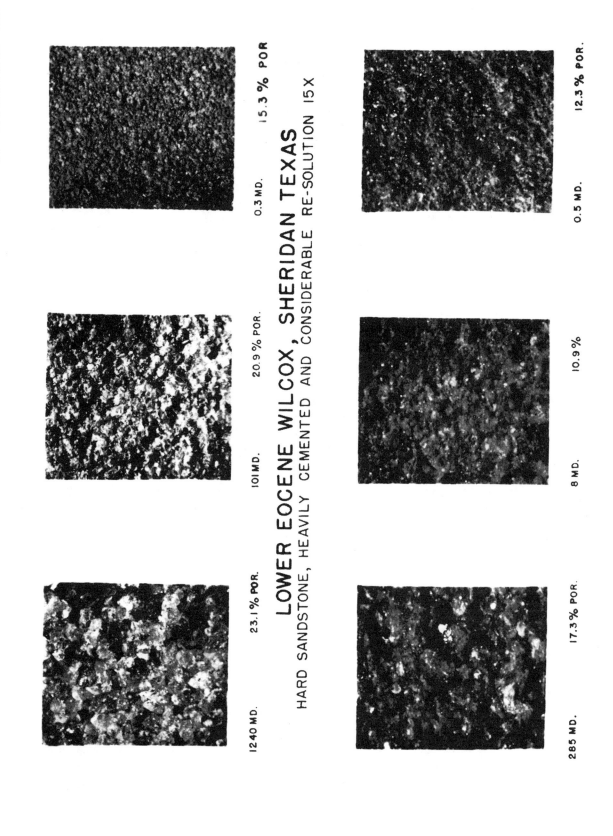

UPPER EOCENE WILCOX, MERCY TEXAS

FRIABLE SANDSTONE, POORLY SORTED GRAINS GRADING TO SHALY SANDSTONE 15X

0.3 MD.   15.3 % POR

20.9 % POR.

101 MD.

23.1 % POR.

1240 MD.

LOWER EOCENE WILCOX, SHERIDAN TEXAS

HARD SANDSTONE, HEAVILY CEMENTED AND CONSIDERABLE RE-SOLUTION 15X

0.5 MD.   12.3 % POR.

8 MD.   10.9 %

285 MD.   17.3 % POR.

9

NACATOCH SANDSTONE, BELLEVUE LA. 15X
HIGH POROSITY FOR GIVEN PERMEABILITY

1180 MD.   38.4% POR.   126 MD.   39.3% POR.   1.4 MD.   29.7% POR.

SAN ANDRES LIMESTONE, WEST TEXAS 10X

  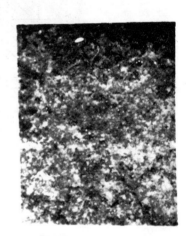

887 MD.   10.9% POR.   182 MD.   20.1% POR.   1.6 MD.   11.5% POR.

DEVONIAN CHERTY LIMESTONE, CROSSETT TEXAS 10X
COMPARATIVELY HIGH POROSITY FOR GIVEN PERMEABILITY

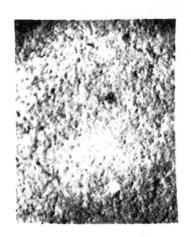 

16.4 MD   35.9% POR.   <0.1 MD.   14.6% POR

trical resistivity, self-potential, and neutron reaction, which can be recorded in a bore hole filled with mud. In order for these indirect measurements to be useful they must be directly related to the physical properties desired (porosity, permeability, and fluid saturation). Correlation of the indirect with the actual physical properties from a macroscopic point of view by actually testing numerous cores has led to the discovery that definite relationships or trends do exist.

ROCK TYPE-SELF POTENTIAL-GROUND WATER SALINITY-PERMEABILITY RELATION

Discussions in the literature have pointed out that the self potential is composed chiefly of two components:[4] (1) the flow potential, and (2) the chemical potential.[5] The flow potential is thought to be a smaller part of the total where

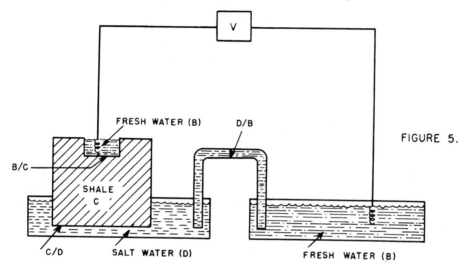

FIG. 5.—Measurement of shale potential.

the ground waters are very saline. Actually, the flow potential in a well can be measured by recording the S.P. with pressure on the well head.

The chemical potential has been expressed by the equation

$$\text{S.P.} = -K \log \frac{R_1}{R_2}$$

where K depends mainly on the type of impermeable rock, and the resistivity ratio is that of the mud filtrate R_1 and formation water R_2. The difference between the S.P. recorded opposite an impermeable shale, for instance, and that recorded opposite an infinitely permeable clean sandstone may be expressed by

[4] Schlumberger and Leonardon, "A New Contribuiton to Subsurface Studies by Means of Electrical Measurements in Drill Holes," *Trans. A.I.M.E.*, Vol. 110, Geophysical Prospecting (1934), p. 273.

[5] M. R. J. Wyllie, "A Quantitative Analysis of the Electrochemical Component of the S.P. Curve," *Jour. Petrol. Tech.*, Vol. 1, No. 1 (January, 1949), p. 17.

this equation. This difference can be measured in the laboratory by placing salt water at one end of a piece of shale and fresh water at the other end. Actually, in the laboratory it is difficult to measure the voltage thus generated by placing the electrodes in the salt and fresh water because of electrode potentials; therefore, the set-up shown in Figure 5 is used. It will be seen that each of the electrodes is now in the same solution, thus cancelling the electrode potential so bothersome in laboratory measurements. Note further that the potential recording at V is the "shale potential" plus the liquid contact potential (D/B). The recorded potential is the result of:

B/C fresh water-shale contact

C/D shale-salt water contact

D/B salt water-fresh water liquid junction contact.

Shale cores obtained from different formations have been tested by using different amounts of dissolved salts in the salt water. It appears that K in the fore-

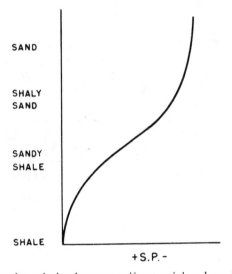

Fig. 6.—Qualitative relation between self potential and sand-shale section.

going equation takes on different values for different shales. For example, shales of the Eocene Wilcox have a K value of the order of 60, while that of some shallow Pennsylvanian shales may be as low as 25.

When the permeability of the formation is not high, however, the S.P. recorded opposite it in a bore hole is somewhat less. In the case of a sand and shale section, the relation between type of formation and self potential may be expressed by Figure 6. Actually, in some cases, the ordinate in Figure 6 may be replaced qualitatively, at least, with a permeability scale. For example, Figure 7-A shows the S.P. curve recorded opposite a section of the Eocene Wilcox formation. The interval 10,800 to 11,800 feet is a sandstone and shale section. The

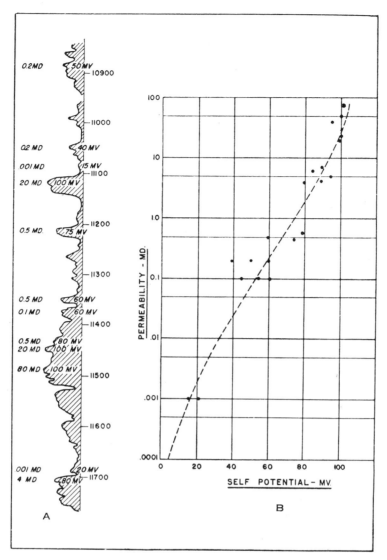

FIG. 7.—Relation between self potential and permeability of Eocene-Wilcox sandstone, Mercy, Texas.

sandstone is more or less the same type throughout, being poorly sorted, well cemented, and hard. Increasing amounts of argillaceous material are noted with decreasing permeabilities. The formation water is the same throughout this section, the mud in the bore hole at the time of the survey was uniform, and all the formations are water-bearing. The formation was extensively cored and the average permeability is indicated opposite the layers from which sufficient cores were recovered and analyzed.

The magnitude of the self-potential for the various layers is plotted *versus* the permeability in Figure 7-B. Note the apparent semi-logarithmic relation (similar to that shown in Figure 6).

It may be noted that there is considerable scattering of points and the relation between S.P. and permeability is qualitative in nature.

ROCK TYPE-ELECTRICAL RESISTIVITY-POROSITY-WATER SATURATION RELATION

The electrical resistivity of rocks when the pores are saturated with brine may be expressed by:[6]

$$R_0 = FR_w$$

where R_0 is the resistivity of the rock when saturated with brine (over about 10 grams per liter dissolved salts): F is the formation resistivity factor; and R_w is the resistivity of the brine. F is found to be related to the porosity and the type of rock.

$$F = f^{-m} \text{ or } R_0 = R_w f^{-m}$$

where f is the porosity of the rock and m is related to the type of rock.

Again it must be remembered that this equation represents a trend or an average line through a number of measured values (Fig. 8 and 9).

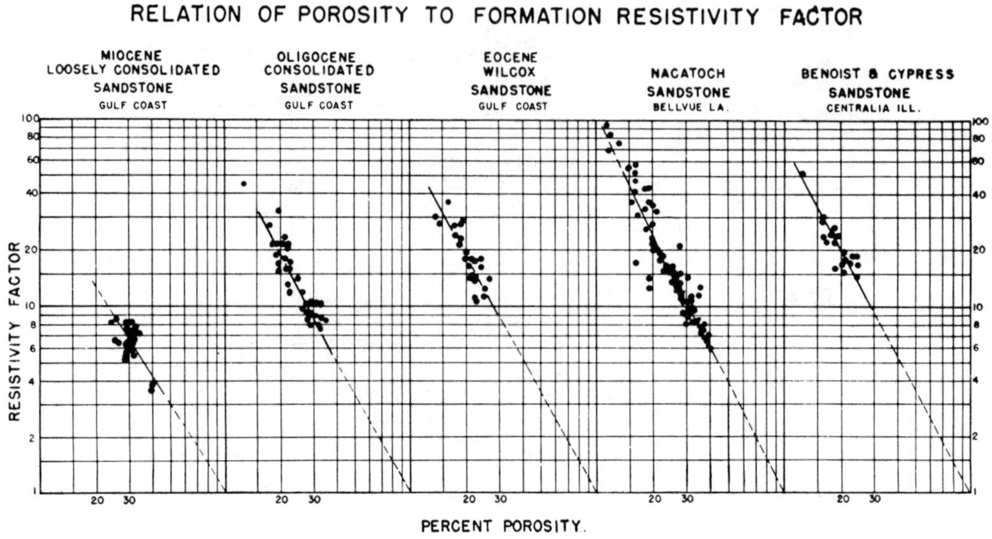

FIG. 8.—Resistivity factors of various sandstones *versus* porosity.

The electrical resistivity of a rock when hydrocarbon-bearing may be expressed by:

$$R = R_0 S^{-n}$$

where S is the fraction of the voids filled with brine; R is the resistivity of the hydrocarbon-bearing rock; and n depends, apparently, on the type of rock.

[6] G. E. Archie, "The Electrical Resistivity Log as an Aid in Determining some Reservoir Characteristics," *Trans. A.I.M.E.*, Vol. 146 (1942), p. 54.

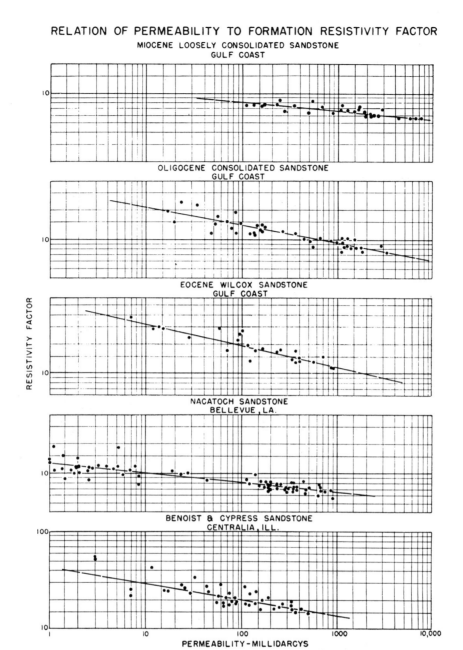

FIG. 9.—Resistivity factors *versus* permeability.

Laboratory results on artificially saturated loose sand packs indicate $n = 2$, (Figure 10-a), whereas measurements on consolidated sandstones naturally saturated (as withdrawn from the well) indicate $n = 1.9$ (Fig. 10-b). The fore-going equation holds for water saturations down to about 10 per cent. Below this value there is some indication that the interconnection between the water is no

15

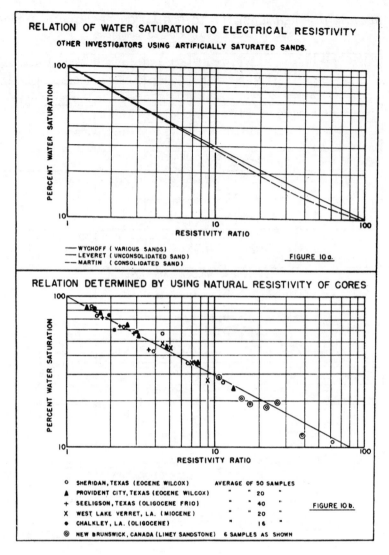

RELATION OF WATER SATURATION TO ELECTRICAL RESISTIVITY

OTHER INVESTIGATORS USING ARTIFICIALLY SATURATED SANDS.

— WYCHOFF (VARIOUS SANDS)
— LEVERET (UNCONSOLIDATED SAND)
-- MARTIN (CONSOLIDATED SAND)

FIGURE 10 a.

RELATION DETERMINED BY USING NATURAL RESISTIVITY OF CORES

		AVERAGE OF 50 SAMPLES	
O	SHERIDAN, TEXAS (EOCENE WILCOX)		
▲	PROVIDENT CITY, TEXAS (EOCENE WILCOX)	" " 20 "	
+	SEELIGSON, TEXAS (OLIGOCENE FRIO)	" " 40 "	FIGURE 10 b.
X	WEST LAKE VERRET, LA. (MIOCENE)	" " 20 "	
●	CHALKLEY, LA. (OLIGOCENE)	" " 16 "	
◎	NEW BRUNSWICK, CANADA (LIMEY SANDSTONE)	6 SAMPLES AS SHOWN	

FIG. 10.—Relation of water saturation to electrical resistivity.

longer uniformly continuous and the resistivity increases more rapidly than indicated by the equation.

Actually, therefore, it may be said that the electrical resistivity of a hydrocarbon-bearing rock depends on the porosity, brine saturation, salinity of the brine, and type of rock.

ROCK TYPE-NEUTRON REACTION-HYDROGEN CONTENT-POROSITY RELATION

Several papers have been written on this subject, and it appears that the reaction from neutron bombardment of formations depends to a large extent on their hydrogen content.[7] The effect of rock type is apparently not known. The porosity must be considered, for this determines the actual void space available to contain hydrogen-bearing fluids. The relative relation between total porosity and response to neutron bombardment, using present field methods, is illustrated in Figure 11. The limestone zones are oil-bearing. Each point represents an average of several feet where the porosity was relatively uniform. The relation would be different if these zones were dry gas-bearing, for the hydrogen density of dry gas in the pores is much lower.

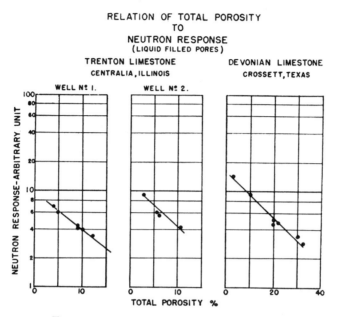

FIG. 11.—Response to neutron bombardment.

SUMMARY OF INDIRECT PROPERTIES

Figure 12 illustrates diagrammatically how the various indirect properties are related to the desired reservoir properties. Some relationships are rigid and quantitative, while others are not easily predicted and are considered only qualitative. After a qualitative relation is studied further, other factors may be discovered which, when incorporated in the equation, put the relationship on a more quantitative basis. For example, the qualitative relation between porosity and permeability is made more quantitative by introducing pore size or the capillary-pressure curve. The solid lines in the chart indicate what are now known to be quantitative relationships, while the broken lines indicate qualitative relationships.

Figure 12 shows how intricately the various properties are related; all relations

[7] R. E. Fearon, "Neutron Bombardment of Formations," *Oil Weekly*, Vol. 118, No. 2 (1945), p. 38; also "Nucleonics," Vol. 4, No. 6 (June, 1949), p. 30.

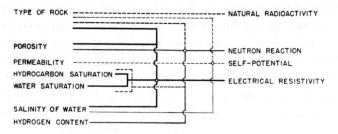

FIG. 12.

PETROPHYSICAL SYSTEM

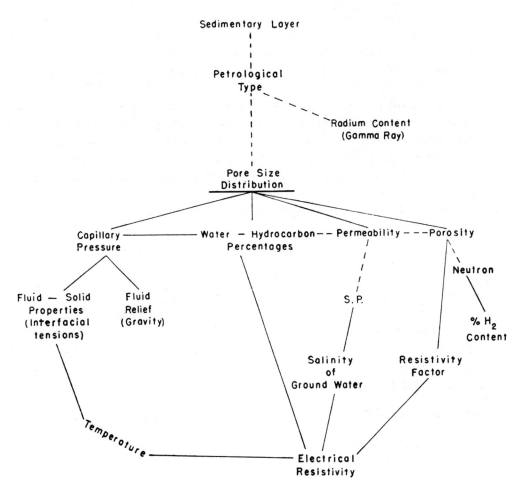

FIG. 13.

18

are tied to the type of rock. The diagram shows the many possibilities that can arise in attempting to unravel the interrelationships in order to evaluate a hydrocarbon accumulation *in situ*. In actual practice, further complications arise due to practical difficulties, economic considerations, and the personal equation. For example, the presence of the bore hole itself, its geometry, and the fact that the bore hole must be filled with mud in order to drill the hole brings up many problems.[8] The layers penetrated are not infinitely thick; therefore, boundary effects (thin layer effect) must be calculated and applied to all indirect measurements.[9] Also, mud filtrate contaminates the permeable layers near the bore hole and it too must be considered.[10] In fact, these practical complications are commonly the most difficult to interpret in attempting to detect and evaluate deposits penetrated in a bore hole.

CONCLUSIONS

In conclusion, a tentative petrophysical system, from the macroscopic viewpoint, is presented in Figure 13 for illustrative purposes. The system revolves mainly around pore-size distribution which defines the capillary-pressure curve, permeability, and porosity. The pore-size distribution does not necessarily define the type of rock, for actually several types of rock may have essentially the same pore-size distribution.

It is not meant that the mineral composition of the rock should be neglected in a study of these relationships. It must be recognized that the type of clay minerals present, for example, will no doubt play a greater role in future study.

It should also be mentioned that lithologic description of rocks is important. In fact, it should be broadened to express pore-size distribution as well as mineral distribution. It is felt that this is very important because drill-cuttings sample-logging (formation type) really is an integral part of this outline.

The relations between rock characteristics should be thought of as trends. Actually, these may be expressed by mathematical formulae; however, the formulae can not be applied in a rigid manner, as is done when considering the properties of homogeneous materials. It must be kept in mind that appreciable deviations from the average trend may occur. The less uniform the data, the less rigid will be the average relation, for some permeable rocks are more heterogeneous than others. The generally uniform types of permeable rocks are sandstones, oölitic limestones, and the so-called granular-appearing dolomites. The less uniform types are the so-called vesicular, vugular, or cavernous, or even fractured limestones and dolomites.

[8] Schlumberger Company pamphlet, *Resistivity Departure Curves* (May, 1947).

[9] H. G. Doll, "The S.P. Log: Theoretical Analysis and Principles of Interpretation," *Trans. A.I.M.E.*, Petroleum Branch, Vol. 179, p. 146.

[10] Schlumberger Company pamphlet, *op. cit.*

Reproduced with permission of the Geological Association of Canada. Published as general introduction in *Facies Models*, R.G. Walker, editor, Geoscience Canada, Reprint Series No. 1, 1979, 1984.

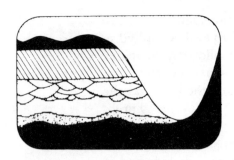

General Introduction: Facies, Facies Sequences and Facies Models

ROGER G. WALKER
Department of Geology
McMaster University
Hamilton, Ontario L8S 4M1

INTRODUCTION

In this paper, I will comment briefly on three concepts – facies, facies sequence and facies models. The intent is to simplify and de-mystify, and hence return some meaning to those misused terms. "facies", and "model". The first part of the bibliography, "basic sources of information", lists with annotations the major texts and monographs on sedimentary environments and facies.

FACIES

The term "facies" was introduced into geology by Nicholaus Steno (1669). It meant the entire aspect of a part of the earth's surface during a certain interval of geological time (Teichert, 1958). The word itself is derived from the latin *facia* or *facies*, implying the external appearance, or look of something. The modern usage was introduced by Gressly (1838), who used the term to imply the sum total of the lithological and paleontological aspects of a stratigraphic unit. Translations of Gressly's extended definition are given by Teichert (1958) and Middleton (1978).

Unfortunately, the term has been used in many different ways since 1838. In particular, arguments have focussed on: 1) whether the term implies an abstract set of characteristics, as opposed to the rock body itself; 2) whether the term should refer only to "areally restricted parts of a designated stratigraphic unit" (Moore, 1949), or also to stratigraphically unconfined rock bodies (as originally used by Gressly and other European workers); and 3) whether the term should be purely descriptive (e.g., "black mudstone facies") or also interpretive (e.g., "fluvial facies").

Succinct discussions of these problems have been given by Middleton (1978) and Reading (1978) – I will use the term in a concrete sense rather than abstractly implying only a set of characteristics, and will use it in a stratigraphically unconfined way. Middleton (1978) has also given the most useful modern working definition of the term, noting that:

> "the more common (modern) usage is exemplified by de Raaf *et al.* (1965) who subdivided a group of three formations into a cyclical repetition of a number of facies distinguished by "lithological, structural and organic aspects detectable in the field". The facies may be given informal designations ("Facies A", etc.) or brief descriptive designations (e.g., "laminated siltstone facies") and it is understood that they are units that will ultimately be given an environmental interpretation; but the facies definition is itself quite objective and based on the total field aspect of the rocks themselves... The key to the interpretation of facies is to combine observations made on their spatial relations and internal characteristics (lithology and sedimentary structures) with comparative information from other well-studied stratigraphic units, and particularly from studies of modern sedimentary environments."

DEFINING FACIES

Many problems concerning the interpretation of depositional environments can be handled without the formal definition of facies. Where the method is invaluable is in stratigraphic sequences where apparently similar facies are repeated many times over (de Raaf *et al.*, 1965; Cant and Walker, 1976).

Subdivision of a rock body into constituent facies (or units of similar *aspect*) is essentially a classification procedure, and the *degree* of subdivision must first and foremost be governed by the *objectives of the study*. If the objective is the routine description and interpretation of a particular stratigraphic unit, a fairly broad facies subdivision may suffice. However, if the objective is more detailed, perhaps the refinement of an existing facies model or the establishment of an entirely new model, then facies subdivision in the field will almost certainly be more detailed.

The *scale of subdivision* is dependent not only upon one's objectives, but on the time available, and the abundance of physical and biological structures in the rocks. A thick sequence of massive mudstones will be difficult to subdivide into facies, but a similar thickness of interbedded sandstones and shales (with abundant and varied examples of ripples, cross bedding and trace fossils) might be subdivisible into a large number of distinct facies. As a general rule, I would advocate erring on the side of oversubdividing in the field – facies can always be recombined in the laboratory, but a crude field subdivision cannot be refined in the lab.

Subdivision of a body of rock into facies ideally should not be attempted until one is thoroughly familiar with the rock body. Only then will it be apparent how much variability there is, and how many different facies must be defined to describe the unit. In the field, most facies studies have relied on distinctive combinations of sedimentary and organic structures (e.g., de Raaf *et al.*, 1965; Williams and Rust, 1969; Cant and Walker, 1976). Statistical methods can also be used to define facies, especially where there is considerable agreement among workers as to the important quantifiable, descriptive parameters. In carbonate rocks, percentages of different organic constituents, and percentages of micrite and/or sparry calcite have been used as input to cluster and factor analyses, with the resulting groupings of samples (in Q mode) being interpreted as facies (Imbrie and Purdy, 1962; Klovan, 1964; Harbaugh and Demirmen, 1964; see also Chapter 7 of the book by Harbaugh and Merriam, 1968, on Computer Applications in Stratigraphic Analysis - Classification Systems). Unfortunately, statistical methods are unsuited to clastic rocks, where most of the important information (sedimentary and biological structures) cannot readily be quantified. Readers unfamiliar with the process of subdividing rock bodies into facies

Figure 1

Cardium Formation, facies 1 massive dark mudstones (from Walker, 1983). For comparison with Figures 2, 3 and 4, note absence of silty or sandy laminae, and absence of recognizable burrow forms. Core from well 10-33-34-6W5, 7851 feet (2293.0 m), Caroline Field, Alberta. Scale in cm.

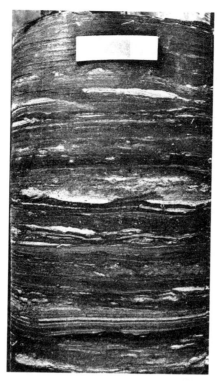

Figure 2

Cardium Formation, facies 2 laminated dark mudstones (from Walker, 1983). Note presence of sharp-based, delicately laminated silty layers (absent in Fig. 1), which are not pervasively bioturbated (compare with Fig. 3). Core from well 8-25-34-5W5, 2098.4 m, between Caroline and Garrington Fields, Alberta. Scale in cm.

Figure 3

Cardium Formation, facies 4 pervasively bioturbated muddy sandstones (from Walker, 1983). Note total bioturbation of silty and sandy layers (compare with Fig. 2), and presence of a few distinct burrow forms - these are better developed in Figure 4. Core from well 10-17-34-7W5, 8390 feet (2557.3 m), between Caroline and Ricinus Fields, Alberta. Scale in cm.

should consult the papers listed in the annotated bibliography, to see how the general principles briefly discussed here can be applied in practise. As one brief example, consider the mudstones and siltstones shown in Figures 1 to 4 from the Upper Cretaceous Cardium Formation of Alberta (Walker, 1983). If one's objective is a detailed study of the hydrocarbon-bearing Cardium sandstones, the examples in Figures 1 to 4 could probably be lumped together as "mudstone or siltstone". But there are clear descriptive differences, involving presence of silty laminations, degree of general bioturbation, and preservation of specific burrow forms. It has turned out that mudstones of Figure 1 only overlie the Cardium "B sand", and mudstones of Figure 2 only overlie the "A sand". Detailed facies subdivision thus happened to define two regional marker horizons (Walker, 1983), which lumping all the mudstones together would not have done.

FACIES SEQUENCE

It was pointed out by Middleton (1978) that "it is understood that (facies) will ultimately be given an environmental interpretation". However, many, if not most, facies defined in the field have ambiguous interpretations – a cross-bedded sandstone facies, for example, could be formed in a meandering or braided river, a tidal channel, an offshore area dominated by alongshore currents, or on an open shelf dominated by tidal currents. Many facies defined in

Figure 4 ▶

Cardium Formation, facies 5 bioturbated sandstones (from Walker, 1983). Note excellent development of burrow forms (compare with Figure 3), including prominent Z-shaped Zoophycos burrow, and small vertical tube at top (Conichnus conicus), with later burrowing by Chondrites (white circles/-ovals). Core from well 10-20-37-7W5, 2294.1 m, between Caroline and Garrington Fields, Alberta. Scale in cm.

the field may at first suggest no interpretation at all. The key to interpretation is to analyze all of the facies communally, in context. The sequence in which they occur thus contributes as much information as the facies themselves.

The relationship between depositional environments in space, and the resulting stratigraphic sequences developed through time as a result of transgressions and regressions, was first emphasized by Johannes Walther, in his Law of the Correlation of Facies (Walther, 1894, p. 979 — see Middleton, 1973). Walther stated that "it is a basic statement of far-reaching significance that only those facies and facies areas can be superimposed primarily which can be observed beside each other at the present time". Careful application of the law, therefore, suggests that in a vertical sequence, a *gradational* transition from one facies to another implies that the two facies represent environments that once were adjacent laterally. The dangers of applying the Law in a gross way to stratigraphic sequences with cyclic repetitions of facies have been emphasized by Middleton (1973, p. 983).

The importance of clearly defining gradational facies boundaries in vertical section as opposed to sharp or erosive boundaries, has been emphasized by de Raaf *et al.* (1965) and Reading (1978, p. 5). If boundaries are sharp or erosional, there is no way of knowing whether two vertically adjacent facies represent

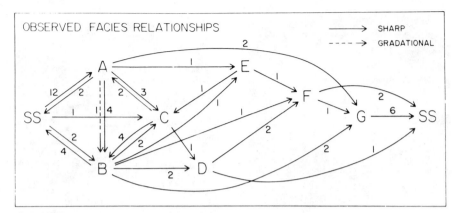

Figure 6
Facies relationship diagram for Battery Point section shown in Figure 8. Numbers indicate the observed number of facies transitions. From Cant and Walker, 1976.

environments that once were laterally adjacent. Indeed, sharp breaks between facies, especially if marked by thin bioturbated horizons implying non-deposition (Fig. 5), may signify fundamental changes in depositional environment and the beginnings of new cycles of sedimentation (see de Raaf *et al.*, 1965, and Walker and Harms, 1971, for examples of sharp facies relationships accompanied by bioturbation).

The first formal documentation of the quantitative relationships between facies was published by de Raaf *et al.* (1965; Fig. 5) in a diagram resembling the web of a demented spider. Note that sharp and gradational boundaries have been carefully distinguished. Note also that there are two "spurs" off the main trend of the web (black mudstone to oscillatory 1, and silty streak to sandy streak). These spurs imply that for the purposes of facies transitions, the facies at the end of the spur is completely contained within another facies (e.g., sandy streak within silty streak). This in turn suggests that facies were oversubdivided in the field, and that (for example) sandy streak is a subset of silty streak and could be combined with silty streak for interpretive purposes.

The spider's web is now termed a "facies relationship diagram" – examples are shown in Figures 5 and 6. As geologists have become more concerned with facies transitions, they have sought methods for simplifying the facies relationship diagram to remove the "noise". In essence, methods have involved converting the *numbers* of transitions (Figs. 5 and 6) to observed *probabilities* of transitions (see Walker, 1979, Fig. 2). The observed probabilities

are then compared with the probabilities that would apply if all the transitions between facies were *random*. It has been argued that those transitions which occur a lot more commonly than random must have some geological significance.

The problem is to derive a matrix of random probabilities. The method used by Selley (1970), Miall (1973), Cant and Walker (1976) and Walker (1979) is statistically incorrect. In the field, it is assumed that one cannot recognize a transition from one facies to itself. Consequently, a matrix of transition probabilities must have "structurally empty cells" along its main diagonal, where the transition from, say, facies A to facies A cannot be recognized in the field and therefore appears in the matrix as zero. However, Carr (1982) has pointed out that "zeros cannot result from a simple independent random process". Consequently, methods for deriving a random matrix based on absolute facies abundances (as explained by Walker, 1979, in the first edition of *Facies Models*) are incorrect.

There is not space here to explain the more complex methods of Markov chain analysis that must now be used, and the reader is referred to the work of Carr (1982) and Powers and Easterling (1982). Another problem of the "old" method, which involved substracting the random probabilities from observed probabilities, was that there was no way of evaluating the differences statistically. This aspect of facies analysis has been improved by Harper (1984) and is explained in "Improved Methods of Facies Sequence Analysis" (this volume). It applies to entries in the

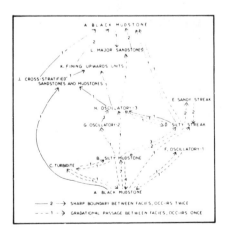

Figure 5
Facies relationship diagram for Carboniferous Abbotsham Formation, North Devon, England. Arrows show nature of transitions, and numbers indicate observed numbers of transitions. This is the first published facies relationship diagram. From de Raaf et al., 1965.

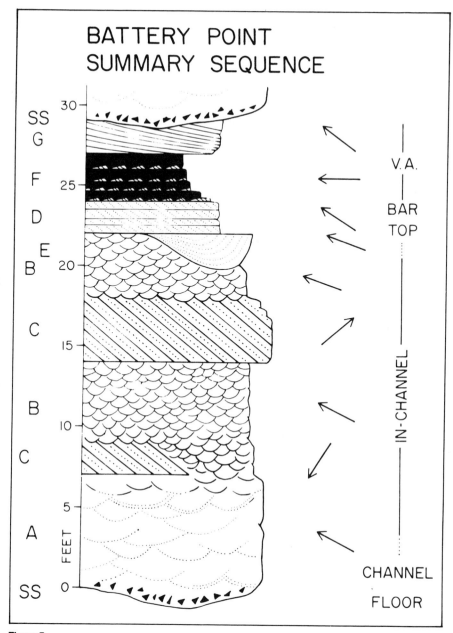

BATTERY POINT SUMMARY SEQUENCE

SS
G
F
D
E
B
C
B
C
A
SS

30
25
20
15
10
5
0

FEET

V.A.

BAR

TOP

IN-CHANNEL

CHANNEL

FLOOR

Figure 7
Summary facies sequence expressed as a vertical section. This has the advantage of visual appeal, and allows the facies to be *drawn to their observed average thickness. Battery Point Formation, Quebec. From Cant and Walker, 1976.*

observed-minus-random matrix that are different from zero, and assumes that a statistically valid random matrix has been derived.

Using Battery Point data from Cant and Walker (1976), Harper ("Improved Methods of Facies Sequence Analysis", this volume) has produced a set of facies transitions where the null hypothesis that the transitions occurred at random can be rejected at a given level of significance. For most of the transitions, that level of significance is less than 0.1 (Harper, this volume, Fig.

1); for E to F, and F to SS the level of significance must be set at 0.13 in order to reject the null hypothesis.

Harper's Figure 1 can be regarded as a simplified facies relationship diagram, or a "distillation" of the Battery Point data. Geologists are most accustomed to seeing transitions of this type expressed as a vertical stratigraphic sequence, and one version of the Battery Point data is shown in Figure 7. This is the original Cant and Walker (1976) version, and has *not* been corrected for the statistical problems dis-

cussed above. It should be compared with the raw Battery Point data (Figure 6) and with Harper's simplified facies relationship diagram (Figure 1 of "Improved Methods of Facies Sequence Analysis", this volume). Clearly, the transitions included in a "summary diagram" will depend on the arbitrarily set level of significance that one accepts. By gradually relaxing the level from, say, 0.1 to 0.2, one can attempt to evaluate the *geological* significance of the transitions judged to be different from random. The problems of statistical versus geological significance have been examined in the discussion of Selley's paper (1970, p. 575-581).

The columnar method of presenting the data shows not only the facies *sequence* but also the mean *thickness* of each facies (calculated from the raw data). This is one way in which data can be "distilled" into summary sequences, or "models", as discussed below.

It is now important to distinguish between a single facies sequence, and repeated sequences (or cycles). The summary sequence diagram in Figure 7, with the suggested basic interpretations, established the probably fluvial origin of the Battery Point Sandstone. The scoured surface SS can then be interpreted as the fundamental boundary between cycles, and hence individual cycles can be defined on the original complete stratigraphic section (Fig. 8). Using the summary stratigraphic sequence (Fig. 7) as an idealization of all of the Battery Point sequences (Fig. 8), each individual cycle can be compared with the summary to identify points in common and points of difference. The reader may do this with the sequences in Figures 7 and 8.

FACIES MODELS
The construction and use of facies models continues to be one of the most active areas in the general field of stratigraphy, as is demonstrated by several new books in the field (see bibliography). This emphasis is not new; many of the ideas were embodied in Dunbar and Rodgers' *Principles of Stratigraphy* in 1957, and were based on studies dating back to Gressly and Walther in the 19th Century (Middleton, 1973). Walther (1893, quoted by Middleton, 1973, p. 981) "explained that the most satisfying genetic explanations of ancient phenomena were by analogy with modern

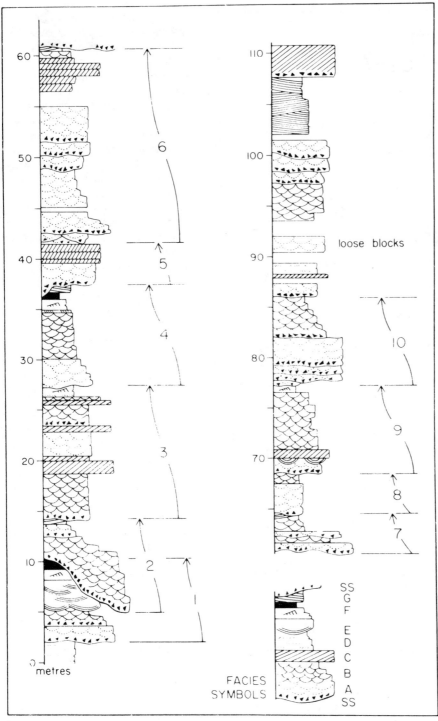

Figure 8

Measured section of the Lower Devonian Battery Point Sandstone near Gaspe.

Quebec. Numbers refer to individual channel-fill sequences. From Cant and Walker, 1976.

diagrams, and as graphs and equations. Examples of all of these are given in "Sandy Fluvial Deposits" (this volume). The term model here has a generality that goes beyond a single study of one formation. The final facies relationship diagram and its stratigraphic section (Fig. 7) are only local summaries, not general models for fluvial deposits. But when the Battery Point facies relationship diagram is compared and contrasted with the facies relationship diagrams from other ancient braided river deposits, and then data from modern braided rivers is incorporated (e.g., Cant, 1978), the points in common between all of these studies begin to assume a generality that can be termed a *model*.

A facies model could thus be defined as a general summary of a specific sedimentary environment, written in terms that make the summary useable in at least four different ways. The basis of the summary consists of many studies of both ancient rocks and recent sediments: the rapidly increasing data base is due at least partly to the large number of recent sediment studies in the last 20 years. The increased need for the models is due to the increasing amount of prediction that geologists are making from a limited local data base. This prediction may concern subsurface sandstone geometry in hydrocarbon reservoirs, the association of mineral deposits with specific sedimentary environments (for example, uraniferous conglomerates), or the movement of modern sand bars in shallow water (Bay of Fundy, tidal power). In all cases, a limited amount of local information plus the guidance of a well-understood facies model results in potentially important predictions about that local environment.

Our aim as geologists is partly to identify different environments in ancient rocks, and also to understand the range of processes that can operate within these environments. We must also be sure of why we want to identify environments in the first place. Is it to provide a name showing that we have thought about the origin of the unit we have mapped ("the Ordovician Cloridorme Formation consists of deep water turbidites"), or is it to provide a framework for further thought? It is the latter – the framework for further thought – that in my mind separates the

geological processes". The study of modern environments and processes was termed the "ontological method" by Walther, who observed that "only the ontological method can save us from stratigraphy" (Walther, in Middleton, 1973, p. 883). Facies models similarly link modern and ancient observations

into coherent syntheses, and their importance at the present time is due to an increasing need for the models, and a rapidly increasing data base on which the models are formulated.

In this volume, facies models are expressed in several different ways —as idealized sequences of facies, as block

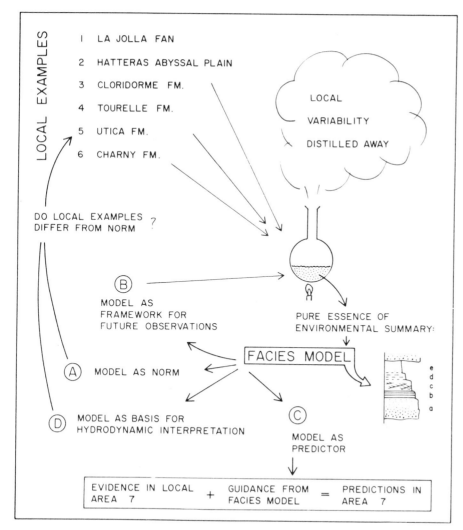

Figure 9

Distillation of a general facies model from various local examples, and its use as a NORM, FRAMEWORK for OBSERVATIONS, PREDICTOR, and BASIS for INTERPRETATION. See text for details.

art of recognizing environments from the art of FACIES ANALYSIS and FACIES MODELLING. The meaning and implication of these two terms will become apparent below.

FACIES MODELS – CONSTRUCTION AND USE

The principles, methods and motives of facies analysis are shown in Figure 9, using turbidites as an example. The principles, of course, apply to all environments. We begin by assuming that if enough modern turbidites can be studied in cores, and if enough ancient turbidites can be studied in the field, we may be able to make some *general* statements about turbidites, rather than statements about only one particular example.

The process of extracting the general information is shown diagramatically in

Figure 9, where numbers 1 and 2 represent recent sediment studies (cores from, say, La Jolla fan and Hatteras abyssal plain) and numbers 3 through 6 represent studies of ancient turbidites (for example, the Cloridorme and Tourelle Formations of Gaspe, the Utica Formation at Montmorency Falls, and the Charny Formation around Quebec City). The entire wealth of information on modern and ancient turbidites can then be distilled, boiling away the local details, but distilling and concentrating the important features that they have in common into a general summary of turbidites. If we distill enough individual turbidites, we can end up with a perfect "essence of turbidite" – now called the Bouma model. But what is the essence of any local example and what is its "noise"? Which aspects do we dismiss and which do we

extract and consider important? Answering these questions involves experience, judgment, knowledge and argument among sedimentologists, and the answers also involve the ultimate purpose of the environmental synthesis and summary. Some of the different methods for "distilling" the examples will become apparent in the papers in this volume. Facies relationship diagrams could be used if the same facies can be recognized in many different examples. Indeed, "standard" facies classifications have been proposed for turbidite (Mutti and Ricci Lucchi, 1972; Walker, 1978) and braided fluvial (Miall, 1977) environments. More commonly, models are still derived by qualitative comparison and contrast, rather than strict quantitative distillation.

I pointed out earlier that the difference between the summary of an environment and a facies model perhaps depends mainly on the use to which the summary is put. As well as being a summary, a FACIES MODEL must fulfill four other important functions:

1) it must act as a *norm*, for purposes of comparison;
2) it must act as a *framework* and *guide* for future observations;
3) it must act as a *predictor* in new geological situations; and
4) it must act as an integrated *basis for interpretation* of the environment or system that it represents.

Figure 9 has been constructed to illustrate these various functions. Using the example of the turbidite model, the numbers 1 through 6 indicate various local studies of modern and ancient turbidites. There is a constant feedback between examples – in this way the sedimentologist exercises his judgment in defining the features in common and identifying "local irregularities". This is the "distillation" process that allows the environmental summary (that will act as a facies model) to be set up.

Having constructed the facies model, it must act first as a norm (Fig. 9, A) with which individual examples can be compared. Without a norm, we are unable to say whether example 5 of Figure 9 contains any unusual features. In this example, Utica Formation turbidites at Montmorency Falls are very thin, silty, and many beds do not begin with division A of the Bouma model (Fig. 9); they begin with division B or C. Because of the existence of the norm (Bouma model), we can ask questions about example 5 that we could not otherwise have asked, and whole new avenues of productive thought can be opened up this way. Thus there is a constant feedback between a model and its individual examples – the more examples and the more distillation, the better the norm will be, and the more we must be forced into explaining local variations.

The second function of the facies model is to set up a framework for future observations (Fig. 9B). In as much as the model summarizes all the important descriptive features of the system, geologists know that similar information must be recorded when working with a new example. In Figure 9, this would include the detailed characteristics and thicknesses of the five Bouma divisions. Although the framework ensures that this information is recorded wherever possible, it can also act to blind the unwary, who might ignore some evidence because it is not clearly spelled out by the model. This leads to imprecise interpretations, and would cause a freeze on any further improvement of the facies model – hence the feedack arrow (Fig. 9B) implying that all future observations must in turn be distilled to better define the general model.

The third function of a model is to act as a predictor in new geological situations (Fig. 9C). This is hard to illustrate on the small scale of an individual Bouma bed, so let us imagine that we have a generalized facies model for automobiles – four wheels, hood, trunk, doors, etc. The new discovery of an *in situ* radiator by itself might be interesting, but without other information, one might be able to say little more than "nice radiator". With a general model, which ideally expresses the relationship of all the parts of the system, we should be able to predict the rest of the car from the discovery of a radiator. Or we might be able to predict other parts of a submarine fan from one thickening-upward prograding lobe sequence. This is obviously a vitally important aspect of facies modelling, and good surface or subsurface prediction from limited data can save unnecessary exploration guesswork and potentially vast sums of money.

The fourth major function of a facies model is to act as an integrated basis for interpretation (Fig. 9D). Again, it is important to eliminate "noise" before looking for a general interpretation, and hence, there should be feedback between the interpretation and the individual examples (Fig. 9D). This is indicated by the feedback arrow to example 5 (Fig. 9), implying the question "does the interpretation of example 5 differ from the idealized hydrodynamic interpretation?" If there is a difference (and there is), we can again ask questions that could not be asked if we had not used the facies model to formulate a general interpretation. This usage of the facies model is demonstrated particularly well by the Bouma sequence for turbidites, as discussed later in this volume.

The turbidite example of Figure 9 illustrates another point, namely that facies models can exist on different scales. The Bouma sequence for individual turbidite beds is a small scale example, but when turbidites are studied as groups of related beds, the system as a whole is referred to a large scale submarine fan model.

The turbidite/submarine fan example has been discussed above because it is reasonably well understood, and because it illustrates the four functions of a facies model (Fig. 9). Some of the other models discussed in this volume are less well understood – because the environmental summary is weaker, so the functioning of the model is weaker. I emphasize that the construction and functioning of facies models is essentially similar for all environments, and that the turbidite example was discussed above to make the general statements about facies models a little more specific.

Just as there can never be any absolute classification of depositional environments, so there will be differing numbers and types of facies models. As very large scale systems are studied in more detail (e.g., submarine fans), models for sub-components of the system may emerge, such as depositional suprafan lobes, or channel-levee complexes on fans. However, it is probably safest at the moment to emphasize and develop the generality of existing models, rather than encouraging the proliferation of more and more very restricted models. The reason for this suggestion is that given one piece of new information, such as an *in situ* radiator, one might make fairly safe generalizations about automobiles in general. But with many different types of automobile models, one may have problems about assigning the new data to the correct model (is the radiator a Chevrolet or Ford?), and hence run the risk of incorrect predictions. But ultimately, as our understanding improves, subdivision of broad models will be both possible and desirable, as in the case of braided and sandy fluvial models; river-, wave- and tide-dominated deltaic models; and storm- and tide-dominated shallow marine models.

BASIC SOURCES OF INFORMATION

This list is not intended to be complete, but highlights some of the more recent and more important books on depositional environment, facies and facies models. The list is roughly in the order of increasing scope and complexity of coverage of the subject, with Selley as a good place to start, and Reading as the most complete and detailed source.

Selley, R.C., 1970. Ancient sedimentary environments. Ithaca, N.Y., Cornell University Press, 237 p.
Selley introduces the volume as "not a work for the specialist sedimentologist, but an introductory survey for readers with a basic knowledge of geology". The book achieves this end very well – it summarizes, it leans on classical examples, and it very briefly indicates the economic implications (oil, gas, minerals) of some of the environments. This volume is a good place to start.

Blatt, H., Middleton, G.V., and Murray, R.C., 1980. Origin of sedimentary rocks, Second Edition. Englewood Cliffs, N.J., Prentice Hall, 782 p.
Chapter 19, on facies models has been greatly expanded in the second edition, and now summarizes concisely the general principles of facies and facies analysis, and reviews all important depositional environments.

Allen, J.R.L., 1970. Physical processes of sedimentation. New York, American Elsevier, 248 p.
Chapter 11 (p. 439-543) is a review of sand bodies and environments written at a fuller and more technical level than Selley (1970), or Blatt, Middleton and Murray (1980). It considers Alluvial, Deltaic, Estuarine, Tidal Flat, Beach and Barrier, Marine Shelf, Turbidite and Aeolian environments, with separate remarks on sand body prediction. Useful follow-up reading after Selley and Blatt, Middleton and Murray in that order.

Galloway, W.E., and Hobday, D.K., 1983. Terrigenous clastic depositional systems. New York, Springer Verlag, 423 p.
This new volume also covers most important depositional environments, but in more detail than the books listed above. It deliberately is slanted toward economic applications and hence gives a different perspective from all the other books in this list.

Rigby, J.K., and Hamblin, W.K., eds., 1972. Recognition of ancient sedimentary environments. Society of Economic Paleontologists and Mineralogists, Special Publication 16, 340 p.
Contains separate papers on many important environments written at a technical level. Many of the papers are disappointing as reviews but there are excellent contributions on Alluvial Fans, Fluvial Paleochannels, Barrier Coastlines and Shorelines. Most of the authors present their environmental summaries but do not attempt to use them as models.

Reineck, H.E., and Singh, I.B., 1973. Depositional sedimentary environments. New York, Springer Verlag, 439 p.
Pages 160-439 are devoted to summaries of many modern environments. Coverage is at the graduate student – professional sedimentologist level, but is patchy and rather uncritical. Vast reference lists are given, but it is hard to single out the very important papers from the trivial. The emphasis on modern environments is useful, but the book should not be used until one is at least somewhat familiar with specific environments.

Scholle, P.A., and Spearing, D.R., eds., 1982. Sandstone depositional environments. American Association of Petroleum Geologists, Memoir 31, 410 p.
This abundantly illustrated volume contains 12 papers reviewing major depositional environments. Most are good, some excellent, one or two are poor. There is no professed overall philosophy to the volume, hence the variability of the contributions. It was suggested in one review that one should read the text of *Facies Models* and use the pictures in Memoir 31.

Reading, H.G., ed., 1978. Sedimentary environments and facies. Oxford, Blackwell, 557 p.
Excellent compilation of data on depositional environments and facies models. An indispensable reference, and the best available summary of major depositional environments.

REFERENCES CITED

I have grouped these under three headings, and then listed the references alphabetically.

FACIES

Gressly, A., 1838. Observations geologiques sur le Jura Soleurois. Neue Denkschr, allg. schweiz, Ges. ges. Naturw., v. 2, 1-112.
Gressly's work first established the concept of facies in the geological literature.

Middleton, G.V., 1973, Johannes Walther's Law of the correlation of facies. Geological Society of America Bulletin, v. 84, p. 979-988.
An excellent discussion of the use, misuse and implications of Walther's Law.

Middleton, G.V., 1978. Facies. *In* Fairbridge, R.W., and Bourgeois, J., eds., Encyclopedia of sedimentology. Stroudsburg, Pa., Dowden, Hutchinson and Ross, p. 323-325.
One of the best and most concise statements of the facies concept, discussing the various ways in which the term has been used.

Moore, R.C., 1949. Meaning of facies. *In* Longwell, C.R., ed., Sedimentary facies in geological history. Geological Society of America, Memoir 39, p. 1-34.
This paper is from the first important North American volume on facies. It emphasizes the lateral variations of facies within a designated stratigraphic unit. Historically, an important paper, but now conceptually out of date (or out of fashion).

Teichert, C., 1958. Concepts of facies. Bulletin of the American Association of Petroleum Geologists, v. 42, p. 2718-2744.
This is probably the best single review of the facies concept. It examines the history of the concept, and its influence in Europe, Britain and North America, and Russia. Teichert's twelve conclusions could be modified slightly in light of the last 20 years work, particularly in recent sediments.

Walther, J., 1893-4. Einleitung in die Geologie als historische Wissenschaft. Verlag von Gustav Fischer, Jena, 3 vols., 1055 p.
See Middleton, 1973, for a commentary on the importance of Walther's work.

FACIES SEQUENCES, FACIES MODELS, EXAMPLES

Cant, D.J., 1978. Development of a facies model for sandy braided river sedimentation: comparison of the South Saskatchewan River and the Battery Point Formation. *In* Miall, A.D., ed., Fluvial sedimentology. Canadian Society of Petroleum Geologists, Memoir 5, p. 627-639.
A detailed comparison of ancient sediments and recent sediments, emphasizing facies comparisons and the construction of a facies model.

Cant, D.J., and Walker, R.G., 1976. Development of a braided fluvial facies model for the Devonian Battery Point Sandstone, Quebec. Canadian Journal of Earth Sciences, v. 13, p. 102-119.
Selley's difference matrix is used to help define fluvial cycles in a sandy braided system. This method is no longer statistically sound – see Carr (1982) and Powers and Easterling (1982).

Carr, T.R., 1982. Log-linear models, Markov chains and cyclic sedimentation. Journal of Sedimentary Petrology, v. 52, p. 905-912.
Explains problems of deriving a random matrix, and suggests improved methods for facies sequence analysis using Markov chain analysis.

de Raaf, J.F.M., Reading, H.G., and Walker, R.G., 1965. Cyclic sedimentation in the Lower Westphalian of North Devon, England. Sedimentology, v. 4, p. 1-52.
This paper gives the first published example of a facies relationship diagram and uses the diagram to establish cyclicity in a series of prograding shoreline sediments. Cycles are defined by black mudstones resting on bioturbated sandstones.

Harper, C.W., 1984. Facies models revisited: an improvement to the method advocated by Walker in the General Introduction to Geoscience Canada Reprint Series No. 1. Geoscience Canada, in press.
Gives details of the method used in the next paper in this volume.

Miall, A.D., 1973. Markov chain analysis applied to an ancient alluvial plain succession. Sedimentology, v. 20, p. 347-365.
An introduction to Markov chain methodology with an example from the Devonian Peel Sound Formation of Prince of Wales Island, Arctic Canada. See modifications by Carr (1982) and Powers and Easterling (1982).

Miall, A.D., 1977. A review of the braided river depositional environment. Earth Science Reviews, v. 13, p. 1-62.
Miall suggests a series of "universal" facies that could be used to describe gravelly and sandy braided systems, and shows facies relationship diagrams for several modern and ancient examples. Four "types of depositional profiles" are suggested – these make very useful reference points for future generalizations about fluvial models.

Mutti, E., and Ricci Lucchi, F., 1972. Le torbiditi dell'Appennino settentrionale: introduzione all'analisi di facies. Memorie della Societa Geologica Italiana, v. 11, p. 161-199. Translated into English by T.H. Nilsen (1978), Turbidites of the northern Appennines: Introduction to facies analysis. International Geology Review, v. 20, p. 125-166.
This paper is one of the most influential in turbidite studies in the last 15 years. It proposes a set of seven facies which have

been used successfully in Italy and other parts of the world. Although some of these facies need revising, most turbidite workers accept the idea of a "universal" set of facies that can be used to describe most turbidites, of any age, anywhere.

Powers, D.W., and Easterling, R.G., 1982. Improved methodology for using embedded Markov chains to describe cyclical sediments. Journal of Sedimentary Petrology, v. 56, p. 913-923.
A companion paper to Carr (1982). Emphasizes problems of zero entries in the transition matrix, and explains improved methods for sequence analysis and testing the significance of the results.

Selley, R.C., 1970. Studies of sequence in sediments using a simple mathematical device. Geological Society of London, Quarterly Journal, v. 125, p. 557-581.
The first discussion of the difference matrix, and its possible use in describing and interpreting facies sequences. Contains written discussions of the paper by various authors, some of which are thought-provoking. See improvements by Carr (1982) and Powers and Easterling (1982).

Walker, R.G., 1979. Facies and facies models. 1) General introduction. In Walker, R.G., ed., Facies Models. Geoscience Canada Reprint Series, 1 (first edition), p. 1-7.
The "old" method for calculating random probability matrices is given here. The correct versions are given by Carr (1982) and Powers and Easterling (1982).

Walker, R.G., 1983. Cardium Formation 3. Sedimentology and stratigraphy in the Garrington-Caroline area. Bulletin of Canadian Petroleum Geology, v. 31, p. 213-230.
This paper illustrates how mudstones, siltstones and sandstones can be subdivided into various facies using such criteria as amount and type of bioturbation, presence or absence of silty laminations, grain size, sedimentary structures, nature of bedding contacts and textures of conglomerates.

Walker, R.G., and Harms, J.C., 1971. The "Catskill Delta": a prograding muddy shoreline in central Pennsylvania. Journal of Geology, v. 79, p. 381-399.
Describes cyclic facies sequences that are defined by transgressive bioturbated sandstone horizons.

Williams, P.F., and Rust, B.R., 1969. The sedimentology of a braided river. Journal of Sedimentary Petrology, v. 39, p. 646-679.
A good example of facies definition in a modern gravelly river (the Donjek, Yukon Territory), with definition of facies sequences and expression of a local "model" in terms of block diagrams.

STATISTICAL DEFINITION OF FACIES

Harbaugh, J.W., and Demirmen, F., 1964. Application of factor analysis to petrologic variations of Americus Limestone (Lower Permian), Kansas and Oklahoma. Kansas Geological Survey Special Distribution Publication 15, 40 p.
A good example of factor analysis used to establish facies (termed "phases") in the Permian Americus Limestone (Kansas and Oklahoma). Maps show distribution of the phases, with interpretations of depositional environments.

Harbaugh, J.W., and Merriam, D.F., 1968. Computer applications in stratigraphic analysis. New York, Wiley, 282 p.
Chapter 7 is concerned with classification systems, and gives a good introduction to factor analysis and other techniques. Several useful examples are discussed.

Imbrie, J., and Purdy, E.G., 1962. Classification of modern Bahamian carbonate sediments. In Ham, W.E., ed., Classification of carbonate rocks. American Association of Petroleum Geologists, Memoir 1, p. 253-279.
A good introduction to factor analysis, with an excellent example of how it can be used to define carbonate facies (with data from the Bahama Banks).

Klovan, J.E., 1964. Facies analysis of the Redwater Reef Complex, Alberta, Canada. Bulletin of Canadian Petroleum Geology, v. 12, p. 1-100.
Defines different types of carbonate particles and uses a hierarchal representation technique to classify them.

Reprinted by permission. First published in the *Journal of Petroleum Technology* (July 1977), pp. 851-866.

Predicting Reservoir Rock Geometry and Continuity in Pennsylvanian Reservoirs, Elk City Field, Oklahoma

R. M. Sneider, SPE-AIME, Sneider & Meckel Associates, Inc.

F. H. Richardson, SPE-AIME, Shell Oil Co.

D. D. Paynter, Shell Oil Co.

R. E. Eddy, SPE-AIME, Occidental Petroleum, Inc.

I. A. Wyant, independent

Introduction

Detailed knowledge of the distribution of pore space and barriers influencing fluid flow combined with reservoir engineering data is critical for selecting, planning, and implementing operationally sound supplementary recovery projects. Reservoir engineers have recognized for many years that sandstone reservoirs are not homogeneous or simply layered. However, reservoir analyses commonly do not account adequately for the inhomogeneities or variations in reservoir and nonreservoir rock properties because these variations are not always defined accurately. The increasing understanding of sand-body genesis — how sands are deposited — makes possible accurate delineation of the spatial distribution of pore space and barriers and helps significantly in predicting performance. This paper shows how the knowledge of sand genesis can be used to provide an accurate picture of the reservoir rock system and insight into reservoir performance. This paper is based on a detailed geological and petrophysical study of sandstone and conglomerate reservoirs and associated nonreservoir rocks in a ±500-ft-thick interval in the Elk City field, a 100-million-bbl field in the Anadarko basin in southwestern Oklahoma (Fig. 1). The paper describes (1) the reservoir rock characteristics of the genetic sand-body types that make up the reservoirs, (2) the geometry and continuity of the pore space and barriers to fluid flow, and (3) most useful maps for portraying floodable pay and predicting performance. The study was conducted during the time of declining primary production and while a waterflood was being planned.

The Elk City field produces from Pennsylvanian sandstones and conglomerates interbedded with nonproductive siltstones, shales, and carbonates. The interval studied includes the L and M zones, the two major reservoirs (Fig. 2). The field is an anticline developed by 310 wells on 40-acre spacing covering about 25 sq miles. Fig. 3 is a structure map on the dense marine limestone marker, Marker M, that separates the L and M zones.

Data Control

Most of the 310 wells were logged with a combination of SP log, 8- and 16-in. normal resistivity log, 24-ft lateral log, and microlog. In 16 wells, a gamma-ray neutron log was run instead of the microlog. Cores representing more than 1,700 ft of interval from 26 wells were studied.

Approach and Methods of Study

The study was conducted in four interrelated parts: (1) lithology and petrophysical properties of the cores were studied, and the logs were calibrated with lithology in the cored intervals; (2) genesis of the reservoir and associated rocks was interpreted; (3) correlation was established; and (4) the interval was divided into subzones, and the net sand and sequences of rock types were mapped for each subzone.

In the lithologic study, the mineralogy, grain size,

A geological-petrophysical engineering study of sandstone and conglomerate reservoirs was made to characterize and map pore-space distribution and to help predict fluid flow response of a waterflood. Floodable sand volume is believed to be represented accurately by net permeable sand isopach maps prepared from sand genesis knowledge.

sorting, and sedimentary structures were examined both visually and microscopically. Thin sections were used to determine pore sizes and geometry and the type and amount of pore-filling material. Grain size and sorting were determined either by sieve analyses or with a grain-size/sorting comparator.

For most core samples, core analyses, including porosity and permeability, were available. Porosity and permeability were measured on an additional 75 samples, and capillary-pressure measurements were made on 35 of these.

The methods and techniques used in calibrating logs, correlating and subdividing the stratigraphic interval, and mapping are discussed in later sections.

Lithologic and Petrophysical Properties And Relations

Rock Types

The interval studied consists of conglomerates, sandstones, siltstones, shales, and limestones. Only the conglomerates and sandstones are reservoir rocks.

The sandstones and conglomerates exhibit a wide range in median grain size and sorting. The sandstones range from very fine (0.062 mm in diameter) to very coarse (2.0 mm); most conglomerates range from granule (2.0 to 4.0 mm) to pebble size (4.0 to 64.0 mm). Sorting

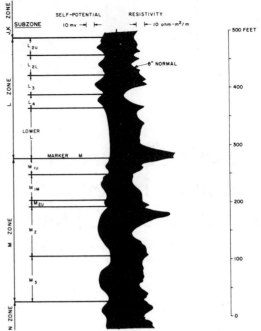

Fig. 2—Idealized log of the L and M zones.

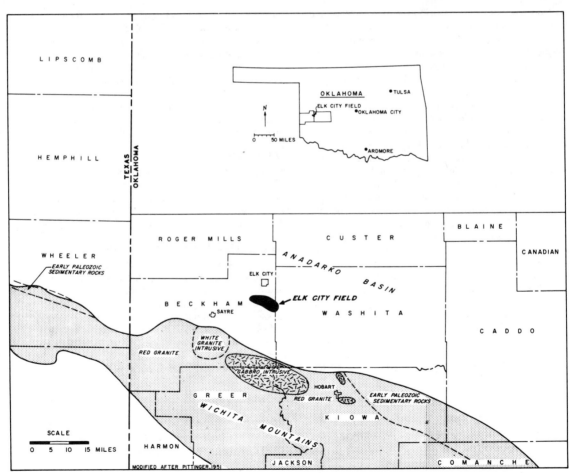

Fig. 1—Map showing the location and geologic setting of the Elk City field, Okla.

30

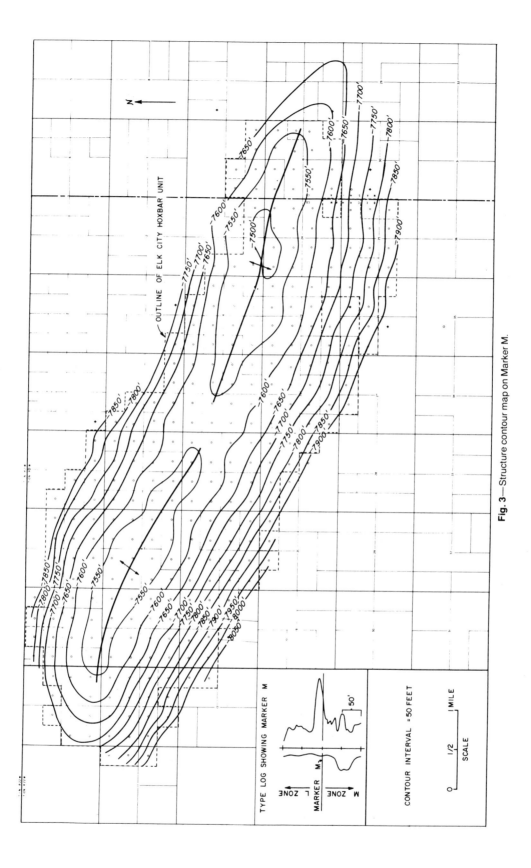

Fig. 3—Structure contour map on Marker M.

31

of the sandstones ranges from very good to moderate. The conglomerates are moderately to very poorly sorted. Although the sorting of these rocks is variable, Fig. 4A shows that grain size and sorting are related. The finer-grained rocks are the best sorted, and the coarser the grain size, the poorer the sorting.

All the sandstones and conglomerates are compacted to some degree. Most are compacted moderately. Most of the sandstones and conglomerates are cemented by silica or calcite. The amount of cement is variable, but most rocks contain less than 7 percent by volume. In a few rocks, the pores are completely or nearly filled by cement.

Siltstones and shales are the predominant nonreservoir rock types. Every graduation, from siltstone with few or no shale intercalations to shale with few or no siltstone intercalations, is observed. The two types most commonly are interbedded. The siltstones and shales are composed of the same minerals and rock particles as the sandstones and conglomerates. The limestones are dense and are composed primarily of clay- and silt-size carbonate particles and sand- and gravel-size fossil debris, quartz, feldspar, and other terrigenous materials.

Pore Space of the Reservoir Rocks And Its Relationship to Texture

Porosity, pore size, and permeability of any sandstone or conglomerate depend primarily on (1) grain size and sorting, and (2) the amount of cementation and compaction. The Elk City reservoir rocks are all compacted, and most contain some cement or pore-filling material. The amount of cement or pore-filling material in most of the

rocks, however, is less than 7 percent by volume. In these rocks, pore space correlates with grain size and sorting.

Porosity. Fig. 4A shows that the finer-grained rocks are the best sorted and that sorting becomes progressively poorer with an increase in median grain size. Fig. 4B, a comparison of size, sorting, and porosity of these rocks, shows that porosity changes for different sizes and sortings, but the change is predictable. The fine-grained rocks have the highest porosities, and with an increase in grain size, sorting is poorer and porosity decreases. The very-fine- and fine-grained sandstones with low porosity contain a high percentage of carbonate detritus and are compacted more than equivalent grain-size rocks with smaller amounts of carbonate detritus. For the conglomerates, the porosities plotted are of the matrix fraction, which is composed of sand. This matrix porosity is higher than the over-all porosity of the conglomerate.

Pore Size. Average pore size correlates with grain size and sorting in the rocks with little or no cement. Estimates of pore size from capillary-pressure curves and pore-size measurements in thin sections show that the fine-grained rocks have predominantly fine and very fine pores, and that with an increase in grain size, the average pore size and the number of medium and large pores increase. Capillary-pressure curves of representative reservoir rock types are shown in Fig. 5.

The general relationships among grain size, sorting, porosity, and pore size are summarized in Table 1.

Permeability. In the relatively uncemented rocks, permeability generally correlates with rock type (grain size

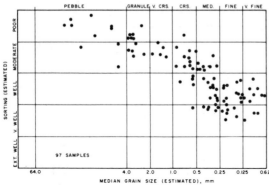

Fig. 4A—Relationship between sorting and grain size in some Elk City sandstones and conglomerates.

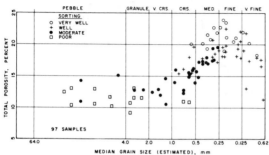

Fig. 4B—Relationship among sorting, grain size, and porosity in some Elk City sandstones and conglomerates.

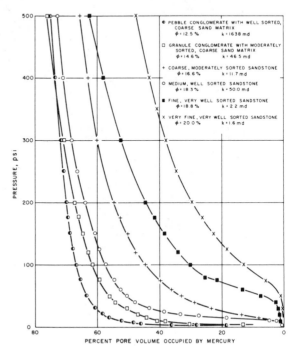

Fig. 5—Capillary-pressure curves of typical reservoir rock types.

32

and sorting). In a later section, we show the empirical relationships between permeability and rock type, and permeability profiles.

Log Calibration

Comparison of the lithologic and core data with the logs shows that rock types and sequences of rock types can be interpreted from a combination of the SP and 8-in. normal resistivity curves and the microlog. Typical relationships between rock types and log characteristics are shown in Fig. 6. Impermeable siltstones, calcareous shales, and limestones are distinguished from the sandstones and conglomerates on the basis of SP development. For the sandstones and conglomerates, which exhibit moderate to well developed SP, grain size is indicated by the 8-in. normal resistivity curve. The fine-grained rocks exhibit low resistivity and, with increase in grain size, the resistivity values increase. Grain-size determinations of the very silty, shaly, or cemented rocks present a problem. The measured resistivity of these rocks is higher than clean or noncemented rocks of equivalent grain size. The microlog is used to determine these rocks. The microlog opposite these rocks shows characteristically "hashy" separation, or no separation.

The reasons for the correlation between grain size and resistivity are apparent from a consideration of the relationships established between rock types and pore space. The fine-grained rocks are generally the best sorted, and have the highest porosity and the largest number of fine pores. This results in low formation-factor and resistivity values. As the grain size increases, sorting generally becomes poorer and porosity and the number of fine pores decreases, giving higher values of formation factor and resistivity. Hydrocarbon saturation has only a minor influence on resistivity values recorded by the 8-in. normal log, since this curve is influenced largely by the zone adjacent to the wellbore, which is invaded by filtrate.

The rock-type/log-response relationships worked out in the cored intervals provided a means for interpreting rock type and sequences of rock types in uncored intervals. Because of the relationships between grain-

TABLE 1—RELATIONSHIPS BETWEEN TEXTURE AND PORE SPACE

Texture		Pore Space	
Grain Size	Sorting	Porosity	Pore Size
Very Fine	Well To Very Well	Highest	Predominantly Fine and Very Fine Pores
↓ Grain Size Increase	↓ Sorting Poorer	↓ Porosity Decrease	↓ Number of Medium and Large Pores and Average Pore Size Increase
Gravel	Poor	Lowest	Large, Medium, and Fine Pores

size/sorting and porosity, permeability, and pore size, the log response also allows one to infer pore-space properties in uncored intervals.

Genetic Types of Sand Bodies

Studies of Recent genetic sand bodies, like those described by Le Blanc,[1] provide the criteria for recognizing similar genetic types in the subsurface. Core studies of the Elk City L and M zones show the reservoirs consist principally of one or more of the following genetic types: barrier bar, alluvial channel, deltaic distributary channel, and deltaic marine fringe. Fig. 7 illustrates schematically the trend and distribution of the different types.

The sand genesis was interpreted from vertical sequences of rock types, lithology (sedimentary structures, grain size, and sorting), nature of the contact with underlying and overlying rock types, fauna, the nature of the lateral boundaries, and the orientation of the sand body relative to depositional strike.

The characteristics of each sand-body type that are important in reservoir analysis are described below. The geological characteristics and criteria used to interpret depositional environments are not described in this paper.

Barrier Bar Deposit

Rock Types and Their Vertical Sequence. The barrier bar deposits are composed of rocks that range from siltstone to pebble conglomerate. As illustrated by Fig. 8, a typical sequence (from bottom to top) is characterized by a progressive upward increase in grain size. The coarsest rocks are coarse sandstones and granule or pebble conglomerates.

The vertical sequences of sedimentary structures are similar to Recent barrier bar deposits. Conglomerates, coarse sandstones, and some fine- and medium-grained sandstones are massive-bedded. The fine- and medium-grained rocks have faintly developed, slightly inclined, or have horizontal bedding. The fine sandstones and siltstones are laminated and ripple-bedded and show evidence of reworking by organisms.

Reason for the Vertical Sequence. The vertical sequence of rock types in the barrier-bar deposits is the result of off-lap sedimentation. As shown by Le Blanc[1] and Bernard et al.,[2] accretion of a barrier bar basinward results in deposition of progressively coarser material

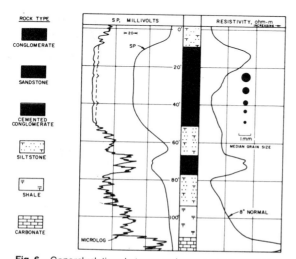

Fig. 6—General relations between rock type and log response.

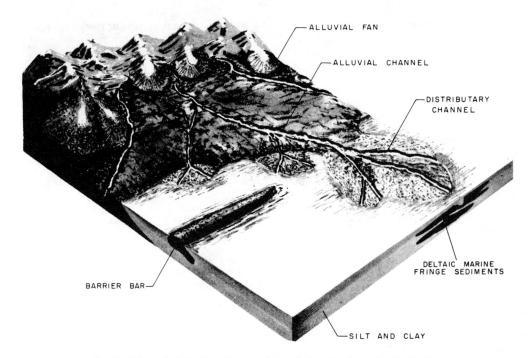

Fig. 7—Schematic illustration of the genetic types of sand bodies in the Elk City field.

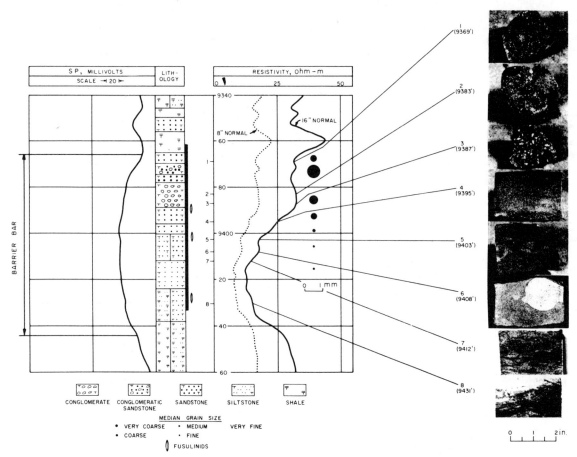

Fig. 8—Vertical sequence of rock types and log response of a barrier bar deposit; Shell, G. Slatten No. 1.

34

over finer material. The grain-size profile will be essentially identical along a trend parallel to the beach.

Log Response. Typical log response to the vertical sequence of rock types is shown in Figs. 8 and 10. The SP log is well developed opposite the sandstones and conglomerates. The short normal resistivity curve reflects the variations in grain size. The upward increase in grain size is shown by a gradual upward increase in resistivity.

Reservoir Trend and Distribution. The barrier bar deposits trend parallel to the depositional strike of the marine strata. The areal distribution and thickness of a portion of a barrier bar deposit are shown in Fig. 9, and cross-sections through the deposits are shown in Fig. 10. In this deposit, as in the other bar deposits,[2] sand thickness and vertical sequences of rock types are remarkably uniform parallel to the strike or trend of the deposit. The coarsest-grained material is concentrated on the landward side and at the top of the deposits. The over-all grain size and sand thickness decrease in a seaward direction across the trend of the deposits. No siltstone or shale bed exists except at or near the base of the sand.

Pore-Space Distribution. The distribution of porosity, permeability, and pore size generally follows the distribution of rock types. The vertical sequence of rock types results in a corresponding vertical sequence of porosity, permeability, and pore-size distribution. In Fig. 11, the permeability profile and porosity-permeability plot for various grain-size classes are presented for a bar sequence. The data show that the fine sandstones have high porosity and low permeability, and that with an increase in grain size, there is a corresponding increase in permeability and a decrease in porosity. Permeability increases vertically upward throughout these deposits and is highest at or near the top of the sequence. Since the sequence of rock types is the same parallel to the trend of the bar deposits, permeability profiles are the same in this direction. Permeability is highest on the shore of the deposit and decreases progressively in a seaward direction. Porosity should be well connected throughout the barrier bar deposits. No shale breaks exist in the sand except near the base; therefore, there should be no permeability barriers to fluid flow.

Alluvial Channel Deposits

The alluvial channel deposits are composed of sandstones and conglomerates deposited by high-gradient braided streams.

Rock Types and Vertical Sequence. The channel deposits are characterized by predominantly coarse-grained rocks and cyclical vertical variations in grain size. Typical rock types and sequences of rock types are shown in Fig. 12. Conglomerates, conglomeratic coarse sandstones, and medium- to coarse-grained sandstones are the dominant rocks. Fine- to very-fine-grained sandstones, siltstones, and silty shales are interbedded with the coarser rocks, but are not abundant quantitatively. The coarser sandstones and conglomerates are moderately to poorly sorted; however, the matrix of some conglomerates is well sorted. Most of the very-fine- to medium-grained sandstones are well to very well sorted. The cyclical vertical variation in rock type, which is so

characteristic of the channel deposits,[3,4] occurs on all scales.

Reason for the Cyclical Vertical Sequence. Sedimentation in braided streams is the result of flooding. Recent braided stream deposits show that rapid vertical and horizontal variations in texture and bedding are characteristic of these deposits. Part of a sequence of sediments is often removed before deposition of the next sequence, so that only parts of sequences are preserved. Cores through a vertical succession of partial sequences would show cyclical variations in rock types.

Log Response. Typical log response of the SP, microlog, and short normal resistivity curves for the channel deposits is shown in Fig. 12. The SP curve is well developed opposite the sandstones and conglomerates. Opposite the silty sandstones, the SP curve is depressed slightly toward the "shale line." In the sandstones and conglomerates, the microlog has positive separation. Opposite intervals that are tightly cemented or very silty, the microlog is hashy and shows no positive separation. Thus, the microlog indicates the intervals in which the resistivity curve cannot be used to interpret grain size.

Reservoir Trend and Distribution. The alluvial channel deposits are linear in trend and are oriented nearly perpendicular to the depositional strike of the marine strata.[1] Fig. 13 is an example of a channel deposit. The cross-sections show that the channel sandstones and conglomerates were deposited in a valley cut into older deposits. Nonchannel strata, particularly the marine siltstones and shales, on either side of the channel correlate very well (compare Well 1-15-14 with Well 1-14-14 in Section LL', Fig. 14). It is apparent, then, that the distribution of channel deposits is controlled by size and shape of the original stream-cut valley. In this example, sand and gravel fill much of the valley; however, silt and clay were also deposited in an appreciable amount in some parts of the valley. Well 1-14-6 penetrates a portion of the valley that was filled with sand and gravel in the lower part and with silt and clay in the upper part. Most of the other siltstone and shale beds within the sandstones are very limited in areal extent.

Pore-Space Distribution. The distribution of pore space in channel deposits generally follows the distribution of rock types. The rapid changes in rock type over short vertical and horizontal distances result in corresponding

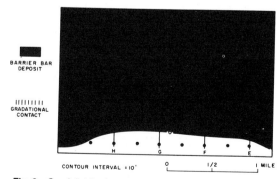

Fig. 9—Sand distribution and thickness in a portion of a barrier bar deposit.

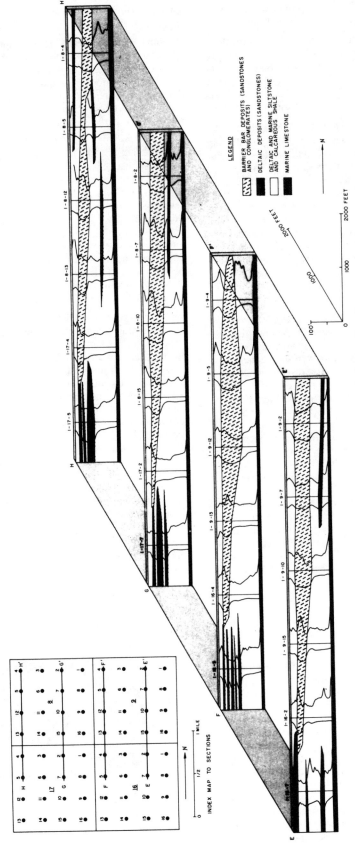

Fig. 10—Cross-sections through a barrier bar deposit.

36

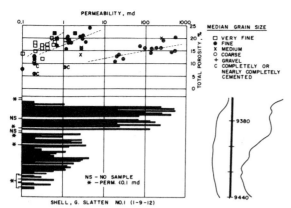

Fig. 11—Plot of porosity vs permeability and vertical permeability profile; Shell, G. Slatten No. 1.

rapid changes in pore space. The vertical-permeability profile of a channel deposit, shown in Fig. 14, illustrates the rapid changes in permeability. Rocks with high permeability are at the top, at the base, or at any position within the interval. Individual rock types probably occur in lenses in these deposits; therefore, variations in permeability in a horizontal direction should be nearly as great as the vertical variations, and permeability should change over short distances.

The channel deposits are primarily coarse-grained rocks, and the permeability of the rocks is relatively high. The sediments deposited near the edges of the channels may be finer grained and may contain a greater abundance of interstitial silt and clay. The permeability of the rocks near the sides of the channel deposits may be low. The rocks adjacent to the edges of the channels, because of their low permeability, may act as a permeability barrier to the flow of fluids. Silt/shale barriers within the channel deposits will not have any appreciable areal extent.

Distributary Channel Deposits

The distributary channel deposits are associated with the deltaic marine-fringe deposits and have the same overall sedimentological and petrophysical properties as the alluvial channels.

Rock Types and Vertical Sequence. The distributary channel deposits are composed principally of fine to very coarse sandstones and conglomeratic sandstones (Fig. 15).

Pore Space Distribution. Fig. 16 shows a vertical-permeability profile of a distributary channel deposit. The higher-permeability rocks can be at any stratigraphic position within the sequence.

Deltaic Marine-Fringe Deposits

This genetic type consists of either one or two stacked genetic units and is believed to be the product of regres-

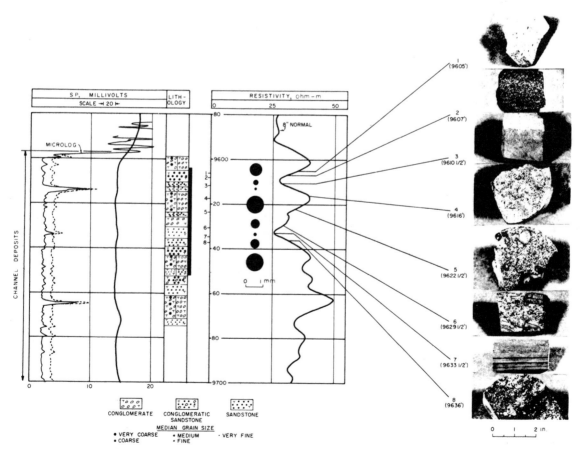

Fig. 12—Vertical sequence of rock types and log response; Shell, B. Pinkerton No. 1.

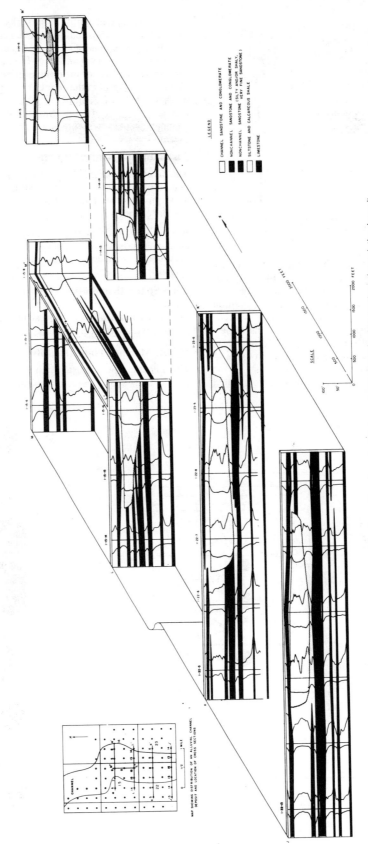

Fig. 13— Areal distribution of a portion of an alluvial channel deposit and cross-sections through the deposit.

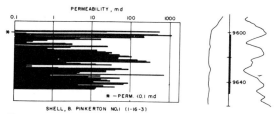

Fig. 14—Vertical-permeability profile; Shell, B. Pinkerton No. 1.

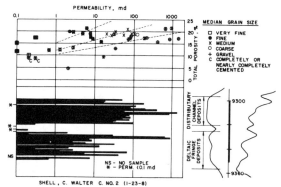

Fig. 16—Plot of porosity vs permeability and vertical-permeability profile; Shell, C. Walter C. No. 2.

sive deltaic sedimentation. The single units are believed to represent deposits from one deltaic advance; the stacked units are a composite of single units.

Rock Types. The deltaic marine-fringe deposits are composed of rocks that range from shale to pebble conglomerate. Typical rocks are shown in Fig. 15. The principal rock types are very-fine- to coarse-grained sandstones and conglomeratic sandstones that contain numerous thin interbeds of shale, siltstone, and carbonaceous material. All rock types contain interstitial silt and clay-size particles in amounts from a few percent up to 25 percent by volume.

Vertical Sequence of Rock Types. The vertical sequence of rock types in these deposits is characterized by (1) interbedding of sandstone with shale and siltstone, (2) a general upward decrease in the number of shale and siltstone beds and in the amount of interstitial clay- and silt-size particles in the sandstones, and (3) a

general upward increase in grain size. The vertical sequence commonly is made up of two fine-to-coarse sequences (Fig. 15). Some deposits have as many as three individual sequences, each showing an upward increase in grain size.

Reservoir Distribution and Boundaries. The deltaic marine-fringe deposits are irregular in geometry and distribution. Toward the lateral and basinward edges of the fringe deposits, the sandstones become progressively finer grained and grade into siltstone and shale.

Pore-Space Distribution. A vertical-permeability profile of a typical fringe deposit is shown in Fig. 16. The

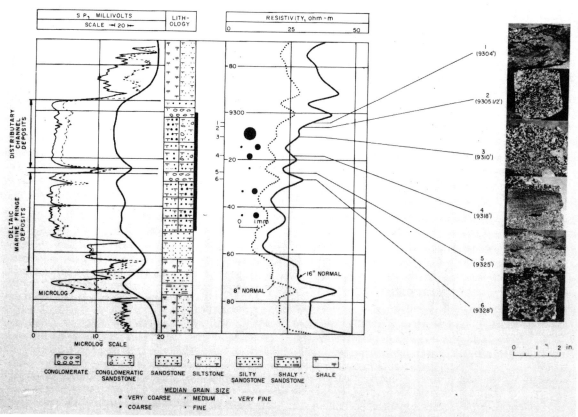

Fig. 15—Vertical sequence of rock types and log response; Shell, C. Walter C. No. 2.

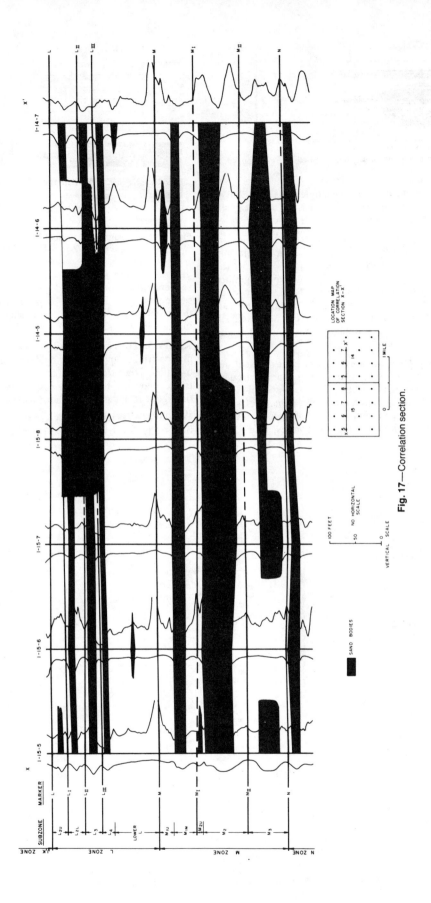

Fig. 17—Correlation section.

profile is characterized by an upward increase in permeability. Where one unit is stacked on another and the vertical sequence of rock types is repeated, the upward increase in permeability is also repeated. The lower part of each sequence is composed of very shaly and/or silty rocks that generally have low permeability. Thus, in deposits where the vertical sequence of rocks is repeated, the permeable rocks in each sequence are separated by impermeable or low-permeability rocks. These tight rocks, although usually very thin (less than 3 ft thick), are widespread.

Toward the lateral or basinward edges of the fringe deposits, the permeability and average pore size of the reservoir rocks decrease gradually as a result of the gradual decrease in grain size and increase in the amount of interstitial clay- and silt-size particles.

Implications of Genetic Sand-Body Identification

Studies of genetic sand bodies in the Recent and in well documented ancient[1] examples demonstrate that lithology and pore space in many types of sands are distributed in a systematic and predictable manner. Identifying the genesis of the sand(s) comprising a reservoir allows one to predict with some degree of confidence the distribution and continuity of pore space and lithology, especially impermeable beds. The next section shows that knowledge of the genesis of sand comprising a reservoir makes correlation and zonation easier and more reliable.

Correlation and Zonation of the Reservoirs

The L and M producing zones are separated vertically by a thin, dense, marine limestone. Each zone consists of bodies of sandstone and conglomerate interbedded with impermeable shale, siltstone, and limestone. Correlation of the sands within each zone was a problem during the primary development of the field and during the initial phase of this study. Most of the correlation problems were resolved, and the sands were correlated in detail, once the genesis of the sands and associated strata was determined.

The following method was used to establish the detailed correlation. Correlation sections similar to Fig. 17 were prepared using the base of the marine limestone (Marker M) as a horizontal datum; the shales and limestones that were identified in cores as marine and that could be recognized on the logs were correlated. Within the marine strata, individual markers, such as tops and bases of limestones, could be traced over much of the field. In wells where a marker was absent, the equivalent stratigraphic position was usually occupied by channel deposits. Markers L_{II} and L_{III} in Fig. 17 are missing in Wells 1-15-8, 1-14-5, and 1-14-6 and are cut out by a channel. The markers within the entire stratigraphic interval were found to be nearly parallel to one another and were assumed to represent nearly horizontal, depositional surfaces or datums. The sands between two markers were then correlated on cross-sections hung on the datum above the sands.

Correlation of the marine strata and the markers within them provided a logical basis for subdividing the L and M producing zones. The subdivisions or subzones established are illustrated in Fig. 17. Each subzone contains a sand accumulation that is separated from underlying and overlying accumulations by impermeable strata over all

or most of the field. In some localities in the field, a sand accumulation of one subzone lies directly on a sand accumulation of an adjacent subzone. The sand accumulations of adjacent subzones are in direct contact with each other where stream valleys cut down through the two subzones and filled with sand and gravel (Fig. 17, Wells 1-15-8 and 1-14-5).

Each sand accumulation is composed of one or more different genetic sand bodies. When two or more different genetic sand bodies make up an accumulation, they are generally laterally equivalent to each other. Most wells through a sand accumulation usually encounter just a single genetic unit or an amplified unit.

The detailed correlation established that the L and M zones together actually are composed of six separate major reservoir subzones and three minor ones. All subzones are separated from one another over all or most of the field by impermeable beds. In almost all cases where two subzones are in contact, their connection is the result of channel downcutting. Although it may not have been necessary to recognize that the L and M zones are actually nine separate reservoirs for efficient primary exploitation, it is necessary for waterflooding. Assuring that water is injected into all floodable reservoirs and that the corresponding intervals are open in the production wells is essential to maximizing waterflood production.

Sand and Pore-Space Distribution
Concepts and Methods Used to Determine And Represent Reservoir Sand and Pore-Space Distribution

One objective of this study was to delineate accurately the distribution of the reservoir sands and their pore space. Before this could be accomplished, however, it was necessary to determine what sands should be mapped together. The sand accumulations of individual subzones appeared to be the logical group to analyze and map, because they are separated from the sand accumulations above and below by impermeable strata over all or most of the field and, therefore, act as separate reservoirs during waterflooding.

Three types of maps were prepared to represent the sand and pore-space distribution of a sand accumulation in each subzone: (1) maps of the distribution of genetic sand bodies combined with the vertical sequence of rock types, (2) thickness maps of net sand, and (3) thickness maps of total microlog separation (permeable sand).

Maps of the distribution of genetic sand bodies were combined with vertical sequences of rock types. Maps like Fig. 18, showing the areal distribution of vertical sequences of rock types with an outline of the genetic sand units, were prepared for each sand accumulation. These maps were prepared from cross-sections (like Figs. 13 and 17). The sequences were determined from cores and, in the uncored intervals, from logs.

In parts of some zones, a sand of one genesis lies directly above, but separated from, sands of different genesis. These areas were designated on the maps. Cross-sections, of course, show the actual sequences and their relationships to one another. On the maps, when the vertical sequence is a hybrid one, only the sequence of the younger sand unit was mapped.

A map of the vertical sequences of rock types, together with a map showing the genesis of the sand bodies, also

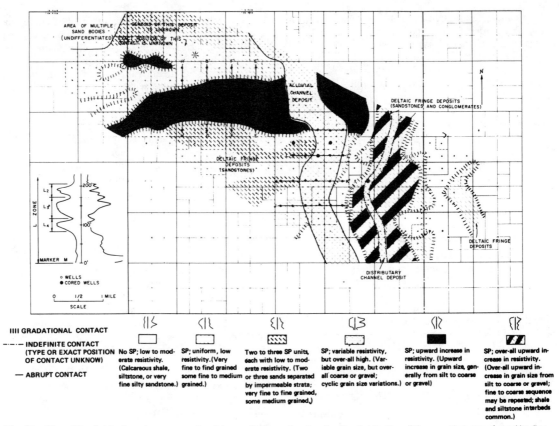

IIII GRADATIONAL CONTACT

—·—·— INDEFINITE CONTACT
(TYPE OR EXACT POSITION
OF CONTACT UNKNOWN)

— ABRUPT CONTACT

No SP; low to moderate resistivity. (Calcareous shale, siltstone, or very fine silty sandstone.)

SP; uniform, low resistivity. (Very fine to find grained some fine to medium grained.)

Two to three SP units, each with low to moderate resistivity. (Two or three sands separated by impermeable strata; very fine to fine grained, some medium grained.)

SP; variable resistivity, but over-all high. (Variable grain size, but over-all coarse or gravel; cyclic grain size variations.)

SP; upward increase in resistivity. (Upward increase in grain size, generally from silt to coarse or gravel)

SP; over-all upward increase in resistivity. (Over-all upward increase in grain size from silt to coarse or gravel; fine to coarse sequence may be repeated; shale and siltstone interbeds common.)

Fig. 18—Map of the distribution of sequence of rock types, Subzone L₃, showing the distribution of the genetic types of sand bodies.

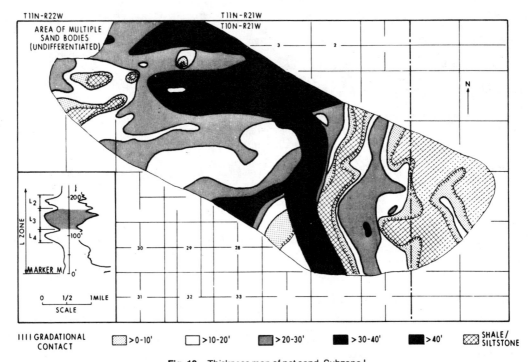

IIII GRADATIONAL CONTACT ▫ >0-10' ▫ >10-20' ▦ >20-30' ■ >30-40' ■ >40' ▨ SHALE/ SILTSTONE

Fig. 19—Thickness map of net sand, Subzone L₃.

represents the vertical and lateral distribution of pore space. In a previous section, we showed that porosity, permeability, and average pore size generally correlate with rock type and that the different genetic types of deposits are characterized by definite vertical and lateral sequences of rock types. Therefore, the distribution of pore space can be inferred if the genesis of a sand body and the distribution of rock types within it are known. An example of the description of pore-space distribution using vertical sequences of rock types on genetic facies maps is discussed later for Subzone L_3.

Thickness Maps of Net Sand. The distribution of "net sand" — sandstone and conglomerate — in each subzone is shown by a thickness map similar to Fig. 19. Net sand includes permeable as well as impermeable sandstone and conglomerate, but not siltstone or other impermeable strata. The number of feet of sandstone and conglomerate in each well was determined from cores and, in the uncored intervals, from electric logs.

The net-sand thickness maps were prepared in the following manner. On cross-sections through the sand accumulation, the amount and vertical distribution of sandstone and conglomerate were plotted for each well. The amount and distribution of net sand between the wells were interpreted from a knowledge of sand genesis and from a consideration of the most probable sand distribution that would result from its deposition. The thickness maps were drawn from the cross-sections. The knowledge of the sand genesis was extremely valuable as a guide in drawing these maps.

Thickness Maps of Microlog Separation (Permeable Sand). "Permeable sand" is defined as sand that is permeable enough to show positive separation on a microlog. From a consideration of the Elk City field rocks and the fluid conditions within them, the microlog should show positive separation opposite all permeable formations. Caliper surveys were run in all wells so that intervals in which the borehole was out of gauge, thus possibly influencing the microlog, could be detected. Comparison of core analyses with micrologs of cored intervals showed that positive microlog separation develops opposite rocks with permeabilities of 0.5 md and greater. Rocks with ±0.5 md and greater are the reservoir rocks.

The distribution of permeable sand is represented by thickness maps of positive microlog separation. For Subzone L_3, the amounts of net sand and permeable sand are essentially identical so the map of permeable sand is the same as Fig. 19.

The permeable-sand maps can be refined to show the number of feet of different rock types (for example, rocks with 1 to 10 md permeability) or potentially high-permeability "thief" layers.

Sand and Pore-Space Distribution in Subzone L_3

The distribution of sand and pore space in each subzone was determined and mapped with the concepts and methods outlined above. The results of the mapping for one sand accumulation — the L_3 — are illustrated in Figs. 18 and 19.

Subzone L_3 was picked because it shows the various combinations of different genetic sand bodies that make up the other individual sand accumulations. Fig. 7, the schematic diagram showing the different sand bodies and their relationship to the mountain front, may be helpful in visualizing the relationships among the different sand bodies.

Subzone L_3 is a composite of four genetic types: barrier bar, alluvial channel, distributary channel, and deltaic fringe deposits.

The *barrier bar deposits* are characterized by rocks that range in grain size from very fine to coarse or gravel. The vertical sequence of rocks is one of uniform upward increase in grain size. Basinward, the rocks become increasingly finer and grade into siltstones. The lower and lateral boundaries are gradational with the adjacent impermeable siltstones and shales. Since the rock types control pore space, porosities of the barrier deposits are relatively high (16 to 24 percent). Permeabilities increase uniformly vertically upward and toward the land (south) and decrease uniformly toward the sea (north). The permeability profile is essentially the same parallel to the sand-body trend and no impermeable barriers should break up the pore space in any direction.

The *alluvial channel deposits* are characterized by predominantly coarse-grained rocks, cyclic vertical grain-size variations, and rapid lateral variations in rock type. The lower and lateral boundaries with adjacent deposits are erosional in nature and, therefore, abrupt. Rocks composing the channel deposits have relatively low porosities (10 to 15 percent), but high permeabilities (75 to 1,500 md). Because of the cyclic variations in rock types, permeability is variable both vertically and laterally. All impermeable strata are very limited in areal extent.

The *distributary channel deposits* are similar to the alluvial channel deposits except that the rocks are not as coarse.

The *delta marine fringe* has an upward-increased grain size and permeability like the barrier bars. However, the thin siltstone/shale beds that cap each cycle are widespread and are effective barriers to vertical flow.

Conclusions

The L and M zone reservoirs are composed of sandstones and conglomerates deposited at the margins of a shallow marine basin adjacent to a mountainous source area.

Pore space (porosity, permeability, and average pore size) is related to grain size and sorting in most of the reservoir rocks. Cementation has a small influence on pore space in most of the sandstones and conglomerates.

Lithology, grain size, sorting, and pore space can be interpreted in uncored intervals from a combination of the SP log, 8- and 16-in. normal resistivity curves, and microlog.

The reservoirs are made up of one or more of the following genetic sand-body types: barrier bar, alluvial channel, distributary channel, and deltaic marine fringe. Each genetic type has characteristic (1) vertical and lateral sequences of rock types (grain size and sorting), sedimentary structures, and bedding types, (2) lateral and vertical boundary relationships with adjacent strata, (3) sand trend, and (4) pore-space distribution and continuity.

Knowledge of the genesis of the sands and associated rocks makes it possible to (1) correlate and subdivide the L and M producing intervals into discrete sand accumulations, and (2) map the distribution of sand and pore space within each sand accumulation.

The methods and concepts used to calibrate lithology and pore space with electric logs and to delineate and map the reservoir sands and pore space should be applicable to problems of predicting and mapping sand and pore-space distribution of many other sandstone reservoirs. Thickness maps of net permeable sand prepared from a knowledge of sand genesis are probably the most accurate means of representing floodable sand volume.

Definitions

A *genetic sand unit* is a sand body (for example, alluvial channel) deposited during a single occurrence of a particular depositional process (for example, flood cycle), and an *amplified sand unit* is an aggradational sand body consisting of superposed sands deposited during reoccurrence of a particular depositional process.

Acknowledgments

We wish to express our appreciation to the managements of Shell Development Co. and Shell Oil Co. for permission to publish this paper. The work leading to the results described was carried out in the Petroleum Engineering Research Dept., Bellaire Research Laboratory, Houston.

Original manuscript received in Society of Petroleum Engineers office July 22, 1976. Paper accepted for publication Jan. 27, 1977. Revised manuscript received April 27, 1977. Paper (SPE 6138) was first presented at the SPE-AIME 51st Annual Fall Technical Conference and Exhibition, held in New Orleans, Oct. 3-6, 1976. © Copyright 1977 American Institute of Mining, Metallurgical, and Petroleum Engineers, Inc.

References

1. Le Blanc, R. J., Sr.: "Distribution and Continuity of Sandstone Reservoirs — Part 1," *J. Pet. Tech.* (July 1977) 776-792.
2. Bernard, H. A., Major, C. F., Parrott, B. S., and Le Blanc, R. J., Sr.: *Recent Sediments of Southeast Texas*, Guidebook No. 11, Bureau of Economic Geology, U. of Texas, Austin (1970) 16-38.
3. Blissenbach, E.: "Geology of Alluvial Fans in Semiarid Regions," *Bull.*, GSA (1954) 175-189.
4. Mackin, J. H.: "Concept of the Graded River," *Bull.*, GSA (1948) 463-512.

JPT

A Stochastic Model for Predicting Variations in Reservoir Rock Properties

D. W. BENNION
JUNIOR MEMBER AIME

J. C. GRIFFITHS

U. OF ALBERTA
CALGARY, ALBERTA, CANADA
PENNSYLVANIA STATE U.
UNIVERSITY PARK, PA.

ABSTRACT

A mathematical model, which does not assume a priori that stratification exists, but was designed to test for the stratification was developed. The model segmented the reservoir horizontally into areas of common variance, then divided it vertically into strata (if strata were present). Next, trend surface techniques were used to determine the lateral extent and variation of each stratum.

The model was tested on two reservoirs (sandstone and limestone) using porosity as the reservoir rock property. The sandstone reservoir contained approximately 60,000 samples from 2,000 cored wells, while the limestone reservoir had approximately 24,000 samples from 430 cored wells.

Within the areas of common variance tested, the model was able to distinguish four separate lithological units or zones in the sandstone reservoir, four in the Marly section and seven in the Vuggy section of the limestone reservoir.

INTRODUCTION

To accurately predict the movement and production of fluids from a reservoir rock, it is necessary that the reservoir engineer have a knowledge of the rock properties and their distribution throughout the reservoir. For this study, a reservoir rock is defined as a solid containing interconnecting holes or voids which occur relatively frequently and are dispersed within the rock in a regular or random manner. The fraction of bulk volume that these interconnecting voids occupy is called porosity. The rock must also possess sufficient permeability to allow fluids to move through it and it may be composed of one or more strata or lithological units. A stratum or lithological unit as defined in this study is a body or volume of reservoir rock whose properties are so distributed that its lateral and vertical extent can be traced throughout or through a portion of the reservoir. Operationally, the existence of strata are question-able unless the variance within strata is less than among strata.

Within the past few years computers have been widely used as a tool for calculating movement and production of fluids from reservoirs. Most prediction models require some type of a mathematical description of certain reservoir rock properties. In the past, most of these models have been relatively simple. Some have assumed vertical variation but no lateral variation, while others have assumed lateral variation but no vertical variation. In general, most of these models have not used functional relationships to predict reservoir rock properties as functions of position.

This study was to develop a model which would predict lateral and vertical variations of reservoir rock properties.

Since samples are usually available only from a small portion of the total reservoir rock, it seemed logical that if measurements from these samples were to be used to infer the properties of the actual reservoir, the data should be treated statistically as a sample from the total population (reservoir). Therefore, the model developed in this study to predict reservoir rock properties was a stochastic model.

The model was tested using porosity (a macroscopic reservoir rock property) measurements from two reservoirs — the first, a sandstone reservoir from which 60,000 samples were obtained; and the second, a limestone reservoir with 24,000 samples.

Several investigators have proposed methods to determine the vertical and areal variation of a reservoir rock property. Stiles,[1] Dykstra and Parsons[2] and Suder and Calhoun[3] have all developed waterflooding prediction techniques which assume vertical variation but no lateral variation. Stiles developed a method to segment a reservoir arbitrarily using frequency distribution. Law[4] suggested that porosity has a normal frequency distribution and that permeability has a log-normal frequency distribution. Elkins and Skov[5] proposed that when a reservoir is segmented vertically, the

Original manuscript received in Society of Petroleum Engineers office July 12, 1965. Revised manuscript received Jan. 11, 1966. Paper (SPE 1187) was presented at SPE Annual Fall Meeting held in Denver, Colo., Oct. 3-6, 1965.

[1]References given at end of paper.

strata should approximate the actual sand deposition rather than be picked from a frequency distribution without regard to spatial origin. Testerman[6] developed a statistical zonation technique which segmented the reservoir vertically using a minimum and maximum variance technique, and then correlated these zones from well to well. Wyllie[7] and Warren[8] have both proposed that the lateral variations are as great as the vertical variations and that variations in reservoir properties may be completely random.

All these investigators used macroscopic variables for their studies. A microscopic study by Dahlberg[9] who used a low rank graywacke[10] found that the petrographic properties had a significant systematic trend in the vertical direction. Techniques for sampling sediments with internal structure have been developed by Griffiths.[11]

DEVELOPMENT OF THE MODEL

To facilitate formulation of a mathematical model capable of predicting reservoir rock properties, some of the factors affecting the sedimentation process must be examined. Most reservoir rocks can be classified into one of three classes: (1) fragmental, (2) amorphous and crystalline precipitates, and (3) a combination of 1 and 2. The first class is usually referred to as a mechanical sediment, the second as a chemical sediment and the third as a hybrid sediment.

Most sediments have been laid down in a marine environment and are of the hybrid type. The sedimentation process is rarely continuous and uniform. Some of the causes for this non-uniformity might be tectonic movements, climatic cycles and changes in the rate of sedimentation. Since most reservoir rocks have been formed in a marine environment, primary structures such as cross-bedding, graded bedding, ripple marks and inclined beds may be present. These are dependent on current velocity and the supply of sedimentary material.

From this brief description of factors which affect the sedimentation process, it is evident that the following two factors must be considered in the model: (1) since the sedimentation process is not usually continuous, reservoir rock properties may vary in the lateral and vertical directions, and (2) primary structures may be present in a reservoir.

The model is composed of three sections. The first segments the reservoir laterally into areas of common variance. The second segments the reservoir vertically into lithological units if they are present. The third predicts the lateral extent and variations of the reservoir rock properties. The model assumes that reservoir rock property Ψ is normally distributed with a mean μ and variance σ^2. Each of the three sections of the model will be described in turn as Steps 1, 2 and 3.

Step 1 segments the reservoir laterally into areas of common variance. Bartlett[12] developed a method which can be used to test the statistical significance of several sample variances. The method tests the hypothesis that $s_1^2 = s_2^2 = s_3^3 \ldots \ldots \ldots = s_k^2 = \sigma^2$. Bartlett showed that under a null hypothesis these are random samples from the same normal population. The ratio H/C has a chi-squared distribution with $k - 1$ degrees of freedom.

$$H = -\sum_{i=1}^{k} f_i \ln \left(s_i^2 / \bar{s}^2 \right) \ldots \ldots \ldots (1)$$

$$C = 1 + \frac{1}{3(k-1)} \sum_{i=1}^{k} \left(\frac{1}{f_i} - \frac{1}{\sum_{i=1}^{k} f_i} \right) \ldots (2)$$

Step 2 segments each of the areas of common variance vertically. A series of cross-sections may be sufficient to pick out the strata or lithological units in a reservoir. If the strata cannot be found visually this does not mean that variation in the reservoir is completely random. It might only mean that the data contain enough *noise* to make it impossible to recognize the pattern. If this is the case, some type of smoothing technique can be used to help eliminate some of the noise. Vistelius[13] has shown that certain reservoir rock properties can be approximated by trigonometric series, so that a Fourier series can be used to help eliminate noise in the vertical profiles. Once data have been smoothed, the profiles can again be visually examined for patterns that may exist across the area or a portion of the area.

A discussion of Fourier series can be found in most advanced calculus and boundary-value problem books, i.e., Churchill[14] and Tolstov.[15] In one dimension, a continuous periodic function can be described by the following equations:

$$f(x) = \frac{A_o}{2} + \sum_{n=1}^{\infty} A_n \cos \frac{\pi n x}{L} + B_n \sin \frac{n \pi x}{L}$$
$$\ldots \ldots \ldots \ldots (3)$$

$$A_n = \frac{1}{L} \int_{-L}^{L} f(x) \cos \frac{\pi n x}{L} \, dx \ldots (4)$$

$$B_n = \frac{1}{L} \int_{-L}^{L} f(x) \sin \frac{\pi n x}{L} \, dx \ldots (5)$$

A Fourier series is periodic by nature, but it can be used to represent non-periodic functions within an interval, outside of which the series would give incorrect periodic results.

Once strata have been obtained, the next step is to determine whether there is any systematic trend in the vertical and lateral directions within each stratum. It is also important to be able to express the lateral extent of each stratum in the form of an equation. This can be accomplished through relationships such as:

$$\Psi = f(x, y, z) \ldots \ldots \ldots \ldots (6)$$

$$\text{Top of stratum} = g(x, y) \ldots \ldots \ldots (7)$$

Base of stratum = $h(x, y)$ . . . . (8)

Thickness of stratum = $r(x, y)$ . . . . . (9)

Trend surface techniques were used to determine these functional relationships. Trend surfaces are described by the mathematical equations in which a dependent variable is a function of its coordinates. Three methods for fitting hyperplanes to equations (Eq. 6) and surfaces to equations (Eqs. 7 through 9) will be explained. They are multiple regression, orthogonal polynomials and Fourier series.

Multiple regression analysis is performed by setting up a model of the type:

$$\psi = \beta_0 + \beta_1 U_1 + \beta_2 U_2 + \cdots + \beta_k U_k + E . \quad . (10)$$

The second method (orthogonal polynomials) has some advantages over multiple regression analysis which uses non-orthogonal polynomials. For instance, Forsythe[16] has shown that, for high order, non-orthogonal polynomial systems, it is difficult to solve for the regression coefficients using a computer without introducing large errors. Also, if it is desired to raise the power of the polynomial using non-orthogonal polynomials, all the regression coefficients must be recalculated. When using orthogonal polynomials, on the other hand, only the coefficient belonging to each new polynomial need be calculated. By using orthogonal polynomials, the following equations can be used to predict a rock property Ψ as a function of its x and y coordinates:

$$\psi_{ij} = \sum_{r=0}^{n} \sum_{k=0}^{m} \beta_{rk} \, x_i^r \, y_j^k \quad . . . . . . (11)$$

The method of generating orthogonal polynomials and obtaining the polynomial coefficients is presented in the Appendix. This method is a combination of the methods proposed by several authors.[16-19]

The third method of fitting surfaces is that of Fourier series. A one-dimensional Fourier series was used to help segment the reservoir vertically. This can easily be expanded to a surface fitting technique as follows:

$$f(x, y) = \sum_{n=0}^{\infty} \sum_{m=0}^{\infty} \lambda_{mn} \left[A_{nm} \cos \left(\frac{\pi n x}{X} - \frac{\pi m y}{Y} \right) \right.$$
$$\left. + B_{nm} \sin \left(\frac{\pi n x}{X} + \frac{\pi m y}{Y} \right) \right] \quad . . . . (12)$$

$$A_{nm} = \frac{1}{XY} \int_0^X \int_0^Y f(x, y) \cos \left(\frac{n\pi x}{X} - \frac{m\pi y}{Y} \right)$$
$$dx \, dy \cdot \quad . . . . (13)$$

$$B_{nm} = \frac{1}{XY} \int_0^X \int_0^Y f(x, y) \sin \left(\frac{n\pi x}{X} - \frac{m\pi y}{Y} \right)$$
$$dx \, dy, \quad . . . . (14)$$

where $X = 1/2$ period in x-direction
$Y = 1/2$ period in y-direction

$$\lambda_{mn} = \begin{cases} 1/4 & m = n = 0 \\ 1/2 & m \neq 0, \, n = 0, \text{ or } m = 0, \, n \neq 0 \\ 1 & m \neq 0, \, n \neq 0 \end{cases}.$$

The above development has been discussed by Tolstov.[15]

Briefly, the model will segment the reservoir laterally, using the principle of common variance; vertically segment the reservoir into strata, if present; and fit surfaces of a reservoir rock property as a function of the coordinates.

The model does not assume a priori that stratification exists, but is designed to test for it. If the model indicates that the reservoir is not stratified, the reservoir is treated as one stratum.

The model is general and can be used on data measured in the laboratory or in situ as long as the scale of measurement is continuous and the frequency is either normal or can be transformed to normality.

TESTING THE MODEL

The model was tested on two reservoirs. The first, a sandstone reservoir with 2,000 cored wells from which 60,000 samples had been taken, consisted of a sand section and a conglomerate section. The latter was so sporadic that it was not tested in the model. The second reservoir, a limestone reservoir with 430 cored wells from which 24,000 samples had been taken, consisted of two sections called the Marly and the Vuggy. Porosity and permeability measurements were available for each sample from both reservoirs.

Frequency distribution and statistical parameters for porosity and log-permeability are presented in Figs. 1 through 4. An examination of these figures indicates that porosity is approximately normally distributed. The log-permeability distributions are skewed to the right and are leptokurtic, with the exception of the conglomerate which is slightly platykurtic. These distributions skew to the right much more than the porosity distribution. Some reasons for this departure from normality have been discussed by Bennion.[20] Since porosity approximated a normal distribution more nearly than

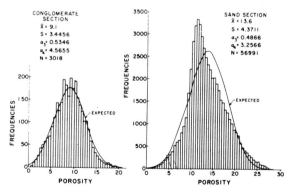

FIG. 1 — POROSITY FREQUENCY DISTRIBUTION, SANDSTONE RESERVOIR.

permeability, it was used to test the model.

Figs. 7 and 8 present results from the first part of the model. Areas enclosed in solid lines have the same variance. Shaded areas do not have common variance. It should be pointed out that these lines are not rigid and each of the areas enclosed within a solid line could perhaps be expanded to include several more wells or sections.

From Figs. 7 and 8 the largest, or one of the largest, areas of common variance was chosen to be tested in the model. In the sandstone reservoir, T. 49, R. 9 was chosen, while in the limestone reservoir the upper left-hand quarter of T. 5, R. 13 was selected.

Porosity profiles in each of the test areas were smoothed using Eqs. 3 through 5. Eqs. 4 and 5 were integrated numerically to obtain the Fourier coefficients. In the sandstone reservoir it was found that eight to eleven coefficients gave a relatively good fit to the data. In the Marly section six to eight coefficients were needed and in the

TABLE 1 — ANALYSIS OF VARIANCE TO TEST DIFFERENCES BETWEEN ZONE MEANS

Source of Variance	Sums of Squares	Degrees of Freedom	Mean Square	F
Sandstone zones	1,547.57	3	515.85667	42.16772**
Within zones	23,084.52	1,887	12.23344	
Total	24,632.09	1,890		
Marly zones	2,475.86	4	618.965	17.1175 **
Within zones	30,012.6	830	36.15975	
Total	32,488.46	834		
Vuggy zones	384.47	6	64.07833	4.0345 **
Within zones	25,032.33	1,576	15.88345	
Total	25,416.80	1,582		

* Significant at the 1 per cent level.

Vuggy section, eight to eleven coefficients were required. The period was taken as the gross thickness of the section.

After observing several cross-sections of the smoothed data, it was apparent that most wells possess certain characteristic minimums and maximums that could be traced throughout the test areas. In each section, one of these characteristic maximums or minimums was chosen and shifted to a common depth.

Fig. 5 contains three cross-sections, one through each of the test areas. Fig. 9 shows the position of the cross-sections within the test areas. From these cross-sections, four zones can be traced throughout the sandstone section, five through the Marly and seven through the Vuggy.

A one-way analysis of variance model was used to test the means of each zone within each section (Table 1). All of the F-tests are significant at the 1 per cent level, indicating that the means are not all equal within a section. While this test indicates that the means are not all equal, it is desirable to know which means are not equal. Tukey[21] has proposed a test, which is fully explained and illustrated by Bennett and Franklin[22] for grouping means (Table 2).

Table 2 shows that although several zones have equal means statistically, with the exception of Marly Zones 4 and 5 no two adjacent zones have equal means. These data give added evidence that the method used to segment the reservoirs vertically has divided them into zones which are significantly different.

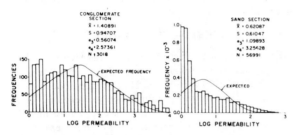

FIG. 2 — LOG PERMEABILITY FREQUENCY DISTRIBUTION, SANDSTONE RESERVOIR.

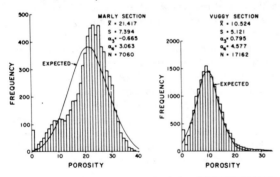

FIG. 3 — POROSITY FREQUENCY DISTRIBUTION, LIMESTONE RESERVOIR.

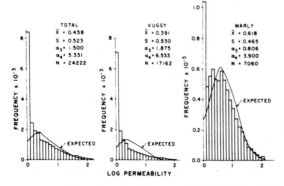

FIG. 4 — LOG PERMEABILITY FREQUENCY DISTRIBUTIONS, LIMESTONE RESERVOIR.

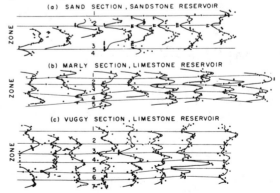

FIG. 5 — CROSS-SECTION OF SMOOTHED POROSITY PROFILES — MEASURED VALUE.

TABLE 2 — MEANS TEST RESULTS

Section	Zones of Equal Means	Zones of Unequal Means
Sandstone	2, 4	1, 3
Marly	1, 3	2
Marly	4, 5	
Vuggy	3, 6	
Vuggy	2, 5, 7	
Vuggy	1, 4	

Marly Zones 4 and 5 could not be recognized in all the wells in the test area and, since the analysis in Table 2 indicates that they have the same mean, they were combined.

First, second and third order, three-dimensional trend surfaces were run using Eq. 10. These surfaces indicated that porosity was not highly correlated with depth. As a result of these analyses it was decided that no simple set of hyperplanes could be fitted to the data. It was then assumed that porosity was random in the vertical direction and the mean value of porosity at each well was used. This reduced the model to two dimensions and allowed the use of two-dimensional, surface-fitting techniques.

In the sandstone reservoir, trend surfaces were run on the top, base and mean porosity of each zone. In the limestone reservoir, they were run on the top, base and gross thickness of each zone. These trend surfaces were calculated using an IBM 7074 computer. Programs to obtain the polynomial coefficient in Eq. 11 and the Fourier coefficients in Eqs. 13 and 14 were written.

Each method generated as many coefficients as desired and contours of the original data, calculated data and residuals were plotted. Since the programs were written for equally spaced data, it was necessary to interpolate to obtain missing values at some of the mesh points. For this study, a linear interpolation program was used. A more accurate method would probably have been to program the calculations

so that they would iterate to obtain the missing values.

After the programs were written, it was found that the orthogonal polynomial program ran 13 times faster than the Fourier series program; the Fourier series program was only used on a few surfaces to show that the technique can be used to fit surfaces.

Fig. 10 shows the contours on the base of Zone 3 in the sand section of the sandstone reservoir and a trend surface on the base of this same zone. In Fig. 11, a Fourier series is used to fit the same surface. In all, 169 cosine terms and an equal number of sine terms were used to calculate this surface. Fig. 12 presents similar data for the top of Zone 3 in the Marly section of the limestone reservoir. Fig. 13 shows an isopach and trend surface for Zone 4 of the Vuggy section. Fig. 14 shows an iso-mean porosity and trend surface for Zone 3 of the sand section. Fig. 6 shows the amount of variance which can be explained by

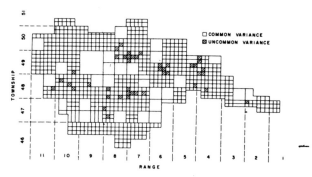

FIG. 7 — AREAS OF COMMON VARIANCE SAND SECTION, SANDSTONE RESERVOIR.

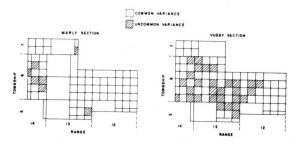

FIG. 8 — AREAS OF COMMON VARIANCE, LIMESTONE RESERVOIR.

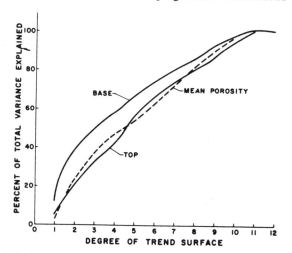

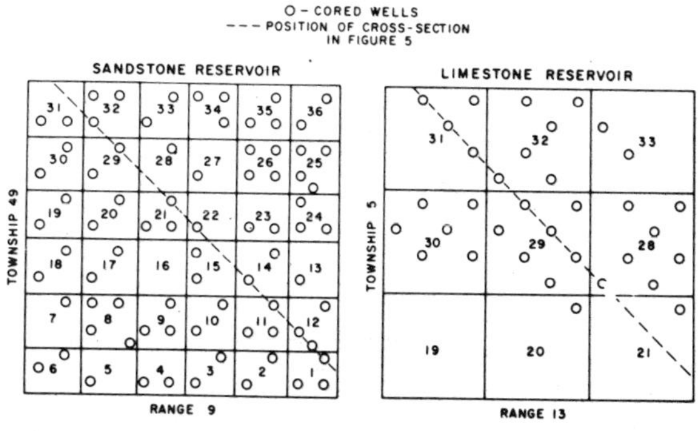

FIG. 6 — PERCENTAGE OF TOTAL VARIANCE EXPLAINED BY TREND SURFACES (ZONE 3, SAND SECTION, SANDSTONE RESERVOIR).

FIG. 9 — MAP OF TEST AREA — O-CORED WELLS (POSITION OF CROSS-SECTION IN FIG. 5).

49

each order trend surface for the top, base and mean porosity in Zone 3 of the sand section.

The order at which surface-fitting is stopped is up to the individual, the main consideration being accuracy of the data. If the data contain a large amount of noise the maximum order of the polynomial should be low enough to prevent fitting the noise. For the sandstone reservoir it was decided that a surface which would explain 80 per cent of variance was sufficient.

This type of model should find wide application in the petroleum industry. The first part of the model can be used to stratify a reservoir if strata are present. This will make it possible to determine how best to sample the reservoir to obtain precise, unbiased estimates of the properties of the reservoir rock. The trend surface part of the model can be used to map structure, to calculate reserves and can also be used with prediction techniques which require reservoir rock properties.

CONCLUSIONS

1. Frequency distributions for porosity were approximately normal but those for log-permeability were generally skewed to the right and leptokurtic.

2. Within the areas of common variance tested, the model was able to distinguish four separate lithological units or zones in the sandstone reservoir, four in the Marly section and seven in the Vuggy section of the limestone reservoir. Therefore, using these techniques, a reservoir may be zoned into relatively homogeneous units and the estimators of the rock properties will be more precise and unbiased.

3. Statistical tests indicate that the zones are significantly different.

4. Within areas tested, the model was able to determine a system of equations for predicting the lateral extent, thickness and mean porosity for each zone.

NOMENCLATURE

ψ = any reservoir rock property
μ = population mean
σ^2 = population variance
s = sample variance
H = first factor of Bartlett's test
C = second factor of Bartlett's test
f = degrees of freedom

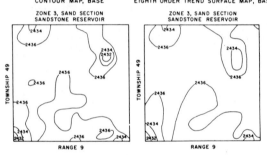

FIG. 10 — MAP OF T. 49, R. 9.

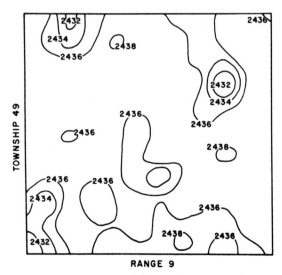

FIG. 11 — FOURIER SERIES TREND SURFACE MAP, BASE (ZONE 3, SAND SECTION, SANDSTONE RESERVOIR).

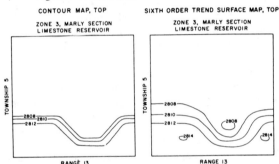

FIG. 12 — CONTOUR MAP OF T. 5, R. 13.

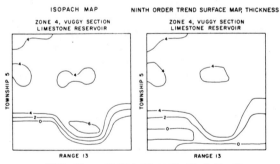

FIG. 13 — ISOPACH MAP, T. 5, R. 13.

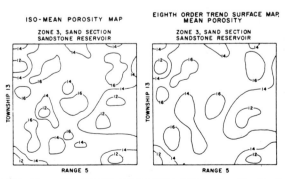

FIG. 14 — ISO-MEAN POROSITY MAP, T. 13, R. 5.

A = Fourier cosine coefficients

B = Fourier sine coefficients

L = one-half of the period

β = polynomial coefficient

U = independent variable

E = error term

n = degree of polynomial in x

m = degree of polynomial in y

x = coordinate in east-west direction

y = coordinate in north-south direction

z = coordinate in vertical direction

ACKNOWLEDGMENTS

The authors wish to thank Pan American Oil Co., Socony Mobil Oil Co. of Canada, Ltd. and Central Del Rio Oil Co. for supplying the data used in this study, and Shell Oil Co. for their financial assistance. We would like to thank Standard Oil Co. of California, Royalite Oil Co. and The Pennsylvania State U. for computer time.

REFERENCES

1. Stiles, W. E.: "Use of Permeability Distribution in Water flood Calculation", *Trans.*, AIME (1949) Vol. 186, 9.

2. Dykstra, H. and Parsons, R. C.: "The Prediction of Oil Recovery by Waterflood", *Secondary Recovery of Oil in the United States*, API, The Lord Baltimore Press (1950).

3. Suder, R. E. and Calhoun, J. C.: "Waterflood Calculations", *Drilling and Production Practices*, API (1949).

4. Law, J.: "Statistical Approach to the Interstitial Heterogeneity of Sand Reservoirs", *Trans.*, AIME, (1944) Vol. 155, 202.

5. Elkins, L. F. and Skov, A. M.: "Some Field Observations of Heterogeneity of Reservoir Rocks and Its Effect on Oil Displacement Efficiency", Paper 282 presented at SPE Production Research Symposium held in Tulsa, Okla. (April, 1962).

6. Testerman, G. D.: "A Statistical Reservoir Zonation Technique", *Trans.*, AIME (1962) Vol. 225, 887.

7. Wyllie, M. R. J.: "Reservoir Mechanics — Stylized Myth or Potential Science", *Jour. Pet. Tech.* (June 1962) 883.

8. Warren, J. E. and Price, H. S.: "Flow in Heterogeneous Porous Media", *Soc. Pet. Eng. Jour.* (1963) Vol. 1, No. 3, 153.

9. Dahlberg, E. C.: "Defining a Gradient in a Sample of Sediments", Paper presented at 50th Annual Meeting of the American Association of Petroleum Geologists, New Orleans, La. (April, 1965).

10. Krynine, P. D.: "The Megascopic Study and Field Classification of Sedimentary Rocks", *Jour. Geol.* (1948) Vol. 56, 130.

11. Griffiths, J. C.: "Experiments in Sampling Sedimentary Rocks", Paper presented at 50th Annual Meeting of the American Association of Petroleum Geologists, New Orleans, La. (April 1965).

12. Bartlett, M. S.: "Properties of Sufficiency and Statistical Tests", *Proc. Royal Society of London* (1937) Vol. 160, 268-282.

13. Vistelius, A. B.: "Sedimentation Time Trend Functions and Their Applications for Correlation of Secondary Deposits", *Jour. of Geol.* (Nov., 1961)

Vol. 69, No. 6, 703.

14. Churchill, R. V.: *Fourier Series and Boundary Value Problems*, McGraw-Hill Book Co., New York, N. Y. (1941).

15. Tolstov, G. P.: *Fourier Series*, Translated from Russian by Richard A. Silverman, Prentice-Hall, Inc., Englewood Cliffs, N. J. (1962).

16. Forsythe, G. E.: "Generation and Use of Orthogonal Polynomials for Data-Fitting with a Digital Computer", *Jour. Soc. Applied Math.*, (June, 1957) Vol. 5, No. 2, 74.

17. DeLury, D. B.: *Values and Integrals of Orthogonal Polynomials up to $n = 26$*, Toronto U. Press (1950).

18. Householder, A. S.: *Principles of Numerical Analysis*, McGraw-Hill Book Co., New York, N. Y. (1953) 221.

19. Stiefel, E. F.: "Kernel Polynomials in Linear Algebra and Their Numerical Applications", *Report*, National Bureau of Standards, Washington, D. C. (1955) 52.

20. Bennion, D. W.: "A Stochastic Model for Predicting Reservoir Rock Properties", PhD Thesis, The Pennsylvania State U. (1965).

21. Tukey, J. W.: "Comparing Individual Means in the Analysis of Variance", *Biometrics* (1949) Vol. 5, 99.

22. Bennett, C. A. and Franklin, N. L.: *Statistical Analysis in Chemistry and the Chemical Industry*, John Wiley and Sons, Inc., New York, N. Y. (1954) 341.

APPENDIX

ORTHOGONAL POLYNOMIALS IN TWO DIMENSIONS

Let P be matrix of orthogonal polynomials in x and Q be a matrix of orthogonal polynomials in y. The $x(i)$ and $y(j)$ points must be equally spaced on a square or rectangular grid with $i = 0, 1, ..., s$ and $j = 0, 1, ..., r$, n is one more than the maximum degree in x and m is one more than the maximum degree in y.

X is a $1 \times r$ vector of the grid point in the x-direction, Y is a $s \times 1$ vector of the grid points in the y-direction. A is a $1 \times n$ matrix whose elements are:

$$a_i = \left[\sum_{\tau=1}^{r} x_\tau P_{i-1,\tau}^2 \right] \Big/ \sum_{\tau=1}^{r} P_{i-1,\tau}^2 \quad \cdot \cdot \quad (1\text{-}A)$$

B is a $1 \times n$ matrix whose elements are:

$$b_i = \left[\sum_{\tau=i}^{r} P_{i-1,\tau}^2 \right] \Big/ \sum_{\tau=1}^{r} P_{i-2,\tau}^2 \quad \cdot \cdot \cdot \quad (2\text{-}A)$$

The elements of the P matrix are:

$$P_{i,\tau} = \left[\left[x_\tau - a_i \right] P_{i-1,\tau} - b_i\, P_{i-2,\tau} \right] \quad \cdot \cdot \quad (3\text{-}A)$$

In the above equations $i = 1, 2, ..., n$

$$\tau = 1, 2, ..., s.$$

Polynomials in the y direction are obtained in a similar manner:

$$e_j = \left[\sum_{k=1}^{s} y_k \, q_{j-1,k}^2 \right] \Big/ \sum_{k=1}^{s} q_{j-1,k}^2 \quad . \; . \; . \; (4\text{-A})$$

$$d_j = \left[\sum_{k=1}^{s} q_{j-1,k}^2 \right] \Big/ \sum_{k=1}^{s} q_{j-2,k}^2 \quad . \; . \; . \; (5\text{-A})$$

$$q_{j,\tau} = \left[y_k - c_j \right] q_{j-1,k} - P_{j-2,k} \quad . \; . \; . \; . \; (6\text{-A})$$

In Eqs. 4-A through 6-A, $j = 1, 2, \ldots m$ and $\tau = 1, 2, \ldots r$. It is not possible to use these equations unless i and j are greater than two.

$$P_{1,\tau} = 1.0 . \; . \; . \; . \; . \; . \; . \; . \; . \; . \; . \; . \; (7\text{-A})$$

$$q_{1,k} = 1.0 . \; . \; . \; . \; . \; . \; . \; . \; . \; . \; . \; (8\text{-A})$$

$$P_{2,\tau} = \sum_{\tau=1}^{r} \left[x_\tau - \bar{x} \right] . \; . \; . \; . \; . \; . \; . \; (9\text{-A})$$

$$q_{2,k} = \sum_{k=1}^{s} \left[y_k - \bar{y} \right] . \; . \; . \; . \; . \; (10\text{-A})$$

$$\bar{x} = \sum_{\tau=1}^{r} x / r . \; . \; . \; . \; . \; . \; . \; . \; (11\text{-A})$$

$$\bar{y} = \sum_{k=1}^{r} y_k / s . \; . \; . \; . \; . \; . \; . \; . \; (12\text{-A})$$

Let Ψ be the matrix of observations of the dependent variable, then the following procedure is used to determine the matrix of the coefficients:

$$[G] = [P][\Psi][Q]^T . \; . \; . \; . \; . \; . \; . \; (13\text{-A})$$

$$f_i = \sum_{\tau=1}^{r} P_i^2 . \; . \; . \; . \; . \; . \; . \; (14\text{-A})$$

$$h_j = \sum_{k=1}^{r} q_{j,k}^2 . \; . \; . \; . \; . \; . \; . \; (15\text{-A})$$

The elements of a matrix Ω are:

$$\omega_{ij} = g_{ij} / (f_i h_j) . \; . \; . \; . \; . \; . \; . \; (16\text{-A})$$

$$[\hat{\psi}] = [P][\Omega][Q]^T . \; . \; . \; . \; . \; . \; . \; (17\text{-A})$$

Eq. 17-A will give a matrix of estimators of Ψ. If a more general system of equations is desired, the following additional steps must be taken:

$$\lambda_{ii} = 1 . \; . \; . \; . \; . \; . \; . \; . \; . \; . \; . \; (18\text{-A})$$

$$a_{ij} = \lambda_{i-1, j-1} . \; . \; . \; . \; . \; . \; . \; (19\text{-A})$$

$$\gamma_{ij} = -b_i \lambda_{i-2} . \; . \; . \; . \; . \; . \; . \; (20\text{-A})$$

$$\lambda_{ij} = \sum_{j=2}^{i-1} -a_i \, \lambda_{i-1, j} + a_{ij} + \gamma_{ij} . \; . \; . \; (21\text{-A})$$

If $j = 1$ $a_{ij} = 0$ $i = 1, 2, \ldots, n$
    $j = i - 1$, $\gamma_{ij} = 0$

$$z_{ii} = 1 . \; . \; . \; . \; . \; . \; . \; . \; . \; . \; (22\text{-A})$$

$$\theta_{ij} = z_{i-1, j-1} . \; . \; . \; . \; . \; . \; . \; (23\text{-A})$$

$$\phi_{ij} = -b_i z_{i-2} . \; . \; . \; . \; . \; . \; . \; (24\text{-A})$$

$$z_{ij} = \sum_{j=2}^{i-1} -c_i \, z_{i-1, j} + \theta_{ij} + \phi_{ij} . \; . \; . \; (25\text{-A})$$

If $j = 1$ $\theta_{ij} = 0$
    $j = i - 1$ $\phi_{ij} = 0$

$$[W] = [\lambda]^T [\Omega] [z] . \; . \; . \; . \; . \; . \; . \; (26\text{-A})$$

W is the matrix of coefficients for the function of x and y. Using the above coefficients, the prediction is of the form:

$$\psi_{kr} = \sum_{i=0}^{n-1} \sum_{j=0}^{m-1} w_{ij} \, x_k^i \, y_r^j . \; . \; . \; . \; . \; (27\text{-A})$$

★ ★ ★

RESERVOIR ENGINEERING

Reservoir Mechanics-Stylized Myth or Potential Science?

M. R. J. WYLLIE
—MEMBER AIME—

GULF RESEARCH & DEVELOPMENT CO.
PITTSBURGH, PA.

The Nature and Task of Reservoir Mechanics

What is reservoir mechanics, or as some less research-minded people might prefer to put it, reservoir engineering? Is it, in the du Pont-inspired words of one of our local radio announcers, a foundation for ''better things for better living through chicanery'', or is it something which we can in good faith describe as a science? Is it an elaborate myth, or is it the fundamental tool which can offset the present inexorable climb in the costs of finding oil in the United States? Let me say that I believe the latter statement to be the true one.

While the reservoir mechanics profession is presently respectable, I must point out that there is danger of its falling into disrepute unless attempts are made to curb present extravagances and to disseminate more widely knowledge concerning its inherent strengths and basic weaknesses. The mood of the oil industry today is not one of expansion—or, at least, not of expansion of personnel, either technical or otherwise. Many companies are carrying on their production activities with a staff which is less than one-half the size of that which existed less than five years ago. Staff reduction has not necessarily reached its nadir. It is imperative, therefore, that the reduction in the quantity of engineering talent be more than offset by an increase in quality and, in particular, by an increase in productivity. A productivity increase can be effected in two ways. The first is by clear thinking, the second by automation. Both are vital, but for the sake of brevity I shall not here consider the latter.

Let us go back to fundamentals. What is the essential job of the reservoir mechanic? Surely it is to get more oil out of the hydrocarbon accumulations which have been so laboriously (and, happily, inexpensively) found in years past. Even allowing for the mystique on which I shall comment later, it appears that (tar sands and oil shales excluded) there may well be in excess of 100-billion bbl of oil which are known to exist in this country but which are presently deemed economically unrecoverable. This 100- billion bbl means something like $300 billion. Even if allowance is made for the conservative view of J. Paul Getty that a billion dollars is not worth what it used to be, this is a tempting target. Certainly it is a target which should be aimed at if, as some figures show, it now costs more than $3.00 to find a new barrel of producible oil.

The recovery of that 100 billion bbl is not going to be easy. But its recovery will be made more difficult if those who control the means to conduct research and the resources to pioneer novel field techniques do not reject specious and ill-defined processes, and avoid being misled by the euphoric misanalysis of pilot studies. If true progress is to be made, it is particularly essential that only the most dispassionate appraisal be made of pilot projects. Too many pilots are only successful because of judicious interpretation; technically, they are failures. It is a sad commentary on the degree of governmental control in our economy that *suppressio veri* or *suggestio falsi* when applied to field trails can sometimes be remunerative. I believe that because of political considerations many novel schemes for increasing recovery are taken to the field at a date far earlier than can be justified technically. Such trials may be not only technical travesties but ones that, by their very nature, may ultimately undermine those who

strive for true technical progress. The danger of a technically unsound trial lies in the cynicism engendered by the too-frequent crying of ''Wolf!''. Eventually, those who have new schemes for oil recovery which are properly understood and have been adequately tested in the laboratory will be impeded.

Surface and the Problem of Oil Recovery

The problem of total oil recovery is not merely one of having enough reservoir energy available; it is primarily the problem of directing that energy. Essentially, the recovery of oil involves the ability to overcome surface forces. In homely terms, the average oil reservoir contains about 1/2 gal of oil/acre of rock surface. Consider that amount of oil spread on a fair-sized parking lot. It might then be considered pretty smart to recover one-half of it if what is recovered only commands $.07/gal (equal, in fact, to the usual state tax on 1 gal of gasoline). Regrettably, of course, it is not smart enough by a factor of about two. More scientifically put, many petroleum-bearing rocks have an envelope surface area of about 500 sq cm/cc of pore volume and a surface area (if measured by the B.E.T. or gas-adsorption technique) of about 0.5 to 1.0 sq m/gm. This surface is small by comparison with those which concern catalyst-minded chemists. Nevertheless, it is sufficiently large to dominate recovery schemes based on the use of surface-active chemicals.

Surfactants and Their Difficulties

For example, Taber has known that about 10^{-3} gm of the nonionic surfactant Triton X-100 is adsorbed per gram of Berea sandstone. In itself, this quantity may not sound large. Nevertheless, in the extraor-

Original manuscript received in Society of Petroleum Engineers Office Feb. 15, 1962. Revised manuscript received May 8, 1962. Paper presented to various local sections of SPE as part of the Society's Distinguished Lecturer series for 1961-62.

53

dinary units in which the industry conducts its business, it amounts to about 2-3/4 tons/acre-ft. Since almost any chemical with any pretensions to molecular sophistication costs about \$.50/lb, even in bulk, it may easily be seen that the cost of the surfactant exceeds the gross value of the 700 bbl of oil/acre-ft that might remain in a Berea sand reservoir after a water flood. In this calculation, no attention has been directed towards an examination of the detailed mechanism whereby surfactants function to increase oil recovery. This is complex in itself. The main purpose, that of diminishing the influence of interfacial forces, seems to be sound. The point made is that, if a surfactant is to find practical use, it is necessary to ensure that residual surfactant adsorption is about 10^{-4} gm/gm of rock, or less. It may be possible to achieve this figure by suitably tailoring the surfactant molecule or by devising some scheme for eluting the surfactant with a much cheaper chemical. There is no indication that any such scheme has been devised. Indeed, there is little evidence that many proponents of surfactant treatment of wells are conscious of the fundamental problem facing them. The penchant that even competent engineers have for accepting a "mysterious-black-box" approach to the solution of their problems is, in many ways, alarming and, I believe, harms their professional standing.

Modern Schemes of Recovery or The Elimination of Surfaces

It was noted previously that the basis for the use of surfactants is the minimization or alteration of the surface forces which, by opposing the squeezing of oil globules through a rock, make difficult the complete recovery of all oil initially present. In the limit, the best way to eliminate the surface forces is to remove the surfaces entirely. This concept is the entire basis of all current schemes for improving oil recovery.

LPG slugs driven by gas or alcohol slugs propelled by water are obvious examples of interface elimination by mutual solubility. Less obvious, but of precisely the same genre, is the much-misunderstood process of in situ combustion. The process may best be viewed as a miscible displacement in which the driv-

ing air is made miscible with the oil and water it pushes by the expedient of vaporizing the air-oil-water interface, thereby generating a single phase. At the same time, viscous fingering is eliminated by the concomitant high rates of molecular diffusion.

It is well known that aspects of the in situ combustions process are not fully understood. In particular, it will never be possible to make a theoretical prediction of the course of recovery by combustion drive until the chemical kinetics of the air-oil interaction are fully grasped. This represents a chemical problem of singular intractability. In addition, the effects of gravity on the combustion process need quantitative elucidation before any conscientious reservoir engineer would be prepared to hazard anything but the most qualitative guess concerning the course a combustion drive might take. This, of course, has not prevented a number of combustion-drive tests from being undertaken. In the light of current inadequate theory, credible generalizations based on these tests have been exiguous.

An imperfect understanding of in situ combustion is not something of which any engineer need be ashamed. The process is, in its details, extraordinarily complex; and, unfortunately, these unknown details may easily dominate the production history and thereby gravely affect the economic consequences of the venture. Nevertheless, there are simple broad generalizations based on elementary heat balances which can be and should be applied at the outset of any proposed application of the combustion process. These heat balances may be conceived to represent limiting cases. Like thermodynamics, they are independent of details of mechanism and, therefore, are broadly applicable.

Perhaps the most important single factor which bears on the possibility of a combustion drive's being economic is the oil content of the formation per cubic foot of reservoir rock. In turn this means that, for any porosity, the oil saturation as a fraction of the pore volume should be high. The porosity also should be as large as possible since the definitive parameter is the product of porosity and saturation, ϕS_o. Now the reason for this requirement is simple and, as noted already, one

that transcends exact mechanistic details. In order that the oil in an oil sand will burn (sustain a chemical reaction leading to the production of water and oxides of carbon), it is now known that a rather high temperature is required—anywhere from perhaps 700° to 1,000° F above the ambient reservoir temperature. Alas, the exact temperature cannot be forecast nor can the nature of the oil cut that is consumed be estimated at present because of our ignorance of the chemical kinetics of the combustion process. But we know we must burn it at a high temperature and, because the oil that burns and the rock are so intimately associated, we cannot burn the one without heating the other. There lies the rub. We do not want to heat the rock, but we cannot avoid so doing. The only source of heat is the oil originally present, so this serves as fuel. Because of the inexorable laws of chemical stoichiometry, the ratio of oil burned by weight to air required by volume is fixed. (The actual ratio depends upon the carbon-to-hydrogen ratio of the oily material consumed.)

The dilemma then is very evident. Only at high porosities and high oil saturations do we meet the best conditions, i.e., a relatively small amount of rock to be heated and, thus, a relatively small amount of oil to be burned to do the heating. Because the oil burned is relatively small, the amount of air that has to be injected is correspondingly small. Also, because the oil saturation is high, the amount of oil left over to be produced is reasonably substantial and the injected air-to-produced oil ratio is economic. This can be seen in Fig. 1.

On the other hand, if the porosity of the reservoir rock is small, the amount of rock to be heated increases and, with it, the amount of oil and air required. Simultaneously, of course, for any oil saturation (even a high one) the gross amount of oil available per cubic foot of rock is reduced by an amount directly proportional to the porosity reduction. The net oil available for production is also dramatically reduced, while the economically dominating injected air-produced oil ratio skyrockets. This is for adiabatic conditions. When heat is lost because the bed being burned is thin or contains shale inclusions, the air–oil ratio worsens. Most field trials have resulted in air–oil ratios some 25 percent worse than those shown in Fig. 1.

The perceptive reader may query the justification for assuming that all heat used to raise the rock temperature is irrevocably lost. Are there no regenerative processes which pick up heat from the rock through which the combustion wave has passed and carry it forward, thereby reducing the heat load downstream and, with it, the amount of oil consumed and the quantity of air to be injected per barrel of oil moved or recovered?

Such a question appears sound. One problem is to move the heat. Some regeneration is obtained because the air that reaches the combustion front attains the rock temperature as it progresses from its point of injection. But a gas is a singularly poor agent for the transfer of heat. Thus, to cool 1 cu ft of rock by 800°F requires 1.5 Mscf of gas. It may surprise those who know of gas-cooled atomic reactors that this figure is almost independent of pressure, as Table 1 shows.

It has been suggested that combustion can be made economic by intermittently burning and moving the heat forward by circulating gas. A calculation shows that under typical adiabatic conditions the gas-oil ratio would be about 11 Mscf/bbl when burning and some 50 Mscf/bbl when circulating gas. Even if the gas circulated is picked up at sufficiently high pressure that only one stage of further compression is required before reinjection, the cost of recirculation is unlikely to be less that $.04/Mscf. Thus, the cost of gas circulated per barrel of oil produced is likely to be about $2.00, while that of air injected is not likely to be more than about $.60 to $.70.

It is clearly more economic to burn. This is particularly so because available experimental data are not in full accord with the assumption that heat regeneration will conserve oil and thus reduce the quantity of air to be injected per barrel of oil produced. This may be expressed in simple terms. It is often assumed that, if the rock ahead of the combustion front could be preheated by bringing up heat from behind the front, the burning temperature would be the same and the quantity of oil burned would be less. On reflection, it becomes evident that this assumption is far from being

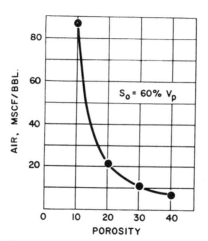

Fig. 1—Cumulative injected air-produced oil ratio as a function of porosity for burning at 1,000°F above reservoir temperature.

TABLE 1—EFFECT OF PRESSURE ON SPECIFIC HEAT AT CONSTANT PRESSURE (C_p) AT 1,260°F

Gas	150 psi	1,500 psi	Per Cent Change
Air	0.2575	0.2618	1.7
Nitrogen	0.2628	0.2674	1.8
Carbon Dioxide	0.27	0.285	5.6

automatically compatible with the chemical kinetics of the process. Some experiments seem to show that the amount of oil burned is independent of the burning temperature. In others, where regeneration of heat is achieved by injection of water, the burning temperature actually decreases slightly and less oil is burned. Thus, heat regeneration by water injection may turn out to be worthwhile. Heat regeneration by gas injection appears to suffer from the fundamental limitation of excessive air–oil ratios.

Pilot Tests and Their Interpretation

It is in the assessment of pilot field tests that some reservoir mechanics seem to display sloppy thinking. Admittedly, this is a difficult subject. How difficult, in fact, can only be adequately gauged by explaining the mechanics of a typical pilot test to, say, a professor of physics in a country devoid of sedimentary basins. When and if such a worthy individual grasps the fact that it is normal practice to conduct a test by injecting one fluid into the center well and recovering another fluid from adjacent wells in an unbounded five-spot, he is likely to be seriously discomposed.

In addition, when the basis for assessing the precise volumetric

content of the test formation is explained, he is quite likely to say that the results of a trial may be in error by several hundred per cent. If your professor is of a sardonic bent, he may well remark that the entire process smacks to him of conducting economically vital research on animal husbandry by using as specimens, on a dark night, the sheep which you think may be in some particular unfenced portion of a large pasture.

The only answer to such views is to emphasize the importance of getting answers and the practical impossibility of proceeding in any other manner. The fact could also be stressed that much excellent thinking has gone into scientific methods of deriving reasonably factual information from pilot tests. In this regard, the recent excellent work by Matthews and his co-workers[2] at Shell in evaluating pilot tests of water flooding need only be cited. Nevertheless, the results of many trials are still reported with a precision which cannot possibly be substantiated if the underlying assumptions are dispassionately reviewed. Others are *ex parte* in outlook or, if this is preferred, so influenced by wishful thinking that they must be to a greater or lesser extent scientifically suspect.

It might be of interest to review briefly a pilot test which, it can be argued, can be evaluated in more than one way.

In this instance, the test was one of miscible displacement using LPG followed by dry gas, and the pattern configuration was an inverted five-spot on a 40-acre spacing. The reservoir oil was above its bubble point. Some 31,000 bbl of propane and butane were injected into the reservoir followed by 696 MMcf of gas. Some 300,000 STB of oil were produced from the test wells. Observations indicated that the flow pattern was that of an isolated, inverted five-spot.

The miscible-displacement process was investigated to determine the volume of the reservoir that was actually swept miscibly. Having regard to the phase behavior of the methane-propane-butane system and the pressure distribution resulting from an injection pressure of 1,500 psi, it appeared that miscibility between oil and LPG was not maintained beyond a radius of 400 ft. Furthermore, the dry gas and LPG were not miscible beyond a radius of 300 ft. Since the injection pressure dropped 200 psi during the first month of gas injection, it was assumed that the lower limit on the radius of miscibility obtained. Beyond this radius, the displacement of oil occurred by a frontal gas-drive mechanism.

It is commonly believed that recovery by frontal gas drive is much less efficient than recovery by miscible displacement. While this is certainly true, the difference between the two processes is sometimes surprisingly small. This is because the complete displacement efficiency of a miscible process is offset, in large measure, by an unfavorable mobility ratio equal to the ratio of the viscosities of the displaced oil and displaced gas. By contrast, the displacement efficiency of an immiscible process is far from unity; but, paradoxically perhaps, just because the process is relatively inefficient, the effective mobility contrast is less unfortunate. Because displacement is incomplete, the relative permeability to the displacing phase is much lower than that to the phase being displaced. Consequently, the effective mobility ratio (obtained by integrating behind the displacing front) never attains very large values. Indeed, in gas-driving or water flooding, the effective mobility ratio before breakthrough never

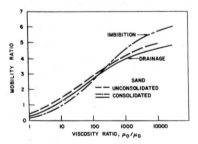

Fig. 2—Mobility ratio at breakthrough for gas- and water-drives.

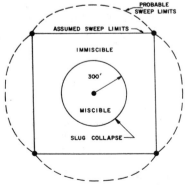

Fig. 3—Radius of miscibility, assumed and probable swept areas.

exceeds 6.0 even though the viscosity ratio of oil to water or gas to oil is 10,000:1. This is shown in Fig. 2. Accordingly, the areal sweep efficiency of an immiscible displacement is superior to that of a miscible one, and this fact must be kept in mind if a pilot test of miscible displacement is to be properly appraised. The effect is shown is Fig. 3.

The calculated results of the test indicate that the true pattern area was probably 60 acres, and not the 40 acres which was assumed on the basis that the miscible-displacement trail was successful and that no early slug breakdown had occurred. It is of interest, and (this is the point) it is possibly significant, that the total recovery experienced in the trial fell within the limits calculated on the basis of breakdown of the LPG slug and the immiscible displacement of oil by gas over an area 50 per cent larger than that rather arbitrarily assumed by the protagonists of a successful miscible displacement, i.e., on the assumption that no breakdown of the slug took place. This is shown in Table 2.

Interpretation Problems—The "Lore of the Core"

Basically, the point here is not that one interpretation is necessarily correct and the other not but, rather, that there are at least two possible explanations for the observed production history. In reservoir engineering, this state of affairs is so common as to be virtually endemic. In very large measure, it stems from the extremely complex physical situation which confronts those who would organize field trials. The fact is that the area in an unbounded pilot in which the trial is actually taking place (and, hence, a vital parameter affecting the volume of oil involved) is frequently uncertain. The three other parameters which affect this volume also are almost always subject to considerable error. These parameters are, of course, the thickness of the formation of interest, its porosity and its interstitial water saturation. In

TABLE 2—FOR RESERVOIR PRESSURE OF 1,207 PSI AND VISCOSITY RATIO OF 60, $2.5 < M < 3.2$

● 262,000 STB < Recovery < 340,000 STB (By LPG and Frontal Gas Drive)
● 199,000 STB < Recovery < 290,000 STB (By Frontal Gas Drive)
● Increase from Miscible Slug ≅ 60,000 STB

Actual Recovery = 300,000 STB

five-spot, the first is generally assessed from the bed thickness noted in each hole and a linear interpolation between holes. Many examples could be given where subsequent in-fill drilling has proved the original gross bed volume computed to be incorrect.

Nonetheless, the average gross thickness of the pay must be held to be one of the more reliable pieces of evidence which is generally available to the reservoir mechanic. It is in the sphere of deriving the net pay thickness and the average porosity, water saturation and permeability that *recherché* things seem to be done to the available data.

As is well known, there are three methods of obtaining reservoir information—by core analysis, logs and special well tests.

Now of the three methods the first, core analysis, seems to be the one in which most faith is placed. The reason for this faith is, it may be argued, partly psychological. In the work of reservoir mechanics, with its many uncertainties and natural variables, there is a tendency to grasp at the solid reality of cores with the fervor that a drowning man grasps at a straw. This instinctive human feeling has in the past also had a genuine scientific basis, for only in quite recent years can it be said that information derived from logs and well tests has reached an acceptable standard of accuracy.

There is no need here to go into the errors which bedevil routine core analysis or the uncertainties which result from the fact that, before any analysis can be made, the core must be abstracted from its normal environment. No one who has faced the problems of analyzing cores from fractured or dirty formations can be very optimistic concerning the accuracy of his results. Yet in spite of all this and in spite of the fact that core recovery is often poor, it seems that the results of core analyses are considered to constitute the ultimate criterion.

For example, it is rather well known that the accuracy of a log is judged by the extent to which it conforms with cores taken from the logged interval. It is really quite extraordinary that this should be so. It means, *in fact, that the log is being asked to conform precisely to the properties of the only portion of the formation which specifically was not logged.* To regard this as being paradoxical is seriously to understate; this is particularly true in the light of modern knowledge of reservoir heterogeneity. If a log is to conform precisely with analyzed cores, it means that a formation must have a layered-cake type of uniformity about the borehole axis, the radial extent of the uniformity varying with the nature of the log considered but being typically several feet. There has been no convincing or comprehensive evidence adduced to justify such a structure as the one universally characteristic of subsurface formations. Indeed, recent investigations of the distribution of heterogeneities made by Atlantic workers on outcrops,[3] and analogous studies at Gulf using the results of special well tests and core data in conjunction with computer studies of the effects of different distributions of heterogeneity, lead to the view that layered-cake homogeneity over significant areas is probably atypical of nature in general. The two schools of thought are illustrated in Fig. 4.

In Fig. 5 a rather interesting comparison is made. This figure shows core analyses of porosity as a function of depth together with a velocity log of the interval cored. It also shows (and this is unique) a synthetic velocity log made by measuring the velocity of the core samples and averaging them in the same manner that the velocity log automatically averages the velocities of the rocks surrounding the void created by the removal of the cores.

It is immediately evident that there is a marked similarity between the actual and synthetic log. It is equally evident that the two are far from identical and that over a number of depth intervals they diverge significantly. As might be expected,

a core porosity-velocity correlation gives better results for the synthetic log than it does for the real one.

The point, of course, is not that this elementary conclusion should be unexpected. Rather, the point is that conventionally it is assumed that cores will always correspond precisely with a log of the cored interval. Indeed, in the example shown, the conventional next step would be to "calibrate" the log of the cored interval with data derived from the only bits of rock not cored. This is an idea which might have appealed to Gilbert and Sullivan—"a paradox, a paradox, a most ingenious paradox"—but it is difficult for a dispassionate engineer to take it seriously. Yet the extraordinary convention of expecting logs to agree with unlogged rocks is now seemingly entrenched in reservoir-engineering lore. Also entrenched and equally difficult to justify in rational argument is the unsophisticated use of air permeabilities measured on core samples to delineate both the distribution of and average of the permeabilities of the uncored rocks not examined in the laboratory. Perhaps the ultimate in the current "lore of the core" is the measurement of relative permeabilities on odd samples and the arbitrary combination of these bits of data to provide what are, allegedly, average reservoir characteristics.

The Holistic Approach to Reservoir Mechanics

Criticism of these practices may perhaps be regarded as too harsh. It is reasonable to ask whether alternatives to them can be suggested. It is held that such alternatives do exist but that their use requires some

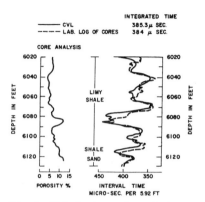

Fig. 5—Velocity log and synthetic velocity log from core analysis.

rethinking by reservoir engineers.

It would seem that a worthwhile approach to the problem of evaluating reservoirs sensibly and simply might have as its basis the philosophy of "holism". This philosophic doctrine was created by the late Field Marshal J. C. Smuts and holds that, in nature, determination is by wholes and not by constituent parts. Translated into reservoir parlance, it would say that what matters is the rock unit and not samples of arbitrary size that may have been taken from it.

This approach may be exemplified particularly by the use of special well tests. Thus, pressure build-up or fall-off analyses give directly the fluid transmissibility of a reservoir as a unit which exists in situ. Such a measurement represents an approach to reality which, as Warren, *et al*, have shown, is often very different from conventional arithmetic averages of permeability measurements made on cores.

On a holistic basis, the optimum method of estimating the average porosity of a formation would be to integrate the appropriate log readings over the interval of interest and to divide this area by the interval selected. It is instructive to apply this technique to the velocity logs shown in Fig. 5. It turns out that the travel time from the actual CVL is 385.3 microseconds and from the synthetic log made on cores, 384.0 microseconds. The percentage difference between these two figures is 0.34 per cent. Actually, such a percentage is less than the precision of either log, and the two travel times may be regarded as being identical. The fact that the two logs are significantly different is detail but show the same transit times may be interpreted as further support for the

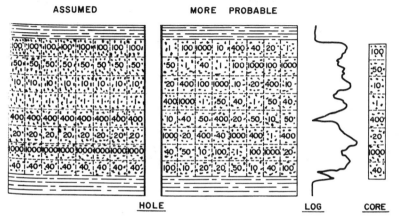

Fig. 4—The "Lore of the Core".

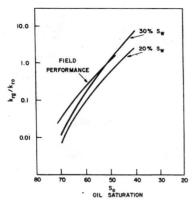

Fig. 6—Divergence between field k_{rg}/k_{ro} and k_{rg}/k_{ro} calculated from "holistic" assumption.

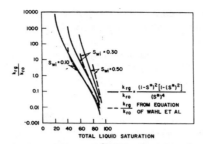

Fig. 7—k_{rg}/k_{ro} calculated for typical sandstones and k_{rg}/k_{ro} based on correlated performance data.

validity of the theory of random block distribution of reservoir heterogeneities.

The holistic approach to saturation and relative permeability estimation would be the following. For saturation, the average resistivity of the formation would be determined and the average formation factor found, either from the average porosity or by integrating under a proximity-log curve. The former method is to be preferred. From the average saturation, average relative-permeability curves for either consolidated or unconsolidated formations can be determined immediately by Corey's excellent approximate method. In this regard, it is pertinent to note that, when the simple holistic scheme is applied to the laboratory and field relative-permeability data of the Rosenwald pool which were presented by Smith and Henderson in 1957, the comparison is reasonably favorable. This is shown in Fig. 6. The divergence between the field and calculated data is in the direction that would be expected if nonuniform saturations exist in the vicinity of the producing wells.

Further evidence for the validity of using the average interstitial water saturation of a formation to characterize its relative-permeability characteristics is given in Fig. 7. Here, a comparison is shown between curves calculated by the Corey method and ones found by Wahl, Mullins and Elfrink of Socony Mobil by correlating field measurements of the gas–oil ratios of producing sandstone reservoirs.

The agreement between the two sets of curves can only be described as remarkable.

The holistic approach specifically excludes from consideration in unitization hearings elaborate arguments concerning cut-off porosities and cut-off permeabilities. It does not seem to be generally known that the cut-off approach has no validity unless a reservoir is ideally stratified and unless the effective mobility ratio is unity. Thus, it can be fairly argued that the present piecemeal approach to reservoir characterization is not only elaborate and complex, but one which is more often than not downright wrong.

The holistic approach, by putting its faith in physical measurements made in situ, is likely to lead to participation factors in water floods or other unitization proceedings which are not only more realistic than those found by present methods, but also ones which are fairer. The influence of the human element will not be eliminated, but it will be sensibly reduced.

The Microcosmic Approach

If for any reason logs and build-up data are unavailable but cores exist, the best approach to determining parameters of interest (at least on the basis of present knowledge) is the following. Find the core with median permeability and use the porosity, irreducible water saturation and relative-permeability characteristics of this core to represent those of the whole reservoir.

In strong contrast to the preferred holistic philosophy, this method may be regarded as the "microcosmic" approach to the determination of reservoir parameters. Its rationale lies in the probablistic reconciliation of core data with gross reservoir measurements that has been carried out at Gulf by Warren, *et al.* These studies have shown that the most probable core properties, based on agreement with gross reservoir properties measured in situ, may be derived from the core having median permeability.

Conclusion and Advice

In the foregoing a lot of ground has been covered. Some of that ground has been covered rather sketchily because of the exigencies of space. The subjects I have chosen to discuss happen to be ones of interest to me. It is an idiosyncratic, not a catholic, choice. Were I to dwell on all the difficulties and problems that beset the reservoir mechanic, I could write almost indefinitely. It is customary in our present parlous and rather earnest civilization for writers of a discussion on this type to end with some sort of message. I find this difficult to do because I am not didactically inclined. Nevertheless, I will try to summarize what I have written in the following terms.

Petroleum engineers and reservoir mechanics are daily engaged in a task of utmost importance and singular complexity. They are, quite literally, dealing with nature and its forces. Nature is a tough cookie. She is not going to be overcome by "gobble-degook" and technical double-talk. Chances are even slimmer if an engineer is pompous or otherwise takes himself too seriously. Nature can be persuaded to cooperate only if she is approached with patience, with humor and with some pretense to intellectual and scientific honesty. In the field of petroleum recovery, she is most readily seduced if she is approached with a sympathy and understanding developed by intelligent, basic laboratory research. That is my Message.

EDITOR'S NOTE: A PICTURE AND BIOGRAPHICAL SKETCH OF M.R.J. WYLLIE WERE PUBLISHED IN THE JULY, 1961 ISSUE OF JOURNAL OF PETROLEUM TECHNOLOGY.

PORE GEOMETRY, PERMEABILITY AND FLUID INTERACTION

Reprinted by permission of the South Texas Geological Society. Published in the *Bulletin of the South Texas Geological Society,* V. 25, No. 6 (February 1985), pp. 35-51.

PRODUCTIVE CHARACTERISTICS OF COMMON RESERVOIR POROSITY TYPES

Edward B. Coalson[1], Dan J. Hartmann[2], John B. Thomas[3]

ABSTRACT

We summarize here the initial results of an effort to catalog the petrophysical characteristics of the most common types of reservoir rocks, and to illustrate how such a catalog can be used to predict the performance of oil fields and the occurrence of subtle traps.

Judging from published descriptions of significant oil and gas reservoirs, nearly all reservoir rock-types are included in a relatively short list: "clean" sandstones; calcite, dolomite, or quartz cemented sandstones; shaly sandstones; solution porous sandstones; oolites and "granular" carbonites; sucrosic dolomite; chalks and diatomites; vuggy/moldic carbonates; and fractured rocks. Each of these rock types contains a characteristic class of pore geometry and thus displays characteristic relationships between porosity, permeability, capillarity, and saturations. Knowledge of these porosity types allows the geologist and engineer to predict zone, well, or field performance, at least qualitatively.

Such a rock-oriented approach was useful in evaluating a Bighorn Basin oil and gas field. Pay zones in the Tensleep (Pennsylvanian) and Phosphoria (Permian) Formations were not recognized without rock typing. Similarly, production of water updip from oil in the Frontier Formation (Cretaceous) was incorrectly explained without reference to the petrophysical attributes of the formation. This latter application is particularly illustrative of the usefulness of the "petrophysical catalog" when cores are not available for analysis.

— —

Editor's Note: This paper was the basis for Dan Hartmann's talk, delivered to the STGS at its luncheon meeting in December, 1984. It was included in the distribution package for the 1984 Denver Petroleum Exhibition and conference (DEPEC), and is published here with DEPEC's permission.

1. Bass Enterprises Production Co., Denver, Colorado
2. Independent geologist, Fredericksburg, Texas
3. Reservoirs, Inc., Denver, Colorado

INTRODUCTION

A key to success in the oil and gas industry is reducing the cost of finding and producing hydrocarbons. Reducing risk is another key. Both these goals can be advanced by perforating departmental "territorial" boundaries via synergistic, multi-discipline exploratory and development projects. We believe that the current growth edge for the industry is the simultaneous application of geological, geophysical, and engineering specialties to a common goal: understanding and predicting, in detail, subsurface traps and reservoir performance (Figure 1).

The success of such combined studies depends in large part on establishing the nature of the reservoir pore system and of fluid flow through it. This emphasis on pore properties is an important common ground between geology and engineering in synergistic studies.

PORE CLASSIFICATION SYSTEM: If one can characterize the pore system in a reservoir, then inferences can be made as to porosity and permeability ranges, "cutoff" water saturation, capillarity, and other useful parameters. The classification system proposed in this paper (Figure 2), and the postulated relationships between pore classes and petrophysical parameters, are based on data generated by the authors and gleaned from the literature, and tempered by many applications during a combined 40 years experience.

In our system, porosity classes are defined first by the "geometry" of the pores and, secondly, by pore size. The pore geometry categories are: intergranular, intercrystalline, vuggy, and fracture (Figure 3). Intergranular porosity is found in rocks composed of non-angular particles, regardless of their origin (clean sandstones, oolites, "fragmental limestones", coquinas, chalks, clean siltstones, pelletal grainstones, etc.). Intercrystalline porosity is found in rocks composed of very angular particles; most typically, these particles are crystals of dolomite in sucrosic dolomites, dolomite or quartz overgrown clasts in heavily-cemented sandstones, or clay particles in shaly sandstones. Very angular clastic grains also may be considered to form intercrystalline porosity. Vuggy pores are those equal to or larger in size than the particles making up the matrix. Fracture porosity results from cracks in the rock, having any size from less than 5 microns to cavern-size. Vugs and fractures may be "linked" (connected) or "dispersed" (not connected).

Macroporosity is porosity visible under 50x magnification, i.e. is larger than about 15 microns. Microporosity is porosity smaller than 5 microns, and is invisible at 50x. Mesoporosity is intermediate in size, and may be just beyond visibility at 50x, appearing as a granular texture to the rock.

Because mesoporosity and microporosity are not visible under the normal microscope, grain-size data are used to infer the presence

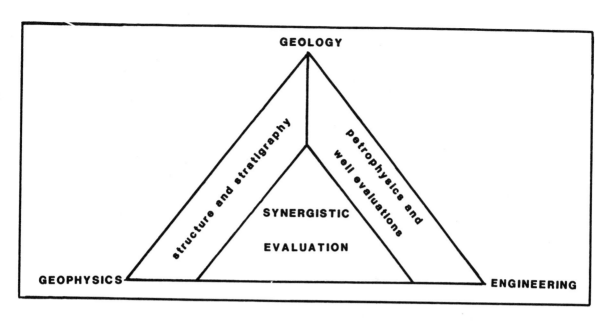

FIG. 1: Areas of cooperative evaluation between petroleum geologists, geophysicists, and engineers

---

FIRST-OBSERVED FEATURE: Pore geometry categories
intergranular - intercrystalline - vuggy - fracture

SECOND-OBSERVED FEATURE: Pore size classes
macroporosity (>15 u) - mesoporosity (15-5 u) - microporosity (<5 u)

example: *"intergranular macroporosity with minor intercrystalline microporosity" (e.g. slightly micritized oolite)*

FIG. 2: Visual (microscopic) system of classifying porosity

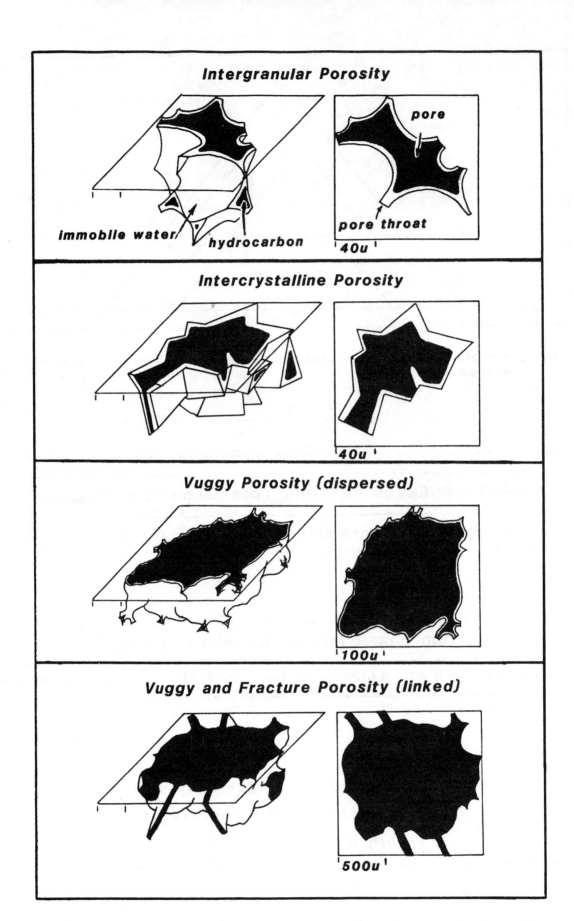

FIG. 3: Categories of pore geometry

of microporosity and mesoporosity. Although it is not necessarily true, microporosity must be assumed to be present when clay or micrite forms an appreciable part of the rock, as in chalks, shaly sandstones, and similar rocks. Mesoporosity should be assumed present when silt-sized particles are abundant.

These parameters, pore size and geometry, determine the pore classes (e.g. "intercrystalline mesoporosity") and often are sufficient to characterize a reservoir, particularly when modifiers are added describing the abundance and degree of connectedness of the pores (e.g. "well-connected, abundant, intercrystalline mesoporosity"). In other cases, the reservoir contains two or more distinctly different pore classes, each of which contributes to the petrophysical characteristics of the formation. These "multi-porosity" rocks can be described by linking class terms, e.g. "vuggy macroporosity with intercrystalline microporosity".

GRAIN SIZES, PORE SIZES, AND BASIC PETROPHYSICS: Each porosity class has characteristic petrophysical qualities, the dominant factor being pore size. A large body of literature documents the general interrelationships between grain size, pore size, porosity, permeability, immobile water saturation, and capillarity for intergranular porosity. The following generalizations, with minor modification, hold true for the other types of pore geometry.

Pore spaces formed between grains of an uncemented reservoir rock are the product of the size, shape, and packing of the grains. When grains are coarse, moderately well-rounded, and orthogonally packed, the porosity is about 33%. Pores are large (radius = 1/5 to 1/6 diameter of the grains) and uniformly distributed; pore throats also are large. Permeability is high and immobile water saturation low.

Smaller grain sizes produce smaller pores and pore throats, and consequently lower permeability and higher immobile water saturation. However, porosity still is about 33% regardless of grain size. Porosity measurements alone, therefore, will not predict the fundamental properties of the pores. On the other hand, by visually classifying the size of the pores in a given reservoir, and the total porosity, one can make reliable, qualitative assumptions about permeability, immobile water saturation, capillarity, and productivity before getting bogged down in a lot of log, core, and test data (Figure 4).

PORE SIZE, CAPILLARITY, AND WATER SATURATION PROFILES: Pore size relates directly to capillarity and water saturation profiles. In Figure 5, the large tube represents large, connected pores with low capillarity, capable of drawing only a short column of water into the tube. Blowing down the tube would displace the water with very little pressure. Smaller pores modelled by the smaller tubes draw a taller column of water. Greater pressure would be required to displace the water. The blotter represents very fine pore systems that draw water very high; one cannot displace it with any easily achievable pressure.

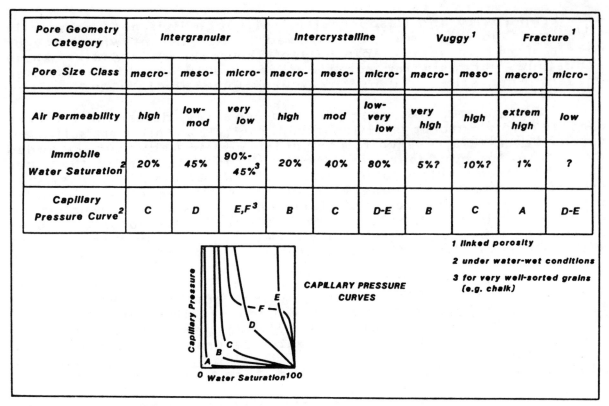

Pore Geometry Category	Intergranular			Intercrystalline			Vuggy [1]		Fracture [1]	
Pore Size Class	macro-	meso-	micro-	macro-	meso-	micro-	macro-	meso-	macro-	micro-
Air Permeability	high	low-mod	very low	high	mod	low-very low	very high	high	extrem high	low
Immobile Water Saturation [2]	20%	45%	90%-45% [3]	20%	40%	80%	5%?	10%?	1%	?
Capillary Pressure Curve [2]	C	D	E,F [3]	B	C	D-E	B	C	A	D-E

1 linked porosity

2 under water-wet conditions

3 for very well-sorted grains (e.g. chalk)

FIG. 4: Typical petrophysical parameters for single-geometry pore classes

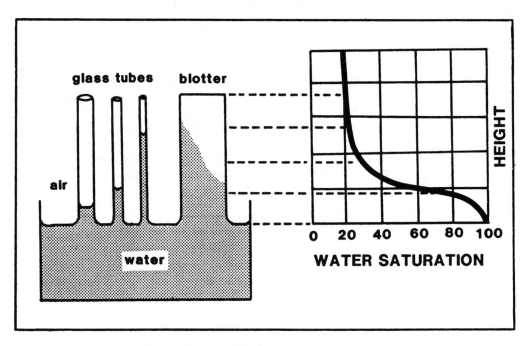

FIG. 5: Pore size, capillarity, and saturation profiles

Capillarity relates directly to the emplacement of oil in a trap. Large-pored rock will accept oil readily at low displacement pressure, i.e. with a short hydrocarbon column. Smaller-pored rock requires a longer hydrocarbon column to displace water. Microporous systems will not accept hydrocarbons without a long hydrocarbon column.

The length of hydrocarbon column possible in a trap is limited by trap closure. If pores are small, closure must be large or else no oil or gas will be emplaced. Conversely, even very slight closures may be charged with hydrocarbons if the pores are large enough.

SUBSURFACE EXAMPLES

CARBONATE RESERVOIR: An example of a reservoir having more than one pore type illustrates the qualitative use of visual pore descriptions to evaluate wells. In Figure 6, rock type A contains abundant vuggy (solution mold) macroporosity connected by inter-crystalline mesoporosity; all fluids must flow through the inter-crystalline pores to reach the vugs. Rock type B contains somewhat less abundant intergranular (oolitic) macroporosity. These are the petrophysical characteristics of each porosity class, based on Figure 4:

ROCK TYPE A — Permeability is low to moderate because the macropores are dispersed and contribute little to permeability.

- Capillarity, controlled by the intercrystalline pores, is moderate; immobile water saturation is lower than for rock type B because of the vuggy macropores.

- Oil or gas in place, if closure is ample to allow emplacement, should be higher than for rock type B.

ROCK TYPE B — Permeability is higher than rock type A.

- Capillarity is moderate because of large pore throats.

- Immobile water saturation is low, about the same as rock type A.

- Oil or gas should be easily emplaced, but original oil or gas in place should be less than rock type A.

- Recovery factor should be better than rock type A.

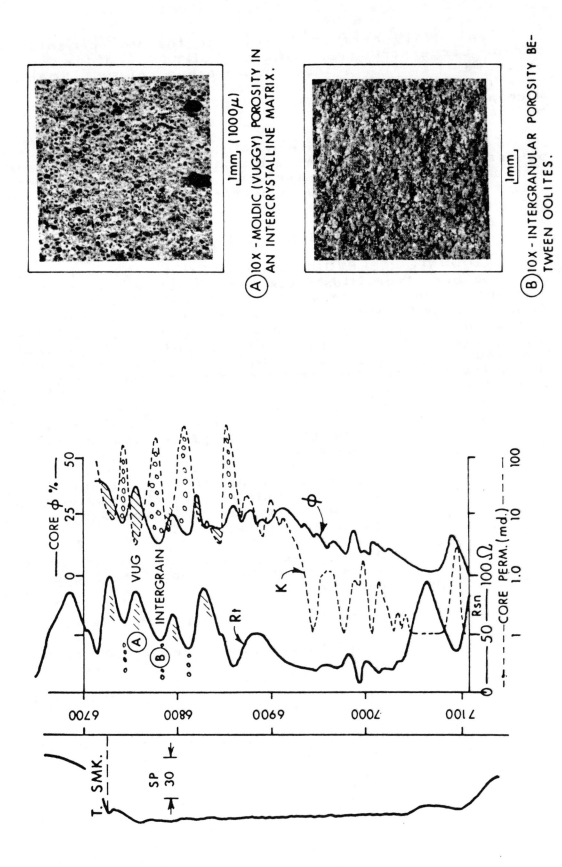

SMACKOVER LEWISVILLE FIELD —
MOBLEY NO. 1 MOORE, LAFAYETTE CO., ARK.

(A) 10X - MOLDIC (VUGGY) POROSITY IN AN INTERCRYSTALLINE MATRIX.

1mm, (1000μ)

(B) 10X - INTERGRANULAR POROSITY BE-TWEEN OOLITES.

1mm

FIG. 6: Petrophysics of a Carbonate Reservoir

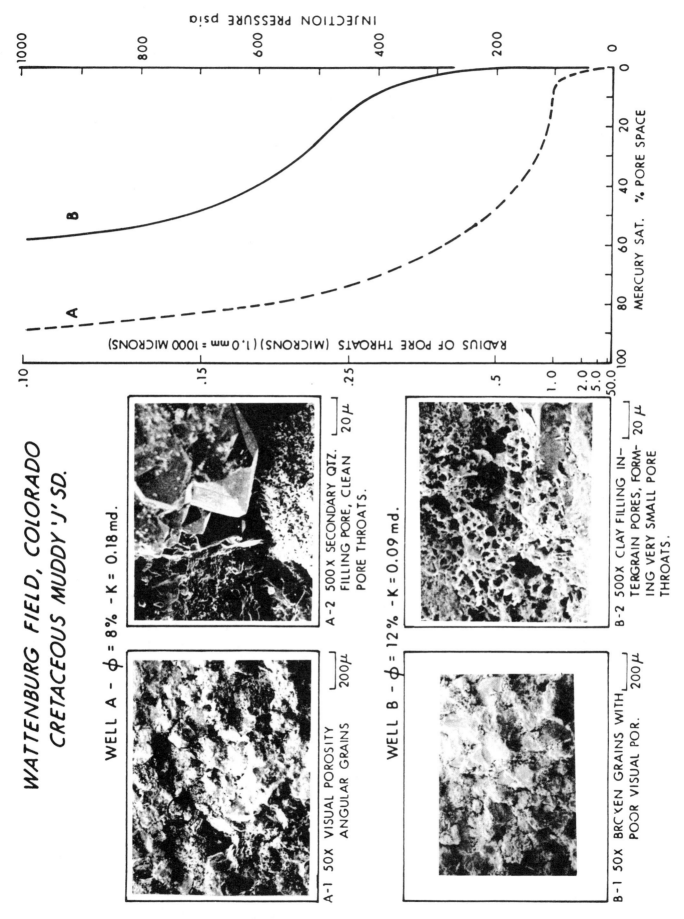

WATTENBURG FIELD, COLORADO
CRETACEOUS MUDDY 'J' SD.

INJECTION PRESSURE psia

MERCURY SAT. % PORE SPACE

RADIUS OF PORE THROATS (MICRONS) (1.0 mm = 1000 MICRONS)

WELL A - ϕ = 8% - K = 0.18 md.

A-1 50X VISUAL POROSITY ANGULAR GRAINS

A-2 500X SECONDARY QTZ. FILLING PORE, CLEAN PORE THROATS.

WELL B - ϕ = 12% - K = 0.09 md.

B-1 50X BROKEN GRAINS WITH POOR VISUAL POR.

B-2 500X CLAY FILLING INTERGRAIN PORES, FORMING VERY SMALL PORE THROATS.

FIG. 7: Petrophysics of a Sandstone Reservoir

67

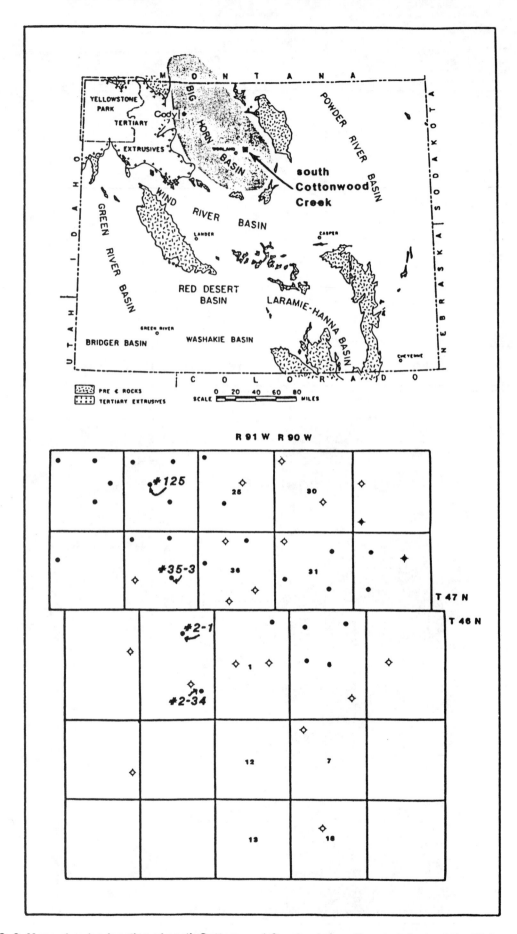

FIG. 8: Maps showing location of south Cottonwood Creek area on the east flank of the Bighorn Basin (after Pedry, 1975) and locations of wells referred to in text

68

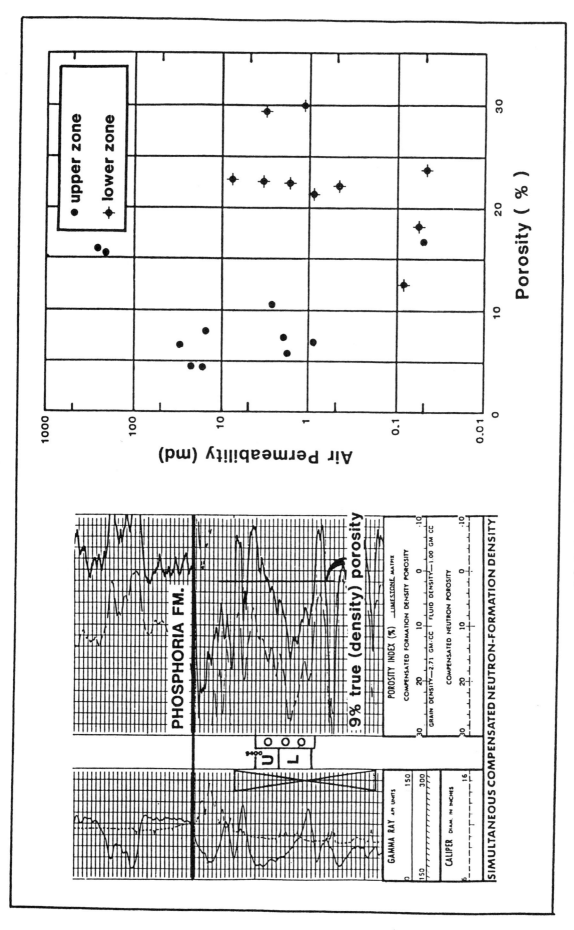

FIG. 9: Log and core data from #2-1 well

The two rock/porosity types occur in 10-20' thick beds throughout the upper 100 feet of the formation. On the logs, type A rocks (vuggy) are the high porosity (30-40%), lower permeability layers, while type B rocks (granular), are the lower porosity (15-20%), higher permeability (>100 md) layers. Resistivity is highest in rock type A, even though it has better porosity. This is due to rock type A having a greater cementation exponent. After applying the proper cementation exponent to the two porosity types, we find that saturation ranges from 20-30% throughout the zone.

INTERCRYSTALLINE POROSITY IN SANDSTONES (Figure 7): Representative samples from two J Sandstone (Cretaceous) wells from the Denver Basin are shown in Figure 7. The extreme magnification required to see the porosity in both samples tells us that the pore system is mesoporous to microporous.

Samples from well A at 500x show intercrystalline mesoporosity and microporosity of a triangular nature, between quartz crystals attached to the grains. The pore throats are small, therefore permeability is low. Based on Figure 4, capillarity and immobile water saturation should be high.

Samples from well B at 500x present the aspect of a honeycomb. The intercrystalline pores are more abundant than in well A, but are even smaller and isolated between clay layers disseminated throughout the intergrain spaces. Though total porosity is greater than well A, permeability is less.

Capillary pressure curves verify these qualitative estimates. Well A requires 100 psi to initiate mercury emplacement, abruptly reaches 40% mercury saturation, then requires significant added pressure to further increase mercury saturation. This well is gas productive at 60% water saturation, but has low rates. It does not produce any water.

Well B requires 400 psi to begin gas emplacement, and does not achieve 40% water saturation until reaching 600 psi. This well would have to be much higher in the trap to achieve the same saturation as well A. It should not be expected to produce at the same rate as well A. In fact, it does not. Similar to well A, well B does not produce water, even at a log-calculated water saturation of 60%.

CASE HISTORIES

Qualitative petrophysical analyses proved useful at Cottonwood Creek field, Wyoming. While standard geologic and engineering approaches misled the operator in at least two of three productive zones, combined geologic and engineering principles were diagnostic.

PHOSPHORIA FORMATION: The #2-1 Cottonwood Creek well was drilled to test Phosphoria (Permian) Dolomite south of the established production at Cottonwood Creek field (Figure 8). In this area, all Phosphoria dolomites were known to be oil-saturated, due to being near the top of an oil column of at least several thousand feet. It was noted that the best wells in the field had 16-18% porosity, with an assumed cutoff porosity of 9%. It was also assumed that permeability was directly proportional to porosity.

The Phosphoria core in the #2-1 well had 19 feet of dolomite that was both oil saturated and porous. The rock was described as "vuggy, fractured dolomite bleeding oil and gas, with excellent porosity." The high oil saturation and good porosity were later confirmed by well logs (see Figure 9). In fact, porosities in the lower 15 feet of the productive bench (9411-26) were extremely high: 20-30%. Apparently, the #2-1 was a very successful well.

Unfortunately, through-pipe tests were disappointing. After perforation and acid stimulation, the well IP'd for only 85 BOPD, declined rapidly, and produced only 18,900 BO in the first four years of production. Company geologists blamed the poor performance on "bad completion"; engineers blamed "bad reservoir rock."

A closer examination of the logs and core analyses showed that permeabilities in the porous lower zone were poor (maximum 10 md) compared to those in the less-porous upper zone, which had permeabilities well above 10 md (Figure 9). The lower dolomite (9411-26) turned out to have intercrystalline microporosity with very large (1-10 cm), dispersed vugs. The upper zone (9399-9408) consisted of vuggy ("fenestral") macroporous dolomite of a type in which the vugs are aligned parallel to bedding and tend to connect well horizontally.

The 15 feet of microsucrosic and very porous dolomite, even though oil-saturated, probably was non-productive. The thinner, fenestral dolomite (8.5 feet) was actually producing most, if not all, of the oil. In fact, essentially all the production was coming from dolomite with less than (not more than) the "cutoff" porosity.

This analysis didn't answer all the questions. The 8.5 feet of fenestral porosity should have produced more oil than recovered from the #2-1, compared to wells in Cottonwood Creek field. It is likely that the cause for low recovery is neither "bad completion" nor "bad reservoir rock", but limited reservoir and thinner pay than originally estimated.

TENSLEEP FORMATION: Because the #2-1 well was located at the top of a seismically-defined anticline (Figure 10), the well was drilled to the Tensleep (Permo-Pennsylvanian) Sandstone. The core and logs (Figure 11) indicated that the Tensleep was oil-saturated but poorly productive or nonproductive due to low porosity and permeability. The well was plugged back to test the Phosphoria.

However, the core permeabilities were high given the low porosity. Why? Later examination of the cores showed the Tensleep to be heavily quartz-cemented, with a few large, smooth-walled, moderately-well connected pores shot through a more-dense matrix. By comparison with similar, heavily quart-cemented Flathead (Cambrian) Sandstones at Lost Soldier field, Wyoming, (Coalson, Hartmann, and Thomas, 1983), the Tensleep in the #2-1 well was re-evaluated as potentially productive. (Conversely, the Tensleep's somewhat lower permeability and the higher viscosity of the Tensleep crude may preclude commercial production). This zone now is slated to be tested through pipe after depletion of the Phosphoria and Frontier Formations.

FRONTIER FORMATION: The Fourth Frontier zone (upper Cretaceous) in the area of the #2-1 well demonstrates that a petrophysical approach to problem solving can be useful even when minimal data are available. In this case, changes in rock quality explain production of water updip from oil.

The South Cottonwood Creek anticline is concordant at Phosphoria (Figure 10) and Fourth Frontier (Upper Cretaceous) levels. Closure on the Frontier is approximately 125 feet.

The #2-1 well encountered the Fourth Frontier at -6910 feet, near the highest point of the structure. The zone was DST'ed and flowed gas at maximum rates of 5700 MCFD. No water was recovered, proving a gas cap is present.

A subsequently-drilled well, the #35-3 Cottonwood Creek, was drilled on the north plunge of the anticline and completed in the Fourth Frontier, pumping 64 BO and 239 BWPD. This well apparently was drilled near the oil-water contact at -7050 feet.

A third well, the #2-34 Cottonwood Creek, was drilled on the south flank of the anticline, and encountered the Fourth Frontier at -6990, updip from the #35-3 well and downdip from the #2-1 well. The operators expected to produce water-free gas or oil. After the Fourth Frontier was acidized, it produced water at the rate of 48 BPD with only traces of oil and gas. In hopes of stimulating into nearby pay, the zone was hydraulically fractured, after which it produced 360 BWPD.

Explanations proposed for this structurally-high water production included 1) a long transition zone, 2) multiple reservoirs within the Fourth Frontier, 3) hydrodynamic tilting of the oil-water contact by water-flow in the Fourth Frontier, and 4) water production from out of zone. Traditional geologic and engineering studies did not yield a solution.

Subsequent evaluations utilized a petrophysical view point. A core cut from the #2-1 well proved to have reasonable porosity and permeability, but less than the next-nearest Frontier core (Amoco, 125 CCU well, Figure 12). This seemed consistent with a uniform thinning and shaling out of the Fourth Frontier southward; in fact, the Fourth Frontier proved to shale out just south of the anticline. These data not only eliminated the possibility of northward water flow in the Fourth Frontier, but suggested

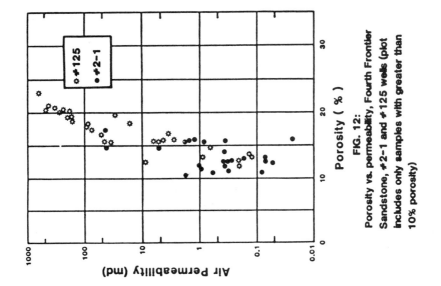

FIG. 12:
Porosity vs. permeability, Fourth Frontier Sandstone, #2-1 and #125 wells (plot includes only samples with greater than 10% porosity)

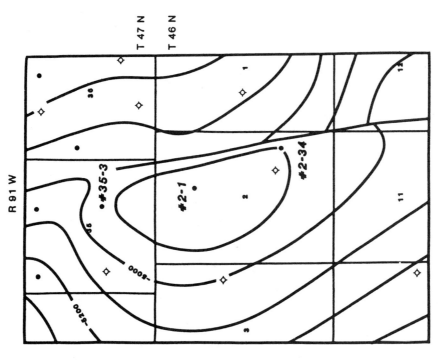

FIG. 10: Phosphoria structure, south Cottonwood Creek area (c.l. 100')

73

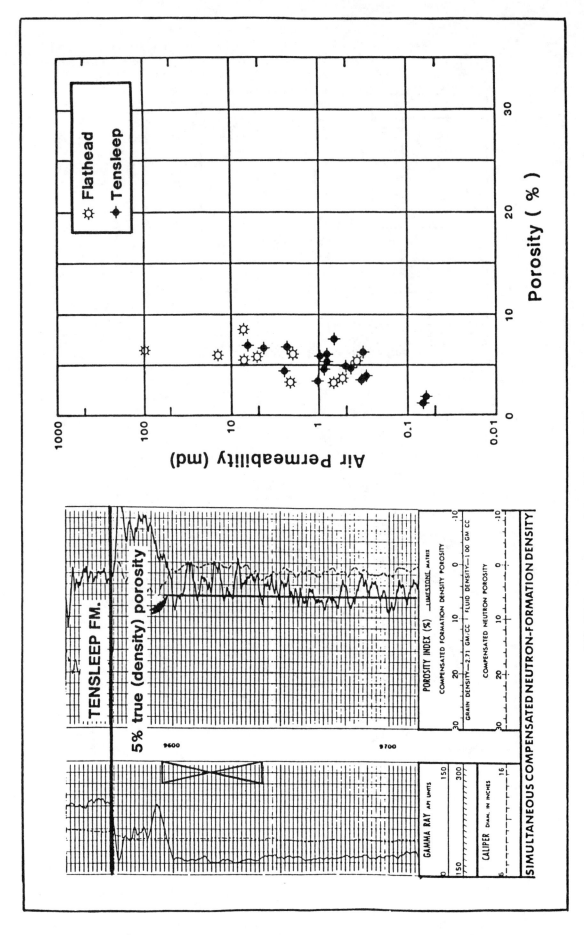

FIG. 11: Log and core data, ≠2-1 well, Tensleep formation

that lateral changes in pore geometry might cause the water production. Westward hydrodynamic tilting seems unlikely because water flow would have to occur upwards along the fault plane bounding the anticline (Figure 10). It was discovered that there are few shale beds within the Fourth Frontier to provide barriers. There was no indication of bad cement to cause suspicion that out-of-zone production was occurring. By process of elimination, changing pore geometry was left as the most likely explanation for the water production. This essentially eliminated the south plunge of the anticline as a prospective area.

CONCLUSION

The qualitative use of petrophysical data to solve everyday oil-field problems is well established. By determining reservoir pore geometry and pore size, the investigator can infer the petrophysics of the formation and qualitatively predict reservoir performance. What is now needed is not a theoretical basis for the approach, but rather: 1) more quantitative data relating rock/pore types to petrophysical parameters, and 2) professionals willing to integrate data of disparate types into unified evaluations of wells, fields, and prospects. We intend to present the quantitative data in later publications in order to aid those interested in better understanding the "container" from which hydrocarbons are extracted: the pore space.

REFERENCES

1. Pedry, John J., 1975, Tensleep fault trap, Cottonwood Creek field, Washakie County, Wyoming, in Geology and Mineral Resources of the Bighorn Basin, Guidebook, Wyo. Geol. Assoc., 27th Ann. Field Conf., pp. 211-219.

2. Coalson, Edward B., Dan J. Hartmann, and John B. Thomas, 1983, Applied Petrophysics in Exploration and Exploitation, privately published, 241 pp.

Petroleum Geologists. Published in the *Bulletin of Canadian Petroleum Geology*, V. 24, No. 2 (June 1976), pp. 225-262.

MERCURY CAPILLARY PRESSURE CURVES AND THE INTEPRETATION OF PORE STRUCTURE AND CAPILLARY BEHAVIOUR IN RESERVOIR ROCKS

N. C. WARDLAW and R. P. TAYLOR[1]

ABSTRACT

Mercury injection and withdrawal capillary pressure curves provide information regarding the efficiency with which a nonwetting phase can be withdrawn from a pore system. Withdrawal (recovery) efficiency varies among samples of differing lithology and can be explained in terms of the geometric aspects of the pore systems, some of which can be interpreted from the capillary pressure curves themselves and others only by direct observation of rocks and resin pore casts. Withdrawal efficiency is defined as the ratio of the volume of mercury withdrawn from a sample at minimum pressure to the volume injected before pressure was reduced.

Some features of capillary pressure curves result from experimental procedure and the influences of sample size and shape in conjunction with effects that are peculiar to the external boundary region of the sample. Features which are not typical of an infinite pore system must be recognized before interpretations are made.

From experiments on rock samples as well as from theoretical consideration of artificial models it is found that withdrawal efficiency tends to increase with increase in initial saturation. Withdrawal efficiency also increases as throat to pore diameter and volume ratios increase, as the homogeneity of throat and pore sizes increases, and as the number of connections (throats) per pore increases. For a particular sample, withdrawal efficiency is also dependent on saturation history.

The initial injection curve combined with the final withdrawal and final re-injection curves for a sample is sufficient to predict withdrawal efficiency from any initial saturation and to construct any desired pair of withdrawal and reinjection curves for intermediate saturations. Experimental determination of the dependency of mercury withdrawal efficiency on saturation history, and the ability to predict withdrawal efficiency for any initial saturation, may be useful in understanding nonwetting phase movements due to capillary effects in hydrocarbon reservoirs.

INTRODUCTION

Mercury is a nonwetting liquid and will enter capillaries or pore spaces only when pressure is applied. As pressure is increased, mercury progressively invades smaller spaces and, provided a sufficiently high pressure is attained, a pore system will become totally saturated. As pressure on the mercury is reduced, progressive ejection or withdrawal from the smaller to the larger spaces occurs.

[1]Department of Geology, University of Calgary, Calgary, Alberta. The authors are grateful to Mr. Lee Helmer of Shell Canada, who made available core samples and mercury injection data for large core samples of Midale dolomite, and Mr. D. P. Andrews of Imperial Oil who provided samples of Becher dolomite. Dr. A. E. Oldershaw is thanked for his assistance in operating the Cambridge Stereoscan 600 scanning electron microscope. The project was funded by the National Research Council of Canada and these funds are gratefully acknowledged. Mr. Lee Helmer and Mr. Stephen Cheshire are thanked for critically reading the manuscript. The authors also have benefitted greatly from discussions with Drs. Frank McCaffery and Norman Morrow.

Purcell (1949) described a method for forcing mercury under pressure into the evacuated pores of a sedimentary rock and for determining the relationship between capillary pressure and volume of mercury injected. Others have studied the ejection or withdrawal of mercury from a mercury saturated pore system as pressure is reduced (Pickell *et al.*, 1966).

Mercury injection and withdrawal capillary pressure curves transformed into dimensionless capillary-pressure functions can be used to predict the behaviour of other fluid pairs, such as oil and water, during pressure changes in the pore systems of reservoir rocks (Dumoré *et al.*, 1974). The mercury injection procedure has the advantage of being simple and rapid compared with direct methods in which typical reservoir fluids are used. A second major use of injection and withdrawal capillary pressure curves is to interpret the size and geometry of pores and throats. Pores may be defined as local enlargements in a pore system which are linked by smaller connecting spaces referred to as throats. Use of the term "throat" does not imply any particular shape.

The porosity of a sample is the "pore volume" expressed as a percentage of the bulk volume. "Pore volume" in this context, appearing on the captions of graphs, includes both pore and throat volume. Where it is necessary to distinguish the volume of pores from the volume of pore-connections or throats this is made clear.

The objectives of this paper are to evaluate various factors that influence the form of mercury injection and withdrawal curves, and to define the extent to which the form of the curves can be used to interpret pore and throat geometry. Following descriptions of the experimental procedures and definition of terms, the paper is divided into three sections:

(1) consideration of features of capillary pressure curves that are related to experimental procedures and to the influences of sample size and shape, in conjunction with effects that are peculiar to the external boundary region of the sample; that is, features of capillary pressure curves that are atypical of infinite pore systems;

(2) consideration of idealized, model pore systems as a means of illustrating how pore and throat geometry influence the form of capillary pressure curves;

(3) presentation and interpretation of injection-withdrawal-reinjection capillary pressure curves for selected sedimentary rocks. Emphasis is placed on the efficiency with which mercury, and by analogy other nonwetting fluids such as oil, can be withdrawn from various types of sedimentary rocks as pressure on the fluid system is reduced.

Apparatus and sample preparation

A Ruska mercury-injection capillary pressure apparatus (Model 1057) was used; the procedure is described in the operating manual and by Purcell (1949).

A cored cylinder of rock, previously cleaned with acetone in a Soxhlet's apparatus and dried in a vacuum oven, is placed in the sample chamber and evacuated to a pressure of 20μ of mercury. Mercury is then raised to a reference mark in a duct leading from the top of the sample chamber.

Nitrogen gas is admitted in increments to increase the pressure on the mercury surrounding the sample, and invasion of the sample by mercury is indicated by recession of the mercury-gas interface from the reference mark. The volume of invading mercury for the pressure increment is then measured by advancing the piston of the displacement pump until the mercury returns to the reference mark. This procedure of incrementally increasing the pressure and advancing the pump is repeated until the final injection pressure is achieved. Pressure is then decreased and the volume of mercury withdrawn from the sample for each pressure decrement is measured.

A pressure-volume correction curve is established for the apparatus by making a run without a sample in the chamber. The volume readings obtained during a sample run are then corrected by making appropriate subtractions determined on the blank run at corresponding pressures. In this way, compression effects and the effects of forcing mercury into small spaces in the interior of the instrument are eliminated.

During the withdrawal procedure, corrections must be made for variation in "backlash" on the pumpwheel. Where withdrawal is followed by reinjection of mercury, it is necessary to follow a similar history of pressure changes on the blank run in order to make appropriate corrections. During a sample run, the chamber is not opened between the withdrawal and reinjection portions of the test.

Temperature changes cause changes in volume of the mercury in the instrument, but would be a source of error only if they were to occur during an individual run. In a room with thermostatically controlled temperatures such problems are minimized.

Definition of terms

For water-oil or gas-oil systems, the displacement of a wetting phase by a nonwetting phase is referred to as drainage and the reverse process is described as imbibition. In the mercury injection procedure, the nonwetting phase is mercury and the evacuated space has air, at a pressure of only 20μ of mercury, and a small mercury vapour pressure related to temperature. Saturation of the sample with mercury is accompanied by compression of residual air, rather than drainage of a wetting phase such as would occur during the displacement of water by oil.

If pressure is reduced following injection, some mercury is expelled from the pore system while a portion is trapped in the system and retained. To avoid confusion, and following the usage of Pickell *et al.* (1966), the terms "injection" and "withdrawal" will be used in discussion of the air-mercury system, and these terms can be thought of as analogous to drainage and imbibition, respectively.

The following terms and symbols are used in describing mercury injection and withdrawal capillary pressure curves (Fig. 1).

Injection curve (I): the relationship of volume of mercury entering the pore system to pressure for the range of pressures from minimum to maximum.

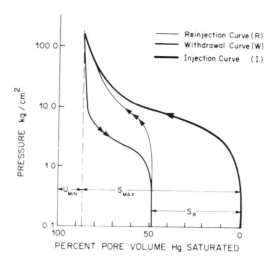

Fig. 1. Capillary pressure diagram illustrating terms defined in the text.

Withdrawal curve (W): the relationship of volume of mercury leaving the pore system to pressure for the range of pressures from maximum to minimum.

Minimum unsaturated pore volume (U_{min}): this is the percentage of the pore volume that is not invaded by mercury at the maximum pressure attained (120 kg/cm²).

The term "irreducible saturation" is inappropriate in this context and should be restricted to situations where wetting phase saturation appears to be independent of further pressure increases on a nonwetting phase. The volume of invaded pore space at maximum pressure is S_{max} ($S_{max} = 100 - U_{min}$) and saturations at intermediate pressures may be designated S_1, S_2, etc.

Residual mercury saturation (S_R): the volume of mercury retained in the pore system when the pressure is reduced from the maximum to the minimum for the system, expressed as a percentage of the total pore volume.

Reinjection curve (R): illustrates the volume of mercury entering the pore system as pressure is increased from the minimum pressure achieved on withdrawal to the maximum pressure attainable with the apparatus.

Mercury can be injected to a saturation less than the maximum, withdrawn to a residual saturation at minimum pressure and then reinjected, with the process being repeated to progressively higher maximum pressures. Two such hysteresis loops are illustrated on Figure 2 with the injection, withdrawal and reinjection portions identified as I_1, W_1 and R_1 on the first loop and I_2, W_2 and R_2 on the second. The residual mercury saturation after the first withdrawal is S_{R1} and after the second S_{R2}. The mercury saturations S_1 and S_2 are those achieved at the maximum pressures of the first injection (I_1) and the second injection (I_2). In the case illustrated, S_2 is equal to ($100 - U_{min}$), since the maximum pressure attainable with the equipment has been reached.

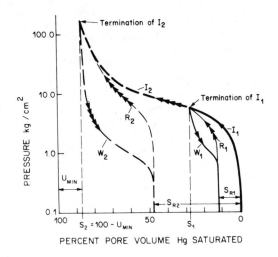

Fig. 2. Two injection-withdrawal hysteresis loops illustrating terminology defined in text. Arrows indicate the sequence of injection-withdrawal events.

Withdrawal efficiency (W_E): is the ratio, expressed as a percentage, of the volume of mercury withdrawn from the sample at minimum pressure to the volume injected before pressure was reduced. That is:

$$W_E = \frac{S - S_R}{S} \times 100$$

Where more than one hysteresis loop has been determined experimentally, corresponding to different degrees of initial saturation (such as S_1 and S_2 on Fig. 2), it is possible to define more than one withdrawal efficiency term (W_{E1} and W_{E2}) and to compare withdrawal efficiencies at various initial saturations for a given sample.

The form of the final withdrawal and reinjection curves appears to be unaffected by previous multiple cycles of injection, withdrawal and reinjection. Similar results are obtained for samples of the same material subjected to one hysteresis cycle as are obtained after six hysteresis cycles.

FEATURES ATYPICAL OF INFINITE PORE SYSTEMS

Some features of capillary pressure curves result from experimental procedure and the influence of sample size and shape, in conjunction with effects peculiar to the external boundary region of the sample. These features, which are not typical of infinite pore systems, should be recognized before making other interpretations from capillary pressure curves.

Inaccuracies of U_{min}

U_{min} is the percentage of the total pore volume that is not occupied by mercury at the highest pressure attained, and tends to be affected by inaccuracies of pore volume measurement. Pore volume commonly is obtained as a difference between bulk volume and grain volume and, if bulk volume is caliper measured or measured in an instrument other than the pressure chamber of the mercury injection apparatus, U_{min} is unlikely to be a true measure of unsaturated pore space.

Bulk volume measurements suffer from errors due to spaces at contact points between the sample and the chamber wall (Taylor and Wardlaw, 1975) and errors due to invasion of exterior pores with mercury. The extent of invasion depends on the size of the exterior pores and the pressure on the mercury. During bulk volume measurements in a pycnometer, mercury is pumped into the pycnometer chamber to cover the sample and enter an overlying duct to a reference mark. Pressure on the mercury in contact with the exterior of the sample depends on duct size and the height of the overlying head of mercury. The effects of both factors on bulk volume measurements must be determined.

The reliability of U_{min} may be judged in the following way. Injection curves A and B (Fig. 3) for two different samples have the same U_{min} but, in case A, the curve runs parallel to the pressure axis over a considerable range of pressures, indicating that no further mercury is entering the sample. One would suspect that U_{min} is nonexistent space created by using a reference pore volume that is too large. In the case of curve B, mercury is still entering the sample up to the maximum pressure, and U_{min} is more likely to represent a true value of pore space too small to be invaded at the maximum pressure attained.

There is a common tendency to overestimate pore volume and therefore U_{min} in small samples or samples of low porosity, because of small spaces at contact points between sample and chamber wall (Taylor and Wardlaw, 1975).

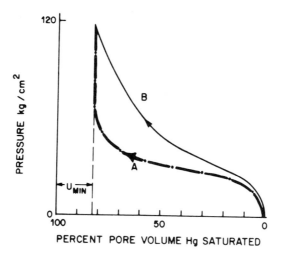

Fig. 3. Injection curves for two samples A and B with similar values of U_{min}. For sample A, the curve is parallel to the pressure axis over a considerable range of pressures suggesting that full saturation may have been reached and the U_{min} is caused by overestimating the reference pore volume. Mercury is still entering the sample at maximum pressure in the case of Curve B and U_{min} probably represents unsaturated pore volume.

Effects of sample size

Pores are accessible through throats and, on the exterior surface of the sample, there are many random intersections of pores which act as artificially enlarged throats. Thus, the pore system at the surface of the sample is not representative of the infinite pore system away from the surface. Sample size and shape influence surface area to volume ratios and therefore can be expected to affect the form of capillary pressure curves. The effects of size and surface area were investigated for samples of Indiana Limestone.

Mercury injection and withdrawal curves were obtained for cored cylinders with bulk volumes from 1.7 to 25 cc (Fig. 4). Except for slight variations in the minimum unsaturated pore volume (U_{min}), due perhaps to small inaccuracies of pore volume measurements, the capillary pressure curves are closely similar, and it can be concluded that sample size has not significantly affected the form of the curves for Indiana Limestone.

A further experiment was conducted with two cylinders of Indiana Limestone of approximately 9 cc bulk volume. One cylinder was sealed with a thin coat of epoxy on all surfaces except for one end, while the other cylinder was left uncoated. Injection-withdrawal curves were obtained, the coated sample being placed with the single uncoated end

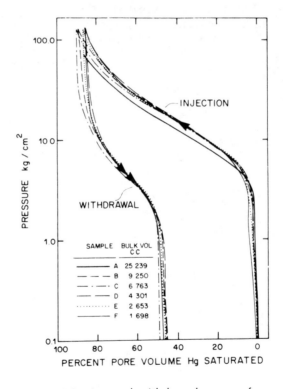

Fig. 4. Capillary pressure injection and withdrawal curves for samples of different size but uniform composition. All samples are from the same block of Indiana Limestone (14% porosity, 2 md permeability). Bulk volumes (cc) are shown on the diagram. Samples supplied by Indiana Limestone Company, Inc., Bedford, Indiana, U.S.A., and desigated as "P and B high porosity limestone".

uppermost in the porosimeter chamber. The curves are closely similar except for the injection curve in the region A-B (Fig. 5), marking the early stages of injection. The rounded shoulder in the region A-B of the uncoated sample must be due to the effects of invasion of surface spaces, including entry via surface-enlarged throats. These effects have been largely eliminated in the coated sample, which has only a small surface for mercury entry and is more representative of an infinite pore system. The curve for the uncoated sample approaches that of the coated sample at 20 per cent saturation which, for the cylinder used, represents the total pore volume in an outer layer 1 mm thick. Presumably, then, the system is not behaving as an infinite pore system until mercury has achieved at least this depth of penetration from the surface.

Entry pressure, displacement pressure and threshold injection pressure are terms used synonymously for the pressure at which significant invasion of a sample with mercury occurs but are usually not precisely defined. As seen from Figure 5, surface throat-enlargement (uncoated sample) allows entry of mercury at lower pressures than would be the case for an infinite pore system (coated sample) and, from comparison of the coated and uncoated samples, it appears that for an uncoated sample of the type illustrated the displacement pressure should be read at about 20 per cent saturation.

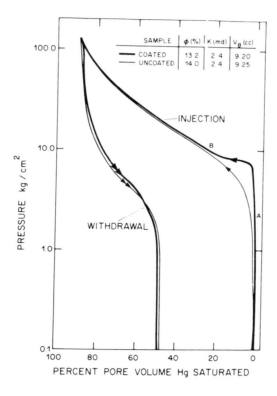

Fig. 5. Capillary pressure injection and withdrawal curves for two similar samples of Indiana Limestone (porosity [φ], permeability [K] and bulk volume V_B indicated on diagram). Both samples are cylinders of 1.86 cm diameter, but one was coated with a film of epoxy except for one end (reducing surface area for entry of mercury) and the other sample was untreated. Gas permeability determined prior to coating.

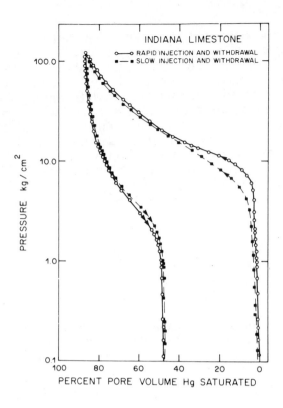

Fig. 6. Capillary pressure curves for two cylinders of Indiana Limestone of approximately 8 cc bulk volume. For one sample injection and withdrawal was done rapidly, allowing only 10 seconds between successive readings. For the second sample, sufficient time was allowed at each pressure change for mercury movements to come to equilibrium.

Rates of injection and withdrawal

Figure 6 illustrates the results of two experiments. In one, mercury was injected and withdrawn rapidly, allowing only ten seconds between pressure increments. In the other, enough time was allowed at each pressure increment for movement of mercury to cease. The effect of rapid injection is to increase the apparent displacement pressure. Less mercury is entering the system for a given pressure because insufficient time has been allowed and equilibrium saturation has not been achieved. For Indiana Limestone, there is little difference between the curves for rapid and slow withdrawal.

Times to equilibrium for each pressure increment were measured during injection and withdrawal from a cylinder of Indiana Limestone which was coated with a thin film of epoxy, so that mercury had access through only one end. It was found that the coated sample, having a smaller surface area to volume ratio, required more time for mercury movement, and that more time was required for mercury withdrawal than for mercury injection.

In summary, errors in U_{min} are likely to be common unless considerable care is taken in determining the appropriate reference pore volume for the injection procedures used.

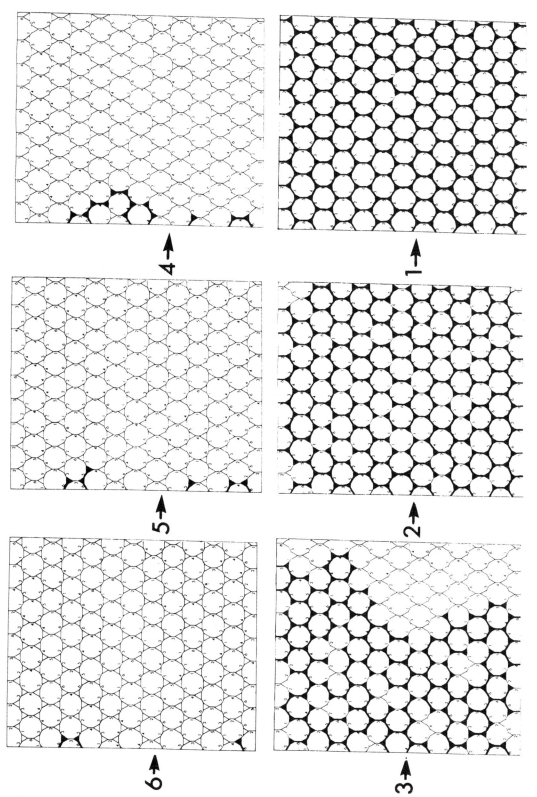

Fig. 7. Model with throat to pore ratio of 1.5 to 1. Throats assigned random numbers from 6 (largest) to 1 (smallest). Nonwetting phase enters the model from the left face only. Successive panels illustrate progressive filling of model as pressure is increased from injection pressure for filling of size 6 throats (low pressure) to that for filling of size 1 throats (high pressure).

Displacement pressures measured on small, uncoated plugs are likely to be lower than those for an infinite pore system because of artificially large surface "throats" (random intersections of pores), which favour access of mercury.

Rapid injection of mercury into a sample, allowing insufficient time for equilibrium, increases the apparent displacement pressure.

PREDICTIONS FROM MODEL PORE SYSTEMS

In this section, idealised models are considered as a basis for interpreting real injection-withdrawal curves which are presented in a following section.

Injection

What aspects of pore geometry influence the form of capillary pressure curves? Access to pores occurs through throats, and it might be thought that the volume of mercury injected for a given pressure increment represents the combined volume of throats and pores that are connected by throats falling within specific size limits. The size limits can be defined in terms of presure limits because the size of space entered is inversely proportional to pressure. In fact, the situation is more complicated because an *accessibility factor* must be considered. Not all pores with throats within the critical size range can be invaded, because they may be shielded by other smaller pores whose displacement pressure has not yet been exceeded.

Figure 7 illustrates a hexagonal network of throats with connecting pores. The pores are of equal size but the throats are of variable size, having been assigned random numbers from 6 (largest) to 1 (smallest). The pressure required to invade the largest throat can be referred to as injection pressure 6. As pressure is increased from injection pressure 6 to injection pressure 1, mercury being allowed to enter the sample via the *left face only*, the progressive invasion of throats and pores can be seen (Fig. 7).

At injection pressure 6, although there are 49 throats of size 6 only 3 of them fill with mercury, the others being inaccessible because they are screened by smaller throats with higher displacement pressures. An *accessibility factor* is defined as the ratio, expressed as a percentage, of throats of a given size invaded by mercury divided by the total number of throats of that size available for invasion. Thus, at injection pressure 6, the accessibility factor for size 6 throats is only 6.1 per cent under the conditions of the model. As pressure is raised and injection proceeds, the accessibility factor for throats and pores of all sizes increases.

Assuming that throat volume is proportional to throat size, an assumption that is valid for sheet-like or lamellar throats (Wardlaw, 1976), and neglecting for the moment the volume of pores, Figure 8A illustrates the actual volume of throats of specified sizes and Figure 8B the apparent volume of throats of given sizes based on the model illustrated in Figure 7. In constructing Figure 8B, it is assumed that at a given injection pressure, say 5, only throats of size 5 or larger are being filled. The accessibility factor makes the larger throats

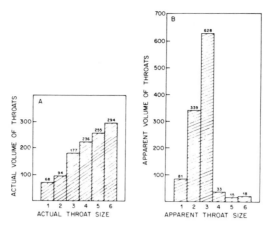

Fig. 8. "A" illustrates the actual volume of throats of various sizes as they occur in the model illustrated in Figure 7 (assuming that volume is proportional to diameter — see text). "B" illustrates the apparent throat size for this model, as it would be interpreted from a capillary pressure curve (Curve A, Fig. 9).

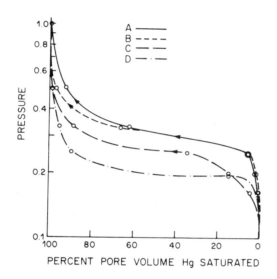

Fig. 9. Curves A, B and C are capillary pressure curves constructed for model with throat to pore ratio of 1.5 illustrated in Figure 7. Pressure units not specified. Diagram to illustrate relative form of curves A to D. Throats have volumes of from 1 to 6 units.

Curve A — for model consisting of throats only (volume of pores neglected). Entry from left face only.

Curve B — for model consisting of throats and pores; throats have volumes of from 1 to 6 units and pores have uniform volumes of 20 units. Entry from left face only.

Curve C — throats and pores model. Curves A and B are with entry from the left face of the model only; Curve C is with entry from all four faces.

Curve D — for model with throat to pore ratio of 3, as illustrated in Figures 10 and 11. As for Curve B model, throats have volumes of from 1 to 6 units and pores have uniform volumes of 20 units. Entry is via the left face of the model only.

appear much less abundant and the distribution as a whole much better sorted than is actually the case.

Since the minimum throat size that can be invaded is inversely proportional to pressure, throats of sizes 1 to 6 can be translated to equivalent relative pressures required for their invasion and, by using the volumes injected for each pressure increment derived from Figure 7, a capillary pressure curve can be constructed (Fig. 9, curve A). This model includes only throats but, assuming that each pore has a volume of 20 units compared with volumes of from 1 to 6 units for the throats, a second capillary pressure curve can be constructed (Fig. 9, curve B). Adding pores to the model, provided they are of uniform size, has surprisingly little effect on the form of the capillary pressure curve.

Rather than allowing entry from the left face of the model only (Fig. 7), entry may be allowed through all four faces. In this case, other factors being the same, the capillary pressure curve has a lower displacement pressure because of greater accessibility related to entry via a larger surface (Fig. 9, curve C). This is the result that was found experimentally by using coated (jacketed) and uncoated (unjacketed) samples (Fig. 5). In the case of the core sample only the low-pressure end of the curve is affected whereas in the model the entire curve is displaced downward, probably because of the small size of the model pore system.

Studies of resin casts of pore systems in dolomites have revealed that tetrahedral pores are common and that they are connected by sheet-like throats which occur at compromise boundaries between crystals (Wardlaw, 1976). Each tetrahedral pore will tend to be linked to 6 sheet-like throats, corresponding to the edges of the tetrahedron. Since each throat is shared by two tetrahedra, the throat to pore ratio will be 3 to 1 (Fig. 10). This compares with a ratio of 1.5 to 1 in the model illustrated in Figure 7. Figure 11A illustrates, in two dimensions, a pore system with throat to pore ratio of 3 being progressively

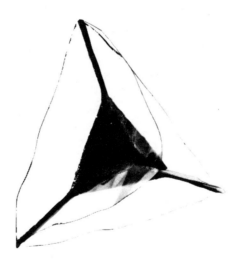

Fig. 10. Six lamellar (sheet-like) throats connecting with a tetrahedral pore. Idealized model but representative of structures commonly found in dolomites (Wardlaw, 1976). Throat to pore ratio of 3 to 1.

A
INJECTION

B
WITHDRAWAL

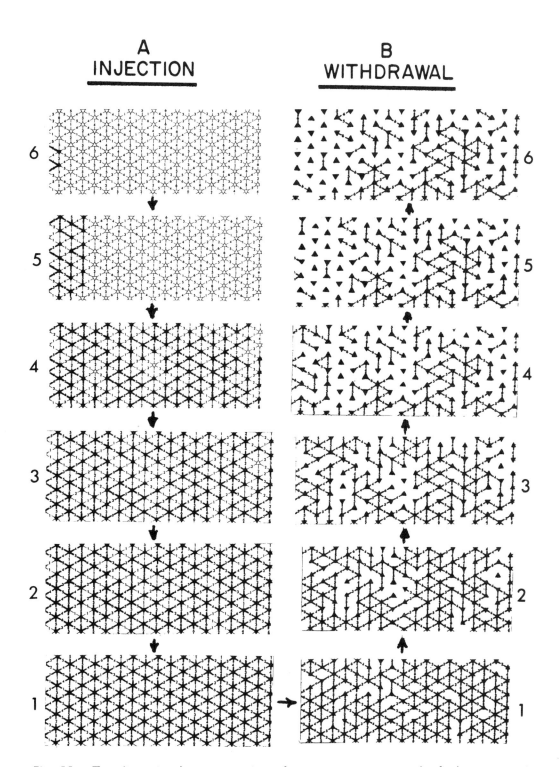

Fig. 11. Two-dimensional representation of pore system composed of elements as in Figure 10. Throat to pore ratio of 3 to 1. Throats assigned random numbers from 6 (largest) to 1 (smallest). Nonwetting phase enters and leaves the model from the left face only. Successive panels "A" illustrate progressive filling of the model as pressure is increased from injection pressure for filling of size 6 throats (low pressure) to that for filling of size 1 throats (high pressure). Successive panels "B" illustrate the withdrawal of nonwetting phase as pressure is reduced from highest pressure (1) to lowest presure (6).

filled via the left-hand face as the injection pressure is raised from 6 to 1. Curve D (Fig. 9) illustrates the corresponding capillary pressure curve that would be obtained for this model.

Curves D and B (Fig. 9) are for models with identical throat size and pore size distributions and with identical experimental conditions. The only difference between these models is the numerical ratio of throats to pores or the number of throats connecting to each pore. Differences between Curves D and B illustrate that the displacement pressure and apparent throat size sorting are influenced not only by throat size distribution but also by the number of throats connecting with each pore, or an accessibility factor. A capillary pressure curve is a function not only of the diameter and volume of conduits but also of the way in which they are connected. In the 1.5 throat to pore model, mercury entering a pore via a throat has a choice of two other throats by which it may continue penetration to the next pore. In the 3 throat to pore model the choice is increased to one of five other throats. Thus, the chances of greater penetration of the pore system at lower pressures are increased in the model with more connections (throats) per pore. This confirms results obtained earlier by Fatt (1956), who considered similar networks. More recently, Dullien (1975) has considered the properties of pore networks and their influence on pore-entry size distributions obtained by mercury porosimetry.

Withdrawal

Capillary forces cause mercury to eject or withdraw from a pore system as pressure is decreased. It is assumed in the model that mercury withdraws from the throats and that the pores remain filled. In reality, some emptying of pores would occur, but is not likely to be great where throats are small in relation to pores. "Snap-off" or breakage of mercury continuity limits the extent to which pores can be emptied. More elaborate models, not attempted here, could allow for variable percentages of nonwetting phase withdrawal from pores during pressure decline.

The relative volume of throats and pores cannot be estimated from an injection curve but, if withdrawal is mainly from throats, this provides a measure of throat and therefore also of pore volumes. Samples with high withdrawal efficiencies are likely to have throat volumes that are large in relation to pore volumes.

Figure 11B is an idealized model illustrating successive stages of mercury withdrawal during pressure reduction. It is constructed with access through the left face of the model only and is based on the assumption that a throat can empty only if:

a. the pressure has been reduced below the threshold pressure for a throat of that size;

b. the throat is connected to the exterior of the sample by a continuous thread of mercury.

Successively larger pores empty as pressure is gradually reduced. Initially, at injection pressure 1 and with the pore space completely filled, there is perfect continuity of mercury. However, as withdrawal proceeds there is progressively less mercury continuity and certain throats can never empty because they become isolated, by withdrawal from surrounding smaller throats, before the pressure falls below their threshold for emptying. Thus, the residual mercury residing in a sample will be at a variety of pressures related to the pressure on the system at the time continuity was broken.

An *accessibility ratio* can be defined for withdrawal as was done for injection, and is equal to the number of throats of a given size emptying for a specified pressure decrement divided by the total number of throats of that size available for emptying. Accessibility decreases as withdrawal proceeds and is therefore less for large throats than for small. Small throats empty more efficiently.

Figure 12 illustrates the injection curve for the model depicted in Figure 11A and also the withdrawal curve for the same model depicted

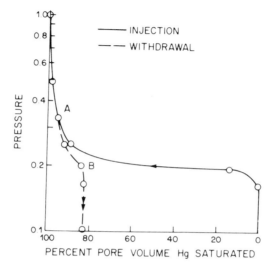

Fig. 12. Injection and withdrawal curves for model with throat to pore ratio of 3 as depicted in Figures 11A and B. Throats have volumes of from 1 to 6 units and pores have uniform volumes of 20 units. Pressure units not specified. Diagram to illustrate relative form of injection and withdrawal curves. No significant withdrawal occurs below point B because of low accessibility.

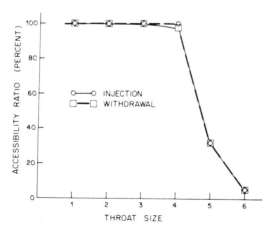

Fig. 13. Accessibility ratios versus throat sizes. Accessibility ratio (defined in the text) increases with decrease in throat size during injection and decreases with increase in throat size during withdrawal. Curves for injection and withdrawal are almost coincident.

in 11B (the injection curve was previously illustrated as curve D in Figure 9). The high-pressure end of the withdrawal curve follows the injection curve to point A below which the departure of the two curves increases and, below point B, there is no further withdrawal of mercury. These relationships can be explained in terms of accessibility ratios during injection and withdrawal. Figure 13 illustrates accessibility ratios, expressed as percentages, for throats of all sizes. At injection pressure 6 only 5.7 per cent of size 6 throats are filled whereas, at injection pressure 4, 100 per cent of size 4 throats are filled (Fig. 13). During withdrawal, when the pressure falls below the threshold for size 1 throats accessibility is 100 per cent and all size 1 throats empty whereas, when the pressure falls below the threshold for size 6 throats, accessibility is only 5.6 per cent, which is the total percentage of size 6 throats that can empty (Fig. 13).

Accessibility increases during injection and decreases during withdrawal. Initially the withdrawal curve follows the injection curve because accessibility is high. As pressure is reduced and mercury withdraws from the throat system, its continuity is broken and accessibility is progressively reduced. No significant withdrawal occurs below point B (Fig. 12) because of low accessibility.

The movements of mercury into and out of the throats and pores of the model in Figure 11 are summarized in Figure 14. The size of throats invaded during injection is inversely proportional to pressure.

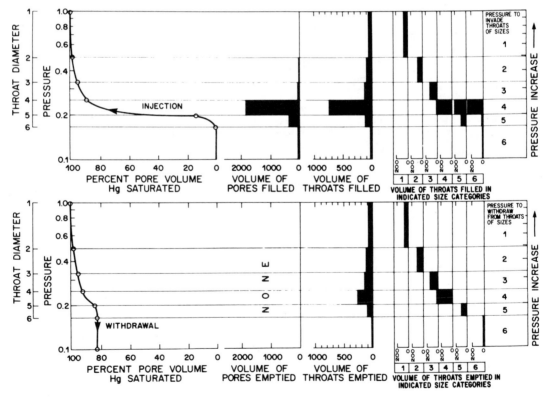

Fig. 14. Summarizing the movements of mercury during injection and withdrawal for the model illustrated in Figure 11. Pressure is represented as the reciprocal of throat diameter. For explanation see text.

To the right of the injection curve, separate histograms show the volume of pores filled and the volume of throats filled for successive pressure increments. Since the pores are assumed to be of equal size, the histograms are similar except for a greater volume of filling at higher pressures due to throats. This is because the throat to pore ratio is 3 to 1 and the pores, which are less numerous than the throats, are all filled before filling of throats is completed. The panels at the far right of the diagram illustrate how, at various pressures (vertical axis), throats of various sizes (horizontal axis) contribute to the total volume injected during successive pressure increments. Figure 14 illustrates that throats of size 6 do not fill in abundance when the threshold pressure for pores of this size is exceeded, but rather at a higher pressure when threshold pressure 4 is exceeded. This is due to low accessibility during the early stages of injection.

Mercury withdrawal in this model is due to the emptying of throats while the pores remain filled. Scales used are the same as for the injection data and it can be seen that at high pressures small throats empty completely because of high accessibility, while at low pressures only a small fraction of the larger throats empty because of low accessibility. For example, the total volume of mercury in size 6 throats is 318 units but only 18 units are withdrawn on pressure reduction. These relationships can be seen by comparing the final columns of data for injection and withdrawal in Figure 14.

In summary, during injection the larger throats tend to fill over a range of pressures as accessibility increases but, on withdrawal, a throat either empties when the pressure falls below the appropriate threshold for that size, or else it never empties because it is already isolated by breakages of the mercury column (snap-off). Thus, low accessibility *delays* the filling of large throats during the early stages of injection, but low accessibility *prevents* the emptying of large throats at low-pressure stages of withdrawal. This generalization would apply if a similar model were made infinitely large and extended to three dimensions.

Saturation and withdrawal efficiency

Under the conditions of injection and withdrawal described for the model in Figures 11 to 14, it is possible to determine the withdrawal efficiency (W_E) beginning at various saturations (S_1, S_2 and S_3 in Fig. 15). Withdrawal efficiency is high (22.7%) for the initial portion of the curve (S_1) because the range of throat sizes invaded is small and accessibility during pressure reduction is high. Since at this stage of saturation there are few continuous threads of mercury, any snap-off would greatly affect withdrawal efficiency. In the model, it is assumed that a size 5 throat can drain into a pore and out through another size 5 throat without snap-off.

Withdrawal from saturation S_2 is less efficient (11.9%) because the range of throat sizes injected is much greater than for S_1 and, during withdrawal, accessibility decreases and the larger throats become disconnected before their threshold for emptying is reached.

Withdrawal from saturation S_3 is more efficient (18.9%) than S_2 (11.9%) because the additional injection of mercury in moving from S_2 to S_3 (Fig. 15) is mainly a filling of small throats (since nearly all available pores have already been filled when stage S_2 saturation is reached) and accessibility is very high during the emptying of these throats (Figs. 13, 14).

In summary, the efficiency of capillary withdrawal of a nonwetting phase (mercury or oil) on pressure reduction increases as:

a) throat to pore volume ratio increases;

b) accessibility increases. Accessibility is related to the number of connections (throats) per pore that are filled with continuous threads of nonwetting fluid and, for any given throat size, tends to increase during injection and to decrease during withdrawal. Accessibility tends to be high for small throats and low for large throats during both injection and withdrawal. For these reasons, withdrawal from high saturations tends to be more efficient than withdrawal from intermediate saturations.

As properties a) and b) increase, threshold injection pressure decreases and withdrawal efficiency increases.

If withdrawal is mainly from throats rather than pores, then withdrawal volume can be used as a measure of throat volume and, by subtraction from total volume, pore volume also can be estimated.

When simulating porous media with two-dimensional network models, such as those presented in Figures 7 and 11, it commonly has been assumed that they can be used to simulate three-dimensional networks. However, as pointed out by Dullien and Chatzis (1976), bicontinua cannot exist in two dimensions and this restricts the usefulness of two-dimensional network models in simulating two-phase flow in porous media. In a two-dimensional model, continuous threads of two fluid phases cannot exist simultaneously from one side of a model (entry face) to the other (exit face), although such bicontinua are possible in a three-dimensional model

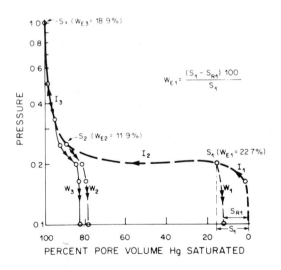

Fig. 15 Withdrawal efficiency [WE] for various initial saturations (S_1, S_2 and S_3). Employing the conditions of injection and withdrawal described for the model in Figures 11 to 14.

and are likely to be present in real porous media. However, the qualitative statements 'a' and 'b' above made for two-dimensional models would apply for three-dimensional models.

HYSTERESIS CURVES AND WITHDRAWAL EFFICIENCY FOR SELECTED ROCKS

For a rock with a particular proportion of water (wetting phase) and oil (nonwetting phase) how much oil can be withdrawn as a result of capillary processes by reducing pressure on the fluid system? In order to predict yield of oil it is not enough to know the proportion of water to oil in the sample, since yield will depend on how that saturation was achieved; for example, by injection of oil displacing water (drainage), by injection followed by withdrawal of some oil (imbibition) or by reinjection of oil following imbibition.

How does withdrawal efficiency (W_E) vary with initial saturation for a variety of rock types? Answers to these questions have been sought by experimenting with mercury as the nonwetting phase and obtaining hysteresis loops from various initial saturations. For a general discussion of capillary pressure hysteresis the reader is referred to Morrow (1970) and Melrose and Brandner (1974).

Trap and drag hysteresis

Figure 16 illustrates an experiment in which mercury was injected (I) into Indiana Limestone to maximum pressure and then withdrawn

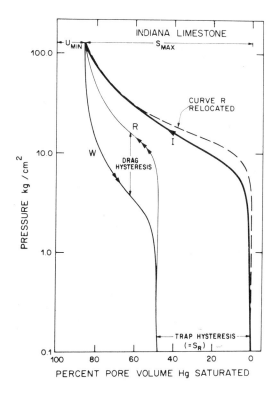

Fig. 16. Terminology as for Figures 1 and 2. Trap hysteresis, drag hysteresis and relocated curve R are discussed in the text.

(W) to minimum pressure and reinjected (R). The process can be repeated through several more withdrawal and reinjection cycles without changing the positions of curves W and R. The original pre-injection state with no mercury in the pore system cannot be regained, even at zero pressure, because discontinuous mercury is trapped in the pore system. This has been referred to as "trap hysteresis" to distinguish it from another effect, "drag hysteresis" (Pickell *et al.*, 1966), which is identified on Figure 16 as the vertical separation of the withdrawal and reinjection curves. For a given saturation, the pressure is higher during reinjection than during withdrawal. This is probably due to differences in contact angle or surface tension between advancing and receding interfaces. The mercury surface is expanding during injection and contracting during withdrawal. Impurities tend to gather on the mercury surface and to become more concentrated there during withdrawal. As Pickell *et al* (1966) pointed out, impurities will affect contact angles and surface tension and may make a major contribution to drag hysteresis. Trap hysteresis is thought to be predominantly a property of the rock and drag hysteresis of the liquid.

Relocation of reinjection curve

The reinjection curve (R) of Figure 16 can be relocated and expressed on a base S_{max} in order to compare its form with the original injection curve (I). The relocated reinjection curve (Fig. 16, broken line) lies above the original injection curve at lower pressures, indicating less injection for a given pressure increment, but comes to join it at higher pressures. This appears to be a general result for other samples studied and can be explained from the theoretical models previously presented. On withdrawal, the residual mercury tends to occupy the larger throats and pores preferentially because of low accessibility ratios in the latter stages of withdrawal. On reinjection, these spaces are then not available for invasion.

Access to the pore system is controlled by throats and, if it is throats that preferentially empty on withdrawal, it is not surprising that the pattern of filling should be generally similar on reinjection to the original filling on injection (*see* Fig. 9, curves A, B).

Residual saturation from various initial saturations

Figure 17 illustrates four hysteresis loops obtained by withdrawal and reinjection from successive initial saturations S_1 to S_4. Residual saturation (S_R) can be plotted against initial saturation (S) as shown in Figure 18. Imagine increasing pressure on the mercury and invading the sample to an initial saturation of 40 per cent. If pressure were then reduced and mercury withdrawn, a residual saturation of 25 per cent would result (Fig. 18). Withdrawal from maximum pressure at S_{max} (85.5%) gives a residual saturation of 48 per cent. It follows that 48 per cent minus 25 per cent (equals 23%) corresponds with the residual saturation, which is associated with the portion of the pore system invaded between 40 per cent saturation and the maximum saturation of 85 per cent. This 23 per cent has been termed *cumulative residual saturation* (Fig. 18). The cumulative residual saturation defined in this way increases from zero at maximum pressure (S_{max}) to 48 per cent at minimum pressure. Cumulative residual saturations computed at intervals of 10 per cent saturation are shown on Figure 19.

96

For each initial saturation there is a corresponding pressure. The pressure corresponding with an initial saturation of 40 per cent is 16 kg per cm² (Fig. 20). The cumulative residual saturation corresponding with this pressure is 23 per cent (Figs. 18, 19) and is plotted on Figure 20 which makes use of the saturation scale. In a similar way, other values for cumulative residual saturation plotted on Figure 19 are replotted on Figure 21.

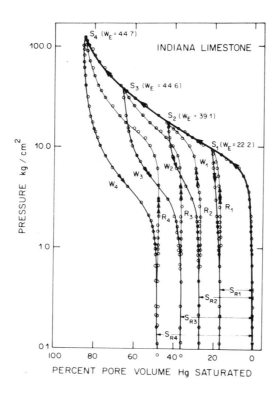

Fig. 17. Hysteresis loops for sample of Indiana Limestone ($\phi = 14.0\%$, K = 1 md) from initial saturations of S_1 to S_4. Injection curves (single arrows), withdrawal curves (double arrows) and reinjection curves (triple arrows). The four residual saturations SR_1 to SR_4 corresponding with initial saturations of S_1 to S_4 are shown, as are withdrawal efficiencies [W_E] which are seen to increase with saturation. Sample location given in caption to Figure 4.

Curve A (broken line) of Figure 21 is obtained by adding the cumulative residual saturation curve (CRC) and initial injection curve (I). This curve exactly follows the experimentally determined final reinjection curve (R_4) and leads to an important general result. Given the final reinjection curve (R_4) it is possible, by subtracting the initial injection curve, to produce a cumulative residual saturation curve (CRC) such as is illustrated in Figure 19. Residual saturations corresponding to any initial saturation can be obtained from this curve. This means that it is not necessary to produce multiple hysteresis loops, such as illustrated on Figure 17, experimentally. Given the final withdrawal curve (W_4) and reinjection curve (R_4), withdrawal and reinjection hysteresis loops for all other initial saturations (such as S_1, S_2 and S_3 in Fig. 17) are predictable within narrow limits by determining the

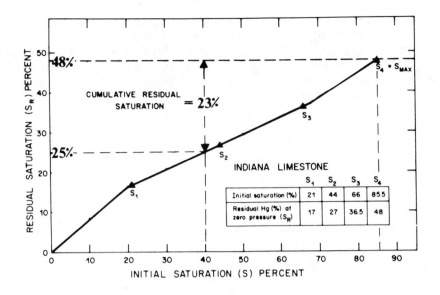

Fig. 18. Residual saturation versus initial saturation from data presented in Figure 17. See text for explanation.

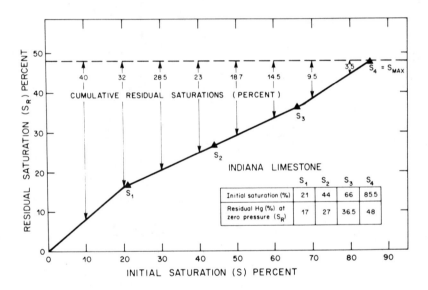

Fig. 19. Residual saturation versus initial saturation from data presented in Figure 17. See text for explanation of cumulative residual saturation.

residual saturation and relocating portions of the final withdrawal and reinjection curves appropriately.

It is clear why the final reinjection curce (R_4) should be equal to the sum of the initial injection curve (I) and the cumulative residual saturation curve (CRC). Mercury in a sample at a specific pressure during reinjection can be considered as being distributed in two quite

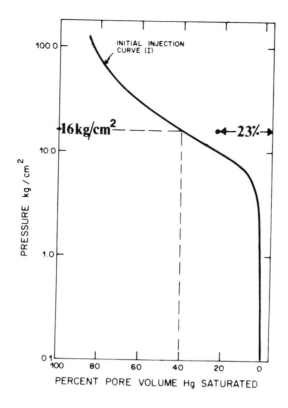

Fig. 20. Initial injection curve (I) for Indiana Limestone is reproduced from Figure 17. At an initial saturation of 40% the pressure is 16 kg/cm². The corresponding cumulative residual saturation (23%) is taken from Figures 18 and 19 and plotted utilizing the saturation scale.

different ways. Mercury occurs in connected networks, the extent of which increases as pressure is increased, and also as disconnected or residual mercury. The residual mercury resides in spaces that have not been reinvaded at the pressure in question because of the small size of connecting throats. The initial injection curve (I) gives the amount of connected mercury and the cumulative residual curve (CRC) the amount of disconnected mercury in the system for any given pressure. Thus, the sum of I and CRC corresponds with the experimentally determined reinjection curve (R).

Residual saturations for initial saturations of S_1, S_2 and S_3 were computed by using the final reinjection curve (R_4) in Figure 17 and by relocating the final withdrawal curve (W_4), withdrawal curves W_1, W_2 and W_3 were constructed (Fig. 22). These curves are almost coincident with the experimentally measured withdrawal curves illustrated in Figure 17, which are reproduced for comparison in Figure 22. Such curves enable predictions to be made regarding future responses to pressure changes, provided the previous saturation history is known. For example, if a sample has 60 per cent saturation achieved by withdrawal from initial saturation S_4, only 11 per cent of the mercury will be recoverable on further pressure reduction. However, if the 60 per cent saturation were achieved by withdrawal from an initial saturation S_3, then 22 per cent of the mercury would be recoverable by reduc-

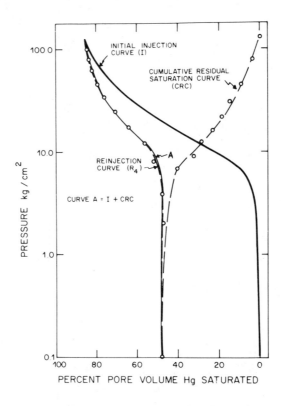

Fig. 21. Cumulative residual saturations obtained from Figure 18 are plotted against appropriate initial saturations and pressures. Adding the cumulative residual saturation curve (CRC) to the initial injection curve (I) gives a curve (I + CRC) which is coincident with the final reinjection curve (R₁ of Fig. 17).

ing pressure on the system. A third possibility is that 60 per cent saturation was achieved by injection without subsequent withdrawal. In this case, residual saturation on pressure reduction can be read from an initial-residual saturation plot, such as Figure 19, and the mercury recoverable by reducing pressure on the system is 26 per cent compared with 22 per cent and 11 per cent in the two previous examples. The required initial-residual saturation plot can be constructed from a single final reinjection curve such as R_4 in Figure 17.

In summary, the initial injection curve and the final reinjection curve provide a simple means whereby residual saturations corresponding with withdrawal from any initial saturation can be computed without the need of experimentally determining capillary pressure hysteresis loops for particular intermediate saturations. Once residual saturation (S_R) is known, withdrawal and reinjection curves can be constructed for any initial saturation by transposing appropriate portions of the final withdrawal and final reinjection curves.

Withdrawal efficiency in selected rocks

Withdrawal efficiency, as previously defined, is the ratio of the volume of mercury withdrawal from a sample at minimum pressure to the

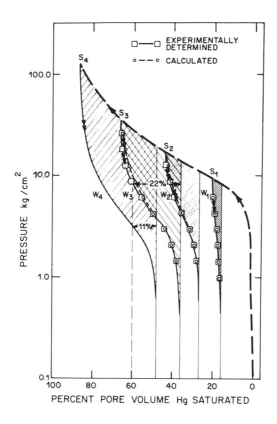

Fig. 22. Heavy broken line is mercury injection curve for Indiana Limestone. W_4 is final withdrawal curve and curves W_3, W_2 and W_1 are shown as experimentally determined (squares) and as calculated (circles) by the method described in the text. Shaded areas represents movable or recoverable mercury.

volume injected before pressure was reduced. For a particular sample, withdrawal efficiency can be studied as a function of initial saturation.

Figures 17 and 23 to 26 illustrate capillary pressure hysteresis loops for limestone, dolomite and sandstone samples. Withdrawal efficiencies (W_E) can be computed from initial (S) and residual (S_R) saturations and, in Figure 27, withdrawal efficiency is shown as a function of initial saturation for five different rock types. For a realistic residual saturation to be obtained, a withdrawal curve should begin at a saturation greater than about 10 per cent to avoid possible sources of error associated with the exterior surface of the sample, and the low-pressure end of the withdrawal curve should have assumed a vertical or near-vertical position, indicating that further decreases of pressure would not cause further withdrawal of mercury. There is an experimental limitation that pressure cannot be reduced to less than the head of mercury that is above the sample in the pycnometer chamber.

Why should Indiana Limestone, Becher dolomite and Midale dolomite yield mercury more efficiently from higher saturations than from lower saturations whereas, for the sample of Rainbow dolomite, ejection efficiency appears to decrease as initial saturation increases (Fig. 27)?

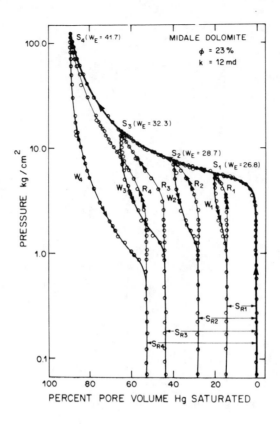

Fig. 23. Pressure-volume hysteresis loops for sample of Midale dolomite from Weyburn field, Saskatchewan (Shell McLane 12-33-33-5W2 at 4739 ft). Porosity is 23% and permeability 12 md. Injection curves (single arrows), withdrawal curves (double arrows) and reinjection curves (triple arrows). The four residual saturations S_{R_1} to S_{R_4} corresponding with initial saturations of S_1 to S_4 are shown, as are withdrawal efficiencies [W_E] Withdrawal efficiency increases with increase of initial saturation.

In the earlier section where idealized models were discussed, the assumption was made that it is throats rather than pores that empty during withdrawal. In the models, residual saturation (S_R) is equal to pore volume plus the volume of mercury in throats that do not empty. These are the larger throats, which become disconnected and isolated by snap-off before a pressure low enough for withdrawal has been achieved (*see* Fig. 14 for summary).

Withdrawal efficiency is relatively high at high pressures in the models because, as previously explained:

a) throats outnumber pores, and it is throats only that are filling and emptying at higher pressures;

b) the accessibility ratio is high at high pressures and, inevitably, decreases during withdrawal as mercury continuity decreases. During withdrawal from low initial saturations, relatively few throats and pores are mercury-filled, and snap-off rapidly reduces continuity of mercury.

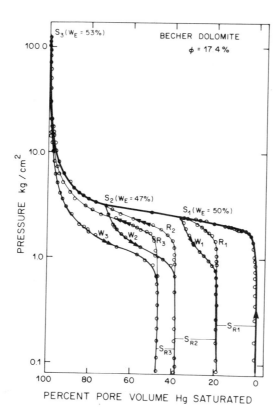

Fig. 24. Pressure-volume hysteresis loops for sample of Becher dolomite from Salina Formation (Imperial Becher No. 75, Becher Field, Ontario at 1875 ft subsurface). Injection curves (single arrows), withdrawal curves (double arrows) and reinjection curves (triple arrows). The three residual saturations S_{R1} to S_{R3} corresponding with initial saturations of S_1 to S_3 are shown as withdrawal efficiencies [W_E].

From these considerations, it is to be expected that withdrawal efficiency would increase with increase in initial saturation. Where withdrawal efficiency decreases with increase in initial saturation, as with the Rainbow dolomite sample and the final point on the Berea Grit curve (Fig. 27), alternative explanations must be sought. Pickell *et al.* (1966, fig. 4) report a similar relationship for a poorly sorted sample of Dalton Sandstone, and suggest that pores that are accessible only through small throats may be invaded at high pressure. A pore to throat volume ratio that increases with decrease in throat size could produce this unexpected relationship, but in the case of data presented here for the Rainbow dolomite and Berea Grit samples (Fig. 27), another explanation is more probable. From Figures 25 and 27 it is seen that withdrawal efficiency for the Berea Grit increases from initial saturation S_2 to S_5 but then decreases for S_6 at the maximum injection pressure. Withdrawal curves W_2, W_3 and W_4 are vertical (parallel to the pressure axis) before the minimum pressure is attained. Further decrease of pressure would not reduce residual saturation (S_R). However, because the "knees" of successive withdrawal curves descend to the left, withdrawal curve W_6 is not asymptotic to the pressure axis at the lowest pressure achieved. Pressure cannot be further reduced in

103

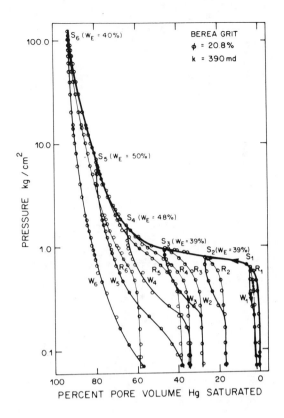

Fig. 25. Pressure-volume hysteresis loops for sample of Berea Sandstone. Samples supplied by Cleveland Quarries Co., Amherst, Ohio, U.S.A. Injection curves (single arrows), withdrawal curves (double arrows) and reinjection curves (triple arrows). Withdrawal-reinjection hysteresis loops for six initial saturations S_1 to S_6 are illustrated and withdrawal efficiencies [W_E] are identified.

the equipment because of the experimntal limitation of mercury head in the duct above the pycnometer chamber, and a final residual saturation cannot be achieved. Were further decrease of pressure possible, a smaller residual saturation and therefore higher withdrawal efficiency would be obtained.

In summary, from experiments (Fig. 27) as well as from theoretical considerations (Fig. 14), it can be seen that withdrawal efficiency for a porous rock tends to increase as initial saturation increases. Where withdrawal efficiency decreases with increase in initial saturation, it should be suspected that this result is caused by a limitation of the experimental procedure rather than a characteristic of the pore system; that is, inability to achieve a low enough pressure in the apparatus for complete withdrawal. Withdrawal is likely to be incomplete for samples that are vuggy or have large conecting throats and pores in addition to finer conduits. Samples of this kind will tend to have injection curves without well-defined plateaus. It was established in an earlier section that the position of the final reinjection curve (R) is given by the sum of the injection curve (I) and a cumulative residual saturation curve (CRC) (*see* Fig. 21). It follows that the horizontal separation of curves R and I provides a measure of the extent of trapped residual

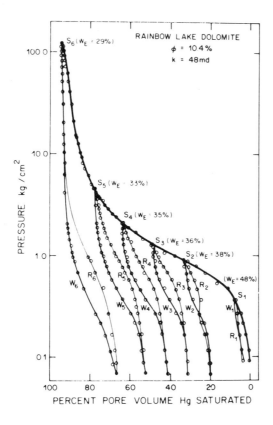

Fig. 26. Pressure-volume hysteresis loops for sample of Devonian dolomite from Rainbow Lake "A" pool, Alberta (Core 12-3-109-8W6 at 5847 ft). Injection curves (single arrows), withdrawal curves (double arrows) and reinjection curves (triple arrows). Withdrawal-reinjection hysteresis loops for six initial saturations S_1 to S_6 are illustrated and withdrawal efficiencies [W_E] are identified.

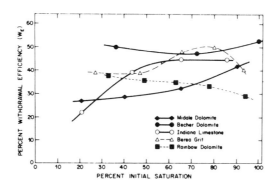

Fig. 27. Withdrawal efficiency increases at higher initial saturations for Indiana Limestone, Midale dolomite and Becher dolomite but not for Rainbow dolomite and Berea Grit. See text for explanation. Withdrawal efficiencies obtained from hysteresis loops on Figure 17 and Figures 23 to 26.

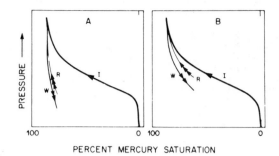

Fig. 28. Injection (I), withdrawal (W) and reinjection (R) curves for two hypothetical cores "A" with relatively low withdrawal efficiency at high saturations and "B" with relatively high withdrawal efficiency at high saturations.

mercury. In case A (Fig. 28), this separation is large, indicating extensive trapping and therefore high residual saturation and low withdrawal efficiency. In B, the separation of I and R is smaller, indicating less trapped mercury, lower residual saturation and higher withdrawal efficiency. In case A it can be expected that the withdrawal curve will fall more steeply than in B, illustrating less withdrawal of mercury for a given pressure drop. Thus, in a qualitative as well as a quantitative way, withdrawal efficiency can be read from the form of the injection, withdrawal and reinjection curves.

Drag hysteresis

Emphasis has been placed on observing and interpreting trap hysteresis and withdrawal efficiency. Finally, some observations are made regarding drag hysteresis. The extent of drag hysteresis is given by the vertical separation of the withdrawal and reinjection portions of a single hysteresis loop (Fig. 16).

It is thought that differences of surface tension and contact angle at advancing and retreating mercury interfaces make an important contribution to drag hysteresis. At an advancing front, new mercury surface is being created and the concentration of surface impurities is reduced while, at a retreating front, surface is contracting and the concentration of impurities is increasing. If surface impurities reduce surface tension and contact angle, it follows that mercury will behave as a more wetting liquid during withdrawal and that the magnitude of drag hysteresis should be related to decrease in surface area. Decrease in surface area will depend on volume of mercury withdrawn as well as on the size of the spaces from which it is withdrawn. Thus, at high pressures where withdrawal is from small throats, the effects should be more pronounced for a given volume. Observations made from the data presented in Figures 17, and 23 to 25 appear to be consistent with these suggestions (Fig. 29).

For a capillary pressure hysteresis loop from any initial saturation, such as S_3 in Figure 29 (inset), it is possible to measure a pressure difference between withdrawal and reinjection curves ($P_2 - P_1$) as a constant volume of withdrawal from the initial saturation (4% of pore

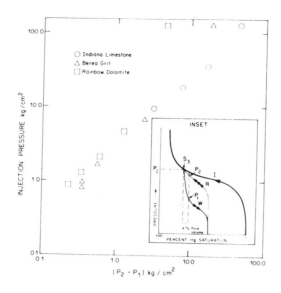

Fig. 29. Illustrates drag hysteresis as given by $P_2 - P_1$ (indicated on inset with P_2 and P_1 arbitrarily selected at 4% pore volume of withdrawal) plotted against the injection pressure (P_1) corresponding with particular stages of saturation (S_1, S_2, S_3 etc.).

volume is arbitrarily selected). Figure 27, then, illustrates drag hysteresis as given by $P_2 - P_1$ plotted against the injection pressure (P_1) corresponding with particular stages of saturation (S_1, S_2, S_3, etc.). Drag hysteresis per unit volume of withdrawal increases at higher pressures. This result may be caused by greater changes in surface area per unit volume at higher pressures, where throats are smaller and area to volume ratios greater.

THROAT AND PORE STRUCTURE FROM SAMPLES AND RESIN CASTS

Electron photomicrographs of the rock samples used in the mercury injection capillary pressure tests (Figs. 17, 23-26) are illustrated in Figures 30 and 31. Each photograph of a sample is accompanied by a photograph of the corresponding resin cast.

Dolomite crystal size is smallest for the Midale samples, intermediate for the Becher samples and largest for the Rainbow samples (Figs. 30A,C,E). Threshold injection pressure decreases as crystal size and, by association, pore-throat size increases. Pore size is related to crystal size because the dolomite crystals tend to maintain a rhombohedral form and compromise boundaries are not extensive (Wardlaw, 1976). Thus, the dolomite samples rank, in order of decreasing threshold injection pressure: Midale, Becher and Rainbow (Figs. 23, 24, 26).

The range of crystal sizes in a sample is important in influencing the slope of the injection curve. Dolomite crystals in the Becher sample are of remarkably uniform size compared with the Midale and Rainbow samples (Fig. 30A,C,E). Greatest variation of crystal size and of pore size occurs in the Rainbow sample, which also has the largest injection curve gradient (compare Figs. 24 and 26 and Figs. 30 A and B with Figs. 30 E and F).

107

ROCK SAMPLE **PORE CAST**

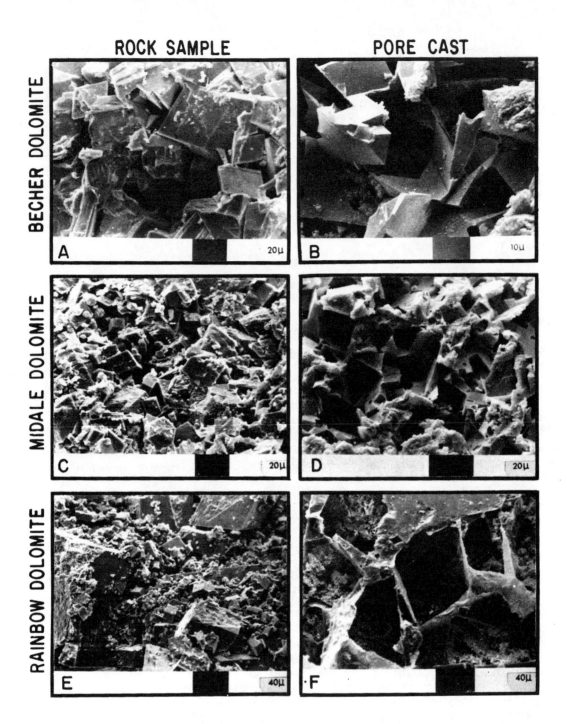

Fig. 30. All illustrations are scanning electron photomicrographs with magnifications given by the bar scales in microns. A, C and E are of dolomite rock samples and B, D and F resin pore casts of the respective rock samples. What appears to be solid in the photographs B, D and F is actually pore space now occupied by injected resin. See text for description.

A and B — Becher dolomite, location given in caption to Fig. 24.
C and D — Midale dolomite, location given in caption to Fig. 23.
E and F — Rainbow dolomite, location given in caption to Fig. 26.

Withdrawal efficiency is uniformly high for a range of initial saturations in the Becher dolomite compared with the Midale and Rainbow samples (Fig. 27). Withdrawal efficiency is thought to increase with increase of throat to pore diameter and volume ratios, the homogeneity of throat and pore sizes and the average number of throats connecting with each pore. Pores in the Becher samples are defined by dolomite crystal faces that are smooth and planar, and throats are large in relation to pores (Fig. 30B). Mercury injected at high pressure, as total saturation is approached, must be invading small recesses and spaces at the corners and edges of larger pores. The uniformity of pore size and the large relative size of throats to pores provide a pore system that has a high withdrawal efficiency (Figs. 24, 27).

There is greater variation of dolomite crystal size in samples of Midale and Rainbow dolomites, and the spaces between larger rhombohedra are filled by a varied assemblage of smaller rhombohedra (Figs. 30C-F). It follows that pore and throat sizes are more varied and that many of the smaller spaces are accessible only via small throats. It is to be expected that snap-off and trapping of residual mercury will be more extensive in these samples than in the pore systems of the Becher dolomite where, at high pressures, it is mainly the recesses of larger pores that are filling.

The Indiana Limestone has fabric selective megapores (Choquette and Pray, 1970) between some of the larger sedimented particles (Figs. 31A, B) but the particles themselves are composed largely of micrite grains with diameters of from 1 to 4μ (Figs. 31C,D). Withdrawal efficiency is low at low saturations but increases to about 45 per cent at high saturations (Fig. 27). The megapores are not sufficiently abundant to be extensively interconnected except via the fine pore networks associated with micrite. Thus, extensive mercury trapping during withdrawal from low saturations may occur. The uniformity of the micrite grain size and pore system may favour more efficient withdrawal from higher saturations.

The Berea grit is composed of relatively large $(100-200\mu$ diameter), well-sorted and partially rounded particles (Fig. 31E). The mercury injection curve (Fig. 25) has a well-defined threshold pressure and flat plateau which extends to about 60 per cent saturation. At higher saturations, the gradient of the curve increases. This may be compared with the Becher dolomite, which also gives a mercury injection curve with flat plateau, but the plateau extends to about 85 per cent saturation before there is a pronounced increase of slope (Fig. 24). Length of the flat portion (plateau) of the injection curve may be related to the volume of recesses associated with the larger pores. Such recesses are more abundant on the highly irregular surfaces of the Berea grains (Figs. 31E,F) than they are associated with the smooth, flat surfaces of dolomite crystals in the Becher samples (Figs. 30A,B). Ejection efficiency is high (50%) for the Berea Grit (Fig. 27).

Some of the spaces between the supporting framework of grains are filled by fine crystal cements or smaller grains, but these are not as abundant as in the Rainbow dolomite sample. Thus, much of the filling at higher pressures may be recess filling of larger pores rather than the filling of small pores acessible only through throats in which snap-off occurs during withdrawal. Accessibility is high and most of the

ROCK SAMPLE   **PORE CAST**

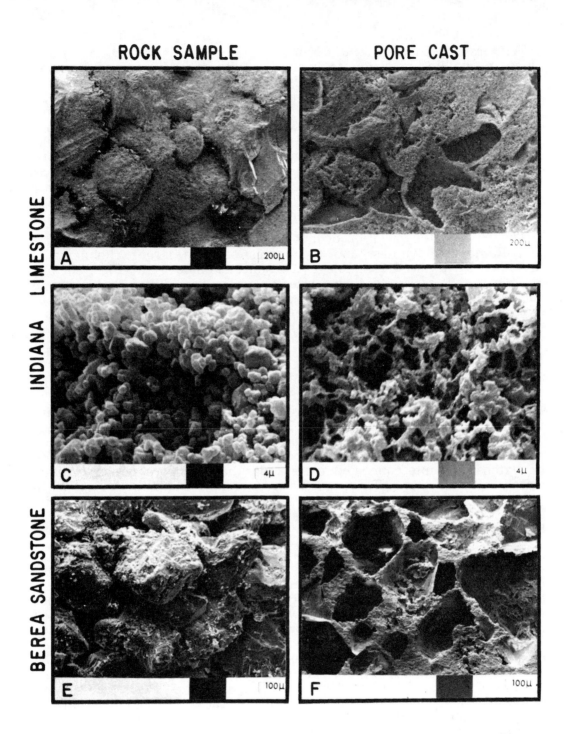

Fig. 31. All illustrations are scanning electron photomicrographs with magnifications given by the bar scales in microns. A, C and E are rock samples and B, D and F are resin pore casts of the respective rock samples. What appears to be solid in photographs B, D and F is actually pore space now occupied by injected resin. See text for description.

A, B, C and D — Indiana Limestone, location given in caption to Fig. 4.

E and F — Berea Sandstone, location given in caption to Fig. 25.

110

pores are connected by sheet-like throats which extend between the grains (Fig. 31F).

CONCLUSIONS

Mercury injection and withdrawal capillary pressure curves provide information regarding the efficiency with which a nonwetting phase can be withdrawn from a pore system during pressure reduction. Withdrawal (recovery) efficiency varies among samples of differing lithology and can be explained in terms of the geometric aspects of the pore systems, some of which can be interpreted from the capillary pressure curves themselves and others only by direct observation of rocks and resin pore casts. Withdrawal efficiency is defined as the ratio, expressed as a percentage, of the volume of mercury withdrawn from a sample at minimum pressure to the volume injected before pressure was reduced.

1. *Features atypical of infinite pore systems*

Some features of capillary pressure curves result from experimental procedure and the influences of sample size and shape in conjunction with effects that are peculiar to the external boundary region of the sample. These features that are not typical of infinite pore systems should be recognized.

a. *Inaccuracies of* U_{min} — U_{min} is the percentage of the total pore volume that is not occupied by mercury at the highest pressure attained (120 kg/cm^2). U_{min}, which is sometimes inappropriately referred to as "irreducible saturation", is commonly overestimated because of inaccuracies in the measurement of pore volumes. Errors are most serious for relatively low-porosity, vuggy samples.

b. *Effects of sample size* — Pores are accessible through throats and, on the exterior surface of the sample, there are many random intersections of pores which act as artificially enlarged throats. Thus, the pore system at the surface of the sample is not representative of the infinite pore system away from the surface. Sample size and shape influence surface area to volume ratios and therefore can be expected to affect the form of capillary pressure curves. Surface effects can be reduced by utilizing large samples or by coating or jacketing samples so that mercury enters through a limited surface. Entry pressure or threshold displacement pressure is larger for a coated sample than for an uncoated sample of similar size, shape and lithology. The coated sample is more representative of the behaviour of an infinite system.

c. *Rates of injection and withdrawal* — Rapid injection of mercury into a sample, allowing insufficient time for equilibrium, increases the apparent threshold displacement pressure. Errors from this source are most serious for large samples with high porosity and low permeability.

2. *Predictions from model pore systems*

Access to pores occurs through throats, and it might be thought that the volume of mercury injected for a given pressure increment represents the combined volume of throats and pores that are connected by throats falling within specific size limits. The size limits can be defined in terms of pressure limits because the size of space entered is inversely proportional to pressure. In fact, the situation is more complicated because an accessibility factor must be considered. Not all pores with throats within the critical size range can be invaded, because they may be shielded by other smaller pores whose displacement pressure has not yet been exceeded. Such effects can be studied conveniently by constructing idealized pore models. An *accessibility factor* is defined as the ratio, expressed as a percentage, of throats of a given size invaded by mercury divided by the total number of throats of that size available for invasion.

From consideration of idealized models, the efficiency with which a nonwetting phase (mercury or oil) will withdraw from a pore system as a result of capillary effects increases as throat to pore volume ratio increases and as accessibility increases. Accessibility is related to the number of connnections (throats) per pore that are filled with continuous threads of nonwetting fluid and, for any given throat size, tends to increase during injection and to decrease during withdrawal. Accessibility tends to be high for small throats and low for large throats during both injection and withdrawal. For these reasons, withdrawal efficiency is higher from high saturations than from low.

It is the size of the connections or throats, not the pores, that controls the relationships of pressure and entry but, in addition to throat size, it is necessary to consider throat to pore accessibility factors. As illustrated in the models presented, any capillary pressure curve will give a very distorted view of actual throat size distribution.

If withdrawal is mainly from "throats" rather than "pores", then withdrawal volume can be used as a measure of "throat" volume and, by subtraction from total void volume, "pore" volume also can be estimated.

3. *Hysteresis curves and withdrawal efficiency for selected rocks*

For a given rock sample, mercury is injected to some initial saturation (I) and then withdrawn (W) to minimum pressure and then reinjected (R). The initial saturation is increased and the process repeated through several hysteresis cycles. In this way, withdrawal efficiency can be studied as a function of initial saturation for a particular sample.

It is found that final reinjection curves provide a simple means whereby residual saturations corresponding with withdrawal from any initial saturation can be computed without the need of determining capillary pressure hysteresis loops experimentally for particular intermediate saturations. The method used is to subtract the initial injection curve from the final reinjection curve to obtain a cumulative residual saturation curve from which residual saturations corresponding with any initial saturation can be read. Once residual saturation (S_R) is known, withdrawal and reinjection curves can be constructed for any initial saturation by transposing the final withdrawal and reinjection curves.

In summary, the initial injection curve combined with the final withdrawal and reinjection curves is sufficient to predict withdrawal efficiency from any initial saturation, and to construct any desired pair of withdrawal and reinjection curves for intermediate saturations.

From experiments as well as from theoretical considerations and artificial models, it can be seen that withdrawal efficiency for a porous rock tends to increase as initial saturation increases. Where withdrawal efficiency decreases with increase in initial saturation, it should be suspected that rather than being a characteristic of the pore system, this results from a limitation of the experimental procedure — an inability to achieve a sufficiently low pressure in the apparatus for complete withdrawal.

Variability of sizes, shapes and arrangements of throats and pores can be studied from rocks and resin casts by utilizing a scanning electron microscope. For the dolomite samples, threshold injection pressure increases as crystal size decreases and the gradient of the central region of the injection curve increases as crystal sorting decreases. Withdrawal efficiency is uniformly high for a range of initial saturations in the Becher dolomite compared with the Midale and Rainbow samples. Withdrawal efficiency is thought to increase with increase of: throat to pore diameter and volume ratios; the homogeneity of throat and pore sizes, and the average number of throats connecting with each pore. During injection (drainage) withdrawal efficiency increases with increase of initial saturation in both idealized models and real porous media. Withdrawal efficiency for a particular saturation also is shown to be dependent on the saturation history of the sample.

REFERENCES

Choquette, P. W. and Pray, L. C., 1970, Geologic nomenclature and classification of porosity in sedimentary carbonates: Am. Assoc. Petroleum Geologists Bull., v. 54, no. 2, p. 207-250.

Dullien, F. A. L., 1975, Effects of pore structure on capillary and flow phenomena in sandstones: J. Can. Petroleum Technology, v. 14, p. 48-55.

—— and Chatzis, J., 1976, Single-phase and two-phase flow in capillary networks, in Symposium on Advances in Petroleum Chemistry, New York, April 1976: Am. Chem. Soc., v. 21, no. 2, p. 243-259.

Dumoré, J. M. and Schols, R. S., 1974, Drainage capillary-pressure functions and the influence of connate water: Soc. Petroleum Engineers J., Oct. issue, p. 437-444.

Fatt, I., 1956, The network model of porous media: Petroleum Trans. A.I.M.E., v. 207, p. 144-159.

Melrose, J. C. and Brandner, C. F., 1974, Role of capillary forces in determining microscopic displacement efficiency for oil recovery by waterflooding: J. Can. Petroleum Technology, v. 13, p. 54-62.

Morrow, N. R., 1970, Physics and thermodynamics of capillary action in porous media: Ind. Eng. Chemistry, v. 62, no. 6, p. 32-56.

Pickell, J. J., Swanson, B. F. and Hickman, W. B., 1966, Application of air-mercury and oil-air capillary pressure data in the study of pore structure and fluid distribution: Soc. Petroleum Engineers J., Mar. issue, p. 55-61.

Purcell, W. R., 1949, Capillary pressures — their measurement using mercury and the calculation of permeability therefrom: Petroleum Trans., AIME, Feb. issue, p. 39-48.

Robinson, R. B., 1966, Classification of reservoir rocks by surface texture: Am. Assoc. Petroleum Geologists Bull., v. 50, no. 3, p. 547-559.

Stout, J. L., 1964, Pore geometry as related to carbonate stratigraphic traps: Am. Assoc. Petroleum Geologists Bull., v. 48, no. 3, p. 329-337.

Taylor, R. P. and Wardlaw, N. C., 1975, Increased precision of porosity measurements using a modified Ruska universal porometer: J. Can. Petroleum Technology, v. 14, no. 2, p. 33-37.

Wardlaw, N. C., 1976, Pore geometry of carbonates as revealed by pore casts and capillary pressure data: Am. Assoc. Petroleum Geologists Bull., v. 60, no. 2, p. 245-257.

Reprinted by permission of the Canadian Society of
Petroleum Geologists. Published in the *Bulletin of Canadian
Petroleum Geology*, V. 26, No. 4 (December 1978) pp.
572-585.

BULLETIN OF CANADIAN PETROLEUM GEOLOGY
VOL. 26, NO. 4 (DEC. 1978), P. 572-585.

ESTIMATION OF RECOVERY EFFICIENCY BY VISUAL OBSERVATION OF PORE SYSTEMS IN RESERVOIR ROCKS

N.C. WARDLAW[1] and J.P. CASSAN[2]

ABSTRACT

Oil-recovery efficiency can be evaluated from the results of relative-permeability tests conducted on core samples but, because of the difficulty and expense of making these tests, typically fewer than fifteen are available for an entire reservoir. There is a need to devise simpler techniques to evaluate the probable recovery efficiency of reservoir rocks in order that a very much larger sample group may be treated.

The efficiency with which oil can be recovered depends on the fluid properties and on the characteristics of the pore system. The most important characteristics of the pore system are thought to be: pore-to-throat size ratio; throat-to-pore coordination number; and type and degree of nonrandom heterogeneity. On the basis of these characteristics, combined with a knowledge of total porosity, an empirical scheme is proposed which enables an approximate estimation of recovery efficiency from visual observations made from resin pore casts of carbonate reservoir rocks. A large number of samples can be examined rapidly by using this scheme, and the chances of properly evaluating the performance of a reservoir are improved. Following such an evaluation, a more representative group of samples could be selected for relative-permeability measurements.

INTRODUCTION

The volume of oil in place in a reservoir can be calculated from a knowledge of average porosity and average water-oil saturation, but this provides little information on the efficiency with which the oil may be displaced and recovered, for example, during a water-flood.

Recovery efficiency can be evaluated from the results of relative-permeability tests conducted on core samples but, because of the difficulty and expense of making these tests, typically fewer than fifteen are available for an entire reservoir. Given the large size and heterogeneous nature of most reservoirs, it is questionable whether the system can be adequately represented by so few test data. There is a need to devise simpler techniques to evaluate the probable recovery efficiency of reservoir rocks so that a very much larger sample group may be treated.

Mercury capillary-pressure tests, performed with drainage and imbibition cycles, provide a measure of nonwetting-phase recovery efficiency (Pickell *et al.*, 1966; Wardlaw and Taylor, 1976). Recovery efficiency is defined as the volume of mercury leaving a sample, as capillary pressure declines to a minimun value, expressed as a percentage of the mercury in the sample at maximum capillary pressure and maximum saturation.

Mercury capillary-pressure tests can be performed more simply and rapidly than relative-permeability tests. In both types of test, the nonwetting phase is trapped during the imbibition cycles as a result of capillary forces and, at residual nonwetting-phase saturation, the nonwetting phase exists entirely as isolated ganglia or blobs. Thus, it can be expected that mercury-recovery efficiency will provide an indication of oil-recovery efficiency in a strongly water-wet system. That this is the case, at least for certain groups of rocks is indicated by the results of experiments by Pickell *et al.* (1966).

---

[1] Department of Geology, University of Calgary, Calgary, Alberta T2N 1N4

[2] Elf-Aquitaine, Boussens, Haute Garrone, France

The authors wish to express their thanks to the management of the Société National Elf Aquitaine (Production) for permission to publish this paper. The senior author also acknowledges grants in aid of research from the Natural Sciences and Engineering Research Council of Canada.

114

The major objective of this paper is to demonstrate that an approximate estimation of recovery efficiency can be made by visual examination of pore casts. Pore networks are conveniently studied by impregnating samples with coloured liquid resins which are allowed to solidify in the pore system. The rock is then dissolved with acid so that a resin pore cast remains in relief. Pore casts can be prepared cheaply and, for a given reservoir, hundreds of samples can be examined by rapid visual techniques. In this study, a comparison was made between mercury-recovery efficiency as determined in capillary-pressure tests and estimations of recovery efficiency made visually from pore casts.

FACTORS AFFECTING RECOVERY EFFICIENCY

Recovery efficiency is affected by the properties of the fluids as well as by the properties of the pore system. For the fluid portion of the system, viscosities, interfacial tensions, fluid densities, wettability, wettability hysteresis and velocity of displacement all may affect recovery. However, the primary purpose of this paper is to consider the effects of pore systems on recovery efficiency, in the presence of a strongly nonwetting phase, where capillary forces are of primary importance in the trapping process. The following aspects are thought to be of major importance:

1) pore-to-throat size ratio

2) throat-to-pore coordination number

3) type and degree of heterogeneity

4) surface roughness

A pore network is made up of larger spaces, referred to as pores, which are connected by smaller spaces or constrictions, which are referred to as throats. It can be demonstrated, by using artificial pore networks in glass, that as the pore-to-throat diameter and volume ratios increase, the recovery efficiency decreases. That is, there is increased capillary trapping of nonwetting phase (Fig. 1). For a given pore-to-throat diameter ratio, it is also possible that recovery efficiency decreases as the absolute sizes of pores and throats decrease, but this has not been investigated experimentally.

Coordination number is defined as the number of throats that connect with each pore, and is a measure of the connectivity of the system. Pores are present at the junctions of throats. For example, in a single hexagonal network the coordination number is three whereas in a triple hexagonal network the coordination number is six (Fig. 1).

Fatt (1956, p. 162) showed, for networks of infinite size, that recovery efficiency increases as coordination number increases. This is true for networks consisting of tubes of a single size as well as those composed of a variety of tube-size sortings (Fatt, 1956, p. 164). From percolation theory applied to residual phases in randomly heterogeneous porous media, it has been shown that nonwetting-phase residual saturation increases; that is, recovery efficiency decreases, as coordination number decreases (Chatzis and Dullien, 1977; Larson et al., 1977). This is true for three-dimensional networks as well as for two-dimensional networks.

In addition to these randomly heterogeneous models, it is also possible to cluster tubes of differing sizes in a nonrandom manner as shown in Figures 2A and B, or to have portions of a network of differing coordination number as shown in Figures 3A and B. Nonrandom heterogeneities of these types are common in reservoir rocks.

In Figure 2A, a discontinuous network of larger tubes, centred at L, is isolated in a continuous network of smaller tubes (S). For simplicity, the network is illustrated as composed of tubes only rather than pores and throats. As capillary pressure falls during imbibition, nonwetting phase withdraws from the smaller spaces first and, in the larger

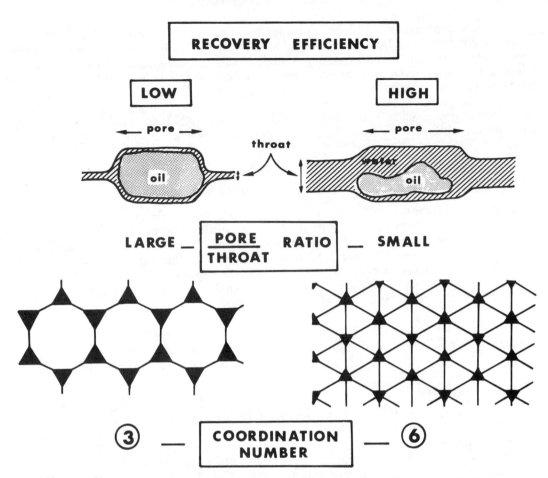

Fig. 1. Illustration of pore-to-throat size contrast and coordination number and their effects on nonwetting-phase recovery efficiency.

tubes, becomes isolated and trapped before the capillary pressure has fallen to a level that would allow withdrawal. Trapping of nonwetting phase in a large tube may be caused by the emptying of a small tube some distance from it.

In Figure 2B, a network of small tubes (S) is isolated in a network of larger tubes (L). Although the fine tubes form an isolated group, withdrawal of nonwetting phase is not impeded because the smaller elements will have emptied before withdrawal from the larger elements begins. Thus, nonwetting-phase continuity will have been maintained. Recovery efficiency will be higher for model 2B than for model 2A and, in these cases, is a function of the nonrandomness of the arrangement of the elements in the network.

Figure 3A illustrates a region with tubes of high coordination number isolated in a region of tubes of low coordination number. In Figure 3B, the region of low coordination number is isolated in a region of high coordination number. If the proportion of the two types of network were the same in the two models, withdrawal efficiency would be higher in model B because here the part of the system with high coordination number and high recovery efficiency forms a continuous phase, and withdrawal of nonwetting phase is not impeded by the presence of a region of low coordination number.

Recognition of nonrandom heterogeneities of these types is important in any classification that relates the pore structure of reservoir rocks to recovery efficiency.

116

In summary, pore-to-throat size ratio and pore-to-throat coordination number (Fig. 1), as well as the arrangement of pores and throats of differing sizes and differing coordination numbers in nonrandom heterogeneities (Figs. 2, 3), are thought to be important in controlling recovery efficiency. It is these aspects that form the basis of the classification which follows.

The surface roughness of pores, although also important, is not included in the classification presented because its effects are less clear. Figures 4 and 5 illustrate contrasting surface roughness in a dolomite sample and a sandstone sample. Surface roughness may affect fluids in the system in several ways.

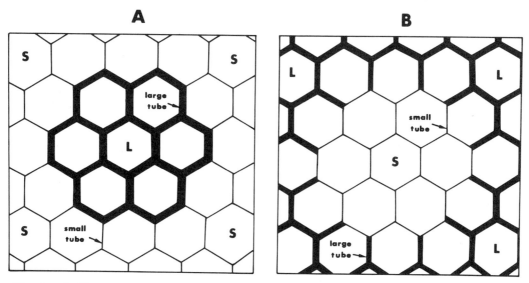

Fig. 2. A. Discontinuous hexagonal network composed of large tubes in a continuous network of small tubes; B. discontinuous hexagonal network of small tubes isolated in a continuous network of large tubes. Recovery efficiency greater for B than for A.

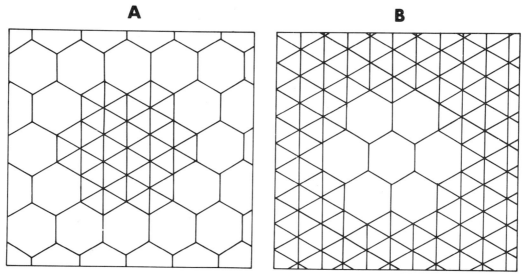

Fig. 3. A. Discontinuous triple hexagonal network (coordination number of 6) in a continuous single hexagonal network (coordination number of 3); B. discontinuous single hexagonal network in a continuous triple hexagonal network. All tubes are the same size. Recovery efficiency greater for B than for A.

Fig. 4. Dolomite crystals with smooth surfaces. Rainbow 'A' pool, Devonian, Western Canada.

Fig. 5. Quartz grains with rough surfaces. Berea sandstone.

Morrow (1970) showed that high irreducible wetting-phase saturations (Swi) in porous media are caused mainly by nonrandom heterogeneities of pore structure. Water is trapped as pendular rings or in clusters of fine pores because these regions lose hydraulic continuity through the medium before the applied pressure is sufficient to effect drainage. However, if the particle surfaces are rough, water may be retained by capillary action in small channels on the otherwise drained surfaces of particles. This water may link regions of pendular water and hydraulic conductivity may be maintained, thus allowing lower values of Swi to be achieved (Morrow, 1971).

Surface roughness is also important in its effects on wettability and, in particular, in accentuating contact-angle hysteresis. The effects on petroleum recovery are discussed by Morrow (1975, 1978).

CLASSIFICATION

The objective of the classification is to present a scheme whereby visual observation made from thin sections and pore casts can be used to predict recovery efficiency for systems in which capillary forces provide the dominant trapping mechanism. Thirty-six carbonate reservoir rocks were studied in detail. They are of widely differing types, with mercury-recovery efficiencies varying from 6 to 60 per cent, and are from the following regions and ages: Permo-Triassic of Iran; Cretaceous of the Congo, France, Iran, Iraq and Italy; Tertiary of Tunisia.

Pore systems are of infinite variety and there is no single attribute that is a sure guide to recovery efficiency. Nevertheless, observation of a relatively small number of attributes allows an approximate estimation of recovery efficiency in many cases. Knowledge of total porosity combined with visual observation of pore structure from pore casts enabled an estimation of recovery efficiency to within ± 10 per cent for 95 per cent of the cases studied. The scheme is simple and requires only a few minutes per sample for the visual examination. Although devised for carbonate rocks, the principles used in the classification are equally applicable for sandstones.

The importance of total porosity is first recognized. Figure 6 illustrates the relationship between mercury-recovery efficiency and porosity for the 36 carbonate samples of this study along with a group of 56 samples from a previous study (Wardlaw, 1976, fig. 2). While there is a significant positive correlation between recovery efficiency and porosity (correlation coefficient 73), there is no correlation with permeability (correlation coefficient 16). Similar results for the relationships between porosity, permeability and recovery efficiency were found for a group of sandstone reservoir rocks.

118

Figure 6 illustrates that as porosity increases, recovery efficiency increases and that, for a given porosity, there is a certain maximum recovery efficiency. Low-porosity rocks invariably have low recovery efficiency but some relatively high porosity rocks also have low recovery efficiency.

Pore-to-throat size ratio and throat-to-pore coordination number probably contribute to these relationships. Wardlaw (1976) explained decreasing recovery efficiency with decreasing porosity as due to increasing pore-to-throat size contrast which accompanies decreasing porosity. Larson (1977, p. 48) reinterpreted this relationship as a topological phenomenon suggesting that, at low porosity, the effective coordination number, or connectivity, of the pore space is low, giving rise to high residual nonwetting-phase saturations and, therefore, low recovery efficiency.

Whatever the explanation, there is a useful empirical relationship that low-porosity rocks (less than 5%) have low recovery efficiencies (commonly less than 20%). However, some high-porosity rocks also have low recoveries. In these cases, the pore-to-throat size contrast is invariably large. Average pore size can be measured directly from pore casts while average throat size is more conveniently estimated from mercury capillary-pressure curves (Wardlaw, 1976).

In the following, idealised models are presented and are then related to examples from reservoir rocks. Four types of pore structure are recognized which occur commonly as discrete regions within a single rock sample. As will be shown later, the specific arrangement and continuity of these regions within the pore network greatly influences recovery.

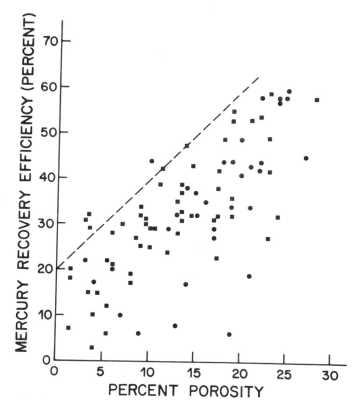

Fig. 6. Relationship between mercury-recovery efficiency and porosity for the 36 carbonate samples of this study (circles) and also for 56 carbonate samples from a previous study (squares) (Wardlaw, 1976, fig. 2). 95 per cent of the values fall below the dashed-line which gives an approximate maximum recovery efficiency for a given porosity.

1. *Regions of high intercrystal porosity.*

These regions are symbolised on Figure 7 by a triple hexagonal network, and by the code XH (intercrystal, high). The use of a triple hexagonal network here (coordination number 6) and that of a single hexagonal network in the following example (coordination number 3), are to indicate a relative difference in coordination number in the two-dimensional diagrammatic representations. Real networks are three-dimensional and the coordination numbers may be much larger than 6.

In the case of high intercrystal porosity, porosity is typically about 20 per cent or greater and is composed of pores of relatively uniform size which have small pore-to-throat size ratios. The pore network occurs between individual crystals which are of relatively uniform size. This type of network is typical of certain micrites (grain size 5-10 μm), microspars (grain size 10-50 μm) and dolomites which have high intercrystal porosity. Recovery efficiency is 55 per cent or greater. Examples of this type of pore network are illustrated (Figs. 8A, 9A,B, 10A,B).

2. *Regions of low intercrystal porosity.*

These regions are symbolised on Figure 7 by a single hexagonal network and by the code XL (intercrystal, low). The porosity is typically 5 per cent or less and composed mainly of spaces at crystal interfaces which are sheet-like or lamellar. These spaces are usually less than 3 μm thick and their other dimensions are related to those of the associated crystals. This is the common type of porosity in regions of more coarsely crystalline calcite or dolomite (spar or dolospar) but may occur also in micrites or other finely crystalline materials of low porosity. Fracture porosity too is of this general type. Recovery is less than 20 per cent, perhaps because of low effective coordination numbers associated with the low total porosity. This type of pore network is illustrated on Figure 8B.

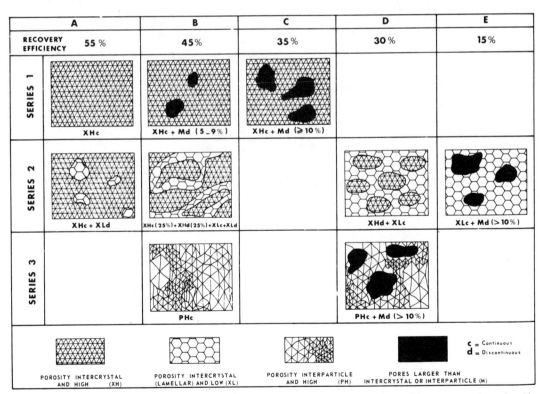

Fig. 7. Classification of some important types of pore networks observed in a reservoir rock with their empirically assigned recovery efficiencies. See text for a discussion.

120

3. *Regions of high interparticle porosity.*

These regions are symbolised on Figure 7 by triple hexagonal networks of variable size and by the code PH (interparticle, high). The porosity is commonly 15 per cent or more and occurs between sedimented particles which commonly are in the sand size-range and somewhat irregular in size and shape. Pore size and shape is less regular than that of intercrystal porosity and this irregularity is symbolised with variably sized network hexagons (Fig. 7). An average recovery efficiency of 45 per cent is assigned for systems of this type. Figures 8C, 11C and 11D illustrate this type of pore system.

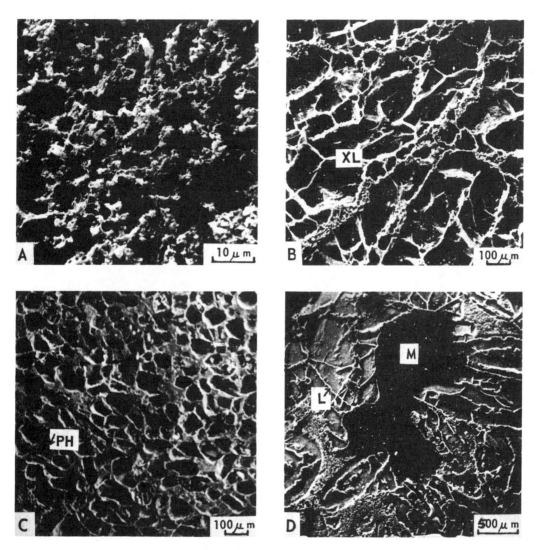

Fig. 8. Electron photomicrographs of resin pore casts. The rocks were impregnated with resin and subsequently dissolved with acid so that the pore system stands in relief as a resin cast. A. High (approximately 20%) intercrystal porosity (symbolised as XH on Fig. 7) in a microspar; B. Low (approximately 3%), lamellar, intercrystal porosity (XL) in a sparite (rock composed of calcite crystals larger than 50 μm). Many of the lamellar pores are seen obliquely, which makes them appear to be thicker than is actually the case; C. High (approximately 20%) interparticle porosity (PH); D. Megapores (M), larger than crystals and particles, are linked by a network of small, lamellar (L) pores at crystal interfaces.

121

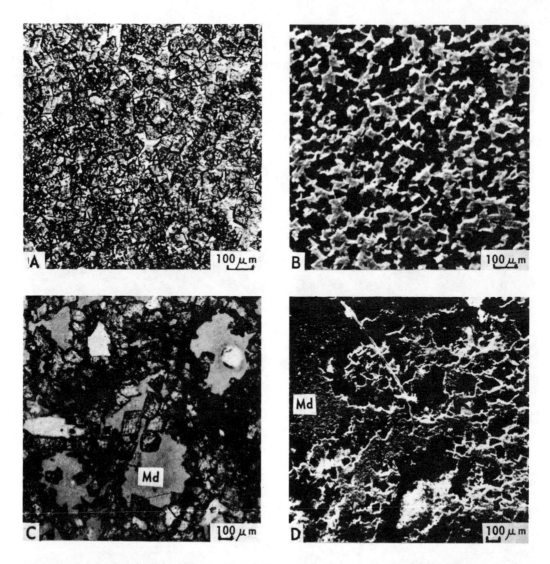

Fig. 9. A and C are photomicrographs of thin sections taken in transmitted light. B and D are electron photomicrographs of resin pore casts in which the host rock has been completely dissolved so that the pore system, cast in resin, stands in relief. A. Dolomite with high (greater than 20%) intercrystal porosity (XH in Fig. 7). Crystal size is relatively uniform as is the intercrystal pore network. Recovery efficiency is 55%; B. pore cast of A; C. dolomite as in A but with the addition of pores (M_d) which are larger than the intercrystal pores and which have relatively large pore-to-throat ratios. Recovery efficiency 35%; D. pore cast of C.

Fig. 10. A, C and E are photomicrographs of thin sections taken in transmitted light. B, D and F are electron photomicrographs of resin pore casts in which the host rock has been dissolved so that the pore system, cast in resin, stands in relief. A. Micrite with high intercrystal porosity (XH of Fig. 7) accompanied by discontinuous patches of spar (light coloured in photograph). Micrite forms a continuous phase. Recovery efficiency 55%; B1 and B2 are pore casts of A at two different magnifications. B2 illustrates the relatively small pore-to-throat size ratio typical of micrites with high intercrystal porosity; C. Micrite (dark coloured) with high intercrystal porosity occurring as discontinuous regions (XH_d) in sparite (light coloured) of low porosity which forms the continuous phase (XL_c). Recovery efficiency 30%; D. pore cast of C; E. dolomite with high intercrystal porosity occurring as discontinuous regions (XH_d) in coarse-grained spar of low porosity which forms the continuous phase (XL_c). Recovery efficiency is 30%; F. pore cast of E. ➤

122

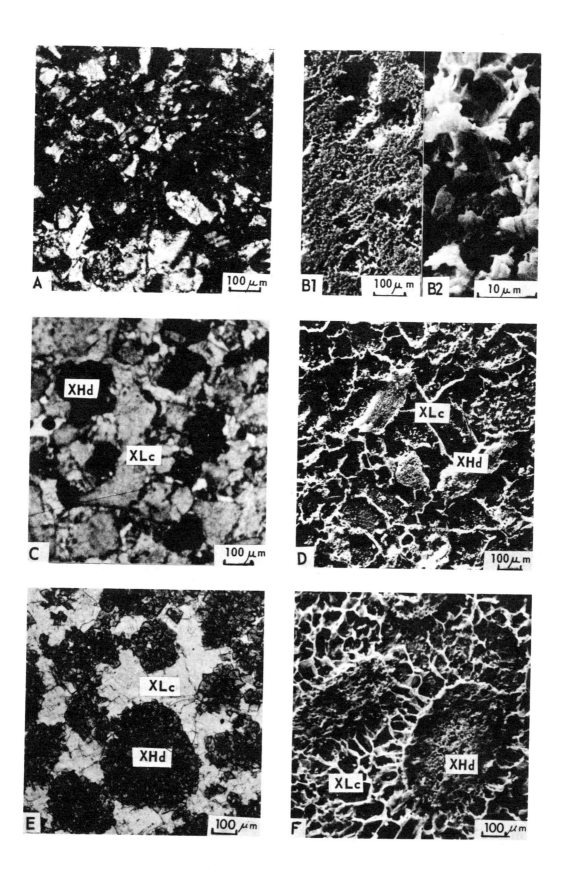

4. Regions of larger pores.

These pores, unlike the intercrystal and interparticle pores, are as large as or larger than the associated crystals or particles and commonly connected by a network of much smaller pores and throats. Porosity of this type is shown on Figure 7 as large black areas and by the code M (mesopore or megapore). The terms mesopore and megapore, as defined by Choquette and Pray (1970), are preferred to the term "vuggy" in order to indicate large size without other connotations of form or origin. Rocks with a large volume of porosity of this type have low recovery efficiencies because of large pore-to-throat size ratios. Examples are illustrated on Figures 8D, 11A and 11B.

Figure 7 illustrates some combinations of the four major types of pore systems described above, and includes an assignment of recovery efficiency that is empirical and based on an examination of pore casts in conjunction with recovery efficiency obtained from mercury capillary-pressure tests. Although 36 samples form the sample group for this study, the approximate recovery factors are assigned on the basis of a much larger sample group of over a hundred.

Series 1 commences, at the left side of the diagram, with member 1A which represents a rock of high intercrystal porosity (20 per cent or more), in which the pores are of relatively uniform size and the pore-to-throat size ratio is small (code XH). An average recovery of 55 per cent is assigned. Member 1B is similar except that between 5 and 9 per cent of the surface area of the sample, or of the volume of the sample, is occupied by larger mesopores or megapores (mainly larger than 250 μm in the samples studied) which are isolated, or discontinuous, in the network of finer spaces. Because the intercrystal porosity is continuous, it is marked with a subscript "c" (XH_c) whereas the larger pores (M), which are not in contact with each other, are marked with a subscript "d" (discontinuous) (M_d). Series 1, member 1C, is similar to member 1B except that the larger pores now occupy more than 10 per cent of the surface or of the volume. The declining recovery efficiency in members 1A, 1B and 1C of series 1 (55 per cent, 45 per cent and 35 per cent, respectively) is caused by the increasing volume of porosity that is of large pore-to-throat size ratio. Figures 9A and 9B are examples of member 1A and Figures 9C and 9D are examples of member 1C of Series 1.

Series 2 member 2A is similar to series 1 member 1A except for the addition of some discontinuous regions of low intercrystal porosity (XL_d). The regions of high intercrystal porosity are continuous (XH_c) and are not affected by the discontinuous regions of low porosity. Recovery is high (55%). A sample of this type is illustrated on Figures 10A and 10B.

In series 2 member 2B, approximately half of the total region of high intercrystal porosity is discontinuous (XH_d) because of the increasing amount and continuity of regions of low intercrystal porosity (XL) which have low recovery efficiency. Total recovery efficiency is estimated as 45 per cent. In member 2D, the regions of high intercrystal porosity are all isolated or discontinuous and the recovery is correspondingly lower (30 per cent). Figures 10C, D, E and F illustrate rocks with systems of this type.

Series 2 member 2E has a similar continuous region of low intercrystal porosity (XL_c), illustrated in member 2D of this series, but the regions of high intercrystal porosity are now occupied by large single pore spaces which are isolated (M_d) and occupy 10 per cent or more of the surface or of the volume. The large pore-to-throat size ratio of the discontinuous large pores combined with the low porosity and low effective coordination number of the continuous regions between the large pores results in very low recovery efficiency (15 per cent). Samples of this type are illustrated in Figures 11A and B.

Series 3 member B is a rock with high interparticle porosity of approximately 20 per cent. The pore system is continuous (PH_c) but less regular than the case of high intercrystal porosity because of irregularity of size and shape of the particles (Figs. 11C, D). An average recovery of 45 per cent is assigned. Series 3 member D is similar to member B but with the addition of 10 per cent or more of pores distinctly larger than the interparticle pores. Recovery efficiency is reduced to 30 per cent because of the addition of a relatively large volume of pores with large pore-to-throat ratios.

The types illustrated in Figure 7 are representatives taken from continuously varying series. A particular rock sample may be composed of any combination of these types in any proportions, and Figure 7 illustrates only certain selected types. The classification is clearly incomplete and should be regarded as only a first step in recognizing certain attributes of natural pore systems that are particularly important in their effect on recovery efficiency.

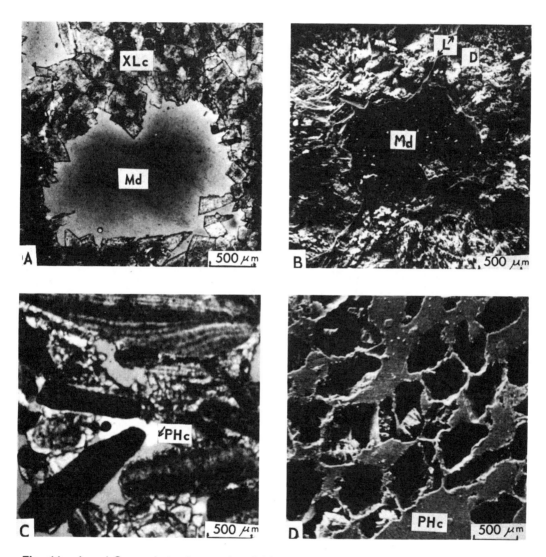

Fig. 11. A and C are photomicrographs of thin sections taken in transmitted light. B and D are electron photomicrographs of resin pore casts in which the host rock has been dissolved so that the pore system, cast in resin, stands in relief. A. Dolomite with low (less than 5%), lamellar, intercrystal porosity forms a continuous phase (XL_c) and is accompanied by megapores which are discontinuous (M_d). Recovery efficieny is 15%; B. pore cast of A. One megapore (M_d) is connected by a few lamellar pores (L) in coarsely crystalline dolomite (D) which is only partly dissolved by acid; C. high (20%) interparticle porosity which is relatively continuous (PH_c). Recovery efficiency is 45%; D. pore cast of C.

125

In summary, samples of high porosity with pores and throats of relatively uniform size and low pore-to-throat size ratio have high recovery efficiencies. The addition of large pores with large pore-to-throat size ratios decreases recovery efficiency (Series 1, Fig. 7).

Samples of low porosity (less than 5%) invariably have low recovery efficiencies (less than 20%), probably because the effective coordination number (connectivity) of the system is low and perhaps also because of a tendency for pore-to-throat size ratio to increase with decrease of porosity.

In a nonrandomly heterogeneous sample, with regions of high porosity and also regions of low porosity, the recovery efficiency is affected adversely if the regions of high porosity are discontinuous; that is, isolated by the regions of low porosity (Series 2, Fig. 7). Thus, the degree and type of nonrandomness are important as well as the effective coordination numbers and pore-to-throat size ratios for the system.

The thin sections and resin casts of the 36 samples were examined and the recovery efficiency for each sample was estimated, by using the classification presented above, without knowledge of the measured recovery efficiencies. Figure 12 presents a graphic plot of the estimated and measured mercury-recovery efficiencies. For more than 95 per cent of the samples, the recovery efficiency was estimated to approximately ±10 per cent. A second observer, not previously familiar with the scheme and without knowledge of measured recoveries, was able to achieve similar results.

The major conclusion is that nonwetting-phase trapping caused by capillary forces is strongly dependent on the characteristics of the pore system of a reservoir rock, and that these characteristics can be identified rapidly by visual observation of thin sections and pore casts. The application is that, in a strongly wetted reservoir of varied lithology, a

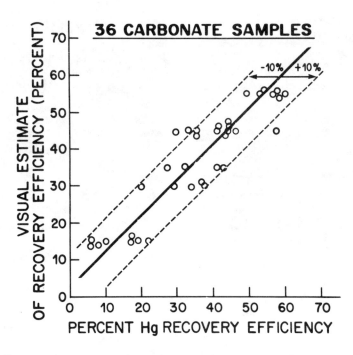

Fig. 12. To illustrate the correspondence between the recovery efficiency estimated from pore casts and mercury-recovery efficiencies measured in drainage-imbibition tests.

large number of samples can be examined and approximate estimates of recovery efficiency can be made, thus improving the chances of properly evaluating the reservoir's performance. Following such an evaluation, a more representative group of samples could be selected for relative-permeability measurements.

Finally, for the sample group studied, there is a significant correlation between mercury-recovery efficiency and porosity but no correlation between recovery efficiency and permeability. A statistically significant estimate of recovery efficiency can be made by using a measurement of total porosity combined with visual observations of pore structure made from pore casts.

REFERENCES

Chatzis, I. and Dullien, F.A.L., 1977, Modelling pore structure by 2-D and 3-D networks with application to sandstones: J. Can. Petroleum Technology, March, p. 1-12.

Choquette, P.W. and Pray, L.C., 1972, Geologic nomenclature and classification of porosity in sedimentary carbonates, in Carbonate Rocks II: Porosity and Classification of Reservoir Rocks: Am. Assoc. Petroleum Geologists, Repr. Ser. 5, p. 154-197.

Fatt, I., 1956, The network model of porous media: Petroleum Trans. A.I.M.E., v. 207, p. 160-181.

Larson, R.G., 1977, Percolation in porous media with application to enhanced oil recovery: M.Sc. Thesis, Univ. of Minnesota, 102p.

———, Scriven, L.E. and Davis, H.T., 1977, Percolation theory of residual phases in porous media: Nature, v. 268, p. 409-413.

Morrow, N.R., 1970, Irreducible wetting phase saturations in porous media: Chem. Eng. Sci., v. 25, p. 1799-1815.

———, 1971, The retention of connate water in hydrocarbon reservoirs: J. Can. Petroleum Technology, March, p. 47-55.

———, 1975, The effects of surface roughness on contact angle with special reference to petroleum recovery: J. Can. Petroleum Technology, Dec., p. 42-53.

———, 1978, Interplay of capillary, viscous and buoyancy forces in the mobilization of residual oil: prepr., 29th Ann. Tech. Mtg., Petroleum Soc. C.I.M., Calgary, 9p.

Pickell, J.J., Swanson, B.F. and Hickman, W.B., 1966, Application of air-mercury and oil-air capillary pressure data in the study of pore structure and fluid distribution: Soc. Petroleum Engineers J., March, p. 55-61.

Wardlaw, N.C., 1976, Pore geometry of carbonate rocks as revealed by pore casts and capillary pressure: Am. Assoc. Petroleum Geologists Bull., v. 60, p. 245-257.

——— and Taylor, R.P., 1976, Mercury capillary pressure curves and the interpretation of pore structure and capillary behaviour in reservoir rocks: Bull. Can. Petroleum Geology, v. 24, p. 225-262.

Reprinted by permission. First published in the *Journal of Petroleum Technology,* V. 34, No. 3 (March 1982), pp. 665-672.

Influence of Common Sedimentary Structures on Fluid Flow in Reservoir Models

K.J. Weber, SPE, Koninklijke/Shell Exploratie en Produktie Laboratorium

Summary

This paper summarizes the state of the art of deriving detailed permeability-distribution models on the basis of cores, sidewall samples, and logs. Reservoir heterogeneities such as clay drapes and intercalations, cross-bedding, sand laminations, slumping and burrowing in various major depositional environments are examined.

Introduction

The rapid increase in the number of secondary- and tertiary-recovery projects has resulted in a greater need for detailed geological modeling. Especially when expensive chemicals are injected into a reservoir, one wants to know how they will be distributed away from the wellbore.

Nearly all reservoirs are heterogeneous to such a degree that realistic reservoir simulation cannot be carried out on the basis of homogeneous prototypes. The overall heterogeneity can be dominated by open or sealing faults, fractures, contrasting lithologies, diagenesis, or sedimentological complexity. The influence of such features is described in a relatively large number of papers, while there are few that have concentrated on the internal heterogeneity within a genetically defined sand body caused by sedimentary structures and intercalations.

There is a fair understanding of the overall shape and continuity of some of the more important genetic reservoir sand-body types. At the other end of the scale, use of the scanning electron microscope (SEM) has revealed much information on pore shape and diagenesis. This paper focuses on the sedimentary heterogeneities on a scale between an entire, genetically defined, sand body and a small volume of matrix.

Although more realistic than glass packs, very homogeneous sandstone for flow and displacement experiments in laboratories—Berea sandstone in the U.S. and Bentheim sandstone in Europe—has been detrimental to the understanding of the influence of medium-scale

heterogeneities. There are few examples in the literature of experiments taking into account sedimentary structures and clay laminations. Pettijohn *et al.*[1] give good annotated references of such literature and also discuss the hierarchical sequence of primary controls on permeability. They state that because cross-bedding with preferred orientation is the dominant structure in many, if not most, sand reservoirs, it is the prime suspect for depositional control of anisotropic horizontal permeability in a reservoir. The use of internal directional structures for predicting the trend or orientation of sand bodies is beginning to be well understood. Numerous examples are given by Potter and Pettijohn.[2]

There is nothing new in the statement that the internal permeability distribution in a sand body will influence fluid flow and sweep efficiency. A literature survey yielded several early papers describing flow experiments on heterogeneous sand models. Mills carried out such experiments as early as 1920.[3] He was particularly interested in the trapping of oil in fine-grained sand lenses embedded in coarser sand during a waterflood. It is stated that there is first a rapid movement of water along the paths of least resistance under the laws of hydraulic flow of liquids, followed by readjustment under the laws of capillarity. Unfortunately, the understanding of the influence of sedimentary structures of the scale used in the experiments has not advanced much beyond this statement of 60 years ago.

A second early (1939) paper worth mentioning is by Illing.[4] He states that the most difficult zone to flush with water is the coarse/fine interface; thus, where there are numerous textural changes in the reservoir rock, it will be increasingly difficult to obtain a high percentage of oil extraction. All such coarse/fine interfaces tend to lock up oil in the strata.

The main purpose of this paper is to encourage the use of more realistic reservoir models in laboratory experiments and simulation studies. The similarity of reservoir deposited under similar conditions is the prime reason that reservoir models can be designed on the basis of relatively limited data. A few statements are made on

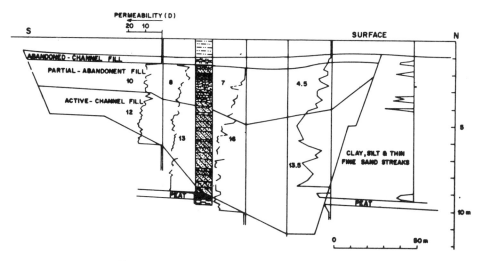

Fig. 1—Distributary channel fill in Holocene Rhone delta.

the relationship between primary sand properties and diagenetic processes. Attention is given to the influence of clay and silt intercalations on vertical fluid flow. The influence of cross-bedding on horizontal fluid flow is discussed and suggestions on how to design realistic reservoir models and what data are required are presented.

Similarity of Sand Bodies Deposited in Similar Environments

If sufficient cores have been taken, it is usually possible to make an accurate assessment of the environment of deposition of a given sand body. This information determines the range of the internal properties of the sand body. It is remarkable how similar sand bodies can be with respect to sedimentary structures and related grain-size variations if the depositional environments match closely. This is also the background of the art of using log shapes in identifying the environment deposition when only logs are available.

Primary differences between sand bodies deposited under similar circumstances are caused mainly by differences in source material and supply rate. However, such differences are usually small in a given area.[5] Thus, when a good statistical data base has been established for a group of associated fields, only limited additional data are needed for making reliable predictions of the internal buildup of another similar reservoir sand body.

The type of bedding of a sand body is determined by hydrodynamic factors and the grain-size distribution of the feed. Excellent publications on the origin of cross-bedding and the occurrence of various cross-bedding types are by Jopling[6] and Reineck and Singh.[7] Jopling gives an excellent explanation of the grain-size variations between foreset lamination and bottomsets. The lower permeability of the bottomsets relative to the foresets is a major reason for permeability anisotropy. Reineck and Singh discuss the relationship of bed form to current velocity and water depth, which forms the background of the similarity in the sedimentary structures found in the same depositional environment.

It is usually possible to split up a reservoir sand body into a number of clearly defined units that internally consist of similarly sorted sand and in which a given sedimentary structure type is dominant. Next, one tries to establish the average properties of these units. Examples of this approach can be found in papers by Harris[8] and Weber et al.[9] Again the same depositional environment leads to the same subdivision in reservoir units. For example, a distributary channel fill is nearly always composed of an active channel fill (mostly consisting of decimeter-size festoon cross-bed sets), a partial-abandonment fill (mainly composed of centimeter-scale cross-bedding), and an abandoned-channel fill of thinly interbedded fine sand, silt, and clay (Fig. 1).

Primary Sand-Body Properties and Diagenesis

Primary sand-body properties are related closely to the grain size of the feed and to the flow regime. Original porosity and permeability of a homogeneous element of a sand body is a function of average grain size and sorting. Beard and Weijl[10] have worked out the influence of texture on porosity and permeability of unconsolidated sand. Permeability-porosity patterns and variations in some Holocene sand bodies were measured by Pryor.[11]

The basic characteristics of deltaic-environment reservoir types are discussed by Sneider et al.,[12] whose article includes a good list of references. The recent book by Reading et al.[13] is also an excellent guide to the various environments and their typical sediments. In general, one can say that especially for deltaic sediments there is now a fairly widespread and well-documented knowledge of the sedimentological processes and the resulting sand-body properties and sedimentary structures in recent deposits.

All sandstone reservoirs have been subjected to a certain degree of loading and often also to various chemical processes. There is a vast amount of literature on diagenetic processes in sandstone. With the widespread use of the SEM[14] and the introduction of the cathodoluminescence technique,[15] recent publications

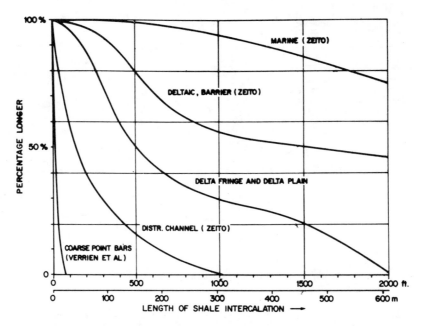

Fig. 2—Continuity of shale (silt) intercalations as a function of depositional environment.

have become more sophisticated. An excellent summary on the origin, diagenesis, and quality of clastic reservoir rocks has been compiled recently by Nagtegaal.[16]

The literature on diagenesis as a whole is somewhat misleading with respect to the average complexity of sandstone matrices. There is much emphasis on exotic clay-mineral configurations and unusual leaching processes. The more pronounced diagenetic alterations are frequently of great importance in deep marginal gas reservoirs but usually much less so in secondary oil-recovery projects at shallow to moderate depth. Fortunately there are numerous cases in which the sands have compacted only in a regular and gradual fashion and in which the chemical processes are restricted to some quartz overgrowth and conversion of feldspar grains into kaolinite. With the present tools it is often possible to work out the train of events leading to the present rock properties and to relate these properties to the original sedimentological characteristics.

The original permeability anisotropy in a sand body is usually quite low if we exclude clay intercalations. Gradually, however, the diagenesis will enhance the permeability differences between the coarse and fine laminae and between the well-sorted and the poorly sorted rock. In our experience the relative changes in porosity and permeability of coarse and fine laminae are often fairly constant over large bodies of rock. Thus, one often finds the same general contrasts in permeability in the reservoir that existed in the original sandstone body but with an enhancement of the ratio between maximum and minimum permeabilities.

Influence of Clay and Silt Intercalations on Vertical Permeability

Continuity of Low-Permeability Intercalations

Vertical fluid flow in a reservoir often is influenced by the presence of a number of shaly intercalations a few centimeters to a few meters thick. Shale breaks with an appreciable continuity will have a negative influence on recovery in reservoirs with partially perforated wells. On the other hand, they may reduce the danger of early gas or water breakthrough. Horizontal flow can be hampered by merging shale breaks or by shale streaks inclined to the average bedding-plane direction.

Shale breaks in reservoirs generally can be detected on logs, but their continuity and petrophysical properties are often unknown. Outcrop studies have shown that geometry, frequency, and lithology of shale and silt intercalations are determined largely by the environment of deposition of the rock. Thus, detailed studies of suitable outcrops of various depositional types should provide a better understanding of subsurface shale-break properties. Only a few such detailed studies have been reported, the best examples being by Zeito[17] and Verrien et al.[18] Analysis of closely drilled wells in recent sediments also can provide useful data on shale continuity.[19] These and other more general outcrop studies indicate that the geometry and distribution of shale and silt breaks in clastic formations show considerable similarities in rock deposited in the following depositional environments.

1. Marine sands of various types, which generally display a remarkable continuity in their usually frequent shale breaks (shale beds extending for several hundred meters are common).

2. Coastal barriers in which the shale breaks are also quite continuous and where they are generally concentrated in the lower half of the sandy interval. Burrowing often provides some permeability across these shale breaks if they are thinner than 10 cm.

3. Delta-fringe and delta-plain sediments contain shale breaks, which are usually less extensive than those of the above depositional environments mainly because of erosion by migrating and nonmigrating rivers and tidal channels.

130

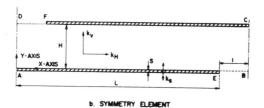

a. DISTRIBUTION OF CLAY INTERCALATIONS

b. SYMMETRY ELEMENT

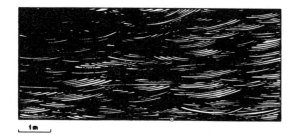

Fig. 3—Pattern of clayey silt intercalations in the active channel fill of a distributary channel.

Fig. 4—Model of clay intercalation distribution in a distributary channel fill.

4. Distributary channel fills in which most shale breaks are short (often less than 10 m), where their most common occurrence is related to festoon cross-bedding and where they often merge, especially perpendicular to the flow direction.

5. Point bars of meandering rivers in which, in addition to the shale-break types found in distributary channel fills, more continuous shale beds may occur along accretion planes. Coarse-grained point bars and braided rivers generally have very short shale and silt intercalations.

An attempt has been made to combine available data on shale continuity into a single graph relating the probability of a shale break extending over a certain length to the depositional environment (Fig. 2). The implications of the graph are that in marine and barrier bar sands, shale breaks tend to split up the reservoir into separate zones with little communication. In the delta fringe and between distributary channel fills, vertical crossflow also is restricted severely. On the other hand, fluvial sand bodies rarely show internal separation on a scale larger than 100 m. The importance of the arrangement of the flow barriers is discussed by Polasek and Hutchinson.[20]

Measuring Vertical Permeability

Much work still has to be done to establish a reliable statistical data base for estimating shale continuity in every type of depositional environment. More information also is needed on the actual in-situ permeability of shale and silt intercalations of various types. In this respect it can be expected that the use of the Repeat Formation Tester (RFT[TM])[21] will provide useful pressure profiles, allowing the computation of vertical conductivities of low-permeability intercalations. Pulse testing[22] also can be applied to detect vertical separation within reservoirs. Kamal[23] describes the use of pressure transients to describe reservoir heterogeneity in a more general way, indicating the advantages of multiwell tests. Direct measurements on preserved samples also can provide insight into vertical conductivities across intercalations, but permeability measurements on low-permeability unconsolidated clayey material are difficult to carry out.

Calculating Vertical Permeability With Models

Of great interest is the common case of a reservoir consisting of cross-bedded sand with discontinuous clay and silt intercalations. In fluvial sand bodies there is often a fairly regular intercalation of short and thin clayey and silty streaks of a brickwork-like pattern (Fig. 3).

Prats[24] calculated the apparent-permeability anisotropy resulting from a repetitive array of impermeable barriers. Richardson et al.[25] describe the effect of small, discontinuous shales on oil recovery through gravity drainage just below the gas-oil contact. In both cases the intercalations are taken to be impermeable and the influence of the sedimentary structures of the sand are not considered. Although in many cases the former assumption may be valid, it is interesting to look into the more complex case of a repetitive array of semipermeable intercalations in a cross-bedded fluvial reservoir with a certain permeability anisotropy caused by the cross-bedding.

The basic model used is shown in Fig. 4. From this configuration a symmetry element ABCD can be considered for which a set of boundary conditions can be imposed on the basis of symmetry and potential considerations. For single-phase flow we have computed the flux (Q_A) through the element ABCD as a function of the dimensionless numbers

$$\frac{Sk_v}{hk_s}, \quad \frac{k_h}{k_v}, \quad \frac{L}{h}, \quad \text{and} \quad \frac{l}{L},$$

where

S = thickness of clay or silt intercalation,
k_v = vertical permeability in anisotropic sand,
k_h = horizontal permeability in anisotropic sand,
k_s = permeability of clay or silt intercalation,
h = vertical spacing of intercalations,
L = length of intercalations, and
l = gap between intercalations at same level.

The anisotropy of the element ABCD can be defined as

$$A = \frac{k_h L P}{Q_A h}$$

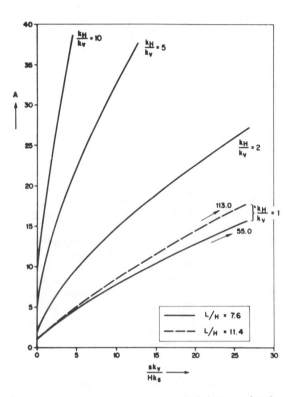

Fig. 5—Effective anisotropy in the vertical plane as a function of Sk_v/hk_s.

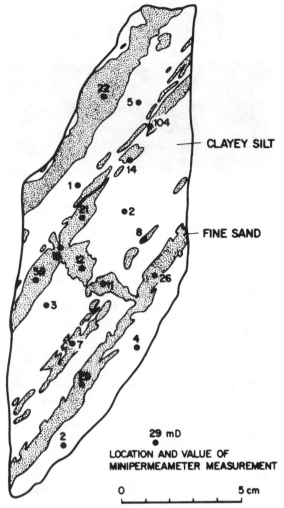

Fig. 6—Permeability distribution in interbedded, burrowed fine sandstone and clayey silt measured on a core with the minipermeameter.

—i.e. the ratio of the total flux Q_A and the flux that would occur for the same potential difference P between CD and AB if the sand were homogeneous with permeability k_h and the clay layers were absent.

In Fig. 5 the calculated anisotropy A has been plotted vs. the dimensionless number Sk_v/hk_s. All the curves are based on a value for l/L of 0.08. The effect of this parameter is small, in the range of 0.05 to 0.2, which is thought to be common in fluvial deposits. It appears that for the most interesting range of values of this number (10 to 30), the anisotropy is influenced most strongly by the parameter k_h/k_v. The geometrical ratio L/h has much less influence within this range of Sk_v/hk_s. This ratio increases in importance with the decrease in clay-layer permeability. If Sk_v/hk_s approaches infinity, the fluid will flow through the sand only, and the anisotropy is roughly proportional to L/h^2. For these conditions the graph presented by Prats is applicable.

For unconsolidated sandstones, k_h/k_v usually lies between one and two. In consolidated sandstones, k_h/k_v

can be much higher in value. The curves in Fig. 5 are thought to cover a wide range of cases.

Influence of Burrowing

Many reservoirs are composed partly of thinly interbedded clay, silt, and sand. In cores, such reservoirs often appear to have no overall vertical permeability. Nevertheless, it is common to find all permeable streaks to be oil-saturated, especially in cases with frequent vertical burrows.

Burrows of this nature occur in a range of environments, especially under marine conditions. It is likely that in many cases the burrows provide the main vertical communication across the thin continuous clay intercalations that we find, for example, in the barrier-foot sediments.

In general, one should first establish, with the aid of cores, whether burrows of a type resulting in vertical-permeability connection occur. Second, one should make a statistical estimate of the number of burrows to

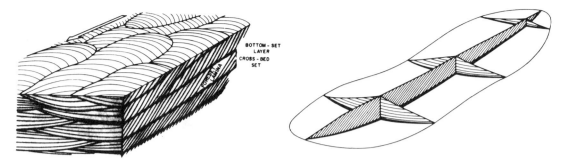

Fig. 7—Festoon cross-bedding.

be expected per square meter of clay intercalation. Third, permeability measurements should be carried out on a number of burrows, for example, with a minipermeameter[26] (Fig. 6). With these data one can make a rough estimate of vertical permeability over an interval of thinly interbedded sand and clay. To give an order of magnitude, a frequency of 25 to 50 burrows per square meter with a diameter of about 2 cm in a series of thin sand and shale beds of similar thickness would result in a level of permeability anisotropy of 50 to 100.

In practice, the bedding often will be disturbed by bioturbation in a more irregular fashion, causing intermixing of clay and sand. This is invariably highly detrimental to the horizontal permeability but, on the other hand, this process may create a certain vertical permeability. A high degree of bioturbation results in a poor but rather homogeneous reservoir.

Horizontal Permeability Anisotropy in Festoon Cross-Bedded Sand Bodies

As an example of the effect of cross-bedding on fluid flow, we have chosen a well-known type of cross-bedding, which is especially common in fluvial deposits (Fig. 7). The general knowledge of the formation of the various cross-bedding types allows us to design models of sedimentary structures that can be used for computing the permeability distribution on the basis of measurements on elements of a cross-bed set. In the case of festoon cross-bedding, the important elements are bottomsets and foreset laminae.

As explained by Jopling,[6] the bottomsets are a mixture of grains rolling down the lee face of the ripple and finer material transported downward in suspension by the eddies created downstream of the ripple crest. The contrast between the foreset laminae depends mainly on water velocity and velocity fluctuations. If the water velocity is constant and high, foresets will be formed continuously from material avalanching down the lee face, and little contrast between the laminae results. If, on the other hand, water flow is variable and around a certain critical velocity where bottom traction transport of the coarser grains is difficult, we find thicker, coarsergrained laminae alternating with thin laminae of fine material mostly deposited from suspension. Sometimes the permeability of these fine laminae is reduced further by the inclusion of mica flakes and clay floccules.

Usually the bottomsets have the lowest permeability of the complex, and cases are known where the foresets retained a reasonable permeability while the bottomsets

had become impermeable.[27] Measuring the permeability distribution parallel and perpendicular to the foresets laminae can be done with the aid of cube-shaped samples[28] or with the minipermeameter.[26]

The minipermeameter in the published configuration and used on a flat rock surface is reasonably accurate over a range of 10 to 5,000 md, provided the rock is clean without a crust of salt, bitumen, or dust.

It should be kept in mind that one measures the permeability of only a very small rock volume—much smaller than the usual permeability plugs. Thus, a large number of measurements is needed to gain insight into the permeability distribution. However, since an individual can carry out about a thousand measurements a day, this is not a limitation. Averaging a number of closely spaced measurements usually provides a more realistic permeability estimate than that obtained from traditional core sampling.

Measurements in the low-permeability range are influenced strongly by any leakage around the tip of the measuring probe, and the accuracy will not be much better than 50 to 100%. The lowest permeability level that can be detected is of the order of 0.5 md. With probes for unconsolidated sands, one can measure with reasonable accuracy up to some 150 darcies. A great advantage of this method is the small degree of disturbance of the sand, in contrast with ordinary measurements that require the use of sampling devices.

The horizontal permeability anisotropy can be measured during in-situ tests like those reported by Hewitt and Morgan.[29] Alternatively, one can construct a prototype model of the cross-bedded formation, measure the various geometrical relationships and permeabilities, and compute the resulting permeability anisotropy. This approach is discussed in some detail by Weber *et al.*[30] and the result is compared with an in-situ fluid-flow experiment.

The formulae used in computing the horizontal and vertical permeability anisotropy are summarized in Fig. 8. With these formulae and the model used, a general graph can be constructed relating horizontal permeability anisotropy in a festoon cross-bedded formation to crossbed set geometry, relative bottomset thickness, ratio of foreset laminae permeability to bottomset permeability, and permeability contrast between the foreset laminae (Fig. 9).

In unconsolidated sands, the ratio of the permeability parallel to the foreset laminae to that perpendicular to the laminae is unlikely to be higher than about three. The

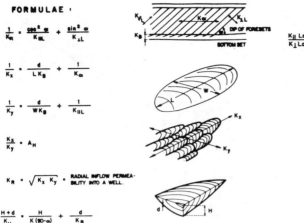

Fig. 8—Formulae used to compute directional permeability in festoon cross-bed sets.

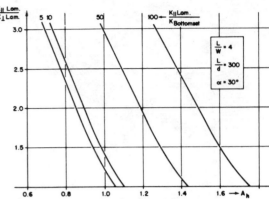

Fig. 9—Relationship between horizontal permeability anisotropy in a festoon cross-bedded formation, and cross-bed set geometry, bottomset thickness, ratio of foreset laminae permeability to bottomset permeability distribution within cross-bed set.

ratio of the permeability parallel to the foresets to the permeability across the bottomsets will rarely be higher than 100. Thus, values of horizontal anisotropy higher than 1.6 or lower than 0.7 are unlikely in unconsolidated sands. In consolidated sandstones, the horizontal anisotropy, of course, can be much larger. Moreover, when numerous shale intercalations are present, the merging of shale breaks in a direction perpendicular to the channel trend is likely to have an effect.

Designing Realistic Reservoir Models

After a thorough core study and calibration of the logs with the aid of core data, the reservoir can be subdivided into sedimentary units with a more or less predictable and correlatable distribution. The internal buildup of such units often can be represented by a not too complex prototype model. This model should be translated into a detailed permeability-distribution model with the aid of a well-designed core-measurement program.

It is usually unnecessary to measure very large numbers of permeabilities, although this is not too time-consuming with a minipermeameter. Our experience is that appreciable packages of cross-bed sets often have comparable grain-size compositions and degrees of diagenesis. Hence, the permeability contrasts are also comparable. Furthermore, it would be an impossible target to aim for very accurate determination of the overall permeability distribution. In general, one should try to establish whether there are certain sedimentary structures present that could have a significant effect on fluid flow. Next, one should focus on these features, and through prototype modeling, the possible range of permeability anisotropy or heterogeneity could be estimated.

On the basis of the previous sections, the example is presented of estimating permeability anisotropy in a distributary channel on the basis of a single cored well (Fig. 10A). The major part of the core consists of festoon cross-bed sets with a few clay drapes, as shown in

Fig. 3. First we measure the permeability of the foreset laminae and bottomsets of a representative series of the cross-bed sets. The pattern of permeability contrast will reveal itself quickly. One can use oriented plugs, cube-shaped samples, or the minipermeameter. Vertical and horizontal plugs also can be used but the resulting permeability values then should be converted with the aid of the formulae given in Fig. 8. With these formulae, the horizontal and the vertical anisotropy in a package of cross-bed sets also can be computed (Fig. 10B).

The next step is to combine cross-bed sets and clay intercalations in a model and compute the overall ratio of horizontal to vertical permeability (Fig. 10C). To this end, it is necessary to measure the vertical permeability across the clay breaks. In the case of these discontinuous clay breaks, the use of large core samples is probably the most reliable approach. With the aid of the graphs in Figs. 5 and 9, it is quite easy to estimate the one-phase permeability anisotropy values. In this paper little attention has been given to multiphase flow but it is obvious that the influence of sedimentary structures is even more important when relative permeabilities are introduced into the calculations.

Conclusions

Knowledge of sedimentology and diagenesis now has reached a stage where realistic prototype models of sand bodies can be designed on the basis of depositional environment and relatively limited core and log data. Such models should be used for computing the possible influence of sedimentary structures on fluid flow. Special core measurements needed for these calculations have to be carried out at an early stage. The resulting permeability-distribution model should be checked against in-situ measurements.

Acknowledgments

This paper represents the result of work done by several research geologists and engineers at Koninklijke/Shell

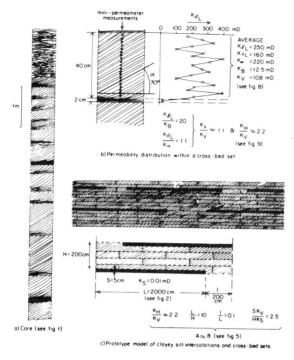

b) Permeability distribution within a cross-bed set

a) Core (see fig 1)

c) Prototype model of clayey silt intercolations and cross-bed sets

Fig. 10—Example of model design and calculation of vertical permeability anisotropy for a distributory channel fill.

Exploratie en Produktie Laboratorium. The author thanks R. Eype and F.C. Wonink for the development of measuring techniques and geological analyses, and R. Barthel for calculating the permeability anisotropy resulting from a combination of clay intercalcations and cross-bedding.

References

1. Pettijohn, F.J., Potter, P.E. and Siever, R.: *Sand and Sandstone*, Springer-Verlag, New York City (1973).
2. Potter, P.E. and Pettijohn, F.J.: *Paleocurrents and Basin Analysis*, Springer-Verlag, New York City (1977).
3. Ambrose, A.W.: "Underground Conditions in Oil Fields," *Bull.*, USBM, Washington, DC (1921) 70–72.
4. Illing, V.C.: "Some Factors in Oil Accumulation," Richard Clay and Co. Ltd., Suffolk, England (1939).
5. Weber, K.J.: "Sedimentological Aspects of Oil Fields in the Niger-Delta," *Geol. en Mijnbouw* (1971) **50**, No. 3, 559–576.
6. Jopling, A.V.: "Origin of Laminae Deposited by the Movement of Ripples Along a Stream Bed: A Laboratory Study," *J. Geol.* (1967) **75**, No. 3, 287–305.
7. Reineck, H.E. and Singh, I.B.: *Depositional Sedimentary Environments With Reference to Terrigenous Clastics*, Springer-Verlag, New York City (1975).
8. Harris, D.G.: "The Role of Geology in Reservoir Simulation Studies," *J. Pet. Tech.* (May 1975) 625–632.
9. Weber, K.J., Klootwijk, P.H., Konieczek, J., and van der Vlugt, W.R.: "Simulation of Water Injection in a Barrier-Bar-Type, Oil-Rim Reservoir in Nigeria," *J. Pet. Tech.* (Nov. 1975) 1555–1565.
10. Beard, D.C. and Weijl, P.K.: "Influence of Texture on Porosity and Permeability of Unconsolidated Sand," *Bull.*, AAPG (1973) 349–369.
11. Pryor, W.A.: "Permeability-Porosity Patterns and Variations in Some Holocene Sand Bodies," *Bull.*, AAPG (1973) 162–189.
12. Sneider, R.M., Tinker, C.N., and Meckel, L.D.: "Deltaic Environment Reservoir Types and Their Characteristics," *J. Pet. Tech.* (Nov. 1978) 1538–1546.
13. *Sedimentary Environments and Facies*, H.G. Reading (ed.), Blackwell Scientific Publications, Oxford (1978).
14. Neasham, J.W.: "Applications of Scanning Electron Microscopy to the Characterisation of Hydrocarbon-Bearing Rocks," *Scanning Electron Microscopy* (March 1977) 101–108.
15. Nickel, E.: "The Present Status of Cathodoluminescence as a Tool in Sedimentology," *Min. Sci. Eng.* (April 1978) 73–100.
16. Nagtegaal, P.J.C.: "Clastic Reservoir Rocks—Origin, Diagenesis and Quality," *Proc.*, Koninklijke Shell Exploration and Production Laboratorium Intl. Meeting on Petroleum Geology, March 18–25, 1980, Beijing, China.
17. Zeito, G.A.: "Interbedding of Shale Breaks and Reservoir Heterogeneities," *J. Pet. Tech.* (Oct. 1965) 1223–1228.
18. Verrien, J.P., Courand, G., and Montadert, L: "Applications of Production Geology Methods to Reservoir Characteristics Analysis From Outcrops Observations," *Proc.*, Seventh World Pet. Cong., Mexico (1967) 425–446.
19. Pryor, W.A. and Fulton, K.: "Geometry of Reservoir-Type Sandbodies in the Holocene Rio Grande Delta and Comparison With Ancient Reservoir Analogs," paper SPE 7045 presented at the Fifth SPE/DOE Enhanced Oil-Recovery Symposium, Tulsa, April 16–19, 1978.
20. Polasek, T.L. and Hutchinson, C.A. Jr.: "Characterisation of Non-Uniformities Within a Sandstone Reservoir From a Fluid Mechanics Standpoint," *Proc.*, Seventh World Pet. Cong., Mexico (1967) 397–407.
21. Smolen, J.J. and Litsey, L.R.: "Formation Evaluation Using Wireline Formation Tester Pressure Data," *J. Pet. Tech.* (Jan. 1979) 25–32.
22. Rijnders, J.P.: "Application of Pulse Test Methods in Oman," *J. Pet. Tech.* (Sept. 1973) 1025–1032.
23. Kamal, M.M.: "Use of Pressure Transients to Describe Reservoir Heterogeneity," *J. Pet. Tech.* (Aug. 1979) 1060–1070.
24. Prats, M.: "The Influence of Oriented Arrays of Thin Impermeable Shale Lenses or of Highly Conductive Natural Fractures on Apparent Permeability Anisotropy," *J. Pet. Tech.* (Oct. 1972) 1219–1221.
25. Richardson, J.G., Harris, D.G., Rossen, R.H., and van Hee, G.: "The Effect of Small Discontinuous Shales on Oil Recovery," *J. Pet. Tech.* (Nov. 1978) 1531–1537.
26. Eype, R. and Weber, K.J.: "Mini-Permeameters for Consolidated and Unconsolidated Sand," *Bull.*, AAPG (1971) 307–309.
27. Roach, C.H. and Thompson, M.E.: "Sedimentary Structures and Localization and Oxidation of Core at the Peanut Mine, Montrose County, Colorado," USGS Professional Paper 320 (1959) 197–202.
28. Fondeur, C.: "Etude Pétrographique Détailée d'un Grès à Structure en Semillets," *Rev. Inst. Français Pétrole* (1964) **19**, 901–920.
29. Hewitt, C.H. and Morgan, J.T.: "The Fry In-Situ Combustion Test—Reservoir Characteristics," *J. Pet. Tech.* (March 1965) 337–342; *Trans.*, AIME, **234**.
30. Weber, K.J., Eype, R., Leijnse, D., and Moens, C.: "Permeability Distribution in a Holocene Distributary Channel-Fill Near Leerdam (The Netherlands)—Permeability Measurements and In-Situ Fluid-Flow Experiment," *Geol. en Mijnbouw* (1972) **51**, No. 1, 53–62.

SI Metric Conversion Factors

in.	× 2.54*	E+00	= cm
ft	× 3.048*	E−01	= m

*Conversion factor is exact. **JPT**

Original manuscript received in Society of Petroleum Engineers office July 31, 1980. Paper accepted for publication Sept. 3, 1981. Revised manuscript received Dec. 11, 1981. Paper (SPE 9247) first presented at the SPE 55th Annual Technical Conference and Exhibition held in Dallas Sept. 21–24, 1980.

Reprinted by permission. First published in the *Journal of Petroleum Technology* (March 1983), pp. 629-637.

Petrophysical Parameters Estimated From Visual Descriptions of Carbonate Rocks: A Field Classification of Carbonate Pore Space

F.J. Lucia, Shell Oil Co.

Summary

The permeability, capillary properties, and m values of carbonate rocks are related to the particle size, amount of interparticle porosity, amount of separate vug porosity, and the presence or absence of touching vugs. Particle size, percent separate vug porosity, and the presence or absence of touching vugs usually can be determined visually. The amount of interparticle porosity is more difficult to determine visually and is done best by subtracting the visual estimate of separate vug porosity from the measure of total porosity obtained from wireline porosity logs or laboratory measurements. In the absence of touching vugs, the permeability, m values, and capillary properties can be estimated if the particle size, percent separate vug porosity, and total porosity are known. No acceptable method has been developed to estimate visually the permeability of touching vugs.

A classification of carbonate porosity is proposed based on the data presented. This classification is intended to be used in the field or for routine laboratory description. Interparticle porosity is classified according to particle size and the dense or porous appearance of the interparticle area. Vuggy porosity is classified according to type of interconnection. Separate vugs are connected through the interparticle pore space and classified by percent porosity. Touching vugs are connected to each other and classified by presence or absence.

Introduction

The role played by the visual description of pore space in carbonate rocks in the field evaluation of a well has changed dramatically over the past 25 to 30 years. The change has been brought about by the development of new and improved logging techniques. The Archie classification, developed in 1952, was the only method at that time to estimate the amount of porosity in uncored wells. The development of porosity wireline tools (neutron, sonic, and density) has provided us with effective ways to measure wellbore porosity. The permeability of a carbonate rock, however, can not be measured directly by wireline tools and it is not directly related to total porosity. Visual descriptions of the pore geometry, therefore, still are needed to estimate permeability. While the Archie classification provides some insight into permeability relationships, new data presented here allow more accurate estimates. In addition, the relationship between pore types and Archie's m value and capillarity can be described.

The role that the visual description of pore space can play in the evaluation is to describe factors that cannot be obtained from logging techniques but that are needed together with the logs to calculate saturations and productive capacities of the reservoir rock. This paper describes the *basic* geologic parameters that control the petrophysical parameters and shows how they are related. More sophisticated studies[1] of specific reservoir rocks may point out other geologic parameters that are of

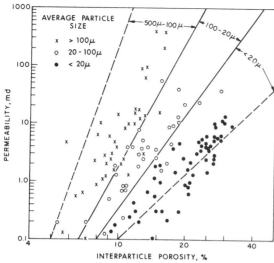

Fig. 1—Porosity/permeability for various particle size groups in uniformly cemented nonvuggy rocks.

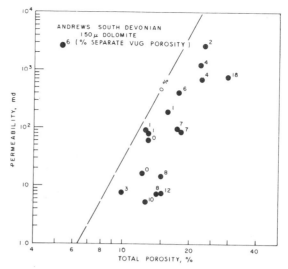

Fig. 2—Reduction in permeability caused by separate-vug porosity.

importance locally. The results presented in this report are based on a large volume of data produced in Shell Development Co. and Shell Oil Co. on the origin of carbonate pore space and the relationships between petrophysical and geological parameters.

Petrophysical/Geological Relationships

First-Order Divisions of Carbonate Porosity

The petrophysical and productive characteristics of a carbonate rock are controlled by two basic pore networks: an interparticle pore network and a vuggy pore network. *Interparticle porosity* can be defined as that pore space located between the particles of the rock that is not significantly larger than the particles. The term "particle" is used here as the general term for grains (multicrystalline particles) and crystals (single-crystal particles). Interparticle porosity between multicrystalline grains is called intergranular porosity and between single-crystal particles is called intercrystalline porosity. *Vuggy porosity* can be defined as that pore space larger than or within the particles of rock and commonly present as leached particles, fractures, and large irregular cavities. The effects of vugs on the petrophysical and productive characteristics of the rocks is related to the type of vug interconnection. Vugs are interconnected in two general ways: (1) through the interparticle pore network (*separate vugs*), and (2) by direct contact with each other (*touching vugs*).

Permeability Relationships

The relationship between permeability and rock texture can be shown by first examining nonvuggy rocks and then determining the effect of separate vugs and touching vugs on the nonvuggy relationships. The geometry of the interparticle pore space is related to the size and shape of the particles and to the amount and distribution of cement, compaction, and interparticle leaching. The size and shape of the particles reflect the largest interparticle

pore geometry possible. This maximum is reduced by either the growth of pore-filling minerals on the particle surfaces or by compacting and breaking of the particles. It is increased by interparticle leaching. The amount of cementation, compaction, and interparticle leaching is reflected in variations in the porosity of the rock. Therefore, the particle size and the amount of interparticle porosity should be the basic factors controlling the permeability of nonvuggy rocks.

The particle size, porosity, and permeability of a large number of nonvuggy carbonate rocks from a number of reservoirs were measured and the results are plotted in Fig. 1. The results show a reasonably good relationship between these three parameters. Therefore, this plot can be used to estimate the permeability of a nonvuggy carbonate rock if the particle size and the amount of interparticle porosity are known.

Deviations from this relationship will occur if the cementation or compaction is distributed erratically. Patchy cementation reduces the permeability of the sample less than does the same amount of cementation uniformly distributed throughout the sample. If the cement is located only in the pore throats, it will be much more damaging to the permeability than if an equal amount of cement were distributed evenly. Sorting and amount and distribution of clay also can modify the relationships. Leaching of interparticle pore space can improve the porosity/permeability relationship.

The effect of adding separate vugs to a nonvuggy rock is to increase the porosity with little or no increase in permeability. Another way to state this is that the permeability of a rock with separate vugs will be less than would be expected if all the porosity were interparticle. To quantify this relationship, several suites of rocks were examined and the percent separate vug porosity was measured visually. The results are shown in Figs. 2 and 3. In general, the rocks with the greatest amount of separate vug porosity are most removed from the porosity/permeability relationship of nonvuggy rocks of the

137

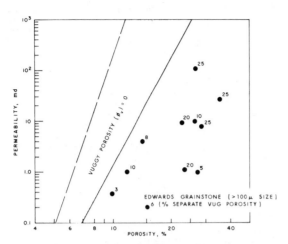

Fig. 3—Reduction in permeability caused by separate-vug porosity formed by dissolution of carbonate grains.

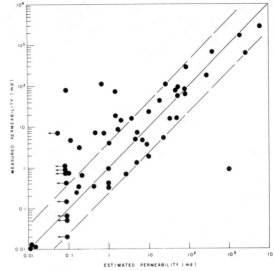

Fig. 4—Accuracy in estimating permeability of rocks with separate vugs.

same particle size. The distance the points are removed from the line is approximately equal to the amount of separate vug porosity. This suggests that the permeability is controlled by the amount of interparticle pore space, and the presence of separate vugs contributes little, if any, to permeability.

The interparticle porosity of the matrix is the porosity of the rock exclusive of the vugs. This is given by the formula

$$\phi'_p = \frac{\phi_T - \phi_v}{1 - \phi_v}, \ldots\ldots\ldots\ldots\ldots\ldots\ldots (1)$$

where ϕ'_p is fractional interparticle porosity of the matrix, ϕ_T is fractional total porosity, and ϕ_v is fractional separate vug porosity. A small error results if the matrix interparticle porosity is estimated by simply subtracting the vuggy porosity from the total porosity. Below a total porosity of 20%, the error is less than 1% porosity. In rocks with greater than 20% porosity and with high separate-vug porosities, the amount of interparticle porosity estimated will be low by 2 or 3% porosity.

The approach, then, to estimating permeability in rocks with separate vugs is as follows.

1. Measure total porosity (wireline logs or core analyses).

2. Measure separate vug porosity (visual estimation).

3. Subtract separate vug porosity from the total porosity to get an approximate interparticle porosity.

4. Estimate the particle size (comparator or ocular micromometer).

5. Use particle size and interparticle porosity to determine the permeability from Fig. 1.

Fig. 4 is a plot of measured permeability against estimated permeability. The total porosity is from core analyses and the particle size and separate-vug porosity

was measured visually. In 70% of the cases, the two methods give results within an order of magnitude of each other. With one exception, the anomalous points have higher measured than estimated permeability and tend to be in the < 1-md range. This may be caused by the presence of small open fractures (touching vugs). It is believed that if the permeability measurements had been conducted under confining pressures, many small fractures would have been closed and the measured permeability would have been closer to the estimated permeability.

The effect of touching vugs on the permeability is to produce a higher permeability than expected if all the porosity were interparticle. Touching-vug pore networks are usually systems of fractures and cavities. Some insight into the geologic parameters that control the permeability of fractures can be gained by converting the following relationships[2] for planar fractures of finite clearance and unit width into a permeability relationship.

$$\Delta p = \frac{12\mu vL}{w^2}, \ldots\ldots\ldots\ldots\ldots\ldots\ldots (2)$$

where

Δp = pressure drop along fracture,
μ = viscosity of the fluid,
v = velocity of the fluid,
L = length of the fracture, and
w = width of the fracture.

This equation can be converted into an equation involving the permeability as follows.

$$v = \frac{q_f}{A_f}, \ldots\ldots\ldots\ldots\ldots\ldots\ldots\ldots\ldots (3)$$

where q_f is the flow rate through the fracture and A_f is

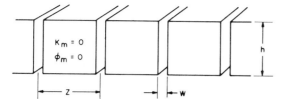

Fig. 5—Model for fracture permeability calculations.

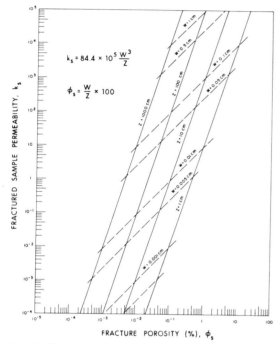

Fig. 6—Theoretical fracture permeability/porosity relationship.

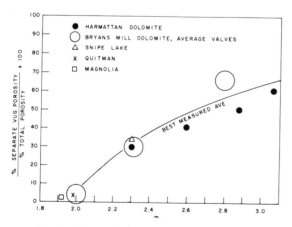

Fig. 7—Measured values of m vs. vug porosity ratio.

the cross-sectional area of the fracture. Therefore, substituting Eq. 3 into Eq. 2,

$$q_f = \frac{\Delta p w^2 A_f}{12 \mu L}.$$

From Darcy's law,

$$q_s = \frac{k_s A_s \Delta p}{\mu L},$$

where q_s is the flow rate through the sample, A_s is the cross-sectional area of the sample, and k_s is the permeability of the sample.

If the matrix permeability is zero and if we assume only one fracture in the sample, then

$$q_f = q_s$$

or

$$k_s = \left(\frac{1}{12}\right)\left(\frac{A_f}{A_s}\right)(w^2).$$

Converting into darcies,

$$k_s = (84.4 \times 10^5)(w^2)\frac{A_f}{A_s}, \quad \dots\dots\dots\dots (4)$$

where

k_s = permeability of *sample*, darcies,
w = width of the fracture, cm (in.),
A_f = cross-sectional area of fracture, cm^2 (sq in.), and
A_s = cross-sectional area of the sample, cm^2 (sq in.).

The model used throughout the rest of the discussion is shown in Fig. 5. The model is composed of a series of planar fractures of constant width w, height h, and spacing Z. The matrix has no porosity or permeability. If the *sample width* is assumed to be equal to the *fracture spacing*, then from Eq. 4,

$$k_s = (84.4 \times 10^5)\frac{w^3}{Z}, \quad \dots\dots\dots\dots (5)$$

and the porosity of the sample is given by

$$\phi_s = \frac{w}{Z}(100). \quad \dots\dots\dots\dots\dots (6)$$

Eq. 5 is plotted in Fig. 6 for values of w from 0.001 to 1 cm (0.0004 to 0.4 in.) and values of Z from 1 cm to 1000 cm (0.4 to 400 in.). The equation is plotted as a function of the porosity of the sample. Since we have assumed that there is no matrix porosity or permeability, k_s and ϕ_s are solely a function of the amount and the geometry of the fracture porosity.

139

The following conclusions can be drawn by inspecting Eq. 5 and Fig. 6.

1. The permeability is related to the third power of the fracture width and inversely related to the fracture spacing. Therefore, the width of the fracture has more control over the permeability than does the fracture spacing.

2. A fracture width of 20 μm (20 microns) in a 2.6-cm (1.02-in.) permeability plug could give 5 md of permeability. Because permeability plugs of dense carbonate may have natural or induced fractures in this width range that could give anomalously high permeability, they always should be run under stress conditions.

3. Fracture porosities of less than 1% can contribute significant permeability.

If we increase the fracture porosity by forming cavities along the fracture, the effect on the permeability will be similar to adding separate vugs to an interparticle pore geometry. This can be seen by examining Darcy's equation,

$$k = \frac{q \mu L}{A \Delta p}.$$

Assuming that the permeability of the cavities is infinite, the pressure drop across each cavity is zero. The effect, then, is to reduce the pressure drop across the sample in proportion to the amount of space occupied by the cavities. If 50% of the fracture length is occupied by cavities, the pressure drop across the sample is effectively $\Delta p/2$ and the permeability is doubled. If the cavity is significantly wider than the fracture, which probably would be the case, the total porosity of the rock would be increased greatly. Therefore, the increase in permeability is much less than would be expected if the porosity had been increased the same amount by increasing the width of the fracture.

The important observation to be made to estimate the permeability of a touching vug pore geometry is the width of the smallest connecting vugs. No predictable relationship has been determined between the size of the cavities or fracture spacing and the size of the connecting fractures. Therefore, to determine the permeability of fractures or of touching-vug pore geometries it is necessary to see the width of the connection fractures. There is no effective method for making this observation.

m-Value Relationships

The m value is given by the well-known Archie formula,

$$\frac{R_o}{R_w} = \phi^{-m}.$$

It is related to the ratio of separate vug porosity to the total porosity; a ratio which is called the *vug porosity ratio*. This relationship is shown in Fig. 7. The vug porosity ratio was measured on a number of samples for which m values had been measured. As shown in Fig. 7, data from the Smackover dolomite in the Bryans Mill field, TX, show an increase in m with an increase in the vug porosity ratio. The separate vugs are oomolds. The same relationship is apparent in the Mississippian dolomite of the Harmattan field, Alta., Canada. Here, however, the separate vugs are leached fossils. Data from the Magnolia and Quitman fields in Texas and the Snipe Lake field in Canada support the trends established by the Bryans Mill and Harmattan fields.

Fig. 7, then, can be used to estimate m values for carbonate rocks with no touching vugs. The percentage of separate vugs usually can be estimated visually. Total porosity should be taken from logs or laboratory measurements.

The effect of touching vugs on the m value is an unknown at this time. However, it is believed that touching vugs will reduce the m value below the value predicted for interpartical porosity.

Estimation of Capillary Properties

The extrapolated entry pressure as determined from the capillary-pressure curve is related to the size of the interconnecting pore space. As stated before, the size of the interconnected pore space in rocks with only interparticle porosity is fundamentally a function of the particle size and the amount of interparticle porosity. Fig. 8 is a plot of the average particle size against the extrapolated entry pressure for nonvuggy rocks. The curve is hyperbolic and becomes asymptotic to entry pressure above 700 kPa (100 psia) and to particle size below 20 μm (20 microns). The amount of interparticle porosity has some effect on the extrapolated entry pressure but has a greater effect on the shape of the curve. An example taken from the Silurian reservoirs in the Cedar Creek (MT) anticline (Fig. 9) shows this effect. The data are from rocks with greater than 6% interparticle porosity and particle sizes between 20 and 50 μm (20 and 50 microns). It can be seen that while the extrapolated entry pressure varies from 276 to 414 kPa (40 to 60 psia), the shape of the curve changes considerably with changing porosity.

Ideally, the presence of separate vugs should not change the extrapolated entry pressure, whereas touching-vug porosity could produce a lower entry pressure than that expected if only interparticle porosity were present. This effect would occur only if the interconnecting vug size was larger than the interconnecting interparticle pore size, which is the usual case. In actual practice, the presence of separate vugs tends to lower the extrapolated entry pressure and alter the shape of the curve. This probably results from the surface pore effect. Thus, the extrapolated entry pressures determined for separate-vug pore geometries tend to be considerably lower than expected for the size of the interparticle pore space.

Conclusion

To estimate petrophysical parameters by visual examination in rocks that have no touching vugs, the amount of interparticle porosity, the particle size, and the amount of separate vug porosity must be determined. From these parameters, estimates of permeability and capillary properties can be made. Archie's exponent m can be estimated from the ratio of separate-vug porosity to total porosity. Also it is important to determine the presence, or absence, of touching-vug pore geometries. If they are present, permeability could be much higher and the

140

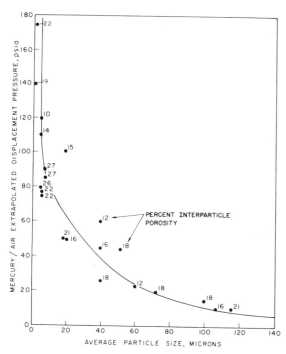

Fig. 8—Relationship between p_d and average particle size for nonvuggy rocks with greater than 0.1 md.

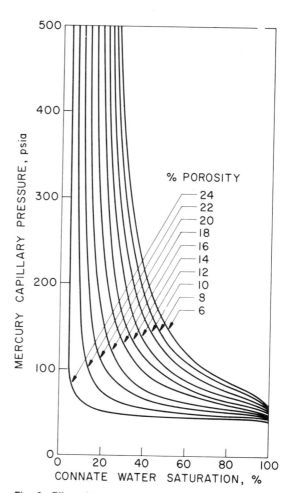

Fig. 9—Effect of interparticle porosity on shape of capillary pressure curve in a 20- to 50-μm dolomite, Silurian, Cedar Creek anticline field, MT.

capillary properties much different than would be predicted if touching vugs were absent. The problem then is to devise methods and techniques of determining these four parameters.

Visual Description of Carbonate Pore Space

Recognition of Interparticle Pore Space

The ease with which interparticle porosity can be recognized depends on the magnification at which the rock is examined and the size and shape of the particles in the rock. The particulate nature of the rock must be determined first, and in carbonate rocks that may not be easy to accomplish. In coarsely grained rocks, the particles usually can be identified even though they have been dolomitized. The interparticle pore space may be very large and observable with the naked eye or very small, requiring microscopic examination. Interparticle porosity can sometimes be recognized by the pore shape—i.e., pores between subrounded grains are often hourglass-shape. However, because of the variety of particle shapes in carbonate rocks, the shapes of interparticle pores vary considerably. This fact was demonstrated

by Dunham[3] with artificially prepared samples containing various kinds of carbonate particles.

In the more finely grained rocks, whether limestones or dolomites, the particles are more difficult to recognize and interparticle pores are more difficult to observe than in the more coarsely grained rocks. Scanning electron microscopy and petrographic techniques often are necessary to observe this porosity. The presence of interparticle porosity in these rocks, however, sometimes can be inferred by observing the texture of the broken surface; a technique described by Archie.[4] Rocks with little or no interparticle porosity are usually compact, hard, resinous, have sharp edges and smooth faces on breaking, and may show feather edges. The presence of interparticle porosity can be inferred by a chalky or a sugary (sucrose) texture.

Visual Estimation of Particle Size

The only method of estimating the particle size is by the visual examination of the rock. It is important because the ability to estimate permeability and capillary properties is primarily dependent on recognition and measure-

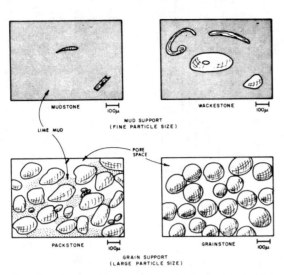

Fig. 10—Diagrams illustrating concept of support in defining particle size in depositional fabric.

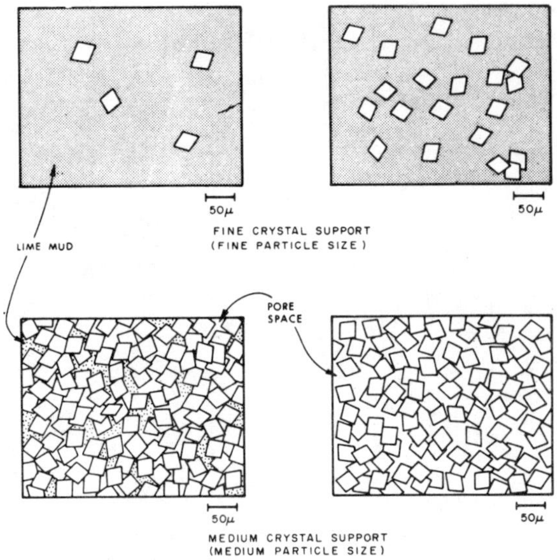

Fig. 11—Diagrams illustrating concept of support in defining particle size in dolomites.

ment of the particle size. The particle size of primary interest is the average particle size of the supporting framework in the rock because the average size of the interparticle pore space basically is controlled by this particle size. As illustrated in Fig. 10, most carbonate rocks are bimodal, being composed of large, sand-size particles and small, silt-to-clay-size particles. If the sand-size particles form a continuous, supporting network (Dunham's packstone or grainstone), their size controls the connected pore size. If the matrix of silt-to-clay-size particles forms a continuous, supporting network (Dunham's wackestone and mudstone), these mud-size particles control the connected pore size.

The same logic applies to large dolomite crystals in a finer limestone or dolomite matrix (Fig. 11). The dolomite crystals form a supporting framework only when they control the connected pore size. However, the particle size that controls the size of the connected pore space in a dolomite is not always the dolomite crystal size. Only when the dolomite crystal size is the same or larger than the sediment particle size is the dolomite crystal size of primary interest. This is usually the case with dolomitized mudstones or wackestones. If the dolomite crystal size is smaller than the sediment particle size, as is usually the case in dolomitized grainstones or packstones, the sediment particle size is more important.

The size of the interparticle porosity in carbonate rocks may be bimodal. If the largest particles in a rock are not single crystals, some of the porosity may be located between the crystals that make up the largest particles. For instance, in a pelleted grainstone the largest pore size is the pore space between the pellets. However, there may be a porosity between the mud-size particles that make up the pellets. A similar situation may be true for a dolomitized grainstone with relict intergranular porosity. Packstones may have interparticle porosity between the mud-size crystals as well as an interparticle pore geometry between the sand-size particles. This type of

porosity is called "microporosity," after Pittman,[5] who showed that large quantities of microporosity can result in unusually high water saturations in productive carbonates.

Visual Recognition of Separate-Vug Porosity

Once the particulate texture of the rock is determined, vuggy porosity usually can be distinguished visually. The best method for recognizing separate vugs, as opposed to touching vugs, is to determine the origin of the vugs. Intrafossil, shelter, and fenestral porosity along with leached particles, leached anhydrite crystals, and leached anhydrite nodules are separate-vug types. These are all classified with interparticle pore types as "fabric selective" by Choquette and Pray.[1] However, they need to be grouped separately to explain their relationship to petrophysical parameters. The origin of the vugs cannot always be determined rapidly. However, all vugs smaller than cutting size are usually separate vugs. The percent separate-vug porosity usually can be estimated visually because separate vugs are large enough to be seen readily with a low-power microscope.

Recognition of Touching Vugs

Touching-vug pore geometries usually consist of interconnected large cavities, channels, and fractures. In general, these are pore geometries classified as "not fabric selective" by Choquett and Pray.[1] Evaluation from cuttings or cores is difficult because touching-vug pore geometries are commonly large. If it is known that a touching vug pore geometry exists in an area, the presence of cuttings with crystal-lined faces may be useful as a criterion for determining their presence in a well. Both interparticle and separate vug porosity, however, also can produce cuttings of this type. Evaluation of fractures in cores is complicated by the possibility of fractures induced by the coring operations.[6]

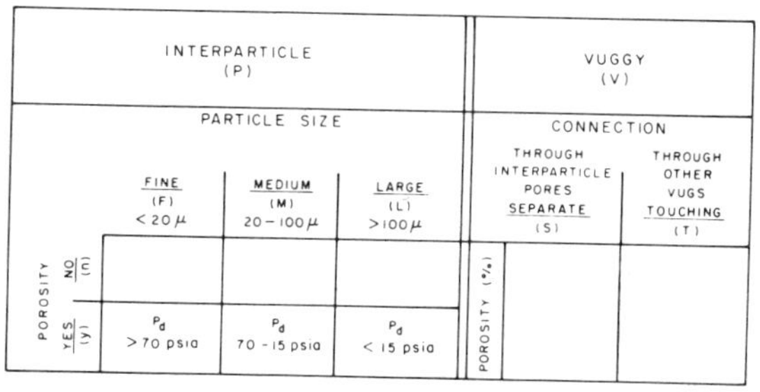

Fig. 12—Classification of carbonate pore space.

TABLE 1—SUGGESTED RECORDING TERMS

Interparticle porosity	P
Fine particle size (< 20 microns)	F
Medium particle size (20 to 100 microns)	M
Large particle size (>100 microns)	L
Yes, interparticle porosity apparent	y
No, interparticle porosity not apparent	n
Vuggy porosity	V
Separate vugs	S
Touching vugs	T

Classification

The proposed classification of carbonate pore space provides a means of grouping rocks with similar petrophysical characteristics and criteria for recognizing these types. As shown in Fig. 12, the first-order division is made between interparticle and vuggy porosity. Interparticle porosity is subdivided further on the basis of particle size and porosity with emphasis on the particle size.

The relationship between the mercury/air extrapolated entry pressures and the average particle size provides a basis for dividing particle size into three groups (Fig. 8). Rocks with <20-μm (<20-micron) particles have high entry pressures, rocks with 20- to 100-μm (20- to 100-micron) particles have intermediate entry pressures, and rocks with >100-μm (>100-micron) particles have low entry pressures. Therefore, size groups of <20, 20 to 100, and >100 μm (<20, 20 to 100, and 100 microns) are considered a petrophysically meaningful breakdown in particle size. These are fine, medium, and large, respectively. Fig. 1 shows that these size groups also give a reasonable partitioning to the porosity/permeability plot.

A visual, meaningful breakdown of the amount of interparticle porosity could not be established. While it is difficult to estimate the amount of interparticle porosity visually, it is relatively easy to distinguish dense rocks from those which have some interparticle porosity by their visual appearance. Rocks appearing dense commonly have <0.1-md permeability and are nonproductive. Therefore, a simple division into tight and porous carbonate seems most appropriate. This is done simply by a "yes" if interparticle porosity appears to be present and a "no" if not present. It should be realized, of course, that no interparticle porosity being evident from the cuttings does not necessarily mean that no interparticle porosity is present in the well.

Vuggy porosity is divided further into separate and touching vugs. An estimate of the percent separate-vug porosity is necessary and usually can be easily obtained visually. The presence or absence of touching vugs is used in the classification. No more detail is provided because of the lack of data on how to evaluate touching-vug pore geometries.

The suggested terms and symbols for recording are shown in Fig. 12 and summarized in Table 1. The classification is recorded as a composite symbol, the first part describing the interparticle pore space and the last part describing the vuggy pore space as shown below.

$$(y, n) (F, M, L) \qquad (S, \%) (T)$$
$$\text{Interparticle} \qquad \text{Vuggy}$$

The following are some examples.

1. "yFS_5" means apparent interparticle porosity, less than 20-μm (20-micron) particle size, with 5% separate vugs.

2. "nMO" indicates no apparent interparticle porosity, 20- to 100-μm (20- to 100-micron) particle size, with no vugs present.

3. "yLS_5T" stands for apparent interparticle porosity, greater than 100-μm (100-micron) particle size, with 5% separate vugs and touching vugs present.

4. "nLT" means no apparent interparticle porosity, greater than 100-μm (100-micron) particle size, with touching vugs present.

Conclusions

The description of carbonate pore space in the manner described in this paper together with a measure of total porosity calculated from wireline logs or laboratory data provides the means by which permeability, m values, and capillarity can be estimated. For example, if an interval is described as FS_5 and the total porosity for that interval is calculated to be 15%, then the interparticle porosity is 10%, and from Fig. 1, the permeability is estimated at 0.1 md. The vug porosity ratio is 0.33, giving an m value of 2.4 from Fig. 7. From Fig. 8, the extrapolated entry pressure for mercury would be greater than 100 psia (700 kPa). On the other hand, an interval described as LS_1 with a total porosity of 15% would have 14% interparticle porosity resulting in a permeability of 100 md. The vug porosity ratio would be about zero, resulting in an m value of 2.0. The pressure at which mercury would first enter this pore network would be about 20 psia (138 kPa).

The recognition of the particulate nature of a carbonate rock is of paramount importance in describing its pore geometry. The distinction between interparticle and vuggy porosity and the selection of the proper particle size is dependent on the ability to interpret the particulate nature of the rock properly. This becomes more difficult with increasing degrees of recrystallization and diagenesis. Nevertheless, pores located between relict sand size depositional particles must be distinguished from vugs if reasonably accurate values are to be determined. The particle size is based on the "support" concept and thus relates to the depositional classification of carbonate rocks as described by Dunham.[3]

By far the most difficult pore geometry to describe effectively is that composed of vugs that are connected to one another, including fractures. It is clear, however, that the presence of touching-vug pore geometries results in a much more productive rock than would be expected. It is also clear that they are difficult to evaluate from wireline logs.

Acknowledgments

The results presented here are based on a large volume of data produced at Shell Development and Shell Oil Co. Particular acknowledgment is made of C.R. Gerling's and R.L. Chuoke's work in explaining the m value relationships. F.E. Coupal first documented the small change in permeability with the addition of separate vugs.

References

1. Choquette, P.W. and Pray, L.C.: "Geologic Nomenclature and Classification of Porosity in Sedimentary Carbonates," *Bull.*, AAPG (1970) **54**, 207–50.
2. Amyx, J.W., Bass, D.M. Jr., and Whiting, R.L.: *Petroleum Reservoir Engineering: Physical Properties*, McGraw-Hill Book Co. Inc., New York City (1960).
3. Dunham, R.J.: "Classification of Carbonate Rocks According to Depositional Texture," *Classification of Carbonate Rocks, A Symposium*, AAPG (1962) 108–21.
4. Archie, G.E.: "Classification of Carbonate Reservoir Rocks and Petrophysical Considerations," *Bull.*, AAPG (1952) **36**, No. 6, 278–98.
5. Pittman, E.D.: "Microporosity in Carbonate Rocks," *Bull.*, AAPG (1971) **55**, No. 10, 1873–81.
6. Kulander, B.R., Barton, C.C., and Dean, S.L.: "The Application of Fractography to Core and Outcrop Fracture Investigations," U.S. DOE (1979) METC/SP-79/3.

SI Metric Conversion Factors

micron	$\times$ 1.0*	E+00	= μm
psi	$\times$ 6.894 757	E+00	= kPa

*Conversion factor is exact.   **JPT**

Original manuscript received in Society of Petroleum Engineers office June 18, 1981. Paper accepted for publication Sept. 5, 1982. Revised manuscript received Dec. 15, 1982. Paper (SPE 10073) first presented at the 1981 SPE Annual Technical Conference and Exhibition held in San Antonio Oct. 5–7.

Reprinted by permission of Parke A. Dickey. Paper presented
at the 24th International Geological Congress, 1972.

Migration of Interstitial Water in Sediments and the Concentration of Petroleum and Useful Minerals

PARKE A. DICKEY,
U.S.A.

ABSTRACT

As sediments compact and recrystallize, large volumes of interstitial water must find their way to the surface. This migration results in the segregation and concentration of useful substances, including oil, gas and minerals. It is therefore a very important process, but little is known about the physical, chemical and geological mechanisms that control it.

The expulsion of water is often temporarily stopped, because relatively impermeable strata are interposed along the paths of migration. Where the sediments are comparatively unconsolidated this situation results in shale diapirs and mud volcanoes. These are common in Tertiary basins where sediments were deposited rapidly. At greater depths and in rocks more greatly altered, the water which is prevented from escaping sustains much of the weight of the overburden, and therefore is found at pressures approximating that weight. It appears that at depths over 10,000 feet these high pressures are common. The suggestion is made that as metamorphism proceeds, the high-pressure interstitial waters rupture the rocks, depositing ore and gangue minerals in cavities which they themselves create. The origin of some veins thus may be the abnormally pressured water escaping from sediments which are undergoing metamorphism.

FLUID PRESSURES IN SEDIMENTS

THE PRESSURE OF interstitial fluid in sediments, usually water but occasionally gas or oil, is measured as pounds per square inch (psi) or kilograms per square centimeter (kg/cm^2). In a continuously permeable stratum containing water, the pressure increases linearly with depth according to the density of the water. It is a general rule that the pressure of the interstial fluid at any depth is approximately that required to sustain a column of water to the surface of the ground; that is, 0.45 psi per foot of depth.

Where the bed has continuity of permeability to the outcrop it may be supposed that the elevation of the outcrop determines the height to which the water will rise in the well bore, (Fig. 1, A). In this simple case the value of the subsurface pressure is determined by the density of the water. However, stagnant subsurface waters are usually salty and their density is greater than 1.00. Furthermore, their concentration usually increases with depth. In this situation the exact value of the hydrostatic pressure potential at any depth is difficult to estimate (Bond and Cartwright, 1970).

In the case of permeable beds which do not outcrop (Fig. 1, B) it is normal for the fluid pressure at any depth to be that required to sustain a column of water to the surface. The reason for this is not clear. It could be supposed that

---

Authors' addresses are given at the back of this book.

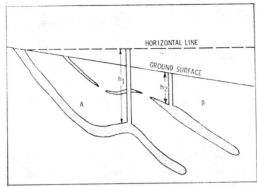

FIGURE 1 — Diagram to show origin of fluid pressure in sediments. In the case of A, the normal pressure in the reservoir is determined by the height of the outcrop above the point of measurement. In the case of B, it is not clear what determines the pressure.

FIGURE 2 — (right) — Pressures of oil reservoirs in Devonian reefs of Alberta plotted against depth. The reefs in the Redwater-Leduc-Rimbey system plot on one line, and those of the Bashaw system on another. The reefs in each system are hydraulically connected, but the two systems are not connected with each other.

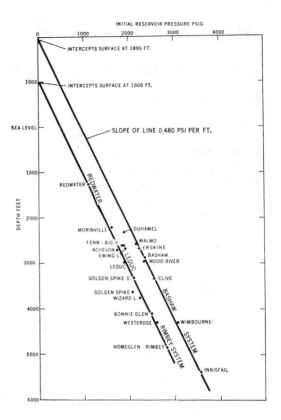

the enclosing shales have small but finite permeability to water, or that thin sandy beds provide continuity of permeability to the outcrop. Then over geologic time the water in the formation would accept more water from above if its pressure were too low, or bleed water upward if it were too high. However, there are so many exceptions to this rule, and so many oil fields with pressures either higher or lower than "normal", that it may be doubted whether compacted shales have any permeability at all to water, at least at right angles to the bedding.

This situation is well illustrated by the actual pressure pattern in the Devonian reef oil pools of Alberta. There are two parallel lines of reefs, 50 miles apart. Figure 2 shows the pressures of the various pools plotted against depth. All the pools on the Leduc-Rimbey trend lie on one line and those on the Bashaw trend lie on another. The slope of both lines is 0.480 psi per foot, which corresponds to a density of water of 1.110. For any given depth the Bashaw system has a pressure about 500 psi higher than the Leduc-Rimbey trend. It may be deduced from this that all the pools in one trend are hydraulically connected with each other, but that the two trends are separated from each other by a hydraulic barrier.

If the lines are extrapolated to zero pressure, the Leduc-Rimbey line indicates a surface outcrop at 1000 feet above sea level, and the Bashaw intercept is 1900 feet above sea level. It may be that the Leduc-Rimbey trend subcrops under permeable Cretaceous beds far to the north. The Bashaw trend, however, loses permeability up-dip before it outcrops. It is not known what determines its pressure.

146

Many fields have abnormally low pressures. They include Gilby in Alberta, Bisti in New Mexico and Wilburton in Oklahoma. All are lenticular sand bodies completely surrounded by shale, and all have only limited amounts of water down-dip from the oil. Obviously, if they had a permeable connection to the outcrop, the pressure would have become normal. The fact that the low pressures exist indicates that some shales, at least, have no permeability to water, even over geologic time.

The reason for the low pressures is unknown, but it may have been caused by the removal of large thicknesses of overburden. Studies of pressure behavior in petroleum reservoirs as fluids are withdrawn show that the pores expand about one part in 7×10^{-6} for each psi drop in pressure. Water has a bulk compressibility of 3×10^{-6} per psi. Thus, removal of the overburden causes the pores to expand more rapidly than the water can expand to fill them, which gives rise to a drop in pressure.

Where a continuously permeable bed outcrops in one place at a high elevation and in another place at a low elevation, meteoric water will enter at the higher outcrop and flow through the bed to come out at the lower. Such flow is usually peripheral, from higher to adjacent lower outcrops. The meteoric water is not usually able to displace the heavier salt water in the deeper parts of sedimentary basins. However, in the Rocky Mountains and other basins in the Cordilleran and Himalayan-Alpine chains it is not uncommon for a stratum to be flushed by meteoric water.

It is a mistake to assume, as has often been done, that differences in fluid potential between separate aquifers or reservoirs indicate flow between them. On the contrary, differences in potential between aquifers can be more logically interpreted as indicating lack of hydraulic connection.

THE MECHANISM OF COMPACTION OF SEDIMENTS

The expulsion of pore water from unconsolidated clays as they compact has been studied by soil mechanics engineers, notably Karl Terzaghi (Terzaghi and Peck, 1968). The process is visualized by means of an imaginary model (Fig. 3). It consists of a cylinder full of water, in which are several plates. The plates are perforated with small holes, and are separated by springs. The holes represent the fact that clays have low permeability to water; the springs represent the fact that the clay minerals have strength and can themselves sustain the weight of the overburden. Between the plates, at various levels, manometers measure the pressure of the water. At equilibrium, before loading, the water in the manometers stands at the level of the top plate.

If, now, a load is suddenly placed on the model, it will cause the springs to compress, and the water to be expelled from between the plates. However, the holes are small, and the water can escape only slowly. This prevents the springs from compressing, and the full weight of the load is carried by the water. There is thus an excess of hydrostatic pressure in the water, which is indicated by the manometers. After a while the water will have bled out of the upper part of the model, and the pressure falls in the upper part but remains high in the lower part. The pressure gradually bleeds off the lower part also, and finally the excess hydrostatic pressure becomes very small. The full weight of the load is now carried by the springs.

The rate at which compaction of clay occurs depends on the permeability of the clay and its volume compressibility. It also depends on the square of the thickness of the bed which is being compacted.

147

As sediments are deposited under water they pile up, increasing the load on those deposited previously, which then compact by the mechanism just described. Sand particles bridge against each other and they compact very little. Shales compact a great deal, losing porosity from near 80 per cent when first deposited to very low values at great depths. This loss in porosity involves the expulsion of large volumes of water and takes a very long time. Thus excess hydrostatic pressures are common in recently deposited sediments.

They were actually measured in the Recent sediments of the Orinoco delta at Pedernales, Venezuela by Kidwell and Hunt (1958). They found (Fig. 4) that the excess pressures increased to 8 psi and more at depths of 120 feet below the surface. They decreased upward to zero at a permeable sandy bed, and also downward to about 4 psi at the pre-Paria unconformity, which is obviously acting as a channel to bleed off the water.

Hedberg (1936) and Athy (1930) measured the changes in shale density with depth. They recognized the importance of the migration of the water in the removal of oil from source rocks. More recently other measurements of shale density have been made, and the values cover a surprisingly large range. At a depth of 6,000 feet on the Louisiana Gulf Coast the average density is 2.3, in Venezuela 2.5, and in Oklahoma 2.6. These densities correspond to porosities cf 0.13, 0.06 and 0.02 respectively. Information from several authors is compiled in Figure 5. Presumably depositional environment, mineralogy and time all affect the density-depth relationship.

Obviously the shales must have permeability to water in order for the water to escape. Few data on the permeability of natural shales are available. Gondouin and Scala (1958) give values between 4×10^{-4} and 2×10^{-6} md. Young et al. (1964) measured values from 8×10^{-4} and 2×10^{-6} md. Undoubtedly the permeability of shales varies widely, depending on differences in mineralogy and on degree of compaction. Miller and Low (1963) believe that water in fine pores behaves as a solid and does not obey Darcy's law. They found that there is a threshold pressure gradient below which no flow occurs in compacted montmorillonite. This may explain the apparent lack of permeability of compacted shales to water.

Bredehoeft and Hanshaw (1968) calculated the rate of expulsion of water from a thick sedimentary section receiving sediments at a constant rate. They

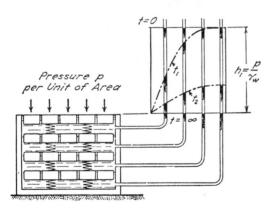

FIGURE 3 — Terzaghi's model of the compaction process. When a load is dropped on the cylinder, the water first carries the full load. As it escapes, the springs sustain the load and the water pressure drops to normal hydrostatic.

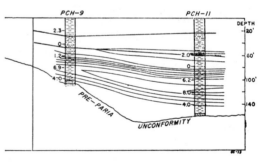

FIGURE 4 — Excess hydrostatic pressures in the recent sediments of the Orinoco river, Pedernales, Venezuela. As the muds compact, pure water is bleeding upward to a permeable bed at depth of about 30 feet, and also downward to the pre-Paria unconformity, which is also permeable laterally (from Kidwell and Hunt).

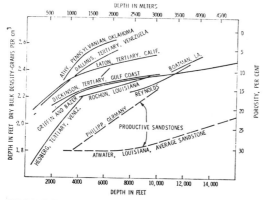

FIGURE 5 — Sediment density and porosity plotted against depth of burial. There is a surprising range of values, resulting from depositional environment and age of the rocks.

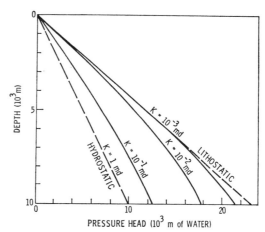

FIGURE 6 — Graphs showing calculated pore pressure versus depth for a situation similar to that of the Gulf Coast. It is assumed that sediments are deposited continually and compact as a result of vertical expulsion of water. If shale permeability is between 10^{-4} and 10^{-7} md, it is obvious that little water escapes vertically. (Modified from Bredehoeft and Hanshaw.)

assumed values comparable to those of the Gulf Coast. Their calculations showed that if the shales had a vertical permeability of 10^{-3} md the water would be expelled extremely slowly so that the pore pressures would remain close to those imposed by the weight of the overburden (lithostatic). For the interstitial pressure to be that of a column of water to the surface (hydrostatic) the sediments would have to have a permeability of 1 md. This is in the range of permeability of sands, and it is doubtful that even in thick sands the gross vertical permeability is as great as 1 md. As a matter of fact, normal pressures in the Gulf Coast down to a depth of about 10,000 feet are all very close to 0.465 psi per foot of depth (1.07 kg/cm²). This is in the part of the section containing abundant sands. Probably the sands connect through thin and silty but relatively permeable stringers to the contemporaneously deposited "massive sand" facies which is found landward of the oil and gas producing zones.

ABNORMALLY HIGH PORE PRESSURES

In the more shaly facies to the seaward, abnormally high pore pressures are very common. Dickinson (1953) showed that extremely high pressures, sometimes almost equal to the weight of the overburden, are found in lenticular sands which are completely surrounded by shale. Obviously the shales must lack permeability to water almost totally. The water in them was unable to escape as sedimentation proceeded. It remained locked in the pores and was forced to carry much of the weight of the overburden.

Hottman and Johnson (1965) showed that the shales containing the sands with abnormal pressures have abnormally high porosity for their depth of burial. This high porosity is reflected in abnormally low electrical resistivity and seismic velocity.

The depth to the first high pressure on the Gulf Coast is quite variable, ranging from less than 4,000 to over 20,000 feet. Often the transition from normal

149

to high pore pressure is very abrupt, taking place in a vertical interval of 100 feet or less. More often it covers an interval of several hundred feet. In general the high pressures occur only where the section is mostly shale and the sands are lenticular. Contemporaneous faulting often controls the boundaries of the high-pressure zones. Apparently the faulting cut off the routes of escape of the water, which must have been mostly parallel to the bedding planes, (Dickey *et al.*, 1969).

High pressures are found in many other areas. The unfortunate blow-out at Santa Barbara, California gave them much notoriety. In the Green River Basin of Wyoming high pressures are found at depths of 10,000 feet both on the west side of the basin in front of the "overthrust belt" and on the east side at the foot of the Wind River mountains. The suggestion has been made that the high pressures originate from laterally directed stresses of tectonic origin (Dickey and Rathbun, 1969). However, it seems improbable that the comparatively unconsolidated shales could transmit horizontal stress. It is more likely that rapid sedimentation trapped the water in shales that had effectively zero permeability to water. Similar situations of high pressures in shaly sections at the foot of mountains occur in the Magdalena Valley of Colombia, in the pre-Caucasus region of Russia, and many other places. Perhaps the most astonishing of these is the Anadarko Basin of Oklahoma, where Pennsylvanian shales contain sand lenses with gas at abnormally high pressures. The water here must have remained locked in the pores since Pennsylvanian time.

Where an impermeable bed A overlies a permeable bed B, as in Figure 7, the stress due to the weight of the overburden, S, is sustained by the stress σ in the skeleton of solid grains, and also by the upward hydrostatic pressure of the pore fluid, p. If the upward pressure p approaches the downward pressure S, then the skeleton pressure σ goes to zero, and the overlying bed A is virtually floating. Hubbert and Rubey (1959) showed that a very small tilt of the contact will cause bed B to slide. Submarine slides can occur on a large scale, resulting in low-angle overthrust faulting. High pore pressures also facilitate small-scale slumping. In the case of a prograding delta, sediments are deposited on a gently sloping bottom. Suppose a bed of fairly rigid and permeable sand a few feet thick is deposited on a mud of low permeability. If the sand is deposited rapidly, the pore pressure in the shale will approach the overburden pressure, and the sand bed may slide down the slope, crumping and rolling up and producing contorted bedding.

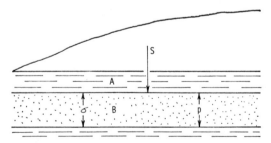

FIGURE 7 — Diagram to illustrate vertical stress in sediments. The weight of the overburden (S) is sustained by the stress in the skeleton of the solid grains (σ) and the pore pressure in the interstitial fluids (p). If the pressure p approaches S, σ drops to near zero, and bed B is practically floating. If the pressure exceeds S, rupture of a bed will occur.

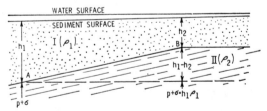

FIGURE 8 — If a heavier, well-compacted sediment overlies a lighter, undercompacted one on a sloping surface, a condition of instability will result. At point B the upward pressure will exceed the downward pressure, and a mud diapir will form.

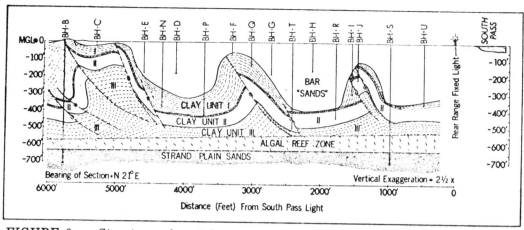

FIGURE 9 — Structure of mud lumps at the mouth of the Mississippi river (from Morgan, *et al.*). Low-density clays (Units I, II and III) have pushed up through the denser bar sands.

Imagine a situation in front of an advancing delta (Fig. 8). Prodelta clays of very low permeability are overlain by fluvial sandy muds of much greater permeability. The sandy muds will compact quickly, reaching a porosity appropriate to their depth of burial. Kidwell and Hunt (1958) found densities of about 1.85 at 140 feet below the bottom, in shales that had not yet compacted completely. Estimates of density at 1000 feet vary from 1.8 to 2.15 (Fig. 5). It is certainly possible that density in the upper bed (I in Fig. 8) could be considerably more than that in the lower bed. The upper bed would have some compressive strength but very little tensile strength. The lower bed, (II) being undercompacted, would have very little strength, and would transmit pressure almost like a fluid.

At point A in Figure 8 the downward stress would be $h_1\rho_1$, where h_1 is the thickness and ρ_1 the density of the upper bed. It is matched by the upward stress $p + \sigma$. At point B the downward pressure is $h_2\rho_1$, and the upward pressure is $h_1\rho_1 - (h_1 - h_2) \rho_2$. If h_1 is greater than h_2, and ρ_1 is greater than ρ_2, then at point B the upward pressure will exceed the downward pressure by $(\rho_1 - \rho_2) (h_1 - h_2)$. The unconsolidated muds will burst upward through bed I, forming a shale diapir.

Such intrusions of soft, undercompacted shale into overlying strata occur at both shallow and deeply buried strata.

At the mouth of the Mississippi River "mud lumps" have appeared continuously successively farther seaward as the river builds its delta outward (Morgan et al., 1968). They form at the rate of one or two per year off the jetties at South Pass, and also off the mouths of other distributaries that are actively depositing large amounts of sediment. Figure 9 shows the structure of the mud lumps as determined by drilling. The soft mud which flowed came from depths of about 500 feet below the bottom.

Shepard reports (1969) larger lumps that he believed to be shale diapirs off the mouth of the Magdalena River, Colombia (Fig. 10). The intrusive shale appears to have come from 400 meters or more below the bottom. Similar lumps have been noted off the constructional slopes near the mouths of rivers in many other places.

Where shale diapirs of this type are found it must be assumed that there is undercompacted shale at depth, which probably contains fluids at abnormally high pressures. Bumps on the sonic reflection profiles resembling shale diapirs should be a warning to drillers that both high pressures and heaving shales can be expected at depth. There is no way to guess at the depth where the dangerous situation may occur. Shale in diapirs may have come from as little as a few feet or as deep as several kilometers. The high-pressure zones are not only directly under the diapirs, but may extend many kilometers laterally away from them.

In the vicinity of Buku, USSR, mud, methane gas and salty water issue from the earth and pile up into mounds tens or even hundreds of meters in height, called mud volcanoes. Occasionally the gas explodes violently. Oil sometimes accompanies the gas. This same area was the first important oil-producing area in Europe, and geologists were impressed by the apparent relation between mud volcanoes and petroleum.

Mud volcanoes occur in many areas of geologically recent sedimentation, notably Trinidad, northwest Colombia, northeast Venezuela, Burma, New Zealand, and the Copper River Valley of Alaska.

Most geologists have ascribed mud volcanoes to pockets of gas at depth which burst their way to the surface, or to tectonic forces related to mountain building (Kugler, 1938; Yakubov, 1959). It seems more probable that they result from a low-density bed of shale in the subsurface, from which the pore water has not been expelled. Such shales are characterized by abnormally high fluid pressures. This relationship was first pointed out by Kalinko (1960). Shale diapirs thus have basically the same cause as salt domes, which they resemble in many respects. Strata on their flanks are often steeply inclined, as they are on the flanks of salt domes and salt anticlines. The water and mud which oozes out originally doubtless contained large amounts of gas in solution. As the water approaches the surface the pressure reduces, and the gas comes out of solution, forming pockets of free gas, which expand as they rise. On reaching the surface they may erupt violently.

In Trinidad, Colombia, Venezuela and Alaska, wells drilled near the mud volcanoes encountered abnormal pressures. No abnormal pressures, however, have been reported from the Baku area. Mud volcanoes should be taken as a warning of high pressures and heaving shales at great depths. Blowouts should be expected when drilling in such areas.

CHEMICAL PROCESSES ASSOCIATED WITH COMPACTION

The expulsion of water from shale is accompanied by mineralogical and chemical changes which are extremely important to applied geology, but they have been little investigated.

The interstitial water in muds changes very little in chemical composition during the first few hundred meters of burial. Many samples have been analyzed from cores from the Deep Sea Drilling Program, and the water composition is almost exactly the same as that of the overlying sea water. In only a few cases has there been notable depletion of the magnesium (Manheim, 1970; Drever, 1971).

At greater depths and longer times of burial the changes in pore water from sea water are profound. Waters produced from porous and permeable sandstones and limestones have usually lost all or almost all their sulfate and bicarbonate, so that practically the only anion is chloride. The calcium is enriched and the magnesium is reduced, so that there is usually 3 to 5 times as much calcium as magnesium. In sea water there is three times as much magnesium as calcium. These waters have been classified by Russian geochemists as "chloride-calcium" brines,

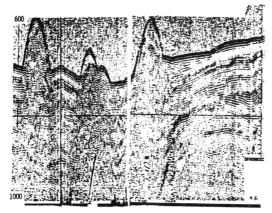

FIGURE 10 — Diapirs, probably shale, coming from depths of 400 m below the sea bottom off the mouth of the Magdalena river, Colombia (from Shepard, 1967).

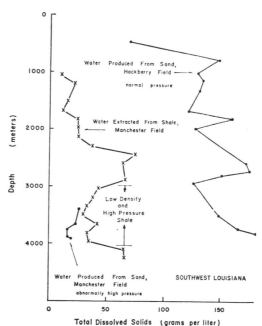

FIGURE 11 — Concentration of interstitial water with depth, southwest Louisiana. The water in the pores of the shale is much less concentrated than that in the adjacent sands. Water in the sands of the high-pressure zone at Manchester is much less concentrated than that in the equivalent normal-pressure sands at Hackberry (from Schmidt, 1971).

(Dickey, 1966). They range in concentration of total solids from less than sea water to almost saturation. Usually there is an increase in concentration with depth, often around 50,000 mg/l per 1000 feet.

The chemistry of the processes by which the sea water is altered and concentrated is very poorly understood. The concentration has been ascribed to reverse osmosis, but this mechanism does not seem to fit the geometry of the actual subsurface situations.

Waters from the abnormally pressured zones conform to the chemical pattern of the chloride-calcium brines, but their concentration is often much less than normal for their depth of burial. This suggests that if the mechanism causing compaction of the shales is inhibited, the process causing the concentration of the water will be inhibited also. At Hackberry, Louisiana, waters produced from normally pressured sand at depth of 3000 to 4000 meters have concentrations ranging from 120,000 to 180,000 mg/l. In the nearby Manchester field, water from equivalent sands at the same depth has abnormally high pressure, and the water concentrations range around 30,000 mg/l, (Schmidt, 1971, Fig. 11).

The water in the pores of shales has seldom been analyzed. It appears to be quite different in composition from the water in the adjacent permeable sands. In Louisiana it has a concentration one-half or one-third that of water in the adjacent sands. Very little calcium or magnesium is present, so that sodium is practically the only cation. It contains sulfate in amounts approximating those of chloride, which is most surprising in view of the absence of sulfate in the sand waters. Bicarbonate is also abundant in the shale waters. Figure 12 is a triangular diagram comparing the composition of waters from shale and sand in the same parts of the section. Unpublished analyses from other areas confirm the large amounts of sulfate in shale waters.

As depth of burial increases mineralogical changes take place in the sediments. In the Gulf Coast Tertiary the clay mineral montmorillonite disappears gradually between 2,500 and 3000 meters (Burst, 1969). It is replaced by mixed layer and illitic clays. This change involves chemical alteration, and also the release of water of crystallization, formerly part of the solid structure of the clay minerals, into the free pore water.

This change seems to be due primarily to temperature, the reaction starting at about 100°C. Consequently it does not seem to be related to the high-pressure zones, which often start at much shallower depths.

Corresponding mineral changes take place in the sandstones, of which the most important is the deposition of secondary silica overgrowths on the quartz grains, which results in a loss of porosity. Rather sparse data differ widely on the amount of porosity loss with depth of burial (Atwater and Miller, 1965; Philipp *et al.*, 1963). Authigenic kaolinite also forms in the pores. In some cases, at least, most of the silica seems to have come from outside the porous sand, in

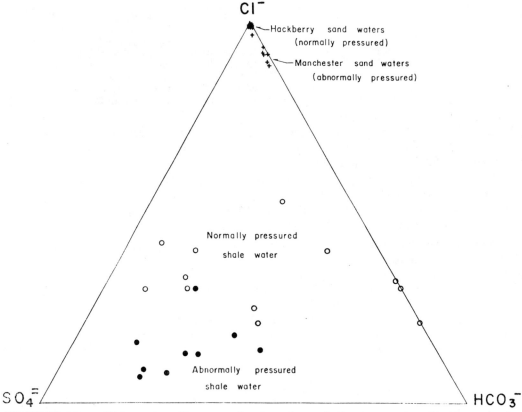

FIGURE 12 — Triangular diagram showing the relative amounts of anions in interstitial waters in southwest Louisiana. The waters in the sands contain almost no bicarbonate and sulfate, but these ions predominate in the shale waters (from Schmidt, 1971).

spite of the common belief that it is formed by recrystallization of the original quartz grains in place. Probably the silica is coming from shales whose pore water is traversing the sand on its way to the surface. Sudden lateral changes in the amount of cementation are often noted. They are probably due to the patterns of water flow.

It is generally believed that petroleum is generated in organic-rich shales, but the mechanism which removes it from the shales and concentrates it in porous reservoir rocks is very poorly understood. It must leave the shale along with the water which is expelled by compaction. There is much geological evidence that oil migrates soon after deposition; for example, contemporaneous structures seem to be more likely to contain oil than structures formed long after compaction and lithification of the sediments.

Recently, however, chemical evidence has been accumulating which suggests that the oil is not generated from the source rocks until their temperature approaches 100°C. This does not occur until the sediments are buried to over 2500 meters. By this time the shales are hard, with a porosity between 10 and 13 per cent, and a permeability to water that is vanishingly low. If oil is not generated until the source rocks have lost practically all of their porosity and permeability, how then is the oil expelled?

The physical form in which the oil migrates is not known. The solubility of light hydrocarbons in pure water is appreciable — several parts per million. However, the heavier hydrocarbons from C_8 up are extremely insoluble (McAuliffe, 1969). Water-wet shale has no permeability at all to immiscible fluids such as gas or oil, so they cannot move as slugs or drops. Peake and Hodgson (1966) report "accommodations" of oil in water up to about 30 parts per million. Cartmill and Dickey (1970) found that such a colloidal suspension was able to pass through water-wet sands, but the tiny droplets coalesced at points where the grain size decreased. Neruchev (1965) offered some evidence that the amount of extractable hydrocarbon decreases in shales for the first few meters away from the reservoir rocks, as if it were removed by some flushing action. This process is the least understood part of the origin and migration of petroleum, and the one on which research is most needed.

As burial proceeds, pressures and temperatures increase. With increasing temperature chemical changes in the solids are accelerated. The organic matter becomes more completely converted to methane and graphite. The clay minerals continue their recrystallization, and finally metamorphism to slates, phyllites and schists takes place. These processes involve a continuing loss of porosity and they also release additional pore water. If the geothermal gradient of deep wells in South Louisiana is extrapolated downward, it appears that a temperature of 400°C will be reached at a depth of 15 kilometers, and that the Tertiary sediments have already begun to metamorphose (Jam L., et al., 1969).

Long before this depth is reached the shales have lost all their permeability to water. How then, does the surplus water generated by metamorphism get out of the rocks? At much shallower depths Darcy-type flow has stopped, as evidenced by the abnormal pressures. Diffusion is extremely slow. Faults in shallower sediments are clearly seals, not channels, because they often form boundaries of the high-pressure zones.

It has long been known that when fluids are injected underground they will rupture the rock, forming their own fissures. The fluid pressures at which rupture occurs vary widely, ranging from more than overburden pressure to only slightly above hydrostatic pressure (Dickey and Andresen, 1950; Hubbert and

Willis, 1957; Eaton, 1969). Oil-producing formations are often ruptured deliberately in order to increase their conductivity to fluids by injecting water or oil containing suspended sand to prop the fissures open. Ruptures also occur accidentally while drilling. If a high-pressure formation is expected, the density of the mud is increased in order to prevent a blow-out. The heavy mud produces a pressure which exceeds that of a column of water to the surface, and may also exceed the critical pressure at which rupture will occur. When it does, cracks form which open and accept the drilling mud. Dickey and Andresen (1950) and Hubbert and Willis (1957) believed that, on the basis of the stress pattern as well as on the behavior of the wells, the cracks had to be predominantly vertical.

It seems obvious now that the interstitial water in deeply buried sediments is usually, if not always, at a pressure close to the weight of the overburden. This pressure may be more than sufficient to burst the rock, and the water can therefore move out through fissures which it forms itself. These fissures will be principally vertical, and will extend upward into zones of lower temperature. The water forming and filling the cracks will be a mixture of salty connate sedimentary pore water and water from the clay minerals released by their recrystallization. It will be hot, and saturated with silica and probably also other minerals. As it forces its way upward it will cool, depositing quartz, feldspar, calcite and other minerals as it goes. It seems likely that many hydrothermal ore veins were formed by sedimentary connate waters, rather than by "juvenile" waters as has often been supposed.

These veins contain metallic minerals including copper, zinc, lead, gold and silver. The waters therefore, by some still unknown process, are able to concentrate and segregate these useful minerals. The process has many points of resemblance to the concentration and segregation of petroleum — the principal difference being that it takes place at a higher temperature.

It has been tacitly supposed that hydrothermal solutions filled pre-existing cavities. However, it seems most unlikely that rocks have enough strength to permit fissures or cavities to remain open at great depths. It is much more probable that the fissures were formed by the solutions themselves, bursting their way upward by the mechanism familiar to oil-well drillers as "lost circulation".

CONCLUSION

The formation of petroleum and mineral deposits, and the diagenesis and metamorphism of rocks, take place in an ambient environment of salty water, which facilitates the processes and participates in the reactions. The chemical composition of subsurface water shows large and significant changes with depth of burial. Dynamic processes of sediment compaction cause the motions of water which segregate and concentrate useful substances such as petroleum and ore minerals. The patterns of water flow determine the location of the petroleum and ore deposits. Consequently, the elucidation of these processes and patterns should provide important principles in the search for such deposits.

REFERENCES

Athy, L. F., 1930. Density, porosity, and compaction of sedimentary rocks. Am. Assoc. Pet. Geol. Bull., 14, p. 1-24.
Atwater, G. I., and Miller, E. E., 1965. The effect of decrease of porosity with depth on future development of oil and gas reserves in South Louisiana: presented at annual meeting. Am. Assoc. of Pet. Geol., New Orleans, 1965.
Bond, D. C., and Cartwright, Keros, 1970. Pressure observations and water densities in aquifers and their relation to problems in gas storage. J. Petrol. Technol., December, 1970, p. 1492-1498.

Bredehoeft, J. D., and Hanshaw, B. B., 1968. On the maintenance of anomalous fluid pressures: I, Thick sedimentary sequences. Geol. Soc. Am. Bull., 79, p. 1097-1106.

Burst, John F., 1969. Diagenesis of Gulf Coast clayey sediments and its possible relation to petroleum migration. Am. Assoc. Pet. Geol. Bull., 53, No. 1, p. 73-93.

Cartmill, John C., and Dickey Parke A., 1970. Flow of a disperse emulsion of crude oil in water in porous media. Am. Assoc. Pet. Geol. Bull., 54, No. 12, p. 2438-2443.

Dallmus, K. F., 1958. Mechanics of basin evolution and its relation to the habitat of oil in the basin: In Weeks, Lewis G., (Editor). Habitat of Oil. Am. Assoc. Pet. Geol., Tulsa, Okla., p. 919.

Dickey, Parke A., 1966. Patterns of chemical composition in deep subsurface waters. Am. Assoc. Pet. Geol. Bull., 50, p. 2472-2478.

————, and Andresen, Kurt H., 1950. Behavior of water input wells: In Torrey, Paul D., (Editor). Secondary Recovery of Oil in the U.S. Am. Pet. Inst.

————,Shriram, Calcutta R., and Paine, W. R., 1968. Abnormal pressures in deep wells of Southwest Louisiana. Science, 160, May 10, p. 609-615.

————, and Rathbun, Fred C., 1969. Abnormal pressures and conductivity anomaly, Northern Green River Basin, Wyoming. The Log Analyst, July-August, 10, No. 4, p. 3-8.

Dickinson, George, 1953. Geological aspects of abnormal reservoir pressures in Gulf Coast Louisiana. Am. Assoc. Pet. Geol. Bull., 37, No. 2, p. 410-432.

Drever, James I., 1971. Magnesium-iron replacement in clay minerals in anoxic marine sediments. Science, 172, p. 1334-1336.

Eaton, B. A., 1969. Fracture gradient prediction and its application in oil-field operations. J. Petrol. Technol., p. 1353-1360.

Gondouin, M., and Scala, C., 1958. Streaming potential and the S P log. Am. Inst. Mining, Metall. and Petrol. Engineers Trans., 213, p. 170-179.

Griffin, David G., and Bazer, Donald A., 1969. A comparison of methods for calculating pore pressures and fracture gradients from shale density measurements using the computer. J. Petrol. Technol., Nov., p. 1463-1473.

Hedberg, Hollis, D., 1926. The effect of gravitational compaction on the structure of sedimentary rocks. Am. Assoc. Pet. Geol. Bull., 10, p. 1035-1072.

————, 1936. Gravitational compaction of clays and shales. Am. J. of Sci., 5th Ser., 31, No. 184, p. 241-287.

Hottman, C. E., and Johnson, R. K., 1965. Estimation of formation pressure from log-derived properties. J. Petrol. Technol., June, p. 717-772.

Hubbert, M. King, and Rubey, William W., 1959. Role of fluid pressure in mechanics of overthrust faulting. Geol. Soc. Am. Bull., 70, p. 115-166.

————, and Willis, D. G., 1957. Mechanics of hydraulic fracturing. Am. Inst. Mining, Metall. and Petrol. Engineers Trans., 210, p. 153-166.

Jam, L. Pedro, Dickey, Parke A., and Tryggvason, Eysteinn, 1969. Subsurface temperature in South Louisiana. Am. Assoc. Pet. Geol. Bull., 53, No. 10, p. 2141-2149.

Kalinko, M. K., 1960. O mekhanizme i usloviyakh gryazevikh vulkanov (On the mechanisms and conditions for mud volcanoes). Trudy VNIGNI, 27, p. 45.

Kidwell, Albert L., and Hunt, John M., 1958. Migration of oil in recent sediments of Pedernales, Venezuela: In Weeks, Lewis G., (Editor). Habitat of Oil. Am. Assoc. Pet. Geol., Tulsa, Okla., p. 790-817.

Kugler, H. G., 1938. Nature and significance of sedimentary volcanism: In Science of Petroleum. Oxford Univ. Press, London, 1, p. 297-299.

Manheim, F. T., Chan, K. M., and Sayles, F. L., 1970. Interstitial water studies on small core samples, Deep Sea Drilling Project, Leg 5[1]: In Initial Reports of the Deep Sea Drilling Project. V., Oct., 1970.

McAuliffe, C., 1969. Determination of dissolved hydrocarbons in subsurface brines. Chem. Geol. 4, No. 1/2, p. 225-234.

Miller, Raymond, J., and Low, Philip F., 1963. Threshold gradient for water flow in clay systems. Soil Sci. Soc. of Am. Proc. No. 6, p. 605-609.

Morgan, James P., Coleman, James M., and Gagliano, S. M., 1968. Mudlumps: diapiric structures in Mississippi Delta sediments: In Braunstein, Jules, and O'Brien, Gerald D., (Editors). Diapirism and Diapirs. Am. Assoc. Pet. Geol., Memoir 8, Tulsa, Okla.

Neruchev, S. G., and Kovalecheva, I. S., 1965. O vliyanii geologicheskikh uslovii na velichinu nefte otdachi materinskikh porod (The effect of geological conditions on the amount of oil given up by source rocks). Akad. Nauk. U.S.S.R. Doklady, 162, No. 4, p. 913-914.

Peake, E., and Hodgson, G. W., 1966. Alkanes in aqueous systems. I. Exploratory investigations on the accommodation of C_{20}-C_{33} n-alkanes in distilled water and occurrence in natural water systems. J. Am. Oil Chem. Soc. 43, No. 4, p. 215-222.

157

Philipp, W., Drang, H. J., Füchtbauer, H., Haddenhorst, H. G., and Jankowsky, W., 1963. The history of migration in the Gifhorn Trough (NW Germany). Sixth World Pet. Congr., Sec. 1, pap. 19, p. 457-481.

Philippi, G. T., 1965. On the depth, time, and mechanism of petroleum generation. Geochimica et Cosmochimica Acta, 29, p. 1021-1049.

Ridd, M. F., 1970. Mud volcanoes in New Zealand. Am. Assoc. Pet. Geol. Bull., No. 4, p. 601-616.

Rochon, R. W., 1967. Relationship of mineral composition of shales to density. Trans. Gulf Coast Assoc. Geol. Soc., 17, p. 135-142.

Schmidt, Gene W., 1971. Interstitial water composition and geochemistry of deep Gulf Coast shales and sands. M.S. thesis, Univ. of Tulsa.

Shepard, Francis P., 1967. Delta-front diapirs off Magdalena River, Colombia, compared with hills off other large deltas. Trans. Gulf Coast Assoc. of Geol. Soc., 17, p. 316-325.

Terzaghi, Karl, and Peck, Ralph B., 1968. Soil mechanics in engineering practice. John Wiley, New York, 2nd ed., p. 84.

Tissot, B., 1969. Premières données sur les mécanismes et la cinétique de la formation du pétrole dans les sédiments: simulation d'un schéma réactionnel sur ordinateur. Rev. de l'Inst. Fr. du Pét., 24, No. 4, p. 470-501.

Yakubov, A. A., 1959. Gryazevie vulkani Azerbaidzhana, ikh genesis i svyaz' s gasoneftynymi mestorozhdeniyami (Mud volcanoes of Azerbaidzhan, their genesis and relation to oil and gas pools). Sovetskaya Geologiya, No. 12.

Yoder, Hatten S., Jr., 1955. Role of water in metamorphism: In Poldevaart, Arie (Editor). The Crust of the Earth. Geol. Soc. of Am. Spec. Pap 62, p. 505-524.

Young, Allen, Low, Philip F., and McLatchie, A. S., 1964. Permeability studies of argillaceous rocks. J. Geophys. Res. 69, No. 20, p. 4237-4245.

Copyright 1979, Society of Petroleum Engineers of AIME.
Reprinted by permission. First published in the *Journal of
Petroleum Technology* (November 1979), pp. 1375-1380.

Some Applications of Scanning Electron Microscopy to the Study of Reservoir Rock

E.D. Pittman, Amoco Production Co.
J.B. Thomas,* Amoco Production Co.

Introduction

This paper discusses some applications of the scanning electron microscope (SEM) to the study of reservoir rocks. The SEM is used extensively to provide qualitative information about the pore geometry of rocks, either directly[1,2] or indirectly through the examination of pore casts.[3-6] Many of these studies were concerned with relating pore geometry to various tests or rock properties; specific applications also are discussed.

The basic principles of the SEM were known in 1935. Experimental instruments were built in Germany and the U.S. before World War II. After the war, the British started a research program that eventually developed a commercial instrument in the early 1960's. Since then, the SEM has become an indispensable scientific tool in all fields of science and industry. Presently, there are approximately 10 SEM manufacturers, with prices ranging from about $17,000 to well over $100,000.

The SEM provides several unique advantages for the geologist or engineer when compared with other types of microscopes: (1) magnification range of $15\times$ to a practical upper limit of about $40,000\times$ for rocks, (2) great depth of field, (3) ease of sample preparation – samples made conductive by a coating of metal applied under vacuum, (4) nondestructive sample study, and (5) three-dimensional image analogous to reflected light requiring no special techniques for interpretation. These features make the SEM ideal for studying irregular rock surfaces.

An energy-dispersive X-ray attachment for the SEM provides qualitative information on chemical elements and aids in identifying minerals.

Evaluation and Importance of Microporosity

Micropores occur in sandstone and carbonate reservoirs and affect fluid flow properties.[7,8] These micropores are seen and evaluated best with an SEM. Microporosity, for convenience, has been defined as pore apertures with a radius of less than 0.5 μm.[7]

In sandstone reservoirs, clay with associated micropores may occur as detrital laminae, pellets (grains) of silt-sized or coarser material, or as authigenic (that is, newly formed or regenerated) material (Fig. 1). In a water-wet sandstone, the micropores and high surface area associated with the clay minerals are conducive to holding significant amounts of bound water that can affect log calculations for water saturation. Because of improved log-calculation techniques, which take into consideration cation-exchange capacities of clay minerals, this is less of a problem than in the past.

Micropores in carbonate reservoirs occur among calcite or dolomite crystals that typically are less than 4 μm in diameter. It is important to consider the

*Now with Reservoirs Inc., Denver.

0149-2136/79/0011-7550$00.25
© 1979 Society of Petroleum Engineers of AIME

The scanning electron microscope (SEM) provides qualitative information about pore geometry through direct observation of a rock or a pore cast of the rock. This aids in understanding reservoir productivity capabilities. The SEM is useful for locating and identifying minerals, particularly clay minerals – an aid when designing drilling and completion programs.

159

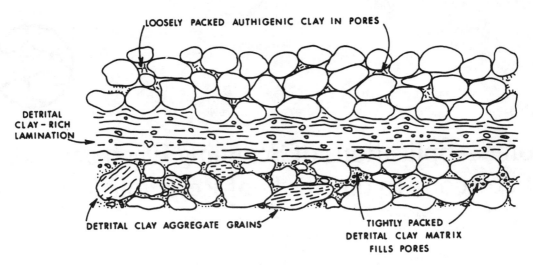

Fig. 1 – Clay minerals in sandstones.

LOOSELY PACKED AUTHIGENIC CLAY IN PORES

DETRITAL CLAY-RICH LAMINATION

DETRITAL CLAY AGGREGATE GRAINS

TIGHTLY PACKED DETRITAL CLAY MATRIX FILLS PORES

relative amounts of macro- and microporosity when evaluating carbonate reservoirs (macro/micro ratio). An almost totally microporous (ratio ≪1) rock in a water-wet system will contain bound water, unless the bouyancy of the hydrocarbon column is great enough to displace the water with oil or gas. Rocks of this type have small pore apertures, low permeability, and require fractures to provide economic flow rates [for example, chalk (Fig. 2)]. When the reservoir has about equal micro- and macroporosity (ratio ≃1) and the macropores are interconnected adequately, it is essential to consider the possibility of abundant bound water in the micorpores (Fig. 3). This can yield high water saturations (based on log calculations) that could give misleading interpretations regarding hydrocarbon production. A bypassed reservoir is a distinct possibility, although such reservoirs would flow little, if any, water. If a reservoir has abundant macroporosity relative to the microporosity (ratio ≫1), there will not be enough bound water to cause problems with log interpretation (Fig. 4).

Evaluation of the Effect of Fluids and Chemical Additives in Enhanced Oil Recovery

The SEM is useful for examining the effect of fluid and chemical additives on rocks during enhanced oil recovery or formation treatment. For example, during laboratory tests for a miscible flood, we discovered that when micellar fluid was mixed with water and flowed through a rock there was a noticeable drop in permeability. We expected a problem existed with the expandable clay minerals in the rock. To evaluate this problem, pairs of side-by-side core plugs were cut. One sample was used for control and the other for a flow test. Examination of the untreated control sample with the SEM showed (Fig. 5) that the rock consists of sand-sized grains cemented by partial grain coatings of the clay mineral smectite (montmorillonite) as well as by calcite. Smectite is an expandable clay that accepts water into the crystal lattice structure, causing the clay to swell. SEM examination of the other sample following a flow test showed that throughout the sample the smectite had reacted to the contact with the water-miscible fluid mixture. The smectite coating on the sand grains had become detached and had migrated into pore throats (Fig. 6), which partially blocked the opening and reduced permeability. The SEM provided visual support for the theory.

Influence of Clay Minerals on Productivity in Sandstone Reservoirs

Reservoir studies using the SEM have shown that

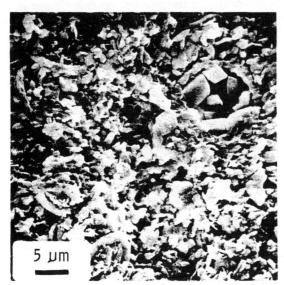

Fig. 2 – SEM view of chalk, showing micropores and absence of macropores; Danian, North Sea; 24% and 0.46-md permeability. (The millidarcy is an informal name for $10^{-3} \mu m^2$.)

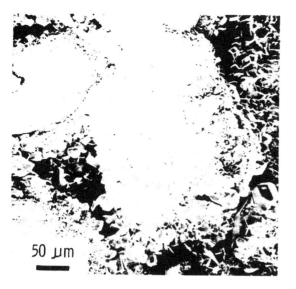

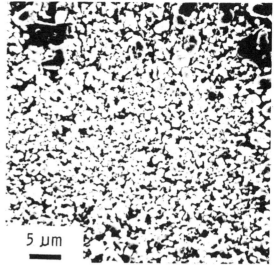

a. Intergranular macropores lined by calcite cement.

b. Enlargement of grain in center of "A," showing micropores among calcite crystals

Fig. 3 – Rodessa limestone; 10% porosity and 4.2-md permeability.

varying distributions and morphologies of clay minerals can be related directly to productivity of sandstone.[9-11] Neasham[12] showed that reservoir quality and mean pore aperture size decreased as follows: (1) discrete, loosely packed particles partially filling the pores (Fig. 7), (2) clay lining pores (Fig. 8), and (3) clay bridging from one sand grain to an adjacent sand grain (Fig. 9).

The Upper Cretaceous Almond and Ericson formation sandstones (Mesaverde Group) of southwestern Wyoming provide good examples of

the influence of clay minerals on porosity and permeability. SEM study of samples from both reservoirs[13] showed that the clay minerals differed in the two reservoir rocks, both in composition and position, within the pore system. The Almond formation reservoir rock typically contained discrete coarse kaolinite euhedra packed in some pores (Fig. 10), whereas in the Ericson formation (a tight gas sand), fibers of authigenic illite line pore throats (Fig. 11). Thus, the two sandstones have similar porosities but markedly different permeabilities.

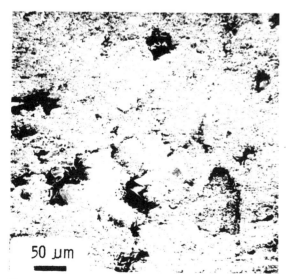

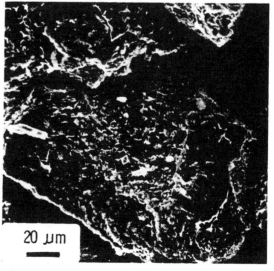

Fig. 4 – Macropores associated with only minor micropores; Rodessa limestone; 7% porosity and 1.2-md permeability.

Fig. 5 – Crinkly coating of smectite clay on sand grains; Second Wall Creek sand; 17.6% porosity and 38-md permeability.

Formation Damage

The type and mode of occurrence of clay minerals in sandstones are important for drilling and completing wells. Information on clays and other minerals can be obtained quickly with the SEM. Authigenic clays, because they commonly occur within pores, readily will contact any introduced fluids. As little as 1 to 2% clay as coatings on the grains of a sandstone could cause formation damage. In contrast, laminae of clay are less affected by introduced fluids, except at the wellbore.

Several papers[14-16] have reported on the role of the SEM during the design of remedial techniques for wells. Three possible types of formation damage caused by clays are (1) migration of loose particles, (2) expandable clays, and (3) acid-treatment side effects.

Loose or loosely attached clay minerals or other fine particles in sandstones are subject to migration during fluid flow. These particles may eventually reach an obstruction, usually a pore throat, and form "brush heaps," which reduce permeability. The clay mineral kaolinite appears particularly susceptible to movement. Reservoirs with potential loose-particle migration problems often are treated chemically to stabilize the particles.

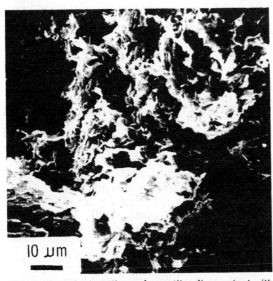

Fig. 6 – Detached coatings of smectite after contact with micellar fluid/water mixture during flow test (compare with Fig. 5).

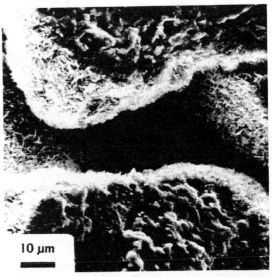

Fig. 8 – Authigenic chlorite lining pores (coating grains) of Berea sand. Chlorite is absent where sand grains were in contact; 14% porosity; permeability not measured.

Fig. 7 – Authigenic kaolinite crystals partially filling pore in sandstone; Frio sand; 20.3% porosity and 80-md permeability.

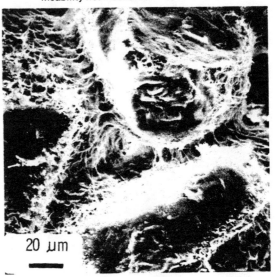

Fig. 9 – Authigenic illite lining pores and bridging gap between sand grains; Rotliegendes sand; 11.3% porosity and 7.7-md permeability

Eolian sandstones such as the Permian Rotliegendes[17] or the Triassic/Jurassic Nugget often have loose or loosely attached, very fine kaolinite that is subject to migration during the production of hydrocarbons. In the Nugget, this kaolinite bridges across grain contacts and is implaced after deposition of the sand through the infiltration of "fines" with downward percolating groundwater (Fig. 12).

The clay minerals smectite, mixed-layer smectite/illite, and degraded illite (potassium-depleted) are all expandable in the presence of fresh water. A sodium-rich smectite is the most expandable, and a degraded illite is the least expandable. An example of expandable clay was discussed previously with regard to the effect of a water/micellar fluid mixture on a smectite-bearing sandstone. Sensitivity to fresh water during drilling normally is controlled by using a KCl mud or an oil-base mud.

Many minerals common in sedimentary rocks are readily soluble in hydrochloric or hydrofluoric acids – the acids commonly used to treat wells. Some of these minerals (the clay mineral chlorite as well as the carbonate minerals siderite, ankerite, and ferroan calcite) are also iron-bearing, which can create a problem. If any of these iron-bearing minerals are dissolved, iron is released to precipitate in the form of a ferric hydroxide gel, which blocks pore throats and lowers permeability. The iron can be controlled using a chelating agent (such as citric acid) with the primary acid. This prevents iron from precipitating. An oxygen scavenger also may be added.

Table 1 summarizes the relative response of the common clay minerals to hydrochloric and hydrofluoric acids. Few quantitative data are available on reaction rates, but we have SEM observations on "before and after" samples of these clays that agree closely with the data presented.

Reservoir Evaluation Using Pore Casts

Pore casts (epoxy resin replicas of pore space) are useful for obtaining good visual perspective of the pore geometry.[3] Pore casts are made by dissolving the minerals of the rock in HCl followed by HF. It is important not to reverse the sequence because if calcite is present it will convert to CaF, which is insoluble in other acids. Delicate pore casts will collapse during dessication. They can be preserved by freezing the pore casts (while still wet from wash water) in liquid nitrogen and removing the ice as vapor using a freeze dryer.* Even delicate pore casts of microporous rocks can be preserved using this technique.

An example of a pore cast from a low-permeability (tight) gas sand is shown in Fig. 13. Pore casts such as this show that tight gas sands typically contain a high percentage of tabular interconnected pore space, while conventional sandstone reservoirs contain more tubular pores.

Pittman[18] showed how SEM examination of pore

*H.D. Winland, oral communication, Amoco Production Co. (1972).

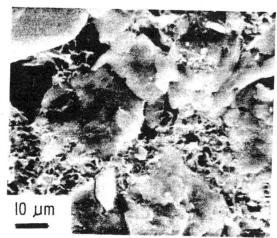

Fig. 11 – Authigenic illite lining pores of Ericson formation sandstone. Gas show while drilling, not tested; 9% porosity, 0.6-md permeability, and 54% calculated water saturation.

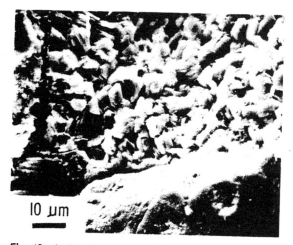

Fig. 10 – Authigenic kaolinite (dickite) filling pore in Almond formation sandstone; initial production flowing 252 BOPD (40.1 m³/d oil) and 9 to 16% porosity (sonic log).

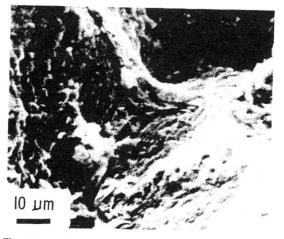

Fig. 12 – Kaolinite, which originated as an infiltration residue from percolating ground water, is attached loosely and subject to migration; Nugget formation; 18% porosity and 19.5-md permeability.

TABLE 1 – RELATIVE SOLUBILITY OF CLAY MINERALS IN COMMON TREATMENT ACIDS

Clay/Acid	Hydrochloric Acid	Hydrofluoric Acid
Kaolinite	S	S
Illite	S	SM
Smectite	S	M
Chlorite*	H	H
Mixed·layer	V	V

S = slightly soluble.
M = moderately soluble.
H = highly soluble.
V = variably soluble.
*Releases iron to system with dissolution.

casts and rocks could be integrated with other data to evaluate pore geometry, which influences the productive capability of sandstone reservoirs

Conclusions

In our opinion, the scanning electron microscope is a valuable tool for qualitative evaluation of reservoir rocks. The SEM provides information on pore geometry through direct observation of the rock or a pore cast of the rock, which aids in formation evaluation and in understanding of productive capabilities. The SEM also can show what minerals are present and where they are located, which is helpful when designing drilling and well completion programs. Clay minerals, which create particular problems in sandstone reservoirs with regard to fluid flow and formation damage, are best evaluated with the SEM.

References

1. Weinbrandt, R.M. and Fatt, I.: "A Scanning Electron Microscope Study of the Pore Structure of Sandstone," *J. Pet. Tech.* (1968) 543-548.
2. Timur, A., Hempkins, W.B., and Weinbrandt, R.M.: "Scanning Electron Microscope Study of Pore Systems in Rocks," *J. Geophys. Res.* (1971) 76, 4932-4947.
3. Pittman, E.D. and Duschatko, R.W.: "Use of Pore Casts and Scanning Electron Microscope to Study Pore Geometry," *J. Sediment. Petrol.* (1970) 40, 1153-1157.
4. Wardlaw, N.C.: "Pore Geometry of Carbonate Rocks as Revealed by Pore Casts and Capillary Pressure," *Bull., AAPG* (1976) 60, 245-257.
5. Wardlaw, N.C. and Taylor, R.P.: "Mercury Capillary Pressure Curves and the Interpretation of Pore Structure and Capillary Behaviour in Reservoir Rocks," *Bull.,* Cdn. Pet. Geologists (1976) 24, 225-262.
6. Swanson, B.F.: "Visualizing Pores and Nonwetting Phase in Porous Rock," *J. Pet. Tech.* (Jan. 1979) 10-18.
7. Pittman, E.D.: "Microporosity in Carbonate Rocks," *Bull., AAPG* (1971) 55, 1873-1878.
8. Keike, E.M. and Hartmann, D.J.: "Scanning Electron Microscope Application to Formation Evaluation," *Trans.,* Gulf Coast Assn. Geological Society (1973) 23, 60-67.
9. Stalder, P.I.: "Influence of Crystallographic Habit and Aggregate Structure of Authigenic Clay Minerals on Sandstone Permeability," *Geol. en Mijnbouw* (1973) 52, 217-220.
10. Gaida, K.H., Ruhl, W. and Zimmerle, W.:

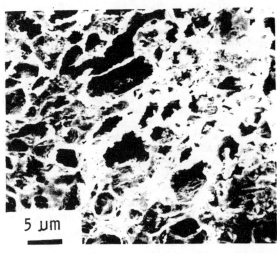

Fig. 13 – Epoxy resin pore cast (replica of void space) for Muddy sand. Note the irregular pore system and presence of tabular pores. This pore cast is more complex at higher magnifications because of micropores associated with authigenid clay minerals; 14.2% porosity and <0.05-md permeability.

"Rasterelektronenmikroskopische Untersuchen des Porenraumes von Sandsteinen," *Erdoel Erdgas Zeitschrift* (1973) 89, 336-343.
11. Wilson, M.D. and Pittman, E.D.: "Authigenic Clays in Sandstones: Recognition and Influence on Reservoir Properties and Paleoenvironmental Analysis," *J. Sediment. Petrol.* (1977) 47, 3-31.
12. Neasham, J.W.: "Applications of Scanning Electron Microscopy to the Characterization of Hydrocarbon-Bearing Rocks," *Scan. Elect. Microscopy* (March 1977) 1, 101-108.
13. Thomas, J.B.: "Diagenetic Sequences in Low-Permeability Argillaceous Sandstones," *J. Geol. Society* (1978) 135, 93-100.
14. Simon, D.E., McDaniel, B.W., and Coon, R.M.: "Evaluation of Fluid pH Effects on Low-Permeability Sandstones," paper SPE 6010 presented at the SPE 51st Annual Fall Technical Conference and Exhibition, New Orleans, Oct. 3-6, 1976.
15. Stout, C.M. and Peters, F.W.: "Preventing Formation Damage During Fracturing of the Clinton Formation," paper SPE 6128 presented at the SPE 51st Annual Fall Technical Conference and Exhibition, New Orleans, Oct. 3-6, 1976.
16. Almon, W.R. and Davies, D.K.: "Understanding Diagenetic Zones Vital," *Oil and Gas J.* (June 6, 1977) 209-216.
17. Glennie, K.W.: "Desert Sedimentary Environments," *Developments in Sedimentology, No. 14,* Elsevier Publishing Co., Amsterdam (1970) 1-222.
18. Pittman, E.D.: "Porosity, Diagenesis and Productive Capability of Sandstone Reservoirs," *Aspects of Diagenesis,* P.A. Scholle and P.R. Schluger, eds., Society of Economic Paleontologists and Mineralogists, Tulsa (1979) Spcl. Publ. No. 26, 159-173.

JPT

Original manuscript received in Society of Petroleum Engineers office Oct. 26, 1978. Paper accepted for publication April 6, 1979. Revised manuscript received Aug. 24, 1979. Paper (SPE 7550) first presented at the SPE 53rd Annual Fall Technical Conference and Exhibition, held in Houston, Oct. 1-3, 1978.

Correlation of k_g/k_o Data with Sandstone Core Characteristics

MARTIN FELSENTHAL
MEMBER AIME

CONTINENTAL OIL CO.
PONCA CITY, OKLA.

INTRODUCTION

Engineers are frequently faced with the problem of having to predict oil recovery from a solution gas drive reservoir in the early life of a field. This is often the time when actual laboratory or field-derived gas-oil permeability (k_g/k_o) data are not yet available. Although excellent collections of k_g/k_o data for various types of rocks may be found in the literature,[1,2,3] a need exists for a more detailed correlation of k_g/k_o data with core properties such as porosity, permeability, clay content or pore-size distribution. It is the purpose of this article to present this type of correlation for sandstone material. The correlation is based on laboratory tests performed on over 300 core samples from 19 reservoirs in the United States and Canada.

EXPERIMENTAL

Cores were analyzed by routine procedures, using samples of conventional size (⅞-in. diameter, ⅞-in long) cut parallel to the bedding planes. Air permeability was measured at 1.67 atm mean pressure in a Hassler-type apparatus. Porosity was determined with the aid of a Ruska bulk volume meter and a Boyle's Law rock volume meter. An evaporation method[4] was used to evaluate irreducible water saturations.

Core samples were saturated with 25,000 ppm NaCl brine. The brine was reduced to the irreducible saturation, and the remainder of the pore space was filled with kerosene. This was followed by multiphase flow tests using the so-called "dispersed feed" technique.[5] In this technique, gas and oil (kerosene and kerosene-saturated air) were dispersed with the aid of a porous Alundum feed head, the face of which contained a waffle-shaped system of Lucite-lined grooves. Oil entered the back side of the feed head and emerged at the raised areas of the front face. Gas was introduced directly into the system of grooves at the front face. The gas-oil mixture then entered the test core which was kept in capillary contact with the feed head. Permeabilities were determined from rate and pressure measurements and saturations from weights. All determinations were performed at equilibrium or steady-state conditions.

Pore-size distributions were determined on selected samples by a mercury injection technique.[6]

GENERAL CORRELATIVE TRENDS

When examining data from a given reservoir, it was

noted that k_g/k_o vs saturation curves became generally less steep as specific air permeability increased, a development that has a favorable effect on oil displacement efficiency by solution gas drive. This trend has also been reported by others in the literature.[7]

An effect of porosity on k_g/k_o data was also noted. This effect was not generally discernible in a study of relative permeability data for a given reservoir but became apparent when data for sandstone reservoirs of similar lithology but differing average porosity were compared. For instance, in a comparison of argillaceous and/or calcareous sandstones from 11 reservoirs ranging in average porosities from 14 to 28 per cent, a definite trend was noted, indicating that for a given permeability k_g/k_o curves became less favorable (i.e., steeper) as porosity increased. A similar trend was observed for a group of comparatively clean sandstones from five reservoirs ranging in average porosities from 15 to 21 per cent.

For a given permeability and porosity, comparatively clean sandstones gave more favorable k_g/k_o curves than argillaceous and/or calcareous sandstone or chert reservoirs. The laboratory studies also showed that least favorable k_g/k_o curves were exhibited by shaly sandstone, conglomerate, and sandstone containing carbon inclusions. Table 1 lists the various source reservoirs classified by lithology types.

To facilitate a correlation of k_g/k_o data with core characteristcis, it was found expedient to work with

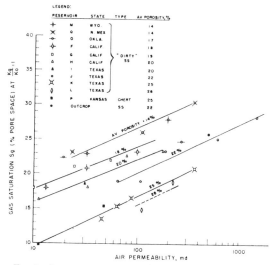

FIG. 1—CORRELATION OF k_g/k_o DATA WITH PERMEABILITY AND POROSITY.

Original manuscript received in Society of Petroleum Engineers office May 15, 1959. Revised manuscript received Aug. 14, 1959.
[1]References given at end of paper.

certain k_g/k_o curve co-ordinates and with averages for groups of six core samples. Fig. 1 illustrates an example correlation of some of these data. Note that the figure also includes a data point for an outcrop sand used in Botset's pioneer gas-liquid flow study.[8]

USE OF CORRELATION GRAPHS

Figs. 2, 3 and 4 present correlations of k_g/k_o curve co-ordinates with core characteristics for various types or classes of cores. Class I represents comparatively clean sandstone, Class II argillaceous and/or calcareous sandstone, and Class III shaly or carbonaceous sandstone and sandy conglomerate. The graphs should be useful for estimating a k_g/k_o curve from permeability, porosity and lithology data when actual k_g/k_o curves for the particular reservoir under study are not yet available. The procedure for using the correlation graphs can be briefly described as follows.

1. With the aid of either Figs. 2, 3, or 4 (depending on the type of rock), co-ordinate points of gas saturation (S_g) at k_g/k_o = 0.01, 0.1, and 1 are selected.

2. The co-ordinate points are plotted on semilogarithmic paper and connected with a smooth k_g/k_o vs S_g curve.

If reservoir lithology is not known, it is suggested that the correlations for Class II cores (Fig. 3) be used as a first approximation.

DISCUSSION OF RESULTS

The standard deviation shown on the correlation graphs refers to the deviation exhibited by individual core samples at k_g/k_o = 1. As the curves converge at

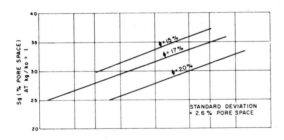

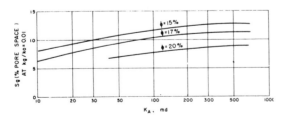

Fig. 2—k_g/k_o Curve Co-ordinates for Comparatively Clean Sandstone (Class I) Based on 97 Core Samples from Five Reservoirs.

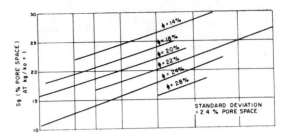

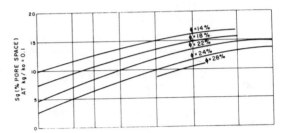

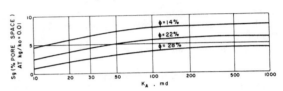

Fig. 3—k_g/k_o Curve Co-ordinates for Argillaceous and/or Calcareous Sandstones (Class II) Based on 161 Core Samples from 11 Reservoirs.

lower k_g/k_o ratios, the standard deviations become correspondingly smaller.

The data of Table 1 indicate that all Class I cores tested to date happen to be well-cemented Cretaceous age rocks originating in or near the Rocky Mountain area.

The majority of cores tested belong in Class II, which generally represents "dirty" sands. Data obtained on cores from a chert reservoir also matched correlations of this class (Fig. 1) as did a few data obtained on intergranular limestone. The reservoirs in this class appear to cover all geological eras and a wide variety of cementation and of geographic origins. It is believed that the majority of reservoirs containing intergranular-type porosity will have Class II characteristics.

Class III cores include the more unusual types, i.e., shaly or highly carbonaceous sand or conglomerate. These cores represent a variety of geologic eras, degrees of cementation and geographic areas of origin.

Class I-A cores occurred least frequently among the samples tested. They represent comparatively clean sandstone possessing micro-stratification, i.e., thin (1 to 2 mm) high permeability layers, a characteristic which was also noted to a lesser extent in Class III cores. In both classes III and I-A, the effect of permeability was greater than for the more homogeneous cores of classes I and II.

The parameters used in the correlations—permeability, porosity and type of sandstone—are actually functions of pore geometry. The correlation graphs show that, for a given type or class of sandstone, k_g/k_o curves become less steep as permeability increases and porosity decreases. This relationship bears a resemblance to the

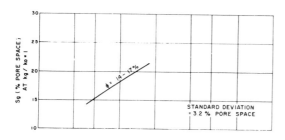

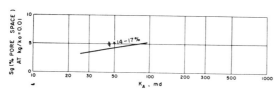

Fig. 4 — k_g/k_o Curve Co-ordinates for Shaly Sandstone, Carbonaceous Sandstone and Sandy Conglomerate (Class III) Based on 14 Core Samples from Three Reservoirs.

Kozeny equation

$$\frac{1}{S_v^2} = t^2 \frac{k}{\phi} ,$$

where S_v = specific surface area on a rock volume basis,

t = tortuosity coefficient,

k = specific permeability,

ϕ = porosity.

From this it is postulated that the steepness of the k_g/k_o curves is in some way related to the specific surface area. Because specific surface area is one of the major factors that determine irreducible interstitial water saturation, it is not surprising that recently developed empirical equations used the latter parameter for predicting k_g/k_o curves.[9] In applying these equations to laboratory data used in this work, a good agreement was obtained for comparatively clean (Class I) cores. For argillaceous and calcareous (Class II) cores, however, the equations gave more optimistic k_g/k_o curves than the correlation curves shown in Fig. 3.

Pore geometry may also be characterized by the distribution of pore sizes. A correlation between this distribution and k_g/k_o characteristics was anticipated for the following reason. Since oil removal by depletion drive proceeds from the larger to the smaller pores, oil displacement should be most efficient (and k_g/k_o curves most favorable) if the pore-size distribution curve has a sharp peak among the larger pores. Such a peak was noted for clean (Class I) cores but not for argillaceous (Class II) cores. As was noted previously, it was the clean (Class I) cores that exhibited the most favorable k_g/k_o curves from the point of view of oil displacement efficiency.

In order to study the effect of pore-size distribution more closely, attempts were made to describe the distribution by a single "number" which would then be correlatable with a key co-ordinate of the k_g/k_o curves. The most successful of these attempts made use of the following procedure.

1. The pore size (microns) and the slope (fraction of pore space $\div$ micron) at the 20 percentile point of the cumulative pore-size distribution curves were determined.

2. Multiplying the two values yielded a number which was correlated with the k_g/k_o curve co-ordinate at k_g/k_o = 1, as illustrated in Fig. 5.

The standard deviation in Fig. 5 is in the same order of magnitude as the deviations from the correlations based on permeability, porosity and lithology (Figs. 2, 3 and 4). This indicates that within the range of values investigated it is possible to estimate k_g/k_o data solely from pore-size data. The correlation shown in Fig. 5 has been used to advantage for loosely consolidated sand that was suitable for pore-size tests but unsuitable for k_g/k_o tests.

Fig. 6 shows a comparison of laboratory data with field-derived data. It may be noted that the laboratory

TABLE 1—SOURCE RESERVOIRS OF k_g/k_o TEST SAMPLES

CLASS I — COMPARATIVELY CLEAN SANDSTONE

	Age	Location	Average Porosity Per cent	Predominant Lithology	Cementation	Grain Size
A	Cret.	Wyo.	15	SS	Very well	Fine to med.
B	Cret.	Wyo.	16	SS	Very well	Fine
C	Cret.	Alberta	17	SS	Well	Fine to med.
D	Cret.	Colo	19	SS	Very well	Fine
E	Cret.	Wyo.	21	SS	Very well	Fine

CLASS II — ARGILLACEOUS AND/OR CALCAREOUS SANDSTONE OR CHERT

	Age	Location	Average Porosity Per cent	Predominant Lithology	Cementation	Grain Size
F	Plio.	Calif.	18	Argill. calc. ss., some sandy cgl.	Well to poor	Fine to coarse
G	Plio.	Calif.	19	Argill. calc. ss., some sandy cgl.	Med.	Fine to coarse
H	Plio.	Calif.	20	Argill. ss., some calc. ss.	Med.	Fine to coarse
I	Olig.	Tex.	20	Argill. ss., calc. ss.	Med. to poor	Med.
J	Olig.	Tex.	22	Calc. ss.	Well	Fine to coarse
K	Olig.	Tex.	25	Argill. calc. ss.	Med.	Fine
L	Olig.	Tex.	28	Argill. calc. ss., some calc. cgl.	Med. to poor	Med.
M	Cret.	Wyo.	14	SS, some carbonaceous inclusions	Very well	Fine
N	Cret.	Wyo.	19	Argill. ss.	Med to poor	Fine
O	Penn.	Okla.	17	Argill. ss.	Med.	Fine
P	Miss.	Kans.	25	Weathered chert, sh. and ls.	Well to poor	Amorphous
Q	Ord.	N. Mex.	14	SS, some argill. ss.	Well to poor	Med.

CLASS III — SHALY OR CARBONACEOUS SANDSTONE OR SANDY CONGLOMERATE

	Age	Location	Average Porosity Per cent	Predominant Lithology	Cementation	Grain Size
M-1	Cret.	Wyo.	14	SS, many carbonaceous inclusions	Very well	Fine
N-1	Cret.	Wyo.	17	Shaly ss.	Med. to poor	Fine
R	Penn.	Tex.	16	Sandy cgl.	Well	Coarse

CLASS I-A — HIGHLY STRATIFIED, COMPARATIVELY CLEAN SANDSTONE

	Age	Location	Average Porosity Per cent	Predominant Lithology	Cementation	Grain Size
C-1	Cret.	Alberta	17	SS, permeability layers few mm thick	Well	Fine to med.
S	Penn.	Wyo.	15	SS, randomly disected by quartz veinlets	Well	Fine

167

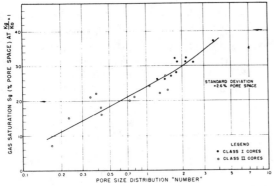

FIG. 5—CORRELATION OF PORE-SIZE DISTRIBUTIONS AND k_g/k_o CHARACTERISTICS OF SANDSTONE CORES.

curves were too conservative because of gravity drainage in one case (Reservoir F). In the remaining cases, for the linear velocity relationship instead of 18,500

SUMMARY AND CONCLUSIONS

Correlation curves between laboratory k_g/k_o data and sandstone core characteristics are presented. The curves should give reasonably accurate k_g/k_o data for predictions of reservoir performance; however, the general complexity of reservoir conditions makes it advisable to review and, if need be, revise these predictions when actual field-derived k_g/k_o data become available.

Within the range of core properties studied (air permeability 10 to 1,500 md, porosity 14 to 28 per cent), k_g/k_o vs saturation curves become less steep as the sand becomes more permeable, cleaner and less porous. This decrease in steepness is generally considered diagnostic of a more favorable oil recovery (on a per cent pore space basis) by solution gas drive.

A correlation between k_g/k_o data and pore-size distribution was also noted; thus the more favorable (less steep) k_g/k_o curves· were generally associated with a pore-size distribution curve having a sharp peak among the larger pores.

ACKNOWLEDGMENTS

The author wishes to express his appreciation to F. R. Conley and G. J. Heuer, Jr., for their helpful criticism and encouragement and to Merle Hutchison and J. W. Quinn for performance of laboratory work. Appreciation is also extended to the management of Continental Oil Co. for permission to publish this paper.

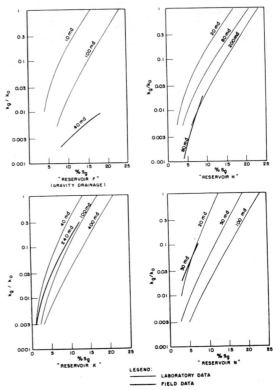

FIG. 6—COMPARISON OF LABORATORY AND FIELD k_g/k_o DATA.

REFERENCES

1. Arps, J. J. and Roberts, T. G.: "Effe Effect of Relative Permeability Ratio, the Oil Gravity, and the Solution Gas-Oil Ratio on the Primary Recovery from a Depletion-Type Reservoir", *Trans.* AIME (1955) **204**, 120.
2. Elkins, L. E.: "The Importance of Injected Gas as a Driving Medium in Limestone Reservoirs as Indicated by Recent Gas-Injection Experiments and Reservoir-Performance History", *Drill. and Prod. Prac.*, API (1946) 160.
3. Patton, E. C., Jr.: "Evaluation of Pressure Maintenance by Gas Injection in Volumetrically Controlled Reservoirs", *Trans.* AIME (1947) **170**, 112.
4. Messer, E. S.: "Interstitial Water Determination by an Evaporation Method", *Trans.* AIME (1951) **192**, 269.
5. Richardson, J. G., Kerver, J. K., Hafford, J. A. and Osoba, J. S.: "Laboratory Determination of Relative Permeability", *Trans.* AIME (1952) **195**, 187.
6. Bucher, H. P., Jr., Felsenthal, M. and Conley, F. R.: "A Simplified Pore-Size Distribution Apparatus", *Jour. Pet. Tech.* (April, 1956) **8**, No. 4, 65.
7. McCord, D. R.: "Performance Predictions Incorporating Gravity Drainage and Gas Cap Pressure Maintenance—LL370 Area, Bolivar Coastal Field", *Trans.* AIME (1953) **198**, 231.
8. Botset, H. G.: "Flow of Gas Liquid Mixtures Through Consolidated Sand", *Trans.* AIME (1940) **136**, 91.
9. Wahl, W. L., Mullins, L. D. and Elfrink, E. B.: "Estimation of Ultimate Recovery from Solution Gas Drive Recoveries", *Trans.* AIME (1958) **213**, 132. ★★★

Effect of Capillary Number and Its Constituents on Two-Phase Relative Permeability Curves

R.A. Fulcher Jr., SPE, ARCO Oil and Gas Co.
Turgay Ertekin, SPE, Pennsylvania State U.
C.D. Stahl, SPE, Pennsylvania State U.

Summary

One primary goal of any enhanced recovery project is to maximize the ability of the fluids to flow through a porous medium (i.e., the reservoir). This paper discusses the effect of capillary number, a dimensionless group describing the ratio of viscous to capillary forces, on two-phase (oil-water) relative permeability curves. Specifically, a series of steady-state relative permeability measurements were carried out to determine whether the capillary number causes changes in the two-phase permeabilities or whether one of its constituents, such as flow velocity, fluid viscosity, or interfacial tension (IFT), is the controlling variable.

For the core tests, run in fired Berea sandstone, a Soltrol 170™ oil/calcium chloride (CaCl$_2$) brine/isopropyl alcohol (IPA)/glycerin system was used. Alcohol was the IFT reducer and glycerin was the wetting-phase viscosifier.

The nonwetting-phase (oil) relative permeability showed little correlation with the capillary number. As IFT decreased below 5.50 dyne/cm [5.50 N/m], the oil permeability increased dramatically. Conversely, as the water viscosity increased, the oil demonstrated less ability to flow. For the wetting-phase (water) relative permeability, the opposite capillary number effect was shown. For both the tension decrease and the viscosity increase (i.e., a capillary number increase) the water permeability increased. However, the water increase was not as great as the increase in the oil curves with an IFT decrease. No velocity effects were noted within the range studied.

Other properties relating to relative permeabilities were also investigated. Both the residual oil saturation (ROS) and the imbibition-drainage hysteresis were found to decrease with an increase in the capillary number. The irreducible water saturation was a function of IFT tension only.

A relative permeability model was developed from the experimental data, based on fluid saturations, IFT, fluid viscosities, and the residual saturations, by using regression analysis. Both phases were modeled for both the imbibition and the drainage processes. These models demonstrated similar or better fits with experimental data of other water- and oil-wet systems, when compared with existing relative permeability models. The applicability of these regression models was tested with the aid of a two-phase reservoir simulator.

Introduction

As world oil reserves dwindle, the need to develop EOR techniques to maximize recovery is of great importance. Methods such as chemical flooding, miscible flooding, and thermal recovery involve altering the mobility and/or the IFT between the displacing and the displaced fluids.

Recovery efficiency was found to be dependent on the capillary number, defined as

$$N_c = \frac{\mu v}{\gamma \phi}. \qquad \dots \dots \dots \dots \dots \dots \dots \dots \dots (1)$$

The viscous forces were defined as the fluid viscosity, flow velocity, and the flow path length. Capillary forces vary with the fluid IFT and the pore geometry of the medium.[1]

Taber defined the capillary number in terms of the pressure drop between two points, the flow length, and the IFT.[2]

$$N_c = \frac{\Delta p}{L \gamma}. \qquad \dots \dots \dots \dots \dots \dots \dots \dots \dots (2)$$

He concluded that as this ratio increased to a value of 5 psi/ft/dyne/cm [0.2 kPa/m/N/m] the ROS was reduced significantly. By decreasing the IFT by using surface-active agents, or by decreasing the path length by altering the field geometry, the capillary number could be increased.

Others have shown similar results. Melrose and Brandner,[3] for example, indicated that as the capillary number rose to a value of 10^{-4}, the microscopic displacement efficiency, which accounts for the residual saturations to both oil and water, increased. The effects of the capillary number on the recovery of residual oil are given by Chatzis and Morrow[4] and by other authors[5] (Fig. 1).

Few studies, however, have shown the effect of the capillary number on the two-phase flow between the residuals. The variables within this group have been researched, but their combined effect on relative permeabilities has been largely ignored.

Several authors have noted that the viscosity ratio of oil and water alters the oil relative permeability but has little effect on that of water.[6-8] Few or no changes by fluid flow velocity were observed, provided that no boundary effects were present during the core tests.[9-11]

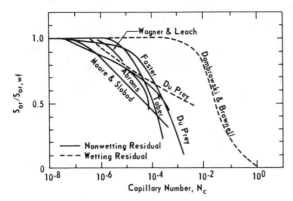

Fig. 1—Recovery of residual oil vs. capillary number.[4]

Studies on oil and gas permeabilities showed that as the IFT decreased by increasing the temperature and the equilibrium pressure between two phases, the relative permeability curves increased and straightened out.[12] The results of other tests on reducing the tension between oil and water indicated that (1) few or no relative permeability alterations occurred for tension above 0.1 dyne/cm [0.1 N/m]; however, larger increases were observed below 0.1 dyne/cm [0.1 N/m] for both phases; (2) the curves tended toward linearity; (3) the imbibition-drainage hysteresis lessened; and (4) the residual saturations to both oil and water decreased.[13-15]

One study did show relative permeability increases with increases in the capillary number; however, these experiments were run using artificial cores of Teflon®,

alumina, and stainless steel rather than reservoir-type rock samples.[16]

Also, with the increased use of mathematical reservoir simulators to predict recovery from different EOR processes, the need to model the various flow properties, such as relative permeability, becomes important. Thus, experimental results need to be applied to empirical or statistical models for use in numerical simulators.

Materials and Experimental Procedure

For this study, a series of steady-state relative permeability tests were carried out at 77°F [25°C] in 2-ft- [0.6-m-] long, 2-in.- [5-cm-] diameter Berea sandstone cores fired at 1,832°F [1000°C]. The cores were baked to prevent large decreases in their absolute permeability because of clay swelling during and between the experimental runs. Cores were reused for several tests before being discarded to minimize differences in porosity, permeability, and lithology. The properties of the cores for each test are given in Table 1. The fluids used were mixtures of Soltrol 170 oil,* 2% CaCl$_2$ brine, IPA, and glycerin, as shown in Table 2. Later, core measurements were taken using Bradford crude oil,** Kendex 0837™,† and 2% sodium chloride (NaCl) brine.

IFT effects were studied with the Soltrol 170, IPA, and CaCl$_2$ brine system. Alcohol acted as the tension reducer. CaCl$_2$ brine was used instead of NaCl brine because it does not form a precipitate with Soltrol and IPA as does NaCl.[17] Glycerin was used as the viscosifying agent to increase the aqueous-phase viscosity.‡

*Phillips Petroleum Co., Bartlesville, OK.
**Pennzoil Co., Bradford, PA.
†Kendall Oil Co., Bradford, PA.
‡Crookston, R., Ehrlich, R., and Bae, J.: private communication, Gulf R&D Co., Pittsburgh (Feb. 13, 1981).

TABLE 1—CORE PROPERTIES

Run	Core	Length (in)	PV (cm)^3	Porosity (%)	Permeability (Pre-run) (md)	Permeability (Post-run) (md)	Change (%)
1	1*	24	276	22.46	241.3	115.1	−52.3
2	2*	12	148	23.96	198.1	136.0	−31.3
3	3**	24	240	19.42	240.6	115.1	−52.2
4	4	24	280	22.48	325.3	278.4	−14.4
5	4	24	280	22.48	278.4	247.8	−11.0
6	4	24	280	22.48	247.8	251.0	+1.3
7	4	24	280	22.48	251.0	219.7	−12.5
8	4	24	280	22.48	219.7	219.7	+0.0
9	4	24	280	22.48	219.7	182.1	−17.1
10	4	24	280	22.48	123.2	128.3	+4.1
11	5	24	287	23.05	365.9	408.2	+11.6
12	5	24	287	23.05	408.2	355.3	−13.0
13	5	24	287	23.05	355.3	353.4	−0.5
14	5	24	287	23.05	353.4	311.1	−12.0
15	5	24	287	23.05	311.1	259.3	−16.7
16	6	24	279	22.40	433.1	433.1	+0.0
17	6	24	279	22.40	433.1	416.4	−3.9
18	6	24	279	22.40	416.4	384.5	−7.7
19	6	24	279	22.40	384.5	—	—
20	7	12	146	23.48	531.6	—	—
21	8	24	330	26.50	388.1	398.5	+2.7
22	9	24	276	22.16	357.7	—	—
23	10*	24	249	20.15	177.3	—	—
24	11*	24	235	19.02	279.1	—	—
25	12*	24	272	22.01	160.2	—	—

*Unfired core.
**Core fired at 250°C.

TABLE 2—PROPERTIES OF THE AQUEOUS AND OLEIC PHASES

Runs	Phase	2% CaCl₂ Brine (%)	Soltrol 170 Oil (%)	IPA (%)	Glycerin (%)	Specific Gravity at 25°C	Viscosity at 25°C (cp)	Refractive Index at 25°C	IFT with Oleic/Aqueous at 25°C (dyne/cm)
1–7,9, 13,21,22	aqueous	100.0	0.0	0.0	0.0	1.007	0.947	1.3385	37.9
8	aqueous	60.0	0.0	40.0	0.0	0.917	2.761	1.3664	5.50
10,11	aqueous	10.9	14.0	75.1	0.0	0.804	2.613	1.3879	0.335
12,16	aqueous	3.9	43.5	52.6	0.0	0.785	2.475	1.3996	0.0389
14	aqueous	40.0	0.0	0.0	60.0	1.156	13.795	1.4231	30.3
15	aqueous	17.5	0.0	0.0	82.5	1.212	128.58	1.4516	29.7
17	aqueous	0.0	9.6	63.5	27.9	0.914	13.636	1.4115	0.454
18	aqueous	0.0	0.1	38.3	61.6	1.069	126.62	1.4418	2.91
19	aqueous	0.0	20.6	62.8	16.6	0.853	6.100	1.4050	0.118
20	aqueous	0.0	0.0	0.0	100.0	1.258	954.00	1.4735	25.3
23,24	aqueous	100.0*	0.0	0.0	0.0	1.009	0.892	1.3356	24.5
25	aqueous	100.0*	0.0	0.0	0.0	1.009	0.892	1.3356	10.8
1–7,9, 12,21,22	oleic	0.0	100.0	0.0	0.0	0.781	2.363	1.4339	37.9
8	oleic	0.2	95.3	4.5	0.0	0.771	2.246	1.4348	5.50
10,11	oleic	0.9	83.2	15.9	0.0	0.774	2.309	1.4277	0.335
12,16	oleic	1.1	71.5	27.4	0.0	0.775	2.352	1.4196	0.0389
14	oleic	0.0	100.0	0.0	0.0	0.781	2.363	1.4339	30.3
15	oleic	0.0	100.0	0.0	0.0	0.781	2.363	1.4339	29.7
17	oleic	0.0	76.1	22.4	1.5	0.785	2.149	1.4207	0.454
18	oleic	0.0	89.7	10.0	0.3	0.781	2.029	1.4264	2.91
19	oleic	0.0	67.0	30.2	2.8	0.791	2.173	1.4168	0.118
20	oleic	0.0	100.0	0.0	0.0	0.781	2.363	1.4335	25.3
23,24	oleic	0.0	100.0**	0.0	0.0	0.814	5.195	1.4504	24.5
25	oleic	0.0	100.0**	0.0	0.0	0.845	11.297	1.4711	10.8

*2% NaCl brine.
**Bradford crude oil.
†Kendex 0837 oil.

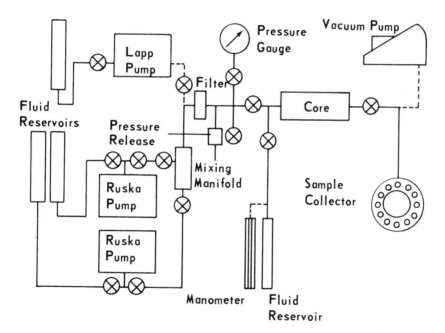

Fig. 2—Relative permeability experimental apparatus.

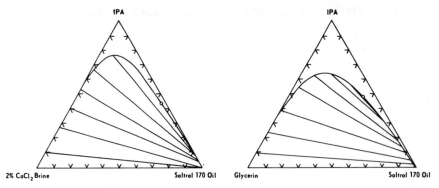

Fig. 3—Equilibrium phase diagrams for Soltrol 1.70/IPA/2% CaCl$_2$ and Soltrol 170/IPA/glycerin systems.

TABLE 3—SUMMARY OF CORE TESTS

Run	System	Flow Rate (cm^3)	IFT (dyne/cm)	Wetting Phase Viscosity (cp)	Capillary Number	S_{or} (%)	S_{wir} (%)	Microscopic Displacement Efficiency (%)
1	SC	160	3.79×10^1	0.947	2.47×10^{-6}	40.4	38.2	34.7
2	SC	160	3.79×10^1	0.947	2.34×10^{-6}	43.5	34.3	33.8
3	SC	200	3.79×10^1	0.947	3.61×10^{-6}	48.9	25.5	34.4
4	SC	200	3.79×10^1	0.947	3.09×10^{-6}	36.3	22.1	53.4
5	SC	80	3.79×10^1	0.947	1.24×10^{-6}	35.1	31.7	48.6
6	SC	120	3.79×10^1	0.947	1.86×10^{-6}	42.3	29.9	39.7
7	SC	160	3.79×10^1	0.947	2.48×10^{-6}	44.1	26.6	39.9
8	SCA	200	5.50×10^0	2.761	6.22×10^{-5}	33.1	39.4	45.4
9	SC	400	3.79×10^1	0.947	6.18×10^{-6}	42.8	34.8	34.4
10	SCA	200	3.35×10^{-1}	2.613	9.66×10^{-4}	0.0	56.3	100.0
11	SCA	200	3.35×10^{-1}	2.613	9.43×10^{-4}	8.9	41.5	84.8
12	SCA	200	3.89×10^{-2}	2.475	7.65×10^{-3}	0.1	33.6	99.0
13	SC	200	3.79×10^1	0.947	3.02×10^{-6}	36.8	33.0	45.1
14	SCG	200	3.03×10^1	13.795	5.50×10^{-5}	38.6	37.0	38.7
15	SCG	200	2.97×10^1	128.58	5.23×10^{-4}	17.1	32.9	74.5
16	SCA	200	3.89×10^{-2}	2.475	7.92×10^{-3}	0.0	32.0	100.0
17	SGA	200	4.54×10^{-1}	13.636	3.74×10^{-3}	2.2	40.2	96.3
18	SGA	200	2.91×10^0	126.62	5.41×10^{-3}	30.6	33.9	53.7
19	SGA	200	1.18×10^{-1}	6.100	6.43×10^{-3}	3.9	30.1	94.4
20	SC	200	2.59×10^1	954.00	4.37×10^{-3}	10.3	36.2	83.8
21	SC	200	3.79×10^1	0.947	2.62×10^{-6}	30.1	40.1	49.8
22	SC	200	3.79×10^1	0.947	3.14×10^{-6}	20.9	38.5	66.0
23	BN	200	2.45×10^1	0.892	4.41×10^{-6}	37.9	30.8	46.4
24	BN*	200	2.45×10^1	5.195	3.13×10^{-5}	39.4	19.7	51.0
25	KN	200	1.08×10^1	0.892	1.06×10^{-5}	33.1	33.0	50.6

*Oil-Wet System.

System Notes:
S - Soltrol 170 oil.
B - Bradford crude oil.
K - Kendex 0837 oil.
C - 2% (by weight) CaCl$_2$ brine.
N - 2% (by weight) NaCl brine.
A - Isopropyl alcohol.
G - Glycerin.

Glycerin mixed completely with the brine and was immiscible with the oil in all proportions. Combined effects of interfacial tension and viscosity were studied by using a Soltrol/IPA/glycerin fluid system.

The experimental apparatus is shown in Fig. 2. The cores were prepared using the Lapp Pulsafeeder™ pump. Relative permeability tests required two Ruska constant-displacement pumps, one for each fluid phase. All fluids were flowed through a 0.4-micron [0.4-μm]

filter before entering the core.

Different procedures were used to prepare the fluids for injection depending on the specific fluid system. For viscosity alterations, the glycerin/CaCl$_2$ brine solutions were mixed separately from the oil. For the IFT and the combined tension-viscosity systems, a tie line, on the ternary diagram, yielding the desired properties was chosen (Fig. 3). A mixture, lying approximately in the center of the tie line, was selected, and the appropriate

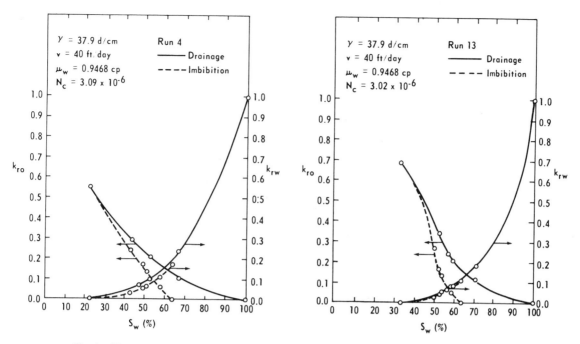

Fig. 4—Oil-water relative permeabilities for the base system and reproducibility of measurements.

amounts of each component were mixed. The resulting solution was shaken, then allowed to sit quietly until complete phase separation occurred.

The Penn State steady-state relative permeability method[19] then was run using the desired fluid system. The experimental technique involved flowing both phases simultaneously through the core and calculating the effective permeabilities with Darcy's law applied to each phase. Fluid saturations were determined by material balance. Drainage curves were found by moving the wetting-phase saturation from 100% to its irreducible value. Imbibition curves then were determined by reversing the saturation direction from irreducible water to the ROS.

To avoid capillary end effects, the criterion given by

Kyte and Rapoport,[20] a 50-psi [345-kPa] pressure differential across the core, was required. Thus, an 80-mL/hr [80-cm³/h] minimum flow rate at a velocity of 16 ft/D [4.9 m/d] was needed for fluids of high IFT (greater than 1.0 dyne/cm [1.0 N/m]), and 200 mL/hr [200 cm³/h] at 40 ft/D [12.2 m/d] was needed for low-tension (less than 1.0 dyne/cm [1.0 N/m]) systems.[21]

Discussion of Results

A summary of the experimental core tests is shown in Table 3. Two control tests were run (Fig. 4) to measure the reproducibility of the procedure, test the viability of using fired Berea cores, and establish a comparative base for the altered fluid runs. Soltrol 170 and 2% CaCl₂ brine were flowed at a total fluid flow rate of 200 mL/hr

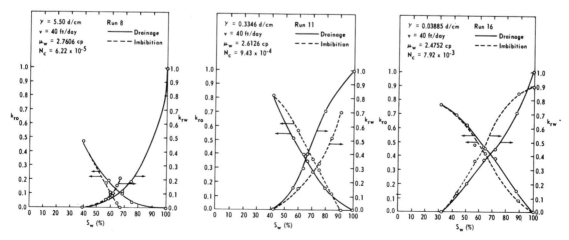

Fig. 5—Behavior of oil/water relative permeabilities at low IFT's.

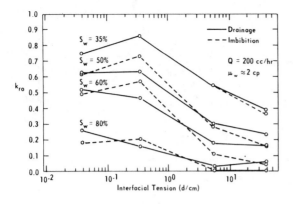

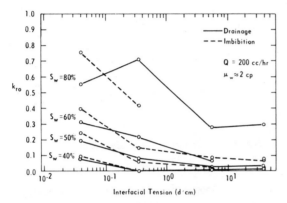

Fig. 6—Variation of oil and water relative permeabilities against interfacial tension for different wetting-phase saturations.

[200 cm³/h] in both imbibition and drainage. These results compared favorably with one another and with those for unfired Berea cores.

Velocity Effects. The initial variable altered was the fluid flow velocity. The rates ranged from 80 mL/hr [80 cm³/h] (16 ft/D [4.9 m/d]), the minimum rate to avoid capillary end effects, to 400 mL/hr [400 cm³/h] (80 ft/D

[24.4 m/d]), the maximum rate attainable with the Ruska pumps. Little or no significant change occurred in the relative permeability curves within this range, which agreed with the work of previous authors.[9-11] Thus, further studies are required to test fluid velocity to attain at least a two-fold order of magnitude increase in the capillary number, although such rates are not seen except around a wellbore.

Oleic/Aqueous IFT Effects. The range of oleic/aqueous IFT's for Soltrol and brine (Fig. 4) was from 37.9 dyne/cm [37.9 N/m] to 0.0389 dyne/cm [0.0389 N/m]. The relative permeability curves for low-tension systems are shown in Fig. 5. The total flow velocity was maintained at 40 ft/D [12 m/d] and the fluid viscosities were kept at approximately 2.0 cp [0.002 Pa·s]. At a tension of 5.50 dyne/cm [5.50 N/m], a slight increase was observed in the water permeabilities but none in the oil values. The former result may have been attributed to the slight increase in the brine viscosity from 0.947 to 2.761 cp [9.47×10^{-4} to 27.61×10^{-4} Pa·s], as is discussed in the next section. At 0.335 dyne/cm [0.335 N/m], both sets of curves showed large increases, indicating less resistance to flow for each phase. At the lowest attainable tension for the oil/brine/IPA system (0.0389 dyne/cm [0.0389 N/m]), further increases in the permeabilities were noted. Also, the curves started to approach linearity, as was reported for fluids of zero IFT.[22] To further illustrate the tension effects, the relative permeabilities of both the oleic and the aqueous phases at various aqueous saturation values are given in Fig. 6. For both drainage and imbibition, larger increase occurred below 5.50 dyne/cm [5.50 N/m], indicating a critical point when fluid began to move more easily.

Wetting-Phase Viscosity Effects. The wetting-phase viscosity range was from that of $CaCl_2$ brine (0.947 cp [9.47×10^{-4} Pa·s]) to that of glycerin (954.0 cp [0.954 Pa·s]). The fluid velocity was 40 ft/D [12 m/d] and the IFT was maintained at approximately 30 dyne/cm [30 N/m]. The relative permeability curves for four viscosity values are shown in Fig. 4 (for brine) and in Fig. 7. As the wetting phase (aqueous) viscosity increased, its

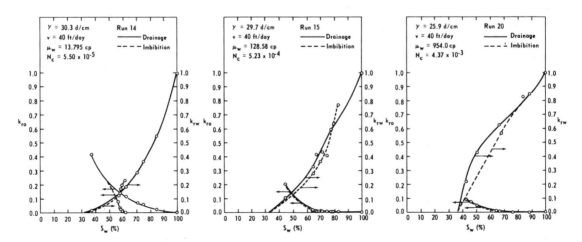

Fig. 7—Behavior of oil/water relative permeabilities at different aqueous-phase viscosities.

174

permeability curves also increased and tended toward linearity. However, the oleic (Soltrol 170) curves decreased in approximately the same order of magnitude as the water curves increased. These effects are seen in Fig. 8, where the relative permeabilities for different saturations are plotted against aqueous viscosity values. Note that caution must be exercised when applying these results since the wettability of the glycerin/brine systems must be verified as strongly water-wet.

Oil-Water Relative Permeabilities at Different Capillary Numbers. To demonstrate the capillary number effect, tests were carried out to measure the results of combining IFT reductions and aqueous viscosity increases (Fig. 9). For a low-tension system of 0.118 dyne/cm [0.118 N/m] and a wetting-phase viscosity of 6.10 cp [61.0×10^{-4} Pa·s], the curves showed near linearity, but the oleic values were slightly less than those for the 0.335-dyne/cm [0.335-N/m] run for tension effects alone. This observation indicated an interaction between the aqueous phase viscosity and the IFT. For a system of higher tension (0.454 dyne/cm [0.454 N/m]) and higher viscosity (13.64 cp [136.4×10^{-4} Pa·s]), the aqueous curves remained relatively unchanged but the oleic curves showed further decreases. However, the oil permeabilities still were larger than those for the 13.80-cp [138.0×10^{-4}-Pa·s] run at approximately 30 dyne/cm [30 N/m] (Fig. 7). At combinations of even larger variable values, the IFT effect was almost negligible.

How the capillary number affected the oil-water relative permeabilities is seen in Fig. 10 for the imbibition process at a 50% wetting-phase saturation. The specific variable within the capillary number determined the direction of the oil curves. As the aqueous-phase viscosity increased, the oleic permeability values decreased. As the tension decreased, the curves shifted upward. Conversely, as the capillary number increased, so did the ability of the wetting phase to flow. The correlation was not a perfect fit because changes from the viscosity were larger than those from the IFT. Thus, the capillary number may not be a true predictor of the wetting-phase permeability. Similar effects were observed in both the oleic and the aqueous drainage curves.[21]

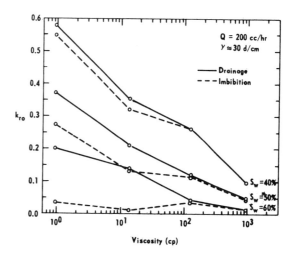

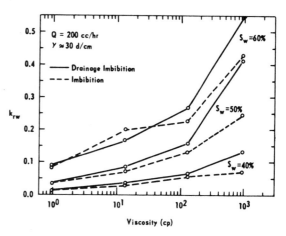

Fig. 8—Variation of oil and water relative permeabilities against aqueous-phase viscosity for different wetting-phase saturations.

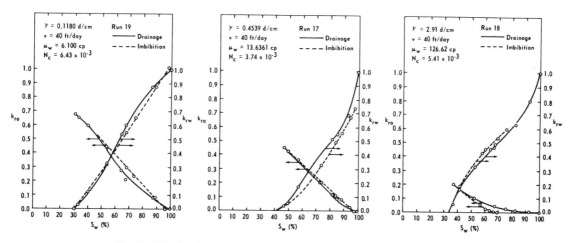

Fig. 9—Behavior of oil/water relative permeabilities at different capillary numbers.

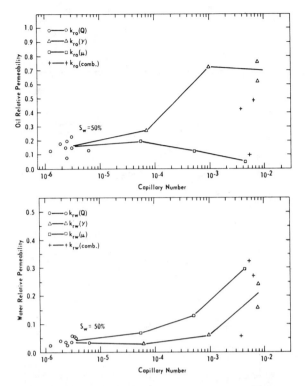

Fig. 10—Imbibition oil and water relative permeabilities as functions of capillary number at 50% water saturation.

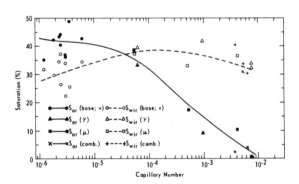

Fig. 11—Experimental residual saturations as functions of the capillary number.

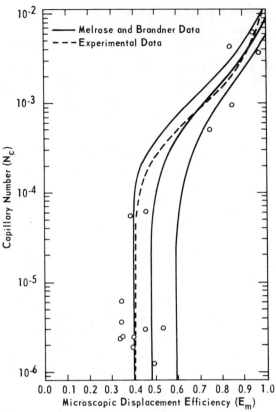

Fig. 12—Microscopic displacement efficiency as a function of capillary number.

Residual Saturation Changes. The effects on the residual saturations of both phases also were noted. The ROS showed a large reduction from 40% to 0% for both decreases in the IFT and increases in the wetting-phase viscosity, corresponding to a capillary number increase from 10^{-6} to 10^{-2}, respectively. As the tension decreased to 5.50 dyne/cm [5.50 N/m], the irreducible water saturation showed very little change, but at the lowest-tension value of 0.0389 dyne/cm [0.0389 N/m] it decreased back to 32%. The residual saturations as functions of the capillary number are given in Fig. 11, with the irreducible water saturation showing changes because of tension changes only. As a check on our experimental

results, the microscopic displacement efficiency, defined as

$$E_m = \frac{1 - S_{or} - S_{wir}}{1 - S_{wir}}, \quad \dots\dots\dots\dots\dots (3)$$

was plotted and compared with the results of Melrose and Brandner[3] (Fig. 12). Although E_m increased at approximately one order of magnitude lower for the experimental capillary numbers than those reported, the data fell within the same general range.

As the capillary number increased, no matter what variable was being altered, the imbibition-drainage hysteresis decreased but never disappeared entirely. This result held true for both the aqueous and oleic phases.[21]

Relative Permeability Model. To develop a relative permeability model based on the experimental results, the Minitab II[©][23] statistical computation system was used. This system uses regression analysis to determine the best coefficients for an equation and the statistical parameters to evaluate the function.

The following functional forms were found for both imbibition and drainage. For the oil (nonwetting-phase) relative permeabilities,

$$k_{ro(dr)} = AS_*{}^{(B + C \ln \gamma)} \left(\frac{\mu_w}{\mu_o} \right)^D \quad \dots\dots\dots\dots (4)$$

176

TABLE 4—SUMMARY OF RELATIVE PERMEABILITY MODEL COEFFICIENTS

Model	$S*$	A	B	C	D	$r^2(\%)*$	MSER**	$F†$	$\hat{r}$ plot‡
$k_{ro(dr)}$	S_o	0.72899	1.2861	0.08043	-0.37932	97.2	1020.8	4.04	r
$k_{ro(im)}$	$\dfrac{S_o - S_{or}}{1 - S_{or}}$	1.56878	1.33874	0.09187	0.08528	96.2	511.2	4.13	r
$k_{rw1(dr)}$	$\dfrac{S_w - S_{wir}}{1 - S_{wir}}$	0.70216	1.25579	0.0	-0.074482	97.9	2070.7	4.89	r
$k_{rw1(im)}$	$\dfrac{S_w - S_{wir}}{1 - S_{wir}}$	0.61135	1.25875	0.0	-0.070812	98.0	1704.1	4.92	r
$k_{rw2(dr)}$	$\dfrac{S_w - S_{wir}}{1 - S_{wir}}$	0.70340	0.66596	0.0	-0.071513	97.8	1613.8	4.89	r
$K_{rw2(im)}$	$\dfrac{S_w - S_{wir}}{1 - S_{wir}}$	0.61135	0.69580	0.0	-0.068221	97.4	1659.5	4.92	r
S_{or}	—	$+5.846 \times 10^{-1}$	$+2.96 \times 10^{-1}$	$+4.62 \times 10^{-2}$	$+1.8855 \times 10^{-3}$	96.6	122.3	5.74	r
S_{wir}	—	$+4.0214 \times 10^{-1}$	$+3.976 \times 10^{-3}$	-7.065×10^{-3}	0.0	62.4	6.64	4.46	r

*Correlation coefficient for linear regression.
**Mean square error ratio.
†F-distribution.
‡r = random scatter of $\hat{r}$ values.

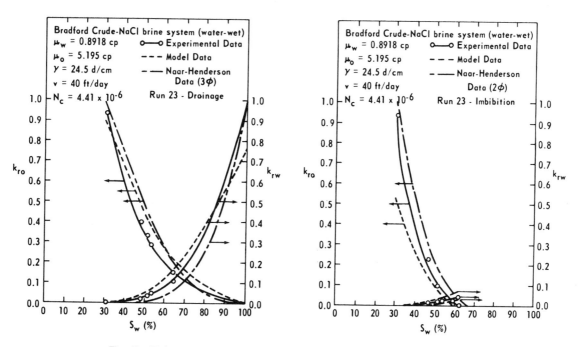

Fig. 13—Drainage and imbibition relative permeability curves for water-wet system.

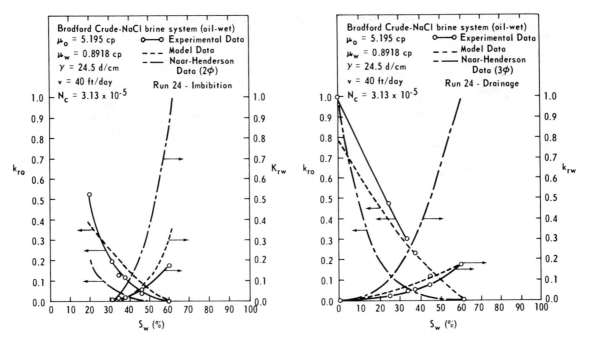

Fig. 14—Drainage and imbibition relative permeability curves for oil-wet system.

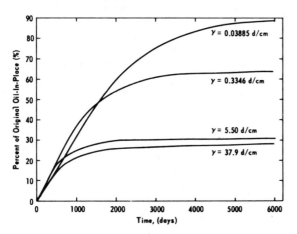

Fig. 15—Recovery vs. time and interfacial changes.

and

$$k_{ro(im)} = AS_*^{[B+C \ln \gamma + D \ln(\mu_w/\mu_o)]}, \quad \dots\dots\dots (5)$$

where S_* is a normalized variable based on the fluid saturation and the residual saturations.

The brine (wetting-phase) relative permeabilities were modeled as functions of both the individual variables and the capillary number, although those found by the variables yielded a slightly better curve-fit.

$$k_{rw1} = AS_*^{[B+D \ln(\mu_w/\mu/\gamma)]} \quad \dots\dots\dots\dots (6)$$

$$k_{rw2} = AS_*^{[B+D \ln(N_c)]} \quad \dots\dots\dots\dots (7)$$

The ROS was modeled as a function of the capillary number and the irreducible water saturation was a function of IFT.

$$S_{or} = A + B[\ln(N_c)] + C[\ln(N_c)]^2 + D[\ln(N_c)]^3 . \quad \dots (8)$$

$$S_{wir} = A + B[\ln(\gamma)] + C[\ln(\gamma)]^2 . \quad \dots\dots\dots\dots (9)$$

The values of S_*, the regression constants A, B, C, and D, and the statistical curve-fit parameters are given in Table 4.

To evaluate the regression models, three experimental tests were run using different oil/brine systems. The test curves were compared with those predicted from the regression model and with those from the equations developed by Naar and Henderson.[24] Both a water-wet and an oil-wet Bradford crude oil/2% NaCl brine system and a Kendex 0837/2% NaCl brine system were tested. The Bradford crude water-wet runs for both drainage and imbibition are given in Fig. 13. In both cases, the regression model yielded slightly larger water relative permeability values than the experimental data. For oil relative permeability, the model showed slightly larger values for drainage but lower values for imbibition. Similar results were observed for the Kendex/brine system.[21] For the oil-wet runs (Fig. 14), the model gave much improved fits over the Naar-Henderson equations for both phases in both drainage and imbibition.

Application of the Proposed Relative Permeability Model to a Reservoir Simulator. The reservoir simulator used was that developed to model two-phase, two-dimensional (Cartesian coordinate system) flow.[25] The problem was set up by the implicit pressure, explicit saturation (IMPES) method and was solved using the general band algorithm.

The runs were carried out in a 40-acre [162 000-m²] five-spot pattern with 3-darcy permeability, 35% porosity, and 58% initial oil saturation. A horizontal 6×6 grid system was established with block lengths and widths of 110 ft [34 m] and a 30-ft [9-m] thickness. Initial reservoir temperature and pressure were 80°F [27°C] and 2,000 psia [13 789 kPa], respectively.

Fig. 15 illustrates the effect on the total oil recovery by the IFT expressed as a percentage of the original oil in place (OOIP) for approximately 16 years. For higher-tension floods, production leveled at 1,000 days. For 0.335 dyne/cm [0.335 N/m], recovery slowed at 3,500 days while production at the minimum tension was still increasing at 6,000 days. Total recovery improved from approximately 30% to 89% OOIP. The aforementioned results of the simulation runs clearly indicate the important role of a low-tension flood in achieving a high mobilization efficiency, and more importantly, how influential the relative permeability characteristics are in the recovery performance.

Conclusions

The following are the conclusions based on the experimental and computational observations.

1. The nonwetting (oil) relative permeabilities were found to be functions of the IFT and the viscosity variables individually rather than a function of the capillary number.

2. The wetting (brine) relative permeabilities behaved as functions of the capillary number but were better modeled using the individual variables.

3. Insignificant IFT effects were observed on both k_{ro} and k_{rw} until a value of 2.0 dyne/cm [2.0 N/m] was obtained. Below this value, the relative permeabilities increased with decreasing tension.

4. Increases in aqueous- (wetting-) phase viscosity yielded reductions in k_{ro} and increases in k_{rw}, provided that the glycerin systems remained strongly water-wet.

5. At very low IFT values, the relative permeability curves straightened out and approached the theoretical X-shape present at zero tension. For large aqueous viscosities, only the water curves behaved in a similar manner.

6. Low tensions (<5.50 dyne/cm [<5.50 N/m]) and viscosity values in the range between 2.0 and 13.6 cp [0.002 and 0.0136 Pa·s] interacted to yield the "X" appearance for the relative permeability curves. However, above 5.50 dyne/cm [5.50 N/m], tension seemed to have little effect on k_{ro} or k_{rw}.

7. No rate effects were observed for the limited range within this study (16 to 80 ft/D [4.9 to 24 m/d]).

8. As the capillary number increased to 0.01, the ROS decreased from approximately 40% to zero.

9. The irreducible water saturation showed no consistent change above a tension of 5.50 dyne/cm [5.50 N/m], but decreased to 32% at 0.0389 dyne/cm [0.0389 N/m].

10. For increases in the capillary number, the imbibition-drainage hysteresis was reduced for both k_{ro} and k_{rw}.

11. Mathematical relative permeability models were developed from the experimental data. These models yielded similar results with experimental data for different fluid systems.

12. Proposed relative permeability models were tested with the aid of a two-phase reservoir simulator. Results of the simulation studies showed that production increased from 30% to 89% OOIP as IFT decreased from 37.9 dyne/cm to 0.0389 dyne/cm [37.9 N/m to 0.0389 N/m].

Nomenclature

$A, B, C,$
D, S_* = coefficients of variables in linear regression model
E_m = microscopic displacement efficiency, fraction
$k_{ro(dr)}$ = drainage relative permeability to oil, fraction
$k_{ro(im)}$ = imbibition relative permeability to oil, fraction
$k_{rw(dr)}$ = drainage relative permeability to water, fraction
$k_{rw(im)}$ = imbibition relative permeability to water, fraction
k_{rw1} = variable modeled water relative permeability, fraction
k_{rw2} = capillary number modeled water relative permeability, fraction
L = flow length [m]
N_c = capillary number, fraction
p = pressure, psia [kPa]
S_{or} = residual oil saturation, fraction
S_{wir} = irreducible water saturation, fraction
v = flow velocity, in./sec [cm/s]
γ = interfacial tension, dyne/cm [N/m]
μ_o = oil viscosity, cp [Pa·s]
μ_w = water viscosity, cp [Pa·s]
ϕ = porosity

References

1. Moore, T.F. and Slobod, R.C.: "The Effect of Viscosity and Capillarity on the Displacement of Oil by Water," *Producers Monthly* (Aug. 1956) 20–30.
2. Taber, J.J.: "Dynamic and Static Forces Required to Remove a Discontinuous Oil Phase from Porous Media Containing Both Oil and Water," *Soc. Pet. Eng. J.* (March 1969) 3–12.
3. Melrose, J.C. and Brandner, C.F.:."Role of Capillary Forces in Determining Microscopic Displacement Efficiency for Oil Recovery by Waterflooding," *J. Cdn. Pet. Tech.* (Oct.–Dec. 1974) 54–62.
4. Chatzis, I. and Morrow, N.R.: "Correlation of Capillary Number Relationships for Sandstones," *Soc. Pet. Eng. J.* (Oct. 1984) 555–62.
5. Stegemeier, W.: *Oil Recovery by Surfactant and Polymer Flooding,* Academic Press Inc., Washington, DC (1977).
6. Downie, J. and Crane, F.E.: "Effect of Viscosity on Relative Permeability," *Soc. Pet. Eng. J.* (June 1961) 59–60.
7. Odeh, A.S.: "Effect of Viscosity Ratio on Relative Permeability," *J. Pet. Tech.* (Dec. 1959) 346–52; *Trans.* AIME, **216.**
8. Schneider, F.N. and Owens, W.W.: "Steady-State Measurements of Relative Permeability for Polymer-Oil Systems," paper SPE 9408 presented at the 1980 SPE Annual Technical Conference and Exhibition, Dallas, Sept. 21–24.
9. Osoba, J.S. *et al.*: "Laboratory Measurements of Relative Permeability," *Trans.*, AIME (1951) **192,** 47–55.
10. Richardson, J.G.: "Calculation of Waterflood Recovery from Steady-State Relative Permeability Data," *J. Pet. Tech.* (May 1957) 64–66; *Trans.*, AIME, **210.**

11. Sandberg, C.R., Gournay, L.S., and Sippel, R.F.: "The Effect of Fluid-Flow Rate and Viscosity on Laboratory Determinations of Oil-Water Relative Permeabilities," *J. Pet. Tech.* (Feb. 1958) 36–43; *Trans.*, AIME, **213**.

12. Bardon, C. and Longeron, D.G.: "Influence of Very Low Interfacial Tensions on Relative Permeability," *Soc. Pet. Eng. J.* (Oct. 1980) 391–401.

13. Amaefule, J.O. and Handy, L.L.: "The Effect of Interfacial Tensions on Relative Oil/Water Permeabilities on Consolidated Porous Media," *Soc. Pet. Eng. J.* (June 1982) 371–81.

14. Batycky, J.P. *et al.*: "Interpreting Relative Permeability and Wettability for Unsteady-State Displacement Measurements," *Soc. Pet. Eng. J.* (June 1981) 296–308.

15. Lo, H.P.: "The Effect of Interfacial Tension on Oil-Water Relative Permeabilities," Research Report RR–32, Petroleum Recovery Inst., Calgary, Alta., Canada (Nov. 1976) 5–9.

16. Lefebvre du Prey, E.J.: "Factors Affecting Liquid-Liquid Relative Permeabilities of a Consolidated Porous Medium," *Soc. Pet. Eng. J.* (Feb. 1973) 39–47.

17. Taber, J.J., Kamath, I.S.K., and Reed, R.L.: "Mechanism of Alcohol Displacement of Oil from Porous Media," *Soc. Pet. Eng. J.* (Sept. 1961) 195–212; *Trans.*, AIME, **222**.

18. Morse, R.A., Terwilliger, P.L., and Yuster, S.T.: "Relative Permeability Measurements on Small Core Samples," *Producers Monthly* (Aug. 1947)19–25.

19. Kyte, J.R. and Rapoport, L.A.: "Linear Waterflood Behavior and End Effects in Water-Wet Porous Media," *J. Pet. Tech.* (Oct. 1958) 47–50; *Trans.*, AIME **213**.

20. Fulcher, R.A.: "The Effect of the Capillary Number and Its Constituents on Two-Phase Relative Permeabilities," PhD dissertation, Pennsylvania State U., University Park (1982)

21. Batycky, J.P. and McCaffery, F.G.: "Low Interfacial Tension Displacement Studies," paper 78–29–26 presented at the 1978 Petroleum Soc. of C.I.M. Annual Technical Meeting, Calgary, Alta., Canada, June 13–16.

22. Ryan, T.A. Jr.: "Minitab Release 81.1," Computation Center writeup, Dept. of Statistics, Pennsylvania State U., University Park (1981).

23. Naar, J. and Henderson, J.H.: "An Imbibition Model — Its Application to Flow Behavior and the Prediction of Oil Recovery," *Soc. Pet. Eng. J.* (June 1961) 61–70; *Trans.*, AIME, **222**.

24. Ertekin, T.: "Numerical Simulation of the Compaction-Subsidence Phenomena in a Reservoir for Two-Phase Nonisothermal Flow," PhD dissertation, Pennsylvania State U., University Park (1978).

SI Metric Conversion Factors

cp $\times$ 1.0*	E$-$03	= Pa·s
dyne $\times$ 1.0*	E$-$02	= mN
ft $\times$ 3.048*	E$-$01	= m
°F (°F$-$32)/1.8		= °C
in. $\times$ 2.54*	E$+$00	= cm
mL $\times$ 1.0*	E$+$00	= cm^3

*Conversion factor is exact.   **JPT**

Original manuscript received in the Society of Petroleum Engineers office Oct. 5, 1983. Paper accepted for publication June 6, 1984. Revised manuscript received Oct. 31, 1984. Paper (SPE 12170) first presented at the 1983 SPE Annual Technical Conference and Exhibition held in San Francisco Oct. 5–8.

PRESSURES

The American Association of Petroleum Geologists Bulletin
V. 59, No. 6 (June 1975), P. 957-973, 23 Figs., 4 Tables

Abnormal Formation Pressure[1]

JOHN S. BRADLEY[2]

Tulsa, Oklahoma 74102

Abstract Abnormal formation pressure requires a seal; without a seal pressures would equalize to normal hydrostatic. Abnormal pressures originate from several interrelated processes, but temperature change appears to be the principal cause.

Both epeirogenic movements with associated erosion and deposition and long-term changes in climate can alter the temperature of a sealed formation at depth. Abnormal pressure resulting from temperature change caused by change of overburden thickness must be corrected for inherited pressure and change of hydrostatic pressure related to elevation difference. Osmosis, precipitation, or solution by trapped pore fluid and carbonization effects are minor in comparison with temperature effects.

Overburden stress cannot cause abnormally high pressure at present drilling depths. The loss of porosity with depth in all sedimentary rocks appears to be a chemical process rather than mechanical compression.

Pressure differentials between wells may indicate ambiguously either no fluid flow (wherein the pressure difference is maintained by a seal) or flow (wherein the pressure drop is from fluid friction in the permeable medium). A relatively small amount of flow across a seal can equalize pressures, retarding further flow. The fluid expelled as a result of loss of porosity during geologic time also flows at a low rate.

The geologist concerned with pressure problems must be aware of (1) the many variables involved in subsurface pressures, (2) the low number and ambiguity of pressure measurements, (3) the need to establish what constitutes "normal" pressure to determine abnormal pressure, and (4) the possibility of uniqueness in any field situation.

Introduction

Abnormal formation pressures were studied to establish a level of understanding about the mechanisms of their origin and preservation and to clarify the terminology to enhance communications among geologists concerned with pressure problems. Formation pressures are of interest to petroleum geologists because they affect the productivity and, hence, the economic evaluation of reservoirs. This study involved investigation of the causes of abnormal pressures to provide guidelines for early economic evaluation of such pressures and thereby enhance the effectiveness of exploration programs.

Because of the multivariate aspects of the geologic environment, the approach in this study was to establish possible and probable limits on pressures and causes. Examination of specific cases of abnormal pressures was not undertaken because any field example, as a result of the many variables, could be unique.

This paper treats the semantics pertaining to pressures and the possible origins of pressure, the factors leading to both origin and preservation of abnormal pressures, and problems associated with pressures; a "type well" standard for evaluating pressure data is presented. No attempt is made to explain the seismic, well-logging, and drilling problems associated with abnormally pressured sedimentary rocks.

Abnormal formation pressure is a pressure above or below the normal hydrostatic pressure for a given depth. It cannot exist without a seal, for without a seal, the pressures would equalize to hydrostatic. Further discussion of definitions and concepts of pressures and seals is included in an appendix to this paper.

Origin and Preservation of Abnormal Pressure

Several factors contribute to the origin and preservation of abnormal pressure: (1) epeirogenic movements with associated erosion and/or deposition; (2) thermal expansion or contraction of fluids reacting to temperature changes; (3) osmosis; (4) chemical action of pore waters trapped in a sealed formation; (5) carbonization of organic matter in sediments; and (6) buoyancy due to the hydrocarbon column of a reservoir. The processes involved are intricately related, but temperature change appears to be the principal factor with other contributors being minor with respect to the temperature effect. The factor of compressive stress caused by overburden does not contribute to abnormally high pressures at present drilling depths. Transient effects (seismic shock waves, barometric pressure, and tides) influence formation pressures only through relatively short periods of time.

[1]Manuscript received, June 3, 1974; accepted, October 14, 1974. Published by permission of Amoco Production Company.

[2]Amoco Production Company, Research Center.
Special thanks are due to the following editors and readers for their assistance in improving the manuscript: Marthann David, S. T. Martner, F. N. Schneider, D. E. Powley, R. Steinmetz, H. H. Hinch, and H. D. Winland. E. R. Buck assisted in the drafting and mechanics of printing. Additional unpublished reference material from D. E. Powley, G. W. Schmidt, H. H. Hinch, A. H. Jageler, and D. J. Hartmann of the Research Department is gratefully acknowledged.

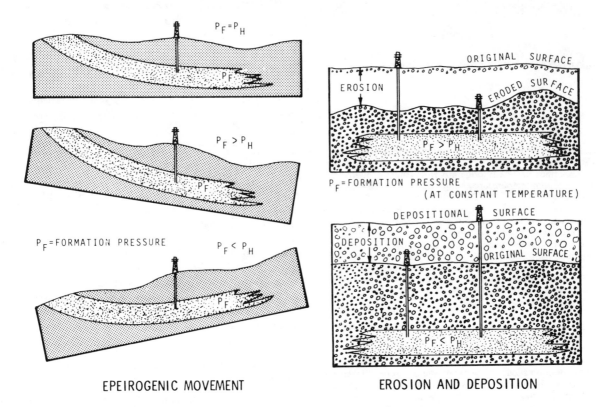

$P_F = P_H$

$P_F > P_H$

$P_F = $FORMATION PRESSURE

$P_F < P_H$

P_F

ORIGINAL SURFACE

EROSION

ERODED SURFACE

$P_F > P_H$

$P_F = $FORMATION PRESSURE
(AT CONSTANT TEMPERATURE)

DEPOSITIONAL SURFACE

DEPOSITION

ORIGINAL SURFACE

$P_F < P_H$

EPEIROGENIC MOVEMENT

EROSION AND DEPOSITION

FIG. 1—Epeirogenic movement with erosion and deposition.

Epeirogenic Movements

Epeirogenic movements may change relative elevation and thus cause abnormal pressures in a formation open to the surface laterally but otherwise sealed (Fig. 1). The formation pressure becomes abnormally high ($P_F > P_H$) if the outcrop is raised and abnormally low ($P_F < P_H$) if the outcrop is deepened. However, abnormal pressures seldom are caused by changes of elevation alone; the associated erosion and deposition resulting from the elevation change also are significant factors. Loss or gain of water-saturated sediments with erosion or deposition alters normal pressure (P_H) for the locality, and the *inherited pressure* (P_F) within a sealed formation becomes abnormally high ($P_F > P_H$) or low ($P_F < P_H$).

Temperature Change

Erosion or deposition accompanying epeirogenic movements alters the temperature of a sealed formation, and long-term changes in climate can affect both temperature gradient and temperature at depth. The effects of volcanoes, dikes, sills, salt domes, permafrost, artesian and hydrothermal activity, and glaciers can alter formation temperatures locally. Subsurface temperatures may change, but temperature gradients are fairly uniform for various rock types (Fig. 2).

If the temperature of a volume of rock and fluid is raised, the volume increases by thermal expansion. The coefficient of thermal expansion of the fluid is greater than that of the rock. If the formation is sealed and either heated or cooled, the pore liquid (in trying to expand or contract, but being unable to pass through the seal) markedly alters the formation pressure. Then, if the seal remains intact and the temperature is maintained, the abnormal pressure is preserved.

The coefficient of linear expansion for sedimentary grains is about 9×10^{-6} volume units per °C (Hodgman, 1957). Assuming uncemented mineral spheres in rhombohedral packing, the pore-volume coefficient of cubic expansion is 5×10^{-6} volume units per °C. Coefficients of cubical expansion of pore fluids are:

Water	200×10^{-6} volume units/°C
Brine	400×10^{-6} volume units/°C
Oil	$1,000 \times 10^{-6}$ volume units/°C
Gas (perfect)	$4,000 \times 10^{-6}$ volume units/°C

Thus, for a temperature change, the volume of a brine will change as much as 80 times the change of pore volume. Hydrocarbons (particularly gases) in the system could cause an even greater volume change.

If a freshwater system is sealed and thus held at constant volume, the pressure changes 125 psi per

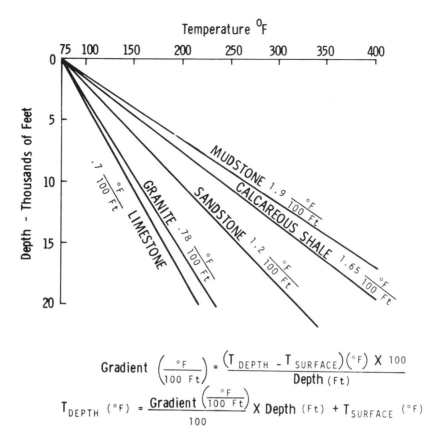

$$\text{Gradient} \left(\frac{°F}{100 \text{ Ft.}} \right) = \frac{(T_{\text{DEPTH}} - T_{\text{SURFACE}})(°F) \times 100}{\text{Depth}_{(\text{Ft})}}$$

$$T_{\text{DEPTH}} (°F) = \frac{\text{Gradient} \left(\frac{°F}{100 \text{ Ft}} \right)}{100} \times \text{Depth}_{(\text{Ft})} + T_{\text{SURFACE}} (°F)$$

FIG. 2—Temperature gradients from wells for various rock types.

°F of temperature change (Smith, 1963). These relations are illustrated in Figure 3. The usual range of interest is at 8,000 to 14,000 ft. If one assumes sufficient geologic time for return of the geothermal gradient to equilibrium at depth, a change in surface temperature of only 8°F could cause a 1,000-psi deviation from normal hydrostatic pressure within a sealed system. A plot of pressure, volume, and temperature for brines is similar to that for fresh water, but the specific volume lines are associated with higher temperatures. For example, for a saturated brine of approximately 300 ppt TDS, the net effect is a 14-percent decrease in pressure gradient, i.e., 110 psi/°F compared with 125 psi/°F for fresh water (Barker, 1972).

The temperature of a sealed formation also is altered by change in overburden with erosion or deposition, so that the true abnormal pressure is a function of the temperature change, the inherited pressure, and the "new" hydrostatic pressure. As shown in Figure 3, with a hydrostatic gradient of 0.5 psi/ft and a temperature gradient of 1.75°F/100 ft, the normal pressure increases at a rate of 0.5 psi/ft of depth change to the seal, below which the abnormal pressure increases at 1.7 psi/ft of depth change in the sealed volume.

$$\Delta G_{PF} - \Delta G_{PH} = G_{\Delta D},$$
$$(0.0175 \text{ °F/ft} \times 125 \text{ psi/°F}) - 0.5 \text{ psi/ft} = 1.7 \text{ psi/ft}.$$

Under these conditions, 588 ft of elevation change causes an abnormal pressure of 1,000 psi.

Osmosis

The mass transfer of water (solvent) through a semipermeable membrane from fresher water (dilute solution) to saltier water (concentrated solution) is termed osmosis (Fig. 4). The pressure seal of a formation can serve as the membrane required for the process. Osmotic pressure is the pressure caused by osmosis and equals the back pressure required to stop osmotic flow (i.e., the flow of water through the membrane which is at a maximum if no back pressure exists). Osmotic flow requires continuing addition of water to the freshwater system and salt to the saltier system. If the pore water within a sealed formation is saltier than the surrounding water, osmosis causes an abnormally high pressure; if the water is fresher

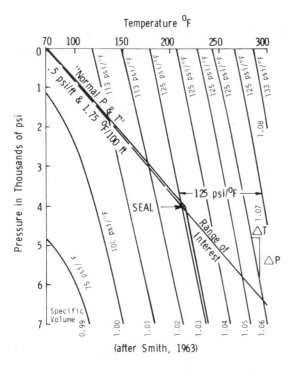

FIG. 3—Pressure-volume-temperature diagram for fresh water.

within the seal, osmosis produces abnormally low pressure. The reverse of this process, ion filtration or reverse osmosis, requires a pressure higher than the osmotic pressure to reverse flow through the membrane. Thus, fresh water is produced while the salt is filtered out of the saltier system which remains high pressured ($P_1 < P_2 + P_F$). The process cannot be effective with a faulty membrane. Ion filtration cannot cause abnormally high pressure, but does require an abnormally high pressure to function.

Theoretical osmotic pressures for various brine concentrations (Fig. 5) can be several thousand psi (Petroleum Research Corp., 1958). However, Young and Low (1965) found that osmotic pressures in cored shale membranes are only 2 to 4 psi. Their findings suggest that effective membranes are absent in shales and that osmosis may not be a major factor in the subsurface pressure systems.

Chemical Effects of Pore Waters

Abnormal pressures could be caused by chemical action of pore waters trapped within a sealed formation, either by solution of a part of the rock matrix to increase porosity or by precipitation of minerals from the pore water to decrease porosity. However, the processes are self-limiting in that no fluid or solid could be added or subtracted through the seal of the system. The limitation

means that any effect of solution or precipitation on pressure must be small. Also, a mechanism to trigger the solution or precipitation would be necessary; if temperature is this mechanism, the minor solution-precipitation effect might be lost in the larger temperature effect. Furthermore, a supersaturated brine is required for mineral deposition (Winkler and Singer, 1972) and such pore brines are uncommon in sedimentary rocks.

One aspect of pore waters of major significance with respect to abnormal pressures involves the dissolved-solids content and its relation to the density of the water. The density is, in turn, a factor of normal hydrostatic pressure. Chemical analyses of produced waters, such as those for the Hackberry and Manchester fields in Louisiana (Figs. 6, 7), usually are not available, but water salinities can be calculated from well logs for use in estimating water densities. On the basis of the data for the Manchester field (Schmidt, 1971), the salinities of waters produced from abnormally high-pressured sandstones in the Gulf Coast appear to be lower than those of the normally pressured zones. However, many examples of similar salinities and even higher salinities in abnormally high-pressured zones (Dickey et al, 1972) render untenable any blanket hypothesis of cause and effect. The hydrostatic pressure plot of the Hackberry field data (Fig. 8) is a curve which demonstrates that the salinity gradient changes with depth. Errors up to several hundred psi can be

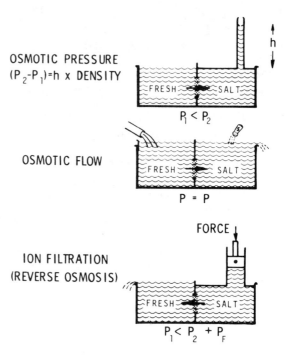

FIG. 4—Osmosis and ion filtration.

184

made by assuming a constant gradient (straight line) which may or may not fit the actual data. For example, the Hackberry data show an error of 140 psi at 15,000 ft if a "normal" gradient of 0.465 psi/ft is assumed, whereas the actual gradients can range from 0.433 psi/ft in surface waters to 0.491 psi/ft at 14,000 ft (Table 4).

Another puzzling factor about the chemical compositions of sediment waters stems from the presumption that initial sedimentation took place in sea water not greatly different in composition from present-day sea water. Thus, in the Manchester field area (Fig. 7), the waters in the sandstones generally become more saline with depth but may become fresher in the abnormally high-pressure zone (Schmidt, 1971). On the other hand, the shale water, if one assumes that the measurements are correct, remains generally at about the same salinity as sea water with increasing depth. Although salinity may remain fairly constant, individual ion concentrations differ radically (Fig. 9). Billings et al (1969) also reported difficulties in explaining chemical changes recorded for sea water to brine and from brine to brine. The great variability of water chemistry is demonstrated

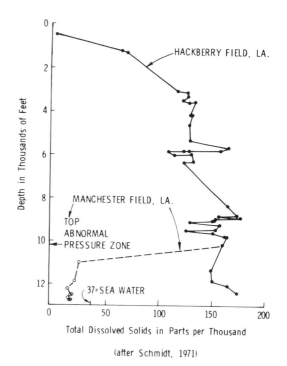

FIG. 6—Composition of waters from sandstones of Hackberry and Manchester fields, Louisiana.

further by a sampling of available data (Table 1). Thus any relations between abnormal pressures and the chemical compositions of brines are not discernible.

Carbonization

Carbonization is the process by which organic materials in sediments are converted by thermal action to hydrocarbons, other volatiles, and carbon.

If the products of the thermal alteration have a greater volume than the original organic matter, they would contribute to an abnormally high pressure in a sealed volume of rock. It would appear that some, if not all, of the gas produced would be dissolved in the oil produced and in the brine. Depending on the type of organic matter and the stage of carbonization, the net volume might be reduced or increased. In any case, the effect of the carbonization with temperature increase would be additive to the large effect caused by the thermal expansion of the brine in the system. Hedberg (1974) emphasized the role of methane generation from organic matter as a probable contributory cause of overpressuring of the fine-grained sediments.

Buoyancy

An unexpectedly high pressure for a depth may be caused by the buoyant effect of a column of

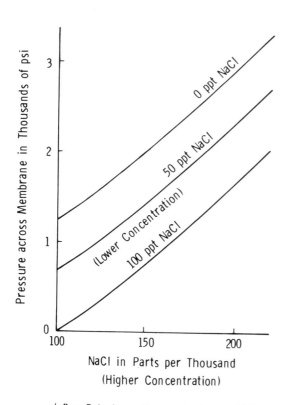

FIG. 5—Theoretical osmotic pressures across semi-permeable membrane for differential brine concentrations.

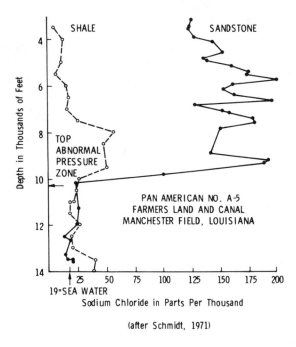

FIG. 7—Composition of waters from sandstones and shales of Manchester field area, Louisiana.

oil and/or gas in a reservoir sealed above and laterally but open at the bottom to hydrostatic pressure (Fig. 10). The buoyant force is the weight of the displaced brine less the weight of the displacing agent; thus, the buoyancy pressure can range from zero for an oil of the same specific gravity as the brine to a maximum limit as the weight of the gas approaches zero. The pressures above hydrostatic which might be expected for various heights of hydrocarbon columns are shown in Figure 11; data for a wet gas and a condensate (55°API) in a heavy brine are plotted. The abnormal pressure caused by buoyancy for varying gradients of brine, oil, and gas is illustrated by the commonly used pressure-depth diagram (Fig. 12).

Compressive Stress

Compressive stress from overburden commonly is cited as a mechanism for "compaction" (which is defined as reduction in volume of sediments, i.e., loss of porosity). However, in popular usage, the term "compaction" has come to mean loss of porosity without regard to cause. In "tectonic compaction," the compressive stress is a result of lateral tectonic forces as opposed to the more common vertical overburden stresses. Compressive stress resulting from overburden (i.e., the weight of sediment in brine) is the vertical lithostatic pressure (P_{LV}). The P_{LV} is relatively low (12,

000 psi at 20,000 ft; see "type well" pressure gradients, Fig. 20) compared with the ultimate strength of the sediment particles in compression (20,000 to 100,000 psi; Clark, 1966).

Removal of vertical compressive stress (i.e., overburden by erosion) allows minor elastic rebound which increases porosity slightly. However, the large loss of porosity in rocks through geologic time is essentially irreversible (Maxwell, 1964).

A reduction of porosity by compressive stress is reported in many publications as causing abnormally high pressures if the sediments involved are sealed. However, it can be demonstrated that compression by overburden cannot cause such pressures at the depths of present drilling capacity. As shown in Figure 13, when a part of water-saturated sediment (1) is sealed off (2), and an overburden of sediment and water is added (3), the extra load is taken by the sediment below the seal. The fluid pressure below the seal remains at the "preloading" hydrostatic pressure(P_{H_1})which has become abnormally low with respect to the hydrostatic pressure directly above the seal. The sediment below the seal continues to absorb the overburden load until the load exceeds the strength of the sediment. When the sediment below the seal fails, a part of the load is assumed by the fluid below the seal. In reality, the system is

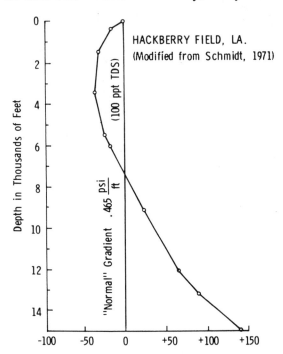

FIG. 8—Deviation of actual hydrostatic gradient from straight-line "normal gradient."

186

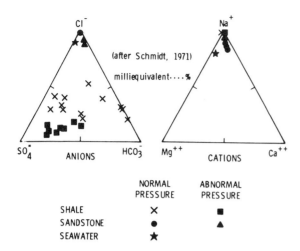

(after Schmidt, 1971)

milliequivalent....%

	NORMAL PRESSURE	ABNORMAL PRESSURE
SHALE	×	■
SANDSTONE	●	▲
SEAWATER	★	

FIG. 9—Ion concentrations in waters from sandstones and shales of Manchester field area, Louisiana.

not so simple because both sediment and water are compressible under the extra load. The compressibilities of water and rock are assumed to be similar, i.e., about 3×10^{-6} volume units/psi (Hodgman, 1957; Clark, 1966). Thus, if 1,000 ft of added overburden increases the pressures by 1,100 psi (600 psi from sediment and 500 psi from brine), the extra load is divided below the seal between the rock and the water proportional to the void ratio (Volume Pores/Volume Rock; see Fig. 19). If the void ratio is 0.1 (porosity: 9.1 percent), the rock carries 990 psi and the water carries 110 psi of the added 1,100-psi load. Thus the pore pressure below the seal is 110 psi above the original ("preload") hydrostatic pressure, but the pore pressure directly above the seal is 500 psi

above its original hydrostatic pressure. The pore pressure below the seal remains abnormally low, showing that compression by overburden in normal-strength rocks cannot cause abnormally high pressures.

Similarly, compressive stress cannot explain the major loss of porosity as actually measured in rocks as their depths increase (Fig. 14). Clays, silica, or calcium carbonate slurries squeezed in a press at pressures comparable with lithostatic pressures compress only slightly (Chilingarian and Rieke, 1968; Ayer, 1971) and porosity is reduced by only a few percent. This type of porosity loss is caused by compression, but in nature, loss of porosity many times greater is caused by other processes. The processes are herein grouped together and termed "lithification" for lack of a better word. "Lithification" is a chemical process which is dependent on reactivity, temperature, surface area, and pressure. In geologic jargon, "limestones are recrystallized, sandstones are cemented, and shales are compacted;" however, it seems probable that all three rock types are subject to the same chemical processes (i.e., "lithification") whereby porosity is reduced.

The dry bulk density of some shales in abnormally high-pressured formations is low compared with that of directly overlying normally pressured sections. However, a complete range of high-normal-low densities is associated with abnormal pressures (D. E. Powley, 1973, personal commun.). Perhaps, then, densities and pressures are not related directly. If grain density remains constant, a low bulk density is equivalent to a high porosity. Thus, if a part of the loss of porosity in shales is caused by chemical processes, high po-

Table 1. Water Chemistry Data

Depth (feet)	Lithology	Salinity TDS °/oo*	Anions °/oo			Cations °/oo			Pressure High-Normal		Remarks
			Cl⁻	HCO₃⁻	SO₄⁼	Na⁺	Ca⁺⁺	Mg⁺⁺	(H)	(N)	
Sea Water	--	37	19	.14	2.7	11	.40	1.3	--		Sea water
9,680	sandstone	82	45	.55	0	31	2.3	.05		N	Dickey et al. (1972)
10,015	sandstone	88	52	.58	0	33	1.6	0	H		
12,350	sandstone	17	8.2	1.5	.18	5.7	.06	.01	H		Manchester field
12,670	sandstone	18	10	1.3	.18	6.6	.14	.02	H		Schmidt (1971)
12,750	sandstone	16	8.5	1.4	.23	5.6	.07	.02	H		
12,865	sandstone	18	9.7	1.7	.17	6.4	.16	.02	H		
3,496	shale	9.7	1.4	5.2	0	2.9	.01	.005		N	Manchester field
6,499	shale	24	4.6	9.5	1.9	8.2	.03	.003		N	Schmidt (1971)
9,470	shale	65	12	9.0	18	26	.07	.008		N	
10,500	shale	38	2.4	9.0	15	11	.02	.02	H		
11,500	shale	29	2.5	8.6	8.3	9.3	.01	.02	H		
12,500	shale	32	2.1	12	8.1	10	.01	.03	H		
1,500	sandstone	68	42	.21	.02	24	1.8	.66		N	W. Hackberry field
3,300	sandstone	128	78	.06	.08	46	2.5	1.1		N	Schmidt (1971)
6,525	sandstone	122	75	.12	0	44	2.7	.76		N	
9,050	sandstone	177	109	.14	0	59	8.0	1.6		N	
9,950	sandstone	162	98	.29	.81	58	3.4	1.0		N	

* TDS °/oo = Total Dissolved Solids, parts per thousand.

187

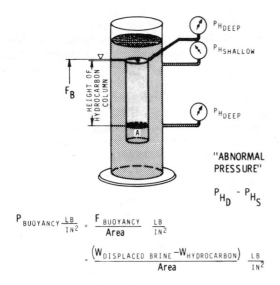

$$P_{BUOYANCY} \frac{LB}{IN^2} = \frac{F_{BUOYANCY}}{Area} \frac{LB}{IN^2}$$

$$= \frac{(W_{DISPLACED\ BRINE} - W_{HYDROCARBON})}{Area} \frac{LB}{IN^2}$$

FIG. 10—"Abnormal" pressure resulting from buoyancy of hydrocarbons in brine.

rosity–low density shales may be caused by suspension of the chemical process. One manner by which the recrystallization and cementation processes could be reduced is a restriction of water flow through shales by some form of "preseal;" such a seal would not necessarily be a pressure seal at inception although a pressure seal might develop later (Bruce, 1973). Thus, the relation between abnormal pressures and low-density shales could depend on the timing and effectiveness of the "preseal" and the seal, and on the subsidence-deposition rates in the basin. However, this explanation probably is also an oversimplification. For example, a sample above the low-density shale with the same density as the low-density shale associated with abnormally high pressures can have different mineralogy (grain density), brine chemistry, shrinkage, texture, and lithification stage. Furthermore, abnormal pressures can be present in sections with normal, high, or low densities. Thus, density and pressure are not directly related, but may result from similar processes operating at different times and during different intervals of time.

Clay Dewatering

Burst (1969) suggested that abnormal pressures could be caused by thermal loss of interlayer water from clay minerals in a sealed formation. If the density of the interlayer water were the same as the density of the pore water, there would be no volume change as the water passed from ineffective porosity (interlayer) to effective porosity (pores). However, Burst calculated that the densi-

ty of the interlayer water can be higher or lower than that of the pore water, depending on the stage of dewatering. If the interlayer-water density is higher than the pore-water density, there will be a net volume increase with expulsion and a resulting increase of pressure within the seal. However, the maximum increase caused by a density difference would be small compared to the pressure increase caused by thermal expansion of the pore water. The density effect, positive or negative, would be additive to the thermal effect.

Transient Effects

The transient phenomena of natural seismic shock waves, barometric pressure, and tides may alter subsurface pressures for different periods of time. Seismic shock waves from various sources can create fairly high pressures through the earth, but they affect only small volumes of fluid, are of relatively short duration, and are not preserved as abnormal pressures, although they might trigger fracturing. Barometric pressure and diurnal tides are low-amplitude transient effects which can affect formation pressures for relatively long periods of time compared with the seismic effects.

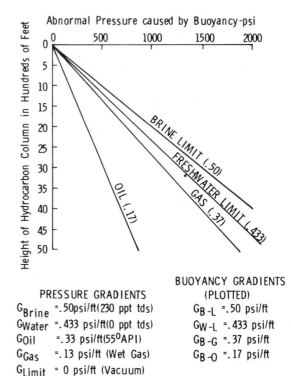

PRESSURE GRADIENTS

G_{Brine} = .50 psi/ft (230 ppt tds)
G_{Water} = .433 psi/ft (0 ppt tds)
G_{Oil} = .33 psi/ft (55°API)
G_{Gas} = .13 psi/ft (Wet Gas)
G_{Limit} = 0 psi/ft (Vacuum)

BUOYANCY GRADIENTS
(PLOTTED)

G_{B-L} = .50 psi/ft
G_{W-L} = .433 psi/ft
G_{B-G} = .37 psi/ft
G_{B-O} = .17 psi/ft

FIG. 11—Abnormal pressure from buoyancy for various heights of hydrocarbon columns.

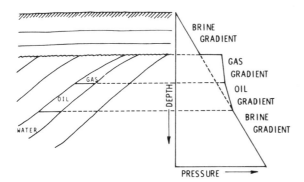

FIG. 12—Gradient diagram showing abnormal pressure resulting from hydrocarbon buoyancy.

PROBLEMS ASSOCIATED WITH INTERPRETING PRESSURES

The geologist concerned with pressure problems has to keep in mind (1) the multivariate aspect of the geologic environment, (2) the possibility of uniqueness in any field situation, and (3) the requirement of establishing a "normal" pressure to determine what is "abnormal" pressure. Ambiguities in interpretations of problems associated with abnormal pressures commonly are a result of the limited precision, types, and numbers of measurements, so that awareness of these limitations is an important factor.

Precision of Pressure Measurements

Pressures usually are measured directly in exploratory wells during drill-stem tests and wireline formation tests, but pressures also may be calculated from mud "kicks" during drilling. Understanding of the accuracy and sensitivity of these measurement procedures is of importance in interpreting pressure data.

In a drill-stem test (Fig. 15), the BT (Bourdon Tube) pressure recorder is considered accurate to approximately ±1 percent of full scale (i.e., ±50 psi on a 5,000 psi gauge—shallow well—or ±100 psi on a 10,000 psi gauge—deep well). Examples of an original DST record and graph are included on Figure 15, and the drill-stem testing sequence is shown in Figure 16. An uncertainty associated with pressure measurements is the origin of flow in the open hole below the packer. If fluid is recovered on a DST, the gauge pressure may be checked by calculating the pressure exerted by the column of fluid recovered (height of column × density of recovered fluid).

Pressure measurements by a wireline formation tester (Fig. 17) are considered accurate to approximately ±2 percent of the range of the instrument; the accuracy also depends on the magni-

tude of the pressures measured. The same BT pressure recorder as that used in the drill-stem test can be run above the sample chamber of the wireline tester to increase the accuracy, but at the expense of utilizing a larger tool.

Formation pressure may be calculated during drilling when the mud "kicks" or "blows out" by estimating the mud weight and the depth of the hole. The accuracy of this method differs widely, depending on the timeliness and exactness of the estimates. True vertical depth (which is less than the hole length) must be considered; pressures for several footage differentials for a 10,000-ft hole are listed in Table 2. Mud weight in the annulus also must be estimated. Formation pressure is calculated by multiplying the mud gradient by depth; thus, a difference of 0.1 lb/gal in mud weight (which is equivalent to 0.05 psi/ft in the mud gradient) can yield an error of 500 psi at 10,000 ft. The estimate of the gradient of the fluid in the annulus appears to be a major item in the accuracy of calculating pressure measurements from mud "kicks." Fluid and gas cutting of the mud, entrapment of air in the mud at the surface, mud compressibilities, and long circulation periods of the mud (several hours in deep wells) can affect the accuracy of density estimates. Pressure gradients and densities for various brines, muds, and oils are listed in Table 3. Accuracy also may be affected if bottom-hole pressure is calculated from wellhead pressures, both flowing and shut in, because of possible error in estimating fluid-column density.

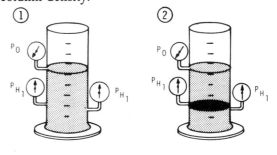

(AT CONSTANT TEMPERATURE)

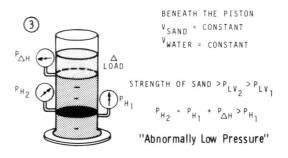

BENEATH THE PISTON
V_{SAND} = CONSTANT
V_{WATER} = CONSTANT

STRENGTH OF SAND $> P_{LV_2} > P_{LV_1}$

$P_{H_2} = P_{H_1} + P_{\triangle H} > P_{H_1}$

"Abnormally Low Pressure"

FIG. 13—Compression by overburden.

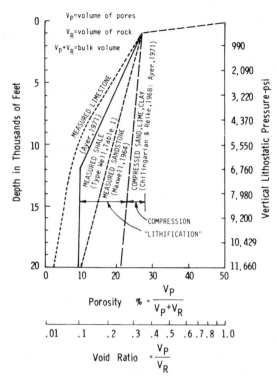

FIG. 14—Loss of porosity with depth for several rock types.

Pressures in Shales

Although the term "high-pressure shales" commonly is used, there never has been a measurement of the pressure in a shale. Where high pressure is measured in a sealed sandstone adjacent to a shale, the pressure in the shale is assumed to be the same as that in the sandstone. In sections in which the sandstone fluid pressures are normal, it is assumed that pressures in the shales also are normal.

Fluid Flow

Natural flow has been measured only in near-surface aquifers. Flow in deeper formations must be inferred from pressure measurements. Measured differences of formation pressure between wells of equal depth and elevation (or pressures corrected for unequal depths and elevations) indicate ambiguously that (1) there is flow, or (2) there is no flow. In the "flow situation," the pressure difference results from a pressure drop of fluid friction in the permeable medium. In the "no-flow situation," the pressure difference is maintained by a seal. Thus, pressure differences alone do not establish the existence of flow.

Flow equalizes pressures, and the volume of fluid which must cross the seal to equalize pressures on either side of the seal is relatively small.

For example, if one assumes a porosity of 10 percent on both sides of a seal, pressure 1,000 psi greater on one side of the seal and a distance of 1 km (3,281 ft) from the seal in both directions (Fig. 18), then a flow across the seal of 15 cc/sq cm will equalize the pressures. If considered through geologic time, this amount of flow is minuscule. Even a distance of 10 km from the seal and a pressure difference across the seal of 6,000 psi would increase the flow required to equalize the pressure only to 900 cc/sq cm. Such flow over the last 10 m.y. (Pliocene-Holocene) would be at an annual rate of 0.0009 cc/sq cm.

Flow also can be caused by movement of the fluid expelled as shales lose porosity. For example, assume a shale bed 1 km (3,281 ft) thick between sandstones that maintain hydrostatic pressure; assume also that the porosity of the shale 11 m.y. ago was 30 percent and is now 10 percent. The maximum flow of fluid will be across the sandstone-shale interface and amount to 11,000 cc/sq cm of surface. If the flow were even through time for 11 m.y., the rate of flow across the sandstone-shale interface would be 0.001 cc/sq cm per year. Such a rate is well within the range of hydraulic conductivities of consolidated clays measured by Olsen (1972).

TYPE WELL FOR ESTABLISHING PRESSURE NORMS

As an aid for evaluating pressure problems in an area, a "type well" standard can be of value in

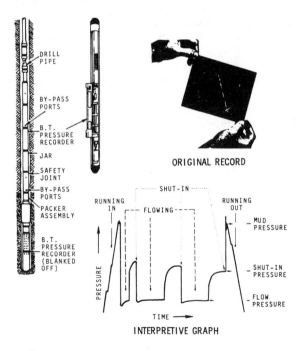

FIG. 15—Drill-stem test pressure measurement.

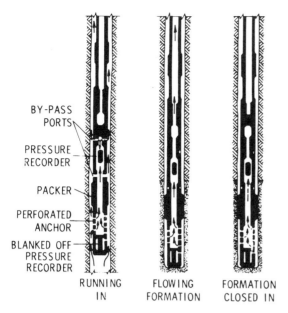

BY-PASS
PORTS

PRESSURE
RECORDER

PACKER

PERFORATED
ANCHOR

BLANKED OFF
PRESSURE
RECORDER

RUNNING
IN   FLOWING
FORMATION   FORMATION
CLOSED IN

FIG. 16—Drill-stem testing sequence.

establishing pressure norms and gradients to compare with "abnormal" parameters which may be encountered in exploratory drilling. A type well of a Gulf Coast model is presented in Figure 19. Reasonable measurements of density, porosity, and salinity are assumed for a sandstone-shale sequence. Various pressure gradients based on these data are plotted in Figure 20. The "type well" parameters are tabulated in Table 4; these or similar calculations can be used as a reference to establish pressure norms for an area.

CONCLUSIONS

Investigation of the mechanisms of origin and preservation of abnormal pressure has led to the following conclusions of possible applicability for geologists concerned with pressure problems encountered in exploration.

1. The primary requirement for existence of abnormal formation pressure is a seal; without a seal, pressures would equalize to hydrostatic. In a single-phase system (water), a pressure difference cannot be maintained unless the rock is completely surrounded by the seal. However, in a multiphase system (water-oil-gas), abnormal pressure can be maintained, because of the buoyant effect of hydrocarbons in water, with the bottom of the system unsealed.

The function of a seal (i.e., maintenance of differential pressures) requires impermeable material separating more permeable material Horizontal seals, top and bottom, may be laminae of impermeable shale or evaporite which, by virtue of mineralogy or sorting, provide a pressure barrier. Vertical (lateral) seals can be formed by fault displacement, by gouge or tear-zone materials from faulting, or by lateral facies change from permeable to impermeable rock types. Seals can be assumed to be thin with respect to both thickness and lateral extent of abnormal pressure zones, as pressure changes laterally across faults or vertically across bedding are abrupt.

2. Temperature change appears to be the principal source of pressure which varies from hydrostatic pressure. The temperature of a sealed formation can be altered by epeirogenic movements with accompanying erosion or deposition, or by a long-term change in climate whereby temperature at depth would be affected. Abnormal pressure from temperature change as a result of additional overburden must be corrected for inherited pressure, and the change in "normal" hydrostatic pressure caused by the elevation difference. Internal pressures in a sealed system held at constant volume will change 110 psi (saturated brine) to 125 psi (fresh water) per °F of temperature change. The presence of volcanoes, dikes, sills, glaciers, permafrost, artesian and hydrothermal activity, and salt domes can alter formation temperature in localized areas.

3. Osmosis, pore-water precipitation or solution, and carbonization are other possible sources of abnormal subsurface pressures, but none is considered to be of major significance in most shales.

Although theoretical osmotic pressures for different brine concentrations can be several thousand psi, measured osmotic pressures on cored shale membranes have been found to be less than 5 psi. Thus, even though osmosis can provide an explanation of certain aspects of salinity and pressure, the effect cannot be considered to be a major factor in subsurface pressure systems.

Any effect of precipitation from trapped pore waters to decrease porosity or solution to increase porosity and thereby increase or reduce pressures in a sealed system must be considered minor and lost in the larger effect of temperature change required to initiate precipitation or solution.

Carbonization, the process by which organic materials in sediments are converted by thermal action to hydrocarbons, other volatiles, and carbon, can cause a volume change within a sealed formation. The volume change could cause abnormal pressure, but the effect would be additive to the large effect of the temperature change on the brine.

4. "Compaction" has been defined as reduction in volume of sediments (loss of porosity) resulting from compressive stress, usually overburden. However, in popular usage, the term has come to

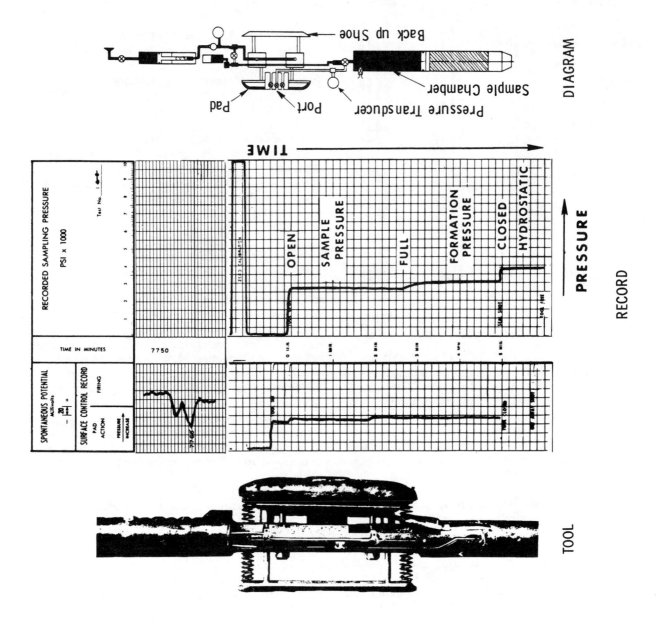

FIG. 17—Wireline formation test tool.

mean loss of porosity without regard to cause. Overburden stress cannot cause abnormally high pressures at present drilling depths. Furthermore, such stresses cannot explain most of the loss of porosity actually measured in rocks as their depths increase. Porosity loss in nature is caused by a group of interrelated chemical processes which are dependent on reactivity, temperature, surface area, and pressure.

Dry bulk densities of shales do not appear to be related directly to pressures. A complete range of "high-normal-low" densities is associated with abnormal pressures. The occasional relation between abnormally high pressures and low-density shales could be dependent on the rate of chemical changes in the shales and on the timing and effectiveness of some form of seal (not necessarily a pressure seal at inception). Such a preseal would restrict water flow through the shales and delay the action of the chemical processes which tend to increase density.

5. An important factor in understanding the limitations of interpreting pressures is assessment of the accuracy and sensitivity of pressure mea-

Table 2.   True Vertical Depth

For a hole 10,000 feet long with - -

Deviation from Vertical	True Depth, ft	ΔD ft [*]	Δ Pressure psi [**]	
			10 lb/gal mud .520 psi/ft	15 lb/gal mud .780 psi/ft
1°	9998.5	1.5	.8	1.2
2°	9993.9	6.1	3.2	4.8
3°	9986.3	13.7	7.1	10.7
4°	9975.6	24.4	12.7	19.0
5°	9961.9	38.1	19.8	29.7

[*]   ΔD = depth differential
[**]   ΔP = pressure differential

surements. Pressures measured during drill-stem testing are considered accurate to ±1 percent (i.e., to ±50 psi on a 5,000 psi gauge—shallow wells—and ±100 psi on a 10,000 psi gauge—deep well). An uncertainty associated with DST pressure measurements is the origin of flow in the open hole below the packer. Pressure measurements by a wireline formation tester are considered to be accurate to within ±2 percent of the range of the instrument, again dependent on the magnitude of the pressures measured. The accuracy of calculating formation pressure from mud "kicks" during drilling can vary widely, depend-

ing on timeliness and exactness of the estimates of mud weight and hole depth.

6. The requirement of establishing "normal" to determine abnormal pressure also must be kept in mind when interpreting formation pressures. Normal hydrostatic pressure is a product of depth and water density. The density of water, in turn, is a function of temperature, pressure, and most important, the dissolved-solids content; all of these change with depth. Errors in estimates of pressure can range up to several hundred psi by assuming a constant salinity gradient. Known variations in salinities of waters produced from

Table 3. Pressure Gradients and Densities for Various Brines, Muds, and Oils[+]

BRINE WEIGHTS (@ 60°F)			MUD WEIGHT			OIL WEIGHT (@ 60°F)			
Density g/cc	Gradient psi/ft	Salinity °/oo [*] TDS ppt	Density lb/gal	Density g/cc	Gradient psi/ft	API Gravity°	Density g/cc	Density lb/gal	Gradient psi/ft
1.00	.433	0	10	1.20	.520	10	1.000	8.328	.433
1.01	.437	13.5	11	1.32	.570	15	.9659	8.044	.418
1.02	.441	27.5	12	1.44	.625	20	.9340	7.778	.404
Sea Water	.444	37	13	1.56	.675	24	.9100	7.578	.394
1.03	.445	41.4	14	1.68	.725	28	.8871	7.387	.384
1.04	.450	55.4	15	1.80	.780	30	.8762	7.296	.380
1.05	.454	69.4	16	1.92	.830	32	.8654	7.206	.375
1.06	.459	83.7	17	2.04	.885	36	.8448	7.034	.366
1.07	.463	98.4	18	2.16	.935	40	.8251	6.870	.357
1.075	.465	100				44	.8063	6.713	.349
1.08	.467	113.2				48	.7883	6.563	.341
1.09	.471	128.3				50	.7796	6.490	.337
1.10	.476	143.5				55	.7587	6.316	.329
1.11	.480	159.5				60	.7389	6.151	.320
1.12	.485	175.8				65	.7201	5.994	.312
1.13	.489	192.4				70	.6852	5.845	.297
1.135	.491	200							
1.14	.493	210.0							
	.500	230.0							

[+]After Levorsen and Berry (1967).
[*]TDS = Total dissolved solids,
  ppt = °/oo = parts per thousand.

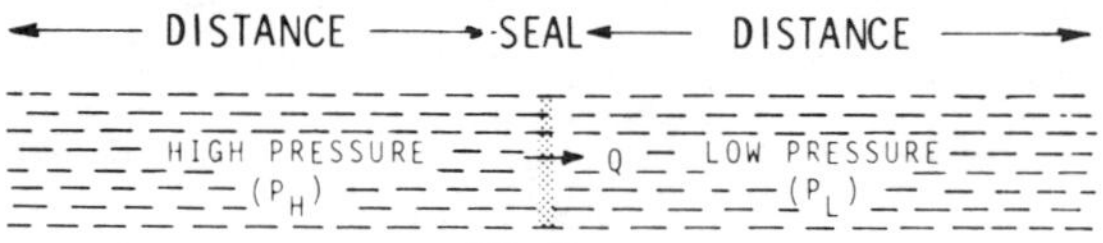

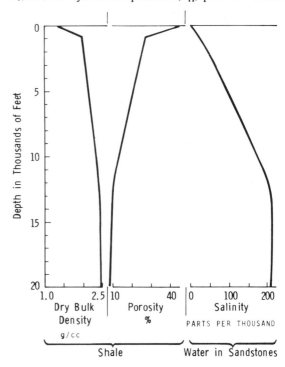

← DISTANCE →•SEAL←— DISTANCE —→

HIGH PRESSURE (P_H) LOW PRESSURE (P_L)

Effective
10 % = Porosity = 10 %

D Kilometers	$P_H - P_L$ psi	Q cm³/cm²
1	1,000	15
2	1,000	30
3	1,000	45
1	2,000	30
1	4,000	60
1	6,000	90
10	6,000	900

1 KILOMETER = 3,281 FT.

FIG. 18—Flow required to equalize pressure.

both normally pressured and abnormally high-pressured zones preclude demonstration of any pattern and render untenable any blanket hypothesis to correlate salinity differentials with pressure variance.

7. Flow within subsurface rocks must be inferred from pressure measurements. Lateral pressure differentials between wells of the same depth and elevation may indicate ambiguously either (a) no flow, wherein pressure difference is maintained by a seal, or (b) flow, wherein pressure difference is indicative of a pressure drop from fluid friction in the permeable medium. Thus, pressure differential alone does not establish the existence of flow.

The amount of flow required to cross a seal and equalize pressures on either side of a seal is relatively small. The rate of flow caused by movement of fluid expelled through geologic time as shales lose porosity will be well within the range of hydraulic conductivities of consolidated clays. Flow will equalize pressures and thus retard further flow unless fluid is added to or lost from the system.

8. The geologist concerned with pressure problems has to keep in mind the large number of variables, the low number and ambiguity of pressure measurements, and the possibility of uniqueness in any field situation.

APPENDIX-- DEFINITIONS AND CONCEPTS

To communicate with the least ambiguity, terms are defined herein to clarify commonly used jargon which often differs from original meanings.

Pressures

Abnormal pressure—A pressure above or below the normal hydrostatic pressure for that depth. *Normal* or *hydrostatic pressure* (P_H) is the pressure equal to that of a "column" of water extending to the surface from a given depth. The shape or size of the column does not affect P_H and, furthermore, P_H is the same for a column of water filled with a porous and permeable medium (e.g., sand). The apparent simplicity of this concept is deceiving in that hydrostatic pressure is a function of the density of the water, which is subject to changes in temperature, pressure, and the dissolved-solids content. Thus, determining a "normal" hydrostatic pressure requires consideration of parameters which may not be measured or even measurable.

Vertical lithostatic pressures (P_{LV})—The pressure of the weight of sediment in water per unit area at that depth (Fig. 21); in some cases this pressure is referred to as "frame pressure" or "matrix pressure." *Horizontal lithostatic pressure* (P_{LH}) results from the sediment having some strength but not being completely rigid; P_{LH} will be somewhat less than P_{LV} and can range from zero (for an ideally rigid rock) to the P_{LV} value for a rock that behaves as a liquid. The P_{LH} value is a function of rock properties. Calculations for an uncemented rhombohedral sphere model (Fig. 22) yield a horizontal lithostatic pressure of 0.7 of the vertical lithostatic pressure ($P_{LH} = 0.7\ P_{LV}$).

Geostatic pressure (P_G)—That pressure in the liquid below a seal which is the sum of the pressure of the water and the pressure of the rock above the seal (Fig. 23), i.e., the hydrostatic pressure (P_H) plus the vertical

FIG. 19—Type well density, porosity, and salinity for Gulf Coast type sandstone-shale sequence.

194

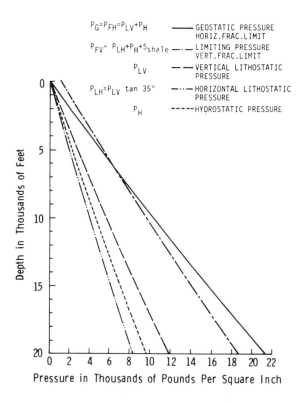

Legend (top left of figure):

$P_G = P_{FH} = P_{LV} + P_H$ ——— GEOSTATIC PRESSURE HORIZ.FRAC.LIMIT

$P_{FV} = P_{LH} + P_H + S_{shale}$ —·— LIMITING PRESSURE VERT.FRAC.LIMIT

P_{LV} — — VERTICAL LITHOSTATIC PRESSURE

$P_{LH} = P_{LV} \tan 35°$ —··— HORIZONTAL LITHOSTATIC PRESSURE

P_H - - - - HYDROSTATIC PRESSURE

FIG. 20—Type well pressure gradients.

lithostatic pressure (P_{Lv}). P_G also may be thought of as the pressure caused by the weight of water plus sediment per unit area.

Seals

The primary requirement for the existence of abnormal formation pressure is a seal; without a seal, the pressures would equalize to hydrostatic. Seals must be envisioned in three dimensions. In a single-phase system (water), a pressure difference cannot be maintained unless the system is completely surrounded by the seal. However, in a multiphase system (water-oil-gas), abnormal pressure can be maintained, because of the buoyant effect of hydrocarbons in water, with the bottom of the system unsealed. Horizontal seals may be laminae of impermeable shale or evaporite which, by virtue of mineralogy and sorting, provide a pressure barrier. Vertical (lateral) seals can be formed by fault displacement, even with throw of less than 1 ft; by gouge or tear-zone materials resulting from faulting; or by lateral facies or diagenetic changes. A seal usually cannot be distinguished visually. The thickness of a seal may be variable, but it can be assumed that a seal is thin with respect to both thickness and lateral extent of an abnormal pressure zone, as pressure changes laterally across faults or vertically across bedding are abrupt (Myer, 1968).

The manner in which a seal is maintained is an enigma. For example, as a sealed volume becomes warmer with deeper burial, the internal fluid pressure can rise sufficiently to fracture the seal. Thus, the high internal

Table 4. Type Well Parameters

Depth thousands of feet	Porosity %	Shale bulk density g/cm³	Salinity TDS °/oo	P_H Hydrostatic pressure psi	Brine gradient psi/ft	P_{LV} Litho-static vertical pressure psi/ft	Litho-static vertical gradient psi/ft	P_G Geostatic pressure psi	Geostatic gradient psi/ft	P_{FV} Fracture pressure (vertical) psi
0	50	1.35	0	0		0		0		1,000
					.443		.539		.982	
2	25	2.03	41	880		989		1,869		2,572
					.452		.551		1.003	
4	22	2.11	84	1,784		2,091		3,875		4,248
					.463		.563		1.026	
6	19	2.19	113	2,710		3,217		5,927		5,962
					.472		.576		1.048	
8	16	2.27	143	3,654		4,369		8,023		7,712
					.480		.590		1.070	
10	13	2.35	176	4,614		5,549		10,163		9,498
					.487		.604		1.091	
12	10	2.43	192	5,588		6,757		12,345		11,318
					.491		.610		1.101	
14	10	2.43	210	6,570		7,977		14,547		13,154
					.494		.611		1.105	
16	9	2.46	210	7,558		9,199		16,757		14,997
					.494		.615		1.109	
18	9	2.46	210	8.546		10,429		18,975		16,846
					.494		.617		1.111	
20	8	2.48	210	9,534		11,663		21,198		18,698
					.494		.632		1.126	

Assumptions: Grain density = 2.7 g/cm³
Gradients @ bottom for 20,000 to 30,000 ft
Fracture pressure $P_{FV} = P_{LH} + P_H + S_{shale}$

26% clay — 2.80 g/cm³
74% silica — 2.65 g/cm³ — 2.7 g/cm³

$P_{LH} = .7\ P_{LV}$

S shale (in tension) = 1000 psi (horizontal) = 0 psi (vertical)

195

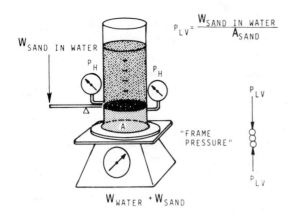

$$P_{LV} = \frac{W_{SAND\ IN\ WATER}}{A_{SAND}}$$

FIG. 21—Vertical lithostatic pressure—P_{LV}.

pressure can "blow down" through fractures to the external pressure. In some areas, particularly some with abnormally low pressures, the seal has been extant during long periods of geologic time and through various regional tectonic episodes. The fact that the seal remains intact also could be an argument for the existence of a healing mechanism which might include precipitation of minerals resulting from release of pressure and elastic closure of a fracture after loss of fluid volume through the seal.

To fracture a seal, the internal pressure would have to exceed the minimum rock stress and the plane of the fracture would be perpendicular to that stress. At shallow depths, the minimum stress is vertical so fracturing

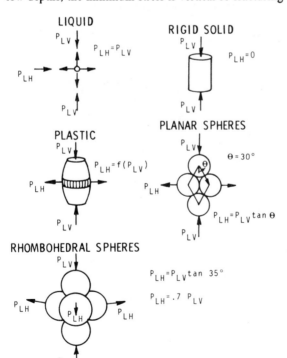

FIG. 22—Horizontal lithostatic pressure—P_{LH}.

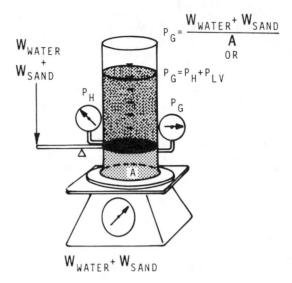

$$P_G = \frac{W_{WATER} + W_{SAND}}{A}$$

OR

$$P_G = P_H + P_{LV}$$

FIG. 23—Geostatic pressure—P_G.

would be horizontal; at great depths, minimum stress is in the horizontal plane and fractures are vertical.

The minimum internal pressure required to fracture a seal is equal to the sum of the hydrostatic pressure (P_H), the horizontal lithostatic pressure (P_{LH}) or minimum rock stress, and the strength of the rock (S_S) in tension. Thus, the mechanism to fracture a seal, i.e., to separate mineral grains, requires that the abnormal internal pressure must (1) balance the hydrostatic pressure (P_H) outside the seal, (2) force the grains apart against the horizontal lithostatic pressure (P_{LH}) and (3) break any cementation between the grains (the strength of the rock, S_S). For example, in Figure 20 the assumptions are (1) the shale strength (S_S) in tension (Clark, 1966) parallel with the bedding is 1,000 psi and perpendicular to the bedding is zero psi; and (2) the horizontal lithostatic pressure (P_{LH}) is 0.7 of the vertical lithostatic pressure (P_{LV}), i.e., $P_{LH} = 0.7\ P_{LV}$ (see Fig. 22). It should be kept in mind that all parameters in the formula for fracturing can vary with such factors as depth, lithology, and salinity. From a similar plot, the pressure at which fracturing of a seal would occur for a given depth and the orientation of the fracture could be predicted for a specific case.

REFERENCES CITED

Ayer, N. J., Jr., 1971, Statistical and petrographic comparison of artificially and naturally compacted carbonate sediments: Univ. Illinois, PhD thesis, 100 p. (available from University Microfilms, Ann Arbor, Michigan).

Barker, C., 1972, Aquathermal pressuring; role of temperature in development of abnormal-pressure zones: AAPG Bull., v. 56, p. 2068-2071.

Billings, G. K., B. Hitchon, and D. R. Shaw, 1969, Geochemistry and origin of formation waters in the western Canada sedimentary basin, 2. Alkali metals: Chem. Geology, v. 4, p. 211-223.

Bruce, C. H., 1973, Pressured shale and related sedi-

196

ment deformation: mechanism for development of regional contemporaneous faults: AAPG Bull., v. 57, p. 878-886.

Burst, J. F., 1969, Diagenesis of Gulf Coast clayey sediments and its possible relation to petroleum migration: AAPG Bull., v. 53, p. 73-93.

Chilingarian, G. V., and H. H. Rieke, III, 1968, Data on consolidation of fine-grained sediments: Jour. Sed. Petrology, v. 38, p. 811-816.

Clark, S. P., ed., 1966, Handbook of physical constants: Geol. Soc. America Mem. 97, 587 p.

Dickey, P. A., A. G. Collins, and I. Fajardo M., 1972, Chemical composition of deep formation waters in southwestern Louisiana: AAPG Bull., v. 56, p. 1530-1533.

Hedberg, H. D., 1974, Relation of methane generation to undercompacted shales, shale diapirs, and mud volcanoes: AAPG Bull., v. 58, p. 661-673.

Hodgman, C. D., ed., 1957, Handbook of chemistry and physics: Cleveland, Ohio, Chemical Rubber Pub. Co., 3213 p.

Levorsen, A. I., and F. A. F. Berry, 1967, Geology of petroleum, 2d ed.: San Francisco, W. H. Freeman, 724 p.

Maxwell, J. C., 1964, Influence of depth, temperature, and geologic age on porosity of quartzose sandstone: AAPG Bull., v. 48, p. 697-709.

Myer, J. D., 1968, Differential pressures: a trapping mechanism in Gulf Coast oil and gas fields: Gulf Coast Assoc. Geol. Socs. Trans., v. 18, p. 56-80.

Olsen, H. W., 1972, Liquid movement through kaolinite under hydraulic, electric, and osmotic gradients: AAPG Bull., v. 56, p. 2022-2028.

Petroleum Research Corporation, 1958, Principles of osmosis applicable to oil hydrology: Research Proj. 12, 66 p.

Schmidt, G. W., 1971, Interstitial water composition and geochemistry of deep Gulf Coast shales and sands: Univ. Tulsa, Master's thesis, 121 p.

Smith, F. G., 1963, Physical geochemistry: Reading, Mass., Addison-Wesley, 624 p.

Winkler, E. M., and P. C. Singer, 1972, Crystallization pressure of sands in stone and concrete: Geol. Soc. America Bull., v. 83, p. 3509-3514.

Young, A., and P. F. Low, 1965, Osmosis in argillaceous rocks: AAPG Bull., v. 49, p. 1004-1007.

The American Association of Petroleum Geologists Bulletin
V. 60, No. 7 (July 1976), P. 1124-1128, 6 Figs.

DISCUSSION

Abnormal Formation Pressure: Discussion[1]

PARKE A. DICKEY[2]
Tulsa, Oklahoma 74104

The paper by Bradley (1975) is comprehensive, thoughtful, carefully prepared, and contains much useful data. Abnormally high formation pressures are of great importance to the oil industry because they constitute an expensive and dangerous hazard in drilling. Many geologic processes, such as overthrusting, diapirism, and hydrothermal vein formation take place in association with and at least partly as a result of abnormally high fluid pressures (Dickey, 1972, p. 13).

Many geologic processes have been postulated to explain abnormal pressures. Bradley dismisses all of them as unimportant, except the increase in temperature with depth of burial and the corresponding volumetric expansion of the pore water.

A series of landmark papers by Shell research people, Hubbert, Dickinson, Hottman, and Johnson, established the compaction of shale resulting from increased depth of burial as the primary cause of abnormally high fluid pressures. Their ideas have been accepted by almost all recent students. The theory of soil science rests on the concept that, if an external load is imposed on a water-saturated porous material, the pressure in the water will be increased, and the material eventually will compact, expelling the water. I feel that the evidence for compaction is so overwhelming, that it cannot be dismissed lightly.

SHALE COMPACTION

The fact that clays compact more than sandstones was recognized in the 1930s. In an effort to work out the structural effects of differential compaction, Athy (1930) measured the density of many shale samples from northeast Oklahoma and plotted it against depth of overburden. He found that the density of shale increases, and porosity decreases, with depth of overburden in a logarithmic relation. He recognized that considerable overburden had been removed. Later Hedberg (1936) determined the density of a set of samples of much younger shales from Venezuela, where little, if any, overburden had been removed. Dickinson (1953) published similar data for the Gulf Coast. Since then much attention has been given to shale density as a means of detecting abnormal pressures, and an abundance of data is available (Griffin and Bazer, 1969).

In the face of this huge amount of data Bradley showed the relation between porosity and depth in a "type well" (his Figs. 14, 19, and Table 4) 0 to 2,000 ft, 25 percent loss of porosity; 2,000 to 12,000 ft, 3 percent loss of porosity per 1,000 ft; and 12,000 to 20,000 ft, 0.25 percent per 1,000 ft. The curves show well-defined doglegs at 2,000 and 12,000 ft.

The data of Hedberg, Dickinson, and others do contain much scatter, which they averaged into a logarithmic relation (Fig. 1). Bradley's type well showed three distinct straight-line parts with different slopes, which contradicts previously published data, and needs some points to substantiate it. It suggests that there are three different mechanisms for the loss of porosity in shales. Bradley argued from his type well that loss of porosity in shale (at depths greater than 2,000 ft) is not due to compaction, but rather to diagenetic recrystallization.

In Figure 1 porosity and density of shales are plotted against depth. Bradley's type well is shown, as well as some of the other data in the literature.

Innumerable cross sections of channel and bar-type sand bodies show draping of overlying beds, which results from the fact that shales compact much more than sands. Sand grains bridge against each other and lock into position, so that imposing a heavy load involves only little movement and some crushing of the grains. Reduction of pore volume in sandstone results mainly from the precipitation of secondary silica and other minerals out of water solution.

Freshly deposited muds, on the other hand, have a very high porosity and their component particles must have a sort of house-of-cards structure. Electron-scanning photomicrographs of Pa-

[1]Discussion received, November 14, 1975; accepted, January 21, 1976.

[2]University of Tulsa.

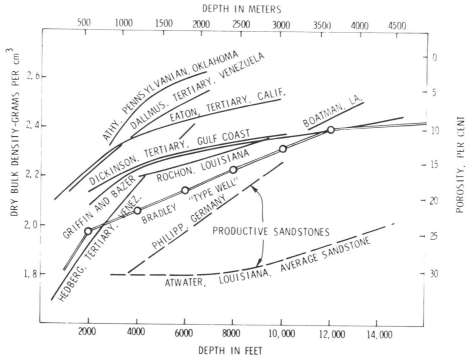

FIG. 1—Density versus depth relation for shales, as reported by various investigators. Porosity is calculated from density assuming a grain density of 2.65 (from Dickey, 1972).

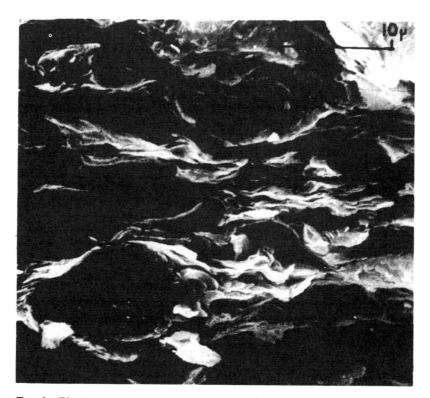

FIG. 2—Electron-scanning photomicrograph of Paleozoic shale from Oklahoma. Texture resembles that of corn flakes with raisins. Bar in upper right corner is 10 μ long (from Dickey, 1975).

leozoic shales show a texture resembling corn flakes. Bradley denied that shales will compact under "lithostatic" pressures of 12,000 psi at 20,000 ft because the ultimate strength of sediment particles in compression is 20,000 to 100,000 psi (Clark, 1966). The photomicrographs (Fig. 2) suggest that no very great pressure will cause bending of the flakes and vertical compaction of the shale.

Effect of Compaction on Fluid Pressures

The usual situation in deep wells is that the fluid pressures are normal hydrostatic down to the sealing bed, and abnormally high below it. In Bradley's Figure 13 he showed a column of sand, filled with water, which contained a horizontal sealing membrane. He wrote that if a load is placed on the column, the sand below the seal will compress elastically according to its bulk modulus, which is about the same as that of water $(3 \times 10^{-6} \text{ vols/vol/psi})$. The extra load will be assumed by the rock and the water proportionally to the relative volume of pores and rock. If the porosity is 10 percent, the pore pressure below the seal will increase but it actually will be lower than the pore pressure above the seal. This cannot be true if there is even a slight amount of crushing or readjustment of the rock particles. This illustration seems to be farfetched and the process unlikely in nature.

Hubbert and Rubey showed (1959, eq. 48, p. 134) that the total overburden pressure at the base of the seal S is supported by the stress in the rock particles σ plus the pressure in the pore fluid p (Fig. 3).

$$S = p + \sigma.$$

In those cases where the fluid pressure p nearly equals the overburden pressure S the stress σ is nearly zero, so the overlying rocks are practically floating. Hubbert and Rubey showed that it is not true that the load S is divided between the rock and the solid according to their void ratio, as previously had been thought (p. 135). They showed that abnormally pressured zones in the Gulf Coast are accompanied by shales with higher porosity than normal for their depth of burial. Using Athy's curve as the situation after the abnormal pressure has bled off, they presented a formula and set of curves relating pore fluid pressure to porosity and depth of burial (Rubey and Hubbert, 1959, Fig. 4, p. 178).

Hottman and Johnson (1965) pointed out that several well-log responses vary with porosity, notably resistivity and sonic velocity. The resistivity of shale shows a regular increase with depth in the Gulf Coast. Where the resistivity of shale becomes abnormally low, the well gives every indi-

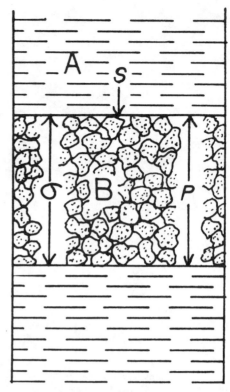

Fig. 3—Diagram to illustrate vertical stress in sediments. Weight of overburden S is sustained by stress in skeleton of solid grains σ and pressure in interstitial fluids p. If pressure in fluid approaches weight of overburden S, stress in solid σ approaches zero and bed A, in this case, practically would be floating. If pressure p exceeds S, rupture of bed A will occur.

cation of high fluid pressures in the adjacent sands. The amount of deviation of shale resistivity from normal will give remarkably accurate estimates of the actual fluid pressures in the sands. Measuring porosity of shale is now an almost universal method of detecting high-pressure zones all over the world. It seems to be successful in Tertiary basins such as those of Nigeria and southeast Asia. It also is used in Mesozoic rocks such as those of the McKenzie delta, the Northwest Shelf of Australia, and the North Sea. In spite of this worldwide application, Bradley wrote that "a complete range of high-normal-low densities is associated with abnormal pressures." However, no examples were given, and the statement needs substantiation in view of the almost universally accepted association of low shale densities and high pressures.

Conclusion

The evidence is impressive that compaction of shale by the weight of overburden is the principal cause of abnormally high fluid-pore pressure. Other phenomena may contribute, but they are

200

based on speculation, and the actual mechanisms have not been worked out fully.

REFERENCES CITED

Athy, L. F., 1930, Density, porosity, and compaction of sedimentary rocks: AAPG Bull., v. 14, p. 1-24.

Bradley, J. S., 1975, Abnormal formation pressure: AAPG Bull., v. 59, p. 957-973.

Clark, S. P., ed., 1966, Handbook of physical constants: Geol. Soc. America Mem. 97, 587 p.

Dickey, P. A., 1972, Migration of interstitial water in sediments and the concentration of petroleum and useful minerals: 24th Internat. Geol. Cong., Montreal, sec. 5, p. 3-16.

——— 1975, Possible primary migration of oil from source rock in oil phase: AAPG Bull., v. 59, p. 337-345.

Dickinson, G., 1953, Geological aspects of abnormal reservoir pressures in Gulf Coast Louisiana: AAPG Bull., v. 37, p. 410-432.

Griffin, D. G., and D. A. Bazer, 1969, A comparison of methods for calculating pore pressures and fracture gradients from shale density measurements using the computer: Jour. Petroleum Technology, v. 21, p. 1463-1474.

Hedberg, H. D., 1936, Gravitational compaction of clays and shales: Am. Jour. Sci., 5th ser., v. 31, p. 241-287.

Hottman, C. E., and R. K. Johnson, 1965, Estimation of formation pressures from log-derived properties: Jour. Petroleum Technology, v. 17, p. 717-722.

Hubbert, M. K., and W. W. Rubey, 1959, Mechanics of fluid-filled porous solids and its application to overthrust faulting, pt. 1 of Role of fluid pressure in mechanics of overthrust faulting: Geol. Soc. America Bull., v. 70, p. 115-166.

Rubey, W. W., and M. K. Hubbert, 1959, Overthrust belt in geosynclinal area of western Wyoming, pt. 2 of Role of fluid pressure in mechanics of overthrust faulting: Geol. Soc. America Bull., v. 70, p. 167-206.

Abnormal Formation Pressure: Reply[1]

JOHN S. BRADLEY[2]

Tulsa, Oklahoma 74102

I want to thank Parke Dickey for his thoughtful discussion of my paper (Dickey, 1976). He discusses my conclusions (Bradley, 1975) concerning: (1) the role of compaction as a cause of abnormally high pressure, (2) the mode of porosity loss of shale with depth, and (3) the relation of shale density to abnormally high pressure.

Muds undoubtedly compact and expel water as Dickey describes until they become grain bound ("lock up") at some porosity between 27 and 30 percent. With further loss of porosity caused by cementation, the sediment becomes a rock. The compressibility of the rock is at least one order of magnitude less than that of the sediment. This rock compressibility at low stresses is predominantly elastic. To squeeze any but minor amounts of water associated with the elastic compressibility out of the rock requires crushing of grains and breaking of cement; i.e., major matrix damage. Stresses required for rock crushing are not available from the lithostatic pressure of overburden at normal drilling depths (Bradley, 1975, Table 4).

Regarding my Figure 13, Dickey states that crushing and readjustment of the rock particles would allow the pore fluid below the piston to take some of the load. I contend that the loads associated with abnormally high pressures are insufficient to crush the rocks. In order to increase the fluid pressure below the piston if the rock is not crushed, either the matrix volume of the rock must be reduced or the volume of the fluid must be increased. Beneath the seal, overburden compresses the rock and fluid at approximately the same rate. Increased temperature, on the other hand, increases the volume of both. However, the fluid volume increases much more rapidly than the rock (pore) volume. If the fluid volume is held constant, the fluid pressure increases.

Loss of porosity in shale with depth occurs linearly with two distinct changes in slope (Bradley, 1975, Fig. 14). The presence of three distinct linear segments suggests that three processes may be involved. Multiwell density-depth data do appear to satisfy a logarithmic curve, but individual well-density plots often show straight-line segments (Fig. 1). Whether the intersections of the segments are curves or knees is debatable, and with the precision of the data, probably indeterminate. The porosity data of Hedberg (1936) also plot as three straight-line segments (Fig. 2) and he described three processes to account for the three segments.

The initial loss of porosity of the sediment (from as high as 80 percent porosity down to ap-

[1]Manuscript received, January 21, 1976; accepted, February 9, 1976. Published by permission of Amoco Production Company.

[2]Amoco Production Company Research Center. Special thanks are due to D. E. Powley, who supplied the density and porosity data, and to R. A. Nelson, who edited this reply.

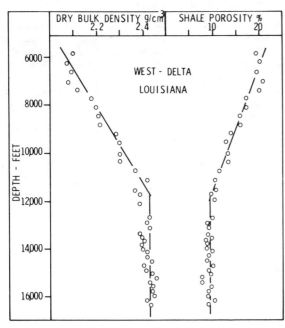

FIG. 1—Shale density and porosity for offshore Louisiana well.

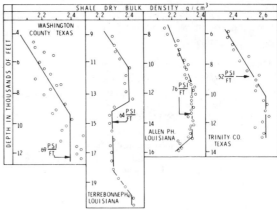

FIG. 3—Abnormally high pressure-depth ratios in high, low, and normal density shale sections.

proximately 30 percent) is an ordering, with water expulsion, of grains with grain movement and packing. This is the compaction or consolidation defined by the soil scientist. At about 30 percent porosity the sediment "locks up," allowing further loss of porosity by recrystallization and cementation. At some porosity near 27 percent we are dealing with a rock (shale), not a sediment (mud). The recrystallization and cementation processes proceed until the shale porosity reaches about 10 percent. From that point to a total depth of 16,000 to 18,000 ft, it is unusual to find more than a few percent further loss of porosity. Without evidence, I would speculate that the second leg represents the loss of effective intergranular porosity and that the third leg represents the mi-

nor loss of porosity caused by clay-interlayer collapse. At some depth (temperature) the shale porosity must approach zero.

I believe that the "house-of-cards" structure of clays is limited to the high-porosity mud phase because there is not enough porosity in the rock phase to allow for such a texture. In the rock it is doubtful that scanning-electron-microscope images represent the true texture of shales at depth, as the clays in the samples are exposed to both heat and vacuum before viewing. It might be noted that the average Gulf Coast shale is not a claystone, but is predominantly quartz.

Abnormally high pressures in sandstones associated with low density shales are widespread but not universal in the Gulf Coast (Fig. 3). Abnormally high pressures also are present in quartzites in low-porosity/high-density quartzitic shales in a well on the Grand Banks. Thus, the conclusion is that abnormal pressures may be associated with a complete range of densities.

To summarize, overburden stress cannot be the sole cause of abnormally high pressures at present drilling depths. Furthermore, such stresses cannot explain the major loss of porosity actually measured in rocks at increasing depth. The dry-bulk densities of shales do not appear to be related directly to fluid pressures. A complete range of densities is associated with abnormal pressures.

I thank Dickey for his perceptive questions and for this opportunity to enlarge upon my arguments and to restate my conclusion.

REFERENCES CITED

Bradley, J. S., 1975, Abnormal formation pressure: AAPG Bull., v. 59, p. 957-973.

Dickey, P. A., 1976, Abnormal formation pressure: discussion: AAPG Bull., v. 60, p. 1124-1127.

Hedberg, H. D., 1936, Gravitational compaction of clays and shales: Am. Jour. Sci., 5th ser., v. 31, p. 241-287.

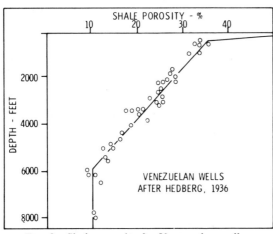

FIG. 2—Shale porosity for Venezuelan wells.

The American Association of Petroleum Geologists Bulletin
V. 61. No. 12 (December 1977). P. 2134-2142, 8 Figs.

Oil and Gas in Reservoirs with Subnormal Pressures[1]

PARKE A. DICKEY[2] and WAYNE C. COX[3]

Abstract Oil and gas are common in stratigraphic traps in structural basins, both deep down near the bottom, and also along the flanks. In many of these traps initial reservoir pressures were subnormal, indicating a lack of permeable connection to the outcrop. Some maps drawn on the potentiometric (piezometric) surface using pressure data from drill-stem tests show clearly the location of the stratigraphic barriers which have trapped the oil; these maps should be used in prospecting. Many giant gas fields with abnormally low initial reservoir pressures are low on the flanks of structural basins. The geologic factors favoring this type of accumulation are not understood.

The cause of the low pressures may be related to removal of overburden, which has resulted in a dilation of the pore volume in the rocks, and a decrease in reservoir temperature.

SUBNORMAL PRESSURE FIELDS

Since the first bottom-hole pressure measuring device was run in Oklahoma in the early 1930s, it has been known that many, if not most, oil and gas reservoirs in the Mid-Continent area had subnormal pressure (Millikan and Sidwell, 1931). That is, the original bottom-hole pressure was less than that necessary to sustain a column of water to the surface. Fresh water exerts a pressure of 0.433 lb per square inch per foot (0.1 kg per square centimeter per meter). Many other low-pressure reservoirs have been reported since. Unfortunately, despite the abundance of low reservoir pressures, and their obvious significance as indicating permeability barriers, few potentiometric-surface maps have been published. The maps that have been published have been contoured as if there were continuous permeability and regional water flow. The very existence of the low pressures, however, can be interpreted better as lack of permeability. A complete seal is required for an aquifer to continue with a subnormal pressure for a long time.

Most low pressure sandstones are stratigraphic traps. The sandstone bodies are limited in extent, and edge water shows little tendency to advance. In some fields there appears to be very little downdip water—the porous and permeable bodies are practically full of oil or gas. The water that is present is salty and typical of connate water in its chemical composition, indicating that it has been removed from the hydrologic cycle for long periods of geologic time.

DECREASING RESERVOIR POTENTIALS (PORE PRESSURE GRADIENTS) DOWNDIP

The most common situation in sedimentary basins is shown in Figure 1. The sandstone body nearest the edge of the basin crops out. Fresh water is present down to a depth determined by the pattern of meteoric-water circulation. At greater depths it becomes salty and changes chemical composition, losing the bicarbonate typical of meteoric waters and taking on the chloride of connate water. Wherever there is continuity of permeability and little or no water movement, the potentiometric surface will be level or nearly level throughout the entire aquifer (i.e., the water-bearing part of the reservoir rock). Oil seldom is found in this part of the sandstone body, because it can migrate easily to the surface and be lost. A closed anticline in this situation could trap oil, but many basins do not have structural closures on their more stable flanks.

Downdip, a sandstone body may be cut by a fault or may pinch out. Farther downdip, another sandstone body may appear at the same, or nearly the same, horizon. This second sandstone body is very apt to contain oil at its updip edge where it is banked against the permeability barrier. The fluid potential in this aquifer commonly is substantially lower than that in the part of the sandstone which crops out.

Still farther downdip the potentials may be even lower. Here it is not unusual to find shoestring sandstones which have very low potentials; they commonly contain almost no water.

VIKING SANDSTONES OF ALBERTA

Figure 2 shows a part of Alberta between T50, R19, W4, and T65, R14, W5. Pressures from drill-stem tests in the Viking have been converted to potentiometric values and plotted by Lynes United Services Ltd. Potentiometric value is the

[1]Manuscript received, December 28, 1976; accepted, May 2, 1977.

[2]University of Tulsa, Tulsa, Oklahoma 74110.

[3]Lynes United Services, Ltd., Calgary, Alberta T2P 2W4, Canada.

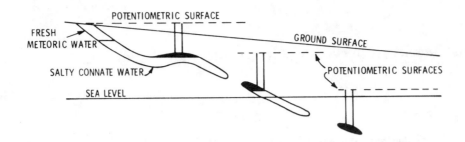

FIG. 1—Schematic cross section showing pattern of fluid potentials common in sedimentary basins. Reservoir pressures are normal just downdip from outcrop; stratigraphic traps deep in basin have subnormal pressures.

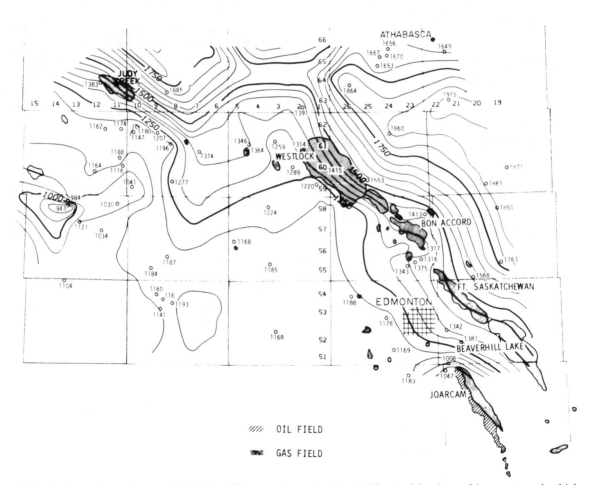

FIG. 2—Potentiometric map of fluids in Viking sandstone in part of Alberta. Map is machine contoured, which assumes continuity of permeability; values are in feet above sea level. Ground surface ranges from 2,000 to 2,500 ft (610 to 762 m) above sea level.

204

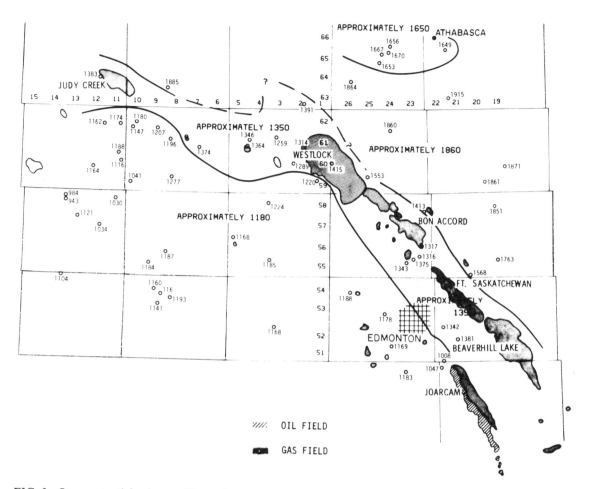

FIG. 3—Same potential values as Figure 2 assuming that there are permeability barriers which separate regions of different fluid potentials.

height in feet above sea level that a column of formation water of a given density would stand, if sustained by the measured reservoir pressure. The density values used are those indicated by actual chemical analyses of the water. The values were contoured by a computer; it was assumed that there is a continuum, that is, hydraulic continuity throughout the whole Viking Formation, although the data seem to indicate lack of continuity.

In the southwest part of the map, where the ground surface elevation is about 2,500 ft (760 m), the potential values are nearly all between 1,150 and 1,180 ft (350 and 360 m) above sea level. These values indicate that there is continuity of permeability throughout this area in the Viking. Farther northeast, between Edmonton and Judy Creek, the contours crowd together between the values of 1,200 and 1,800 ft (365 and 550 m) indicating a low-permeability barrier. Practically all the oil and gas fields in the Viking in the West-

lock area are in this zone of closely spaced contours. In the east part of the map, the potentiometric values are all near 1,850 ft (560 m), although there is a small cluster of values near 1,650 ft (500 m) in the northeast corner around Athabasca. The trend of the barrier shown by the potentiometric map corresponds well to the trend of sandstone bars in the Viking.

There is a decrease in potentiometric surface from northeast to southwest. However, it is far from smooth. Pressure measurements from good drill-stem tests should have a maximum error of 25 psi (1.75 kg/sq cm)—less for recent tests. This amounts to an error of about 50 ft (15 m) in the calculated potential. If we assume that the potential values may be in error by as much as 50 ft (15 m), a line can be drawn which separates a region of about 1,850 ft (560 m) from one of about 1,350 ft (400 m)—a 500-ft (169 m) sudden drop (Fig. 3). This pressure difference may be interpreted as indicating a complete pinchout (or stratigraphic

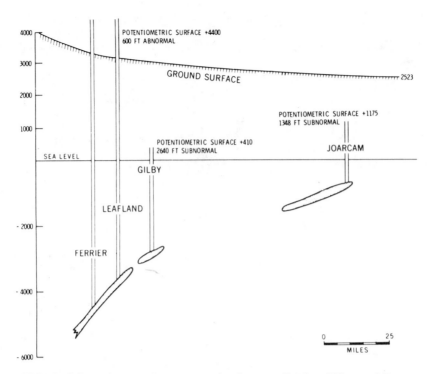

FIG. 4—Schematic west-east cross section between Ferrier, Gilby, and Joarcam oil fields. Bars show height at which column of fresh water would stand if sustained by original reservoir pressure. Ferrier is abnormally high and Gilby is abnormally low, although they produce from same zone.

offset) of the sandstone, thereby separating two different hydraulic systems. The oil and gas clearly are banked against this barrier on the downdip side. Another line can be drawn which separates the 1,150-ft (350 m) value of the southwest part of the map from the 1,350-ft (400 m) values of the oil and gas belt. The lines on Figure 3 have been drawn without reference to the local correlations, lithology, or depositional environment of the Viking sandstones. They could be drawn with much more assurance on the basis of this information.

Still farther downdip is the Gilby field. The initial pressure in the Viking "A" sandstone was 1,445 psi (102 kg/sq cm) which is a potentiometric surface of only 411 ft (125 m) above sea level. The Gilby field is part of the Joffre-Bentley-Gilby trend, a narrow shoestring sandstone in the Viking which had practically no downdip water. About 25 mi (40 km) farther downdip are the Ferrier and Leafland pools. Their initial pressure was 4,170 psi (293 kg/sq cm), which is a potentiometric surface of 4,400 ft (1,340 m)—that is, there is almost 4,000 ft (1,200 m) of hydraulic head gradient over the few kilometers between Ferrier and Gilby (Fig. 4). In the Willesden Green field, the Viking is also high potential. This indicates that there is extremely low permeability to water,

probably zero, parallel with the bedding, in the Viking sandstones between these oil fields.

A map of the potentiometric surface in the Viking was published by the Petroleum Research Corporation in 1961 (Hill et al, 1961; Fig. 5). This map (Fig. 5) shows the extremely low value at Gilby, and the very high value at Ferrier. The arrows indicate that water is flowing from all over central Alberta toward Gilby field, but Gilby contained practically no water, and it seems most unlikely that the water could run out of the system through the shales above or below, across the bedding.

SAN JUAN BASIN, NEW MEXICO AND COLORADO

A very similar situation has been described in the Cretaceous Gallup Sandstone of New Mexico by Berry (1959) and later by McNeal (1961). The potentiometric surface is shown in Figure 6. The sandstone crops out around the south and west margins of the basin, where it is continuous and permeable, but it pinches out downdip toward the center of the basin (Sabins, 1963). Northeast of the pinchout there are two long, narrow, beach-type sandstone bodies full of oil which form the Bisti and Gallegos oil fields.

206

The southwest permeable part of the Gallup Sandstone contains fresh water. The potentiometric contours indicate that the water enters at the high outcrops and flows in a general peripheral direction, emerging at the low outcrops, as shown in Figure 6. (Where there is continuity of permeability, fluids flow perpendicular to the potentiometric contours.) No oil is found in this part of the basin.

In the Bisti field, however, the original reservoir pressure was 570 psi (40 kg/sq cm) too low for the depth, and at Gallegos 720 psi (50 kg/sq cm) too low. To show this discontinuity the potentiometric contours on Figure 6 are crowded together, which indicates a permeability barrier (i.e., there is no connection between the Bisti reservoir and the Gallup aquifer). The water at Bisti and Gallegos is salty which confirms the lack of hydraulic connection with the freshwater area on the southwest.

In the same general area, at a shallower depth, is the great Mesaverde Blanco gas field. The basin is an asymmetric syncline. The gas occurs deep in the basin on the northeast-dipping southwest flank (Fig. 7). Hollenshead and Pritchard (1961) ascribed the trapping of the gas to updip pinchouts of successive benches of the Mesaverde sandstones. During deposition the sea lay on the northeast and the land on the southwest, and the successive benches were laid down as the sea retreated to the northeast. The sandstone crops out all around the rim of the basin, and it is hard to see why the gas did not escape, at least along the strike to the northwest and southeast.

Figure 8 is a map of the potentiometric surface of the pore fluids in the Point Lookout Sandstone

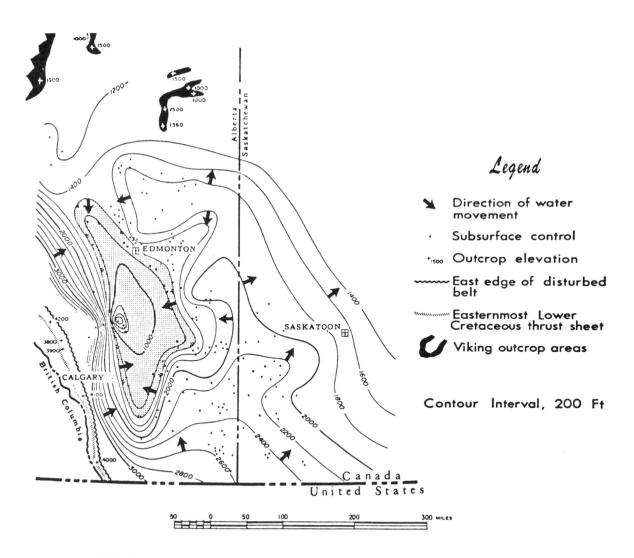

FIG. 5—Viking sandstone potentiometric surface as contoured by Hill et al (1961).

207

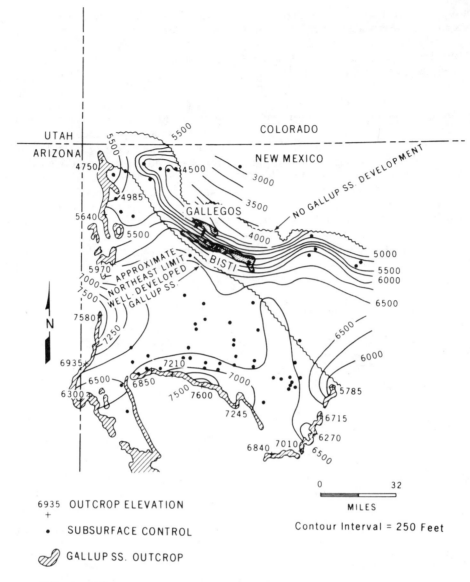

GALLEGOS

BISTI

APPROXIMATE
NORTHEAST LIMIT
WELL DEVELOPED
GALLUP SS.

NO GALLUP SS. DEVELOPMENT

N

6935 OUTCROP ELEVATION
  +
  •   SUBSURFACE CONTROL

      GALLUP SS. OUTCROP

0   32

MILES

Contour Interval = 250 Feet

FIG. 6—Gallup Sandstone potentiometric surface. In southwest part of San Juan basin water in Gallup Sandstone is fresh and flow is mostly peripheral. Potentials are normal. Northeast of pinchout line, reservoir pressures are subnormal and water is salty. Oil is found in stratigraphic traps downdip from permeability barrier. From Berry (1959).

of the Mesaverde Group. Both the Point Lookout and Gallup Sandstones have widespread continuity of permeability in the southwest half of the basin. The potentiometric contours show a curious ring-shaped permeability barrier which surrounds the Blanco gas field. Within the gas field the potentiometric surface is less than 4,000 ft (1,200 m) above sea level whereas outside the field it is normal, that is, about the elevation of the outcrop (about 7,000 ft or 2,100 m above sea level).

The origin of the ring-shaped barrier is a mystery. It has been suggested that the meteoric water which entered at the outcrop caused the montmorillonite clays in the sandstone to swell, thereby destroying its permeability. It was suggested by Berry (1959) and Hill et al (1961) that the gas was retained by a downdip flow of water in the sandstone, parallel with the bedding. This explanation requires the water to penetrate the gas-bearing sandstone and flow through it, and then to run out the bottom across the bedding

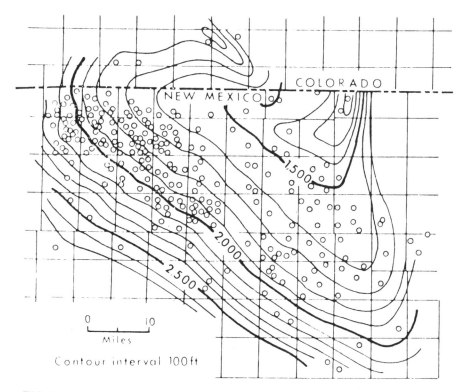

FIG. 7—Structure contours on "green marker horizon" in Blanco gas field. Gas is on southwest flank of syncline. From Hollenshead and Pritchard (1961).

like the water in a washbasin. The Mancos Shale below is several hundred feet thick, so this explanation is not very plausible.

OTHER LOW-PRESSURE FIELDS

The Morrow gas fields of the western Anadarko basin in Oklahoma are low pressured, with a pore-fluid pressure gradient of 0.3 psi per foot of depth (0.07 kg/sq cm/m). These sandstones are notably lenticular (Busch, 1974) and, although they string like beads along the strike, each reservoir covers only 1 or 2 townships (36 or 72 sq mi, 100 or 200 sq km). Each reservoir had a different initial pressure, and shows its own pressure decline with time (Masroua, 1973). This indicates a complete lack of permeable connection between the different reservoirs.

Russell (1972) shows many reservoirs with notably low pressure. Most, if not all of these, are in synclines in lenticular sandstones with no structural control of the oil or gas accumulation.

Silver (1973) recognized the abundance of isolated porous bodies in the deeper parts of many basins. They are mainly oil- or gas-bearing and commonly devoid or almost devoid of water. He called these "isolani," and pointed out that some very large gas fields occur in structural lows. Silver stated that many are "pressure deficient." Silver

called on various compaction and capillary phenomena to account for the presence of oil and gas, the absence of water, and the abnormally low or high pressure, but none of his explanations is entirely satisfactory for the observed conditions. His paper pointed out the great importance of this type of trap, and the need to understand the processes of migration to develop methods of prospecting for them.

It is an interesting question why several great gas fields, like San Juan, Wattenberg, Hugoton, Morrow, and those in the Arkoma basin and the Appalachians, occur on the flank of a synclinal basin, with no structural closure, and with subnormal pressure. We need to understand the geologic situation which gave rise to this type of accumulation to develop better methods of prospecting for them.

ORIGIN OF LOW PRESSURES

There are no obvious reasons why fluids in the bottom of structural basins should have subnormal pressures. Several suggestions have been offered.

Berry (1959) and Hill et al (1961) ascribed the subnormal pressures of the Alberta and San Juan basins to osmotic-pressure differences. Osmotic pressures across semipermeable membranes, in-

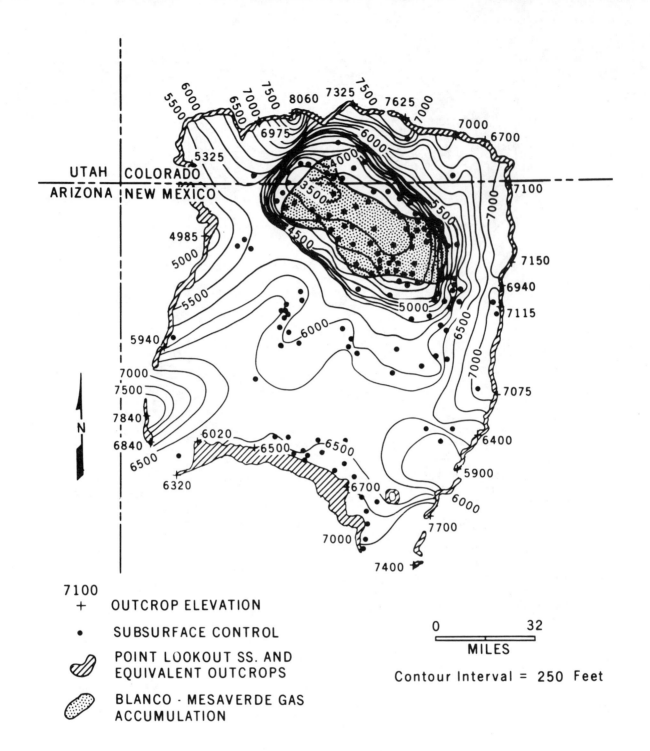

7100
+   OUTCROP ELEVATION

•   SUBSURFACE CONTROL

POINT LOOKOUT SS. AND
EQUIVALENT OUTCROPS

BLANCO - MESAVERDE GAS
ACCUMULATION

0   32
MILES

Contour Interval = 250 Feet

FIG. 8—Point Lookout Sandstone potentiometric surface. Potentials downdip from outcrop are normal. Flow is peripheral, inward at high outcrops and outward at low ones. Great gas field is near bottom of basin and has very low potential. From Berry (1959).

210

cluding shales, can be high—1,000 psi (several hundred kg/sq cm) or more—if there is a large difference in water salinity across the membrane. However, no evidence for adequate salinity differences in the pore water in these areas has ever been published.

Berry (1959) and Hill et al (1961) also assumed continuity of permeability and cross-formation flow of water. However, many low-pressure fields are almost devoid of active edge water and operate under dissolved-gas drive.

Silver (1973) ascribed the low pressures to capillary-pressure differences, but this does not seem to fit the geologic pattern. Theoretically, in water-wet porous media, the pressures should be higher in the gas than in the water.

Russell (1972) pointed out that many low-pressure reservoirs are in well-consolidated sediments which have been uplifted in the geologic past. Fatt (1958) and McLatchie et al (1958) have determined that a sandstone reservoir contracts elastically about 7×10^{-6} pore volumes per pore volumes per psi (5×10^{-5} pv per pv per kg/sq cm) as the internal pressure of the fluid is removed during oil production. This is an elastic compression, and should not be confused with compaction, which is irreversible. It may be supposed that the removal of overburden will cause an elastic dilation of the pores in the sandstones at about the same rate. Shales appear to contract and dilate even more than sandstones. The modulus of compressibility of water is about 3×10^{-6} volumes per volume per psi (2×10^{-5} vol per vol per kg/sq cm). Therefore, as overburden is removed, the pore volume dilates but the interstitial water expands only about half as much as necessary to fill the new pore volume thus created. Consequently, the pressure will drop.

As the pressure in the pore water in the shales drops, water (but not oil or gas) will seep out of the sandstone into the adjacent shales. This will result in low pressure in the oil and gas, and only a small amount of water may remain in the aquifer.

As overburden is removed the subsurface becomes cooler. Water contracts as it cools, and this effect also would cause a drop in pressure in the pore water (Barker, 1972).

The trouble with these theories is that they require that the shales have some permeability to water. However, if they had any permeability, water from the surface would seep down and raise the pressure in the reservoir to normal. The evidence in this paper, notably the contrast between Leafland and Gilby fields, suggests that some shales have no permeability, even parallel with the bedding.

The low pressure sandstones of the Appalachian and Anadarko basins are in areas where considerable overburden has been removed. However, there is no reason to suppose that much overburden has been removed from the San Juan or Alberta basins.

REFERENCES CITED

Barker, C., 1972, Aquathermal pressuring—role of temperature in development of abnormal pressure zones: AAPG Bull., v. 56, p. 2068-2071.

Berry, F. A. F., 1959, Hydrodynamics and chemistry of the Jurassic and Cretaceous Systems in the San Juan basin, northwestern New Mexico and southwestern Colorado: PhD thesis, Stanford Univ.

Busch, D. A., 1974, Stratigraphic traps in sandstones—exploration techniques: AAPG Mem. 21, 174 p.

Fatt, I., 1958, Compressibility of sandstones at low to moderate pressures: AAPG Bull., v. 42, p. 1924-1957.

Hill, G. A., W. A. Colburn, and J. W. Knight, 1961, Reducing oil-finding costs by use of hydrodynamic evaluations, in Petroleum exploration, gambling game or business venture—Inst. Econ. Petroleum Explor., Devel., and Property Evaluation, Internat. Oil and Gas Educational Center: Englewood, N.J., Prentice-Hall, Inc., p. 38-69.

Hollenshead, C. T., and R. L. Pritchard, 1961, Geometry of producing Mesaverde sandstones, San Juan basin, in J. A. Peterson and J. C. Osmond, eds., Geometry of sandstone bodies: AAPG, p. 98-118.

Masroua, L. F., 1973, Patterns of pressure in the Morrow sands of central Oklahoma: Master's thesis, Univ. Tulsa, 78 p.

McLatchie, A. S., R. A. Hemstock, and J. W. Young, 1958, The effective compressibility of reservoir rock and its effect on permeability: AIME Trans., v. 213, p. 386-388.

McNeal, R. P., 1961, Hydrodynamic entrapment of oil and gas in Bisti field, San Juan County, New Mexico: AAPG Bull., v. 45, p. 315-329.

Millikan, C. V., and C. V. Sidwell, 1931, Bottom-hole pressures in oil wells: AIME Trans., v. 92, p. 194-205.

Russell, W. L., 1972, Pressure-depth relations in Appalachian region: AAPG Bull., v. 56, p. 528-536.

Sabins, F. F., Jr., 1963, Anatomy of stratigraphic trap, Bisti field, New Mexico: AAPG Bull., v. 47, p. 193-228.

Silver, C., 1973, Entrapment of petroleum in isolated porous bodies: AAPG Bull., v. 57, p. 726-740.

Reprinted by permission of the American Institute of Mining,
Metallurgical, and Petroleum Engineers. Published in the
Journal of Petroleum Technology, V. 17, No. 6 (1965), pp.
717-722.

Estimation of Formation Pressures from Log-Derived Shale Properties

C. E. HOTTMANN

R. K. JOHNSON
JUNIOR MEMBER AIME

SHELL DEVELOPMENT CO.
HOUSTON, TEX.

SHELL OIL CO.
NEW ORLEANS, LA.

ABSTRACT

Fluid pressure within the pore space of shales can be determined by using data obtained from both acoustic and resistivity logs. The method involves establishing relationships between the common logarithm of shale transit time or shale resistivity and depth for hydrostatic-pressure formations. On a plot of transit time vs depth, a linear relationship is generally observed, whereas on a plot of resistivity vs depth, a nonlinear trend exists. Divergence of observed transit time or resistivity values from those obtained from established normal compaction trends under hydrostatic pressure conditions is a measure of the pore fluid pressure in the shale and, thus, in adjacent isolated permeable formations. This relationship has been empirically established with actual pressure measurements in adjacent permeable formations. The use of these data and this method permits the interpretation of fluid pressure from acoustic and resistivity measurements with an accuracy of approximately 0.04 psi/ft, or about 400 psi at 10,000 ft. The standard deviation for the resistivity method is 0.022 psi/ft, and for the acoustic method 0.020 psi/ft.

Knowledge of the first occurrence of overpressures, and of the precise pressure-depth relationship in a geologic province, enables improvements in drilling techniques, casing programs, completion methods and reservoir evaluations.

INTRODUCTION

GENERAL STATEMENT

Operators engaged in the search for and production of hydrocarbon reserves in Tertiary basins are more and more frequently confronted with complications associated with overpressured (abnormally high fluid pressure) formations. This is particularly true in the Texas-Louisiana Gulf Coast area. The problems associated with these formations are of direct concern to the combined activities of all phases of operations, i.e., geophysical, drilling, geological and petroleum engineering.[1-3] Knowledge of the pressure distribution of a given area of operations would greatly reduce the magnitude of many of these complexities and in some cases would completely eliminate specific problems.

This paper presents techniques developed for estimating formation pressures from interpretations of acoustic and electric log data. Specifically, the acoustical and electrical properties of shales, reflected by conventional acoustic and electrical surveys, can be used to infer certain *reservoir* properties, such as formation pressure, at any level in a well. It has been possible to develop these techniques because of a firm understanding of the basic principles that govern and apply to such overpressured provinces.

NORMAL PRESSURES

Normal pressures refer to formation pressures which are approximately equal to the hydrostatic head of a column of water of equal depth. If the formations were opened to the atmosphere, a column of water from the ground surface to the subsurface formation depth would balance the formation pressure. On the Gulf Coast, the shallow, predominantly sand formations contain fluids which are under hydrostatic pressure. These formations are said to be normally pressured or to have a normal pressure gradient.* Experience has shown that the normal pressure gradient on the Gulf Coast is approximately 0.465 psi/ft of depth.

OVERPRESSURES

Formations with pressures higher than hydrostatic are encountered at varying depths in many areas. These formations are referred to as being abnormally pressured, abnormally high pressured, or overpressured. Formation pressures up to twice the hydrostatic pressure have been observed. These formations require extreme care and much expense to drill and to exploit.

COMPACTION-FLUID PRESSURE RELATIONS

THEORY

The generation of overpressured formations in Tertiary sections of the Gulf Coast and several other Tertiary sedimentary basins is, in general terms, considered to be primarily the result of compaction phenomena.[1] This portion of the paper presents a brief review of the theory which associates compaction and fluid pressure relations, and should thus provide the necessary background for an understanding of the techniques presented. See Hubbert and Rubey[4] for a more comprehensive treatment of this subject.

The theory of the consolidation of a water-saturated clay[5] has been well established by workers in soil mechanics. The concept is explained by a model with perforated metal plates separated by metal springs and water and enclosed in a cylindrical tube. Fig. 1 is a schematic representation of such a model (see Ref. 5, Fig. 27, page 74). The springs simulate communication between clay particles, and the plates simulate the clay particles. Mano-

Original manuscript received in Society of Petroleum Engineers office Jan. 8, 1965. Revised manuscript received April 26, 1965. Paper (SPE 1110) to be presented at 40th Annual SPE Fall Meeting in Denver, Colo., Oct. 3-6, 1965.

[1]References given at end of paper.

*Pressure gradient is defined as p/D, where p is the reservoir pressure at depth D.

meters are used to record the fluid pressure. Upon application of pressure to the uppermost plate, the height of the springs between the plates remains unchanged as long as no water escapes from the system. Thus, in the initial stage the applied pressure is supported entirely by the equal and opposite pressure of the water.

A useful manner of recording this pressure is in terms of the ratio of the fluid pressure p to the total pressure S, which is defined as λ and is symbolically represented by Eq. 1:

$$\lambda = \frac{p}{S} . \quad . \quad . \quad . \quad . \quad . \quad . \quad . \quad . \quad . \quad (1)$$

At conditions for Stage A in Fig. 1, λ has a value of 1; the system is overpressured. As water is allowed to escape from the system, the plates move downward slightly (the system compacts), and the springs carry part of the applied load. As more and more water is allowed to escape from the system, the springs carry a greater share of the load, and λ has a value less than 1. Finally, sufficient water escapes from the system for the springs to attain their compaction equilibrium. At this stage—terminal compaction equilibrium—the applied load is supported jointly by the springs and the water pressure, which is simply hydrostatic. The value of λ is approximately 0.465.

This model is analogous to a clay undergoing essentially uniaxial compaction in response to an axial component of total stress S (overburden pressure), where

$$S = \bar{\rho}_{bw}gD . \quad . \quad . \quad . \quad . \quad . \quad . \quad . \quad . \quad (2)$$

Here, $\bar{\rho}_{bw}$ is the mean value of the water-saturated bulk density of the overlying sediments, g is the acceleration of gravity and D is the depth of burial. Hubbert and Rubey[4] (Eqs. 48, 53 and 75) have demonstrated that the load S is supported jointly by the fluid pressure p and the grain-to-grain bearing strength σ of the clay particles, where

$$\sigma = S - p; \quad . \quad . \quad . \quad . \quad . \quad . \quad . \quad . \quad . \quad (3)$$

σ then is analogous to the support afforded by the springs in the Terzaghi-Peck model.[5]

As stated by Hubbert and Rubey,[4] "The effective stress σ exerted by the porous clay (or by the springs in the model) depends solely upon the degree of compaction of the clay, with σ increasing continuously as compaction increases. A useful measure of the degree of compaction of a clay is its porosity ϕ, defined as the ratio of the pore volume to the total volume. Hence, we may infer that for a given clay there exists for each value of porosity ϕ some maximum value of effective compressive stress σ which the clay can support without further compaction."

From Eqs. 2 and 3 and from the foregoing, we can state that the porosity ϕ at a given burial depth D is dependent upon the fluid pressure p. If the fluid pressure is abnormally high (greater than hydrostatic), the porosity will be abnormally high for a given burial depth.

Considering the conditions as they exist in the geologic column, and applying the previous model concept, factors which can influence the overpressuring of a section are the ratio of shale thickness to sand thickness, the mean formation permeability, the elapsed time since deposition, the rate of deposition and the amount of overburden. Dickinson[1] reported on the first recognized association of the occurrence of overpressures and the relative proportion of sand and shale in the geologic column when he stated, "Abnormal pressures occur commonly in isolated porous reservoir beds in thick shale sections developed below the main sand series". The interrelation of all these parameters controls the compaction of the sediments. Overpressuring can result if compaction is restricted.

In the Tertiary sediments of the Gulf Coast, shale intervals of great thickness are frequently encountered. Many of these intervals are deep-water marine shales containing isolated sands. These sediments have essentially been subjected only to uniaxial compaction, the compressive stress of the overburden. Eqs. 2 and 3 should therefore apply to such a region.

For a shale to compact, fluids must be removed. Sands, which are highly permeable media, act as avenues of fluid escape. These sands may be thought of as pipelines. The near-absence of sands in thick shales reduces the rate of fluid removal from these shales in comparison with thinner shales sandwiched between sands. Fractures and nonsealing faults can also act as avenues of fluid escape, but sands are believed to be the more important avenues. In such shale intervals, the permeability is quite small

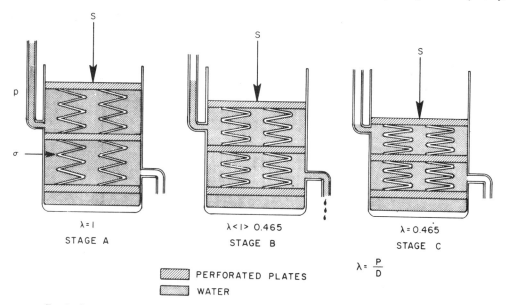

FIG. 1—SCHEMATIC REPRESENTATION OF SHALE COMPACTION (AFTER TERZAGHI AND PECK, REF. 5).

and fluid removal is restricted; thus, the shale fluid pressure will be large for a given burial depth D.

As previously stated, a useful measure of the degree of compaction of a clay is its porosity. Obviously, then, an estimation of clay or shale porosity as a function of depth will reveal the degree of compaction. A section which is "undercompacted" with regard to a given burial depth will be a section whose fluid pressure is abnormally large (in excess of hydrostatic pressure) or overpressured for the burial depth. The data recorded by various logs can be utilized to infer the degree of compaction. Thus, through the interrelated parameters discussed, a practical method of estimating formation pressures is achieved.

METHODS OF ESTIMATING PRESSURE

ACOUSTIC METHOD

The development of the acoustic log several years ago provided a new means for estimating the porosity of sedimentary rocks. The longitudinal acoustic velocity recorded by the various types of acoustic logs may be thought of as primarily a function of porosity and lithology. If a given lithology such as shale is investigated, the acoustic log response will be essentially a response to porosity variations. The change of porosity with depth can be studied in this manner to gain an insight into shale compaction. An investigation of the response of an acoustic log in normally pressured shales indicates a relation between the column logarithm of *shale* travel time $\Delta t_{(sh)}$ and depth. An example of the type of relation for Miocene and Oligocene sediments is presented in Fig. 2.

Fig. 2 illustrates that the travel time decreases (velocity increases) with increasing burial depth. This indicates that porosity decreases as a function of depth. This trend represents the "normal compaction trend" as a function of burial depth, and the fluid pressures exhibited within this normal trend will be hydrostatic.

If intervals of abnormal compaction are penetrated, the resulting data points will diverge from the "normal compaction trend". If overpressured formations are encountered, the data points will diverge from the normal trend toward abnormally high transit times for a given burial depth, since the porosity is higher. Fig. 3 illustrates such data. A lesser degree of compaction is also borne out by bulk density measurements upon shale cores, as illustrated in this figure.

The amount of divergence of a given point from the established "normal compaction trend" has been related to the observed pressure in adjacent *reservoir* formations. Fig. 4 presents a schematic plot of $\Delta t_{(sh)}$ vs depth and the parameters used to determine pressures. The relation between the $\Delta t_{(sh)}$ parameter and pressure for Miocene and Oligocene formations is presented in Fig. 5. Pertinent information used to establish this empirical relation is presented in Table 1. The standard deviation from the line representing the data of Fig. 5 is 0.020 psi/ft.

To estimate the formation pressure of reservoirs from adjacent shale acoustic log data, the following steps are necessary.

1. The "normal compaction trend" for the area of interest is established by plotting the logarithm of $\Delta t_{(sh)}$ vs depth (see Figs. 2 and 3).

2. A similar plot is made for the well in question.

3. The top of the overpressured formations is found by noting the depth at which the plotted points diverge from the trend line.

4. The pressure of a reservoir at any depth is found as follows:

> (a.) The divergence of adjacent shales from the extrapolated normal line is measured (See Fig. 4).
> (b.) From Fig. 5 the fluid pressure gradient (FPG) corresponding to the $\Delta t_{ob(sh)} - \Delta t_{n(sh)}$ value is found.
> (c.) The FPG value is multiplied by the depth to obtain the reservoir pressure.

A pressure gradient profile can be constructed for a well by using the above procedure.

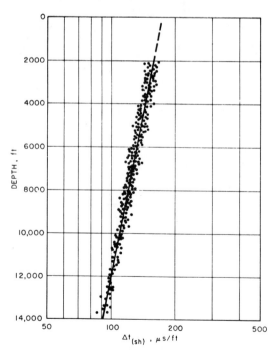

Fig. 2—Shale Travel Time vs Burial Depth for Miocene and Oligocene Shales, Upper Texas and Southern Louisiana Gulf Coast.

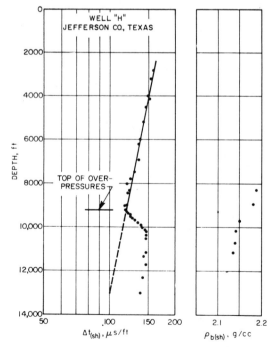

Fig. 3—Shale Travel Time and Bulk Density vs Burial Depth.

Parish or County and State	Well	Depth (ft)	Pressure (psi)	FPG* (psi/ft)	$\Delta t_{ob(sh)} - \Delta t_{n(sh)}$ (microsec/ft)
Terrebonne, La.	1	13,387	11,647	0.87	22
Offshore Lafourche, La.	2	11,000	6,820	0.62	9
Assumption, La.	3	10,820	8,872	0.82	21
Offshore Vermilion, La.	4	11,900	9,996	0.84	27
Offshore Terrebonne, La.	5	13,118	11,281	0.86	27
East Baton Rouge, La.	6	10,980	8,015	0.73	13
St. Martin, La.	7	11,500	6,210	0.54	4
Offshore St. Mary, La.	8	13,350	11,481	0.86	30
Calcasieu, La.	9	11,800	6,608	0.56	7
Offshore St. Mary, La.	10	13,010	10,928	0.84	23
Offshore St. Mary, La.	11	13,825	12,719	0.92	33
Offshore Plaquemines, La.	12	8,874	5,324	0.60	5
Cameron, La.	13	11,115	9,781	0.88	32
Cameron, La.	14	11,435	11,292	0.90	38
Jefferson, Tex.	15	10,890	9,910	0.91	39
Terrebonne, La.	16	11,050	8,951	0.81	21
Offshore Galveston, Tex.	17	11,750	11,398	0.97	56
Chambers, Tex.	18	12,080	9,422	0.78	18

*Formation fluid pressure gradient.

RESISTIVITY METHOD

Logging specialists on the Gulf Coast have for several years observed and recognized that shale resistivity decreases in overpressured zones. This phenomenon has been used in various areas to detect the presence of so-called "sheath" material near salt domes; it has also been considered a qualitative indication of high formation pressure gradients. The next logical step is to determine how shale resistivity can be used to estimate actual formation pressures.

Little is known of the effects of the many factors which influence shale resistivity. It is reasonable that many of the same parameters which influence the resistivity of reservoir rocks will also affect shale resistivities. Among these, the more important are (1) porosity, (2) temperature, (3) salinity of the contained fluid and (4) mineral composition. Rather than attempt to isolate the effect of each factor on shale resistivity, we have investigated the resultant combination of all of these factors. However, we should consider the individual effects (summarized in Table 2) so that we can recognize the problem if we encounter an anomalous situation.

As in the acoustic method, a trend of shale resis-

tivity vs depth for hydrostatic shales is established for a given area. Typical trends of data from hydrostatic pressure sections are illustrated in Fig. 6. These data points were obtained from standard electrical resistivity logs; the amplified short normal device was used because of its readability and because of negligible borehole corrections in the range of resistivities considered. These trends, in a given area, reflect the "normal compaction trend" as a function of depth. If overpressured formations are encountered, the shale resistivity data points diverge from the normal trend toward lower resistivity values, owing to exceptionally high porosity. An example resistivity-depth plot is presented in Fig. 8(a). The degree of divergence of a given point from the established "normal compaction trend" has been related to the observed pressure gradient in adjacent reservoir formations. The pertinent information used to establish this empirical relation is presented in Table 3 and is plotted in Fig. 7. The maximum deviation of the data from the smooth curve in Fig. 7 is approximately 0.08 psi/ft, and the standard deviation is 0.022 psi/ft. Fig. 7 illustrates that an increase in the ratio of extrapolated normally pressured shale resistivity to actual recorded shale resistivity signifies an increase in formation pressure gradients. The trend il-

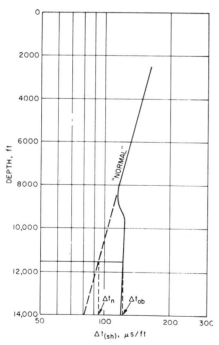

FIG. 4—SCHEMATIC PLOT OF SHALE TRAVEL TIME VS BURIAL DEPTH.

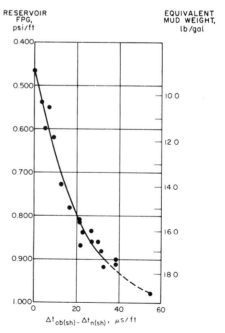

FIG. 5 — RELATION BETWEEN SHALE ACOUSTIC PARAMETER $\Delta t_{ob(sh)} - \Delta t_{n(sh)}$ AND RESERVOIR FLUID PRESSURE GRADIENT (FPG).

215

TABLE 2—EFFECT OF VARIOUS FACTORS UPON RESISTIVITY

Resistivity Increases With	Resistivity Decreases With
Lower Porosity	Higher Porosity
Lower Salinity	Higher Temperature
Lower Temperature	Higher Salinity
High Nonclay Mineral Content	Lower-than-Average Nonclay Fraction

lustrated in Fig. 7 should be considered an example plot and should be used only as a guide until actual pressure and log data are obtained for the particular region under study.

To estimate the formation pressure of reservoirs from adjacent shale resistivity data, the following steps are necessary.

1. The normal "compaction trend" for the area of interest is established by plotting the logarithm of shale resistivity from the amplified short normal device vs depth. (Usually, numerous wells in the area are examined.)

2. A similar plot is made for the well in question.

3. The top of the overpressured formations is found by noting the depth at which the plotted points diverge from the trend line.

4. The pressure gradient of a reservoir at any depth is found as follows:

(a.) The ratio of the extrapolated normal shale resistivity to the observed shale resistivity is determined.

(b.) The fluid pressure gradient (FPG) corresponding to the calculated ratio is found (from Fig. 7).

5. The reservoir pressure is obtained by multiplying the FPG value by the depth.

A pressure gradient profile for the well can be constructed by repeating the above procedure at numerous depths.

LIMITATIONS

The conditions of the borehole and the surrounding disturbed formation will have an influence upon the recordings of both the acoustic and resistivity logs. Generally, these effects can be overcome by employing normal borehole correction procedures. If there is a large temperature disturbance caused by drilling, it may prove necessary to use one of the longer spaced resistivity devices to determine shale resistivity. The caliper survey should be used to determine zones of extreme borehole enlargement which can lead to erroneous shale transit times owing to weak signal and cycle skipping.

In general, the presence of fresh- or brackish-water

TABLE 3—PRESSURE AND SHALE RESISTIVITY RATIOS, OVERPRESSURED MIOCENE-OLIGOCENE WELLS

Parish or County and State	Well	Depth	Pressure (psi)	FPG* (psi/ft)	Shale Resistivity Ratio** (ohm-m)
St. Martin, La.	A	12,400	10,240	0.83	2.60
Cameron, La.	B	10,070	7,500	0.74	1.70
Cameron, La.	B	10,150	8,000	0.79	1.95
	C	13,100	11,600	0.89	4.20
	D	9,370	5,000	0.53	1.15
Offshore	E	12,300	6,350	0.52	1.15
St. Mary, La.	F	12,500	6,440	0.52	1.30
		14,000	11,500	0.82	2.40
Jefferson Davis, La.	G	10,948	7,970	0.73	1.78
	H	10,800	7,600	0.70	1.92
	H	10,750	7,600	0.71	1.77
Cameron, La.	I	12,900	11,000	0.85	3.30
Iberia, La.	J	13,844	7,200	0.52	1.10
		15,353	12,100	0.79	2.30
Lafayette, La.	K	12,600	9,000	0.71	1.60
		12,900	9,000	0.70	1.70
	L	11,750	8,700	0.74	1.60
	M	14,550	10,800	0.74	1.85
Cameron, La.	N	11,070	9,400	0.85	3.90
Terrebonne, La.	O	11,900	8,100	0.68	1.70
		13,600	10,900	0.80	2.35
Jefferson, Tex.	P	10,000	8,750	0.88	3.20
St. Martin, La.	Q	10,800	7,680	0.71	1.60
Cameron, La.	R	12,700	11,150	0.88	2.80
		13,500	11,600	0.86	2.50
		13,950	12,500	0.90	2.75

*Formation fluid pressure gradient.

**Ratio of resistivity of normally pressured shale to observed resistivity of overpressured shale: $R_{n(sh)}/R_{ob(sh)}$.

zones at considerable depths may lead to anomalously high resistivity values and will make it extremely difficult, if not impossible, to use the resistivity method of pressure estimation. The acoustic log data can frequently be used in such an area. Variations in shale clay mineralogy and nonclay constituents impose difficulties upon either technique. Prudent choice of data points can greatly reduce this problem. Care should always be taken to select zones of low SP deflection and uniform resistivity or sonic readings.

The use of the acoustic and resistivity techniques has been most successful in Tertiary Age sediments, particularly those of the Miocene and Oligocene. Success has been achieved in Quaternary and Cretaceous Age sediments with difficulty. In general, for sediments of any age the correlations of acoustic transit time vs depth are more easily established than the trends of shale resistivity vs depth. This is undoubtedly true because fewer parameters influence the acoustic properties as compared with the number that influence the resistivity of shales.

These techniques are limited to areas in which the gen-

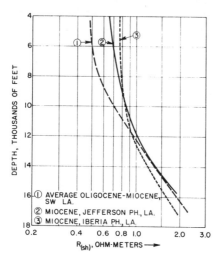

FIG. 6—SHALE RESISTIVITY VS BURIAL DEPTH.

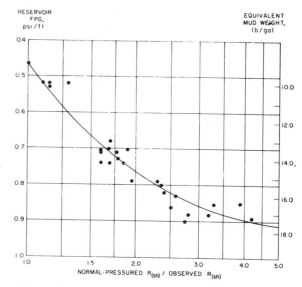

FIG. 7—RELATION BETWEEN SHALE RESISTIVITY PARAMETER $R_{n(sh)}/R_{ob(sh)}$ AND RESERVOIR FLUID PRESSURE GRADIENT (FPG).

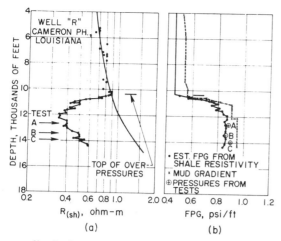

FIG. 8—EXAMPLE OF ESTIMATING PRESSURES FROM RESISTIVITY LOG.

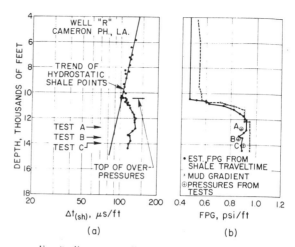

FIG. 9—EXAMPLE OF ESTIMATING PRESSURES FROM ACOUSTIC LOG.

eration of overpressures is primarily the result of compaction processes in response to the stress of overburden.

APPLICATIONS

Knowledge of the reservoir pressure is of considerable engineering value. Such knowledge will provide the means for improving drilling techniques and designing better casing programs and completion methods. From data gathered from surrounding wells, it is possible to predict the probable pressure profile that will be encountered by a drilling well. In addition, this pressure information will enable the reservoir engineer to make more accurate reserve estimates and performance predictions.

EXAMPLE

To illustrate the techniques of estimating formation pressures from shale properties, we have selected a well drilled in Cameron Parish, La. This well penetrated several thousand feet of overpressured sediments, and there are three actual bottom-hole pressure measurements to verify the accuracy of the methods.

In Fig. 8(a), shale resistivity is plotted against depth on semilog grid. The average normal resistivity trend, fitted to the data as discussed in the preceding sections, is shown. The top of overpressures occurs at approximately 10,400 ft, as can be determined by the departure of the observed shale resistivity points from the normal trend. The ratio of the observed resistivity to the "normal trend resistivity" at the same depth is determined at numerous levels. From Fig. 7, the fluid pressure gradient at each level is determined; these data are plotted in Fig. 8(b). For comparison, the mud column pressure gradient used while the well was being drilled is shown. Also, three bottom-hole pressure readings were obtained from tests at 12,700, 13,500 and 13,900 ft, respectively. The calculated pressure gradients are within 0.04 psi/ft of the measured gradients.

Observed shale travel times are plotted against depth in Fig. 9(a). A straight line is drawn through the shallow shale compaction trend. The deviation of observed points from the normal compaction trend occurs at approximately 10,400 ft and designates the top of overpressured formation. The departure of the observed shale transit times from the trend line is measured at numerous levels, and the corresponding pressure gradient is obtained from Fig. 5. The resultant trend of pressure gradient with depth is plotted in Fig. 9(b). As above, the mud column hydrostatic gradient and the measured pressure gradients from tests of the formations are shown

for comparison. Excellent agreement is observed between the estimated pressure gradient and the measured pressures.

CONCLUSIONS

The degree of compaction of a shale in response to an axial component of total stress S depends upon several variables. The fluid pressure can be related to the degree of compaction and burial depth. The degree of compaction can be ascertained from techniques which utilize various logging data; thus, fluid pressures of isolated reservoir rock can be estimated from adjacent shale compaction data.

Shale acoustic and resistivity log data have been extensively and successfully used to estimate reservoir formation pressures. The acoustic log and resistivity methods have an accuracy of fluid pressure predictions of approximately ± 0.04 psi/ft. The standard deviations for the resistivity and acoustic methods are 0.022 and 0.020 psi/ft, respectively.

The examples presented in this paper are from the Gulf Coast, but the principles of the techniques presented here will apply equally well to other Tertiary sedimentary basins in which the principal stress has been the result of overburden weight.

Knowledge of the first occurrence of overpressures, and indeed of the precise pressure-depth relationship in a geologic province, enables improvements in drilling techniques, casing programs, completion methods and reservoir evaluations. From data gathered from surrounding wells, we can predict the probable pressure profile that will be encountered by a drilling well.

REFERENCES

1. Dickinson, G.: "Geological Aspects of Abnormal Reservoir Pressures in the Gulf Coast Region of Louisiana, U.S.A.", *Proc.*, Third World Petroleum Cong., The Hague (1951) 1.

2. Thomeer, J. H. M. A. and Bottema, J. A.: "Increasing Occurrences of Abnormally High Reservoir Pressures in Boreholes and Drilling Problems Resulting Therefrom", *Bull.*, AAPG (1961) **45**, No. 10, 1721, 1730.

3. Mullins, John D.: "Some Problems of Superhigh-Pressure Gas Reservoirs in the Gulf Coast Area", *Jour. Pet. Tech.* (Sept., 1962) 935.

4. Hubbert, M. King and Rubey, W. W.: "Role of Fluid Pressure in Mechanics of Overthrust Faulting. Part I", *Bull., GSA* (Feb., 1959) 70.

5. Terzaghi, Karl and Peck, R. B.: *Soil Mechanics in Engineering Practice*, John Wiley & Sons, Inc., N. Y. (1948) 566. ★★★

Reprinted by permission. First published in the *Journal of Petroleum Technology*, V. 33, No. 10 (October 1981), pp. 1828-1834.

Thistle Field Development

F.T. Nadir,* SPE, BNOC (Development) Ltd.

Summary

A mathematical reservoir model of the Thistle field was used to explain the rapid pressure decline in some early development wells. This paper describes (1) how evidence was gathered to confirm the hypothesis that the Thistle reservoir is divided into three fault blocks and (2) the changes made to the development plan to arrest and reverse the field's pressure and productivity decline.

Introduction

The Thistle field is located about 125 miles (201 km) northeast of the Shetland Islands (Fig. 1) in the very prolific northern part of the Viking graben. The field, discovered in July 1973 with the completion of Well 211/18-2, is mainly in U.K. Offshore Block 211/18 but extends into Block 211/19. The participants in Block 211/18 are the British Natl. Oil Corp. (BNOC), Deminex, Santa Fe, Tricentrol, Burmah Oil, Charterhouse, and Ultramar. Block 211/19 participants are Conoco, Gulf, and BNOC. Blocks 211/18 and 211/19 were awarded, respectively, in the third and fourth rounds of U.K. offshore licenses. BNOC (Development) is operator for the unitized field.

Nadir and Hay[1] presented a comprehensive history of Thistle field in which they discussed the initial development plan and its objectives and compared the reservoir thickness and quality found in the first seven development wells with what was predicted. This paper updates the earlier work by emphasizing reservoir engineering studies conducted to pinpoint the reason for the rapid pressure decline and, hence, production decline. It describes modifications to the development plan that allowed water injection wells to be located correctly to repressurize the reservoir and reverse the trend of declining production.

*Now with LNS Petroleum Consultants Ltd.

Geology

The Thistle field is an easterly dipping Middle Jurassic Brent sandstone reservoir (Fig. 2) with an average net pay thickness of 380 ft (115.8 m). The formation is subdivided into four reservoir sands: D, C, B, and A. Stratigraphy and the reservoir quality were described in detail by Nadir and Hay[1] and updated by Hallett.[2] There have been few changes to the original geological model; the main change is some evidence of erosion between the Sands D and C, less Sand D erosion, and the possibility of syndepositional fault movement. The quality of the sand is slightly better than originally expected.

Reservoir Modeling

Production from Thistle field started in Feb. 1978, and the eighth development well was completed in Aug. 1978 (Fig. 3). The wellhead pressures of the producing wells in the crestal area of the field were declining faster than anticipated, which could have been a result of localized pressure sinks and/or a small or nonexistent aquifer. The completion of Well 08A showed Sand-B pressure to have been depleted by 1,200 psig (8274 kPa), which dispelled any localized pressure sink concepts. The search for a solution started.

A black-oil mathematical reservoir model was used to match field performance. The matching parameter was pressure. Wellhead flowing pressures, measured with dead-weight gauges, were converted to bottomhole flowing pressures by Orkiszewski's method[3] and to equivalent grid-block pressures by Peaceman's method.[4] Bottomhole pressure buildup surveys were used as spot checks to confirm the reliability of the resultant grid-block pressure values. Repeat formation tester (RFT) pressures identified the different pressure regimes in the various sands at a specific point in time at some distance from producing wells.

The initial pressure matching runs concentrated on Sands B and A, using a 1,080 grid block model (12 ×

218

30 × 3); each areal grid block had dimensions of 820 × 820 ft (250 × 250 m). Later pressure matching runs on all the reservoir sands used an 1,800 grid block model (12 × 30 × 5).

Reservoir performance depends on the magnitude of parameters such as horizontal and vertical permeability, fault locations and degree of communication across them, aquifer permeability and size, rock compressibility, and rock quality. By varying these parameters, we obtained two virtually identical pressure matches of the early pressure decline using two different reservoir descriptions; the first had restricted vertical permeability and the second had sealing faults. Fig. 4, showing the performance of Well 04A, is an illustration of the quality of these matches. The faulting model seemed the more plausible because it required no other changes, whereas the restricted vertical permeability

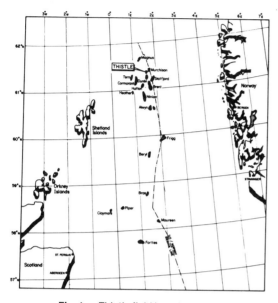

Fig. 1 — Thistle field location map.

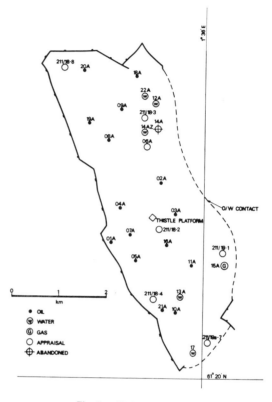

Fig. 3 — Well location map.

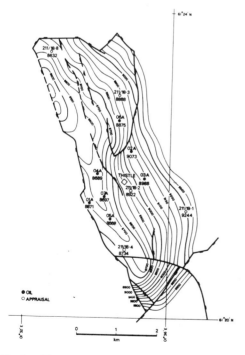

Fig. 2 — Structure contour map on top Brent reservoir.

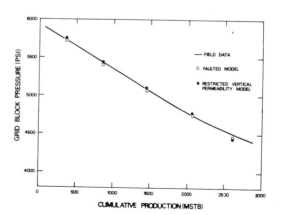

Fig. 4 — Pressure match, Well 04A.

219

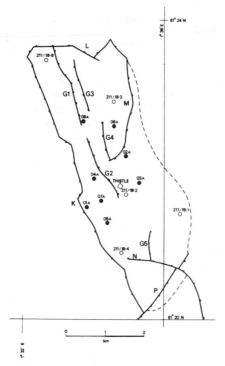

Fig. 5 – Original fault pattern.

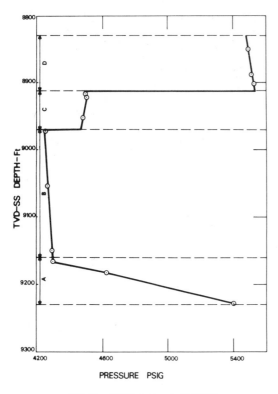

Fig. 6 – RFT pressures, Well 09A.

model required reducing aquifer size, rock porosity, and rock compressibility. We had a second look at the available seismic data to find possible fault locations.

Original Fault Pattern

The major faults (marked K and L on Fig. 5) to the west and north always were assumed sealing. We assumed two more sealing faults: Fault M, with a throw of 200 to 400 ft (61 to 122 m), located in the northeastern part of the field and Fault N to the south, with a throw of about 100 ft (30.5 m). The remainder of the faults were small, about 50 ft (15.2 m). Nevertheless, minor Faults G1, G2, G3, and G4 in the northern part of the field formed a graben and, being an area of disturbance, may have formed a barrier to flow. At that time, Well 09A was being drilled to the west of this graben area and we decided to alter the target location of this well by about 1,650 ft (500 m) eastward to test whether this graben was a possible seal. Well 10A was drilling south of Fault N and would test the assumption that Fault N was sealing.

Well 09A

Well 09A was the first well in the field that encountered significantly less Sand C than anticipated. The initial interpretation was that the well had crossed the central Fault G3. An alternate explanation is that Sand C is depositionally thin as a result of localized erosion. The RFT (Fig. 6) showed that Sand D had depleted by 500 psig (3448 kPa) and Sand C depleted by 1,500 psig (10 342 kPa) from the original reservoir pressure of 6,060 psig (41 782 kPa) at 9,200 ft (2804.2 m) subsea. Sand B was pressure depleted to approximately 4,300 psig (29 647 kPa), which was very similar to Sand B crestal producers. The RFT pressures showed the following.

1. The graben was not forming a continuous barrier to flow.

2. Sand D pressure depletion was probably due to communication to sands below. The amount of vertical permeability was very small since there was a pressure difference of nearly 1,000 psig (6895 kPa) between Sands D and C.

3. There was no significant vertical barrier to flow within Sand B. Parts of Sand A were still at high pressure, showing the poor vertical communication between Sands B and A.

4. Sand C consists of more than one pressure region, showing that the shale stringers within Sand C are continuous and form a partial barrier to flow.

Well 10A

Well 10A was completed at about the same time as Well 09A and showed nearly original reservoir pressure in all sands. The RFT pressures are shown in Fig. 7. Nevertheless, there was some depletion in Sand B of about 120 psi (827 kPa) and in Sand C of some 70 psi (483 kPa). The data showed (1) Sand D was nearly at original reservoir pressure, (2) Sand C showed a reverse gradient indicating downward flow, and (3) Sands B and A were in pressure equilibrium

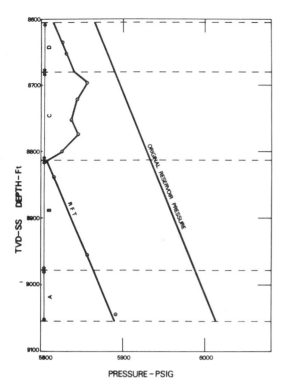

Fig. 7 − RFT pressures, Well 10A.

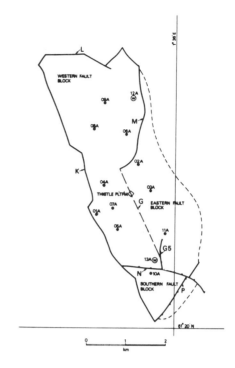

Fig. 8 − Fault pattern, second version.

and were only slightly below original reservoir pressure.

Fault Pattern, Second Version

The invaluable evidence from Wells 09A and 10A produced a second fault pattern (Fig. 8). Faults K, L, and N were proved sealing, and Fault M was assumed to be sealing. There was still a problem as to the location of the central fault. The best pressure match resulted from assuming a sealing fault connecting the graben area to Fault G5 (Fig. 8). Fault blocks were designated as eastern, western, and southern. By then it became obvious that 1978 and 1979 Thistle field production rate would be significantly below expectation and that water injection was required quickly. The rapid pressure decline was complicated by drilling and completion difficulties, which meant that low-deviation water injection wells were needed to minimize drilling times. Furthermore, injection wells should not be too far updip as this may jeopardize ultimate recovery.

Much discussion centered on the location of Wells 11A and 12A − the first because it was too far updip and the second because it had too high a deviation. Finally, locations were agreed on. Well 11A location was chosen so that it could be useful as a Sand B injection well if it ended up in the western fault block or a Sand C producer in the eastern fault block. The oil/water contact was intended to be in Sand B. Well 12A was chosen as an injector in the northern part of the field that was far enough downdip to be the

correct location for an injector whether Fault M was sealing or not.

Well 11A

Unfortunately, Well 11A had a reduced section; only 20 ft (6.1 m) of Sand D and no Sand C. Sand B thickness was normal. The well had crossed a fault downthrown 170 ft (51.8 m) to the west. The RFT pressure (Fig. 9) proved that Sand B was high pressure and that a central fault existed. Further observations include the following.

1. Sand B pressure of 5,500 psig (37 921 kPa) proved the existence of a central barrier to flow between Well 11A and Wells 01A, 04A, and 05A, whose pressures were about 3,800 psig (26 200 kPa).

2. Sand B pressure was 500 psig (3448 kPa) lower than original. Probably, the central fault is not fully sealing, allowing some small pressure communication to the western fault block.

3. Sand D pressure was similar to Well 03A, a Sand D producer, indicating that the fault intersected by the well may not be the central sealing fault.

4. The presence of a fault meant that Sand B was much higher in the well than expected, and the well was completed as a Sand B producer.

Well 12A

Well 12A was also a surprise. It had a full Brent sand section, but the quality of Sand B was much poorer than that of the other wells in the field. The pressure, nevertheless, showed the well to be in the western

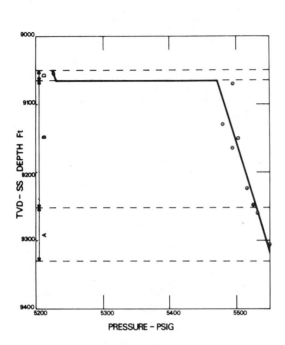

Fig. 9 – RFT pressures, Well 11A.

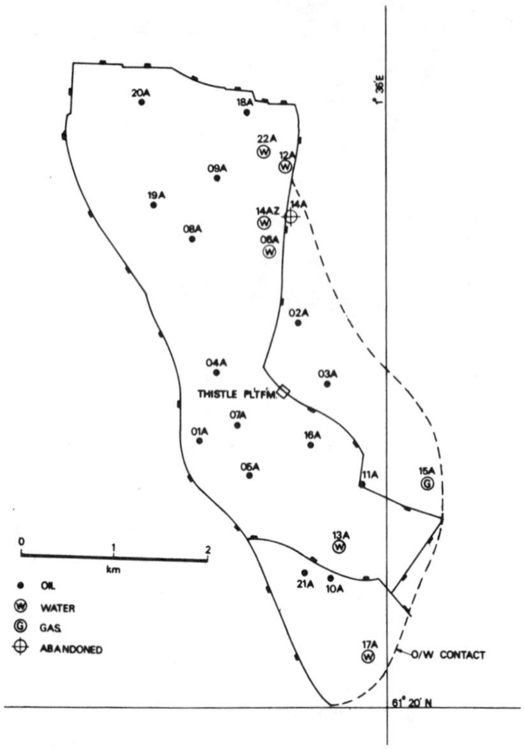

Fig. 10 – Fault pattern, third version.

fault block and, hence, suitable for injection.

Well 13A

A high-volume water injection well in the southern part of the western fault block was needed to replenish the depleted pressure. The location of the central Fault G was still unknown although the pressure matches showed it to be within 820 ft (250 m) from where it is marked in Fig. 8. We picked a location (Fig. 8) close to the western corner of the intersection of Faults N and G5; Fault G5 was the only fault that could be seen with any confidence on the seismic survey in the central part of Thistle field.

The Thistle field production rate was down significantly on budget because of lack of water injection and the longer drilling and completion times. A high-volume water injection well was needed in the western fault block where most of Thistle oil is present.

The well had to be as close as possible to the intersection of Faults G5 and N (Fig. 8) to ensure that no oil is trapped between the injection well and these two faults, which could reduce ultimate oil recovery by several million barrels (about 0.3×10^6 m^3). Although the location chosen was between two high-pressure wells (Wells 10A and 11A) and may have encountered a high-pressure regime, the fault pattern shown on Fig. 8 had to be proved (or disproved) before committing an injection pattern for the western fault block.

The well results were encouraging. The RFT showed low pressures (Fig. 11) and the openhole logs showed good quality sands. The following conclusions could be drawn from the RFT.

1. Sand C pressures were in four regions: the uppermost the same as Sand D, the lowest the same as Sand B, and two intermediate pressures. Most of Sand C was below 4,000 psig (27 580 kPa), showing communication to Well 07A, a Sand C producer.

2. Sand B pressures were all below 3,700 psig (25 512 kPa), which is very close to the pressure in Wells 01A, 04A, and 05A.

3. The upper part of Sand A seems to be in pressure equilibrium with Sand B, while the lower part of Sand A shows much high pressure.

The sand quality in this well is excellent and has been injecting at a restricted rate of some 45,000 B/D (7155 m^3/d). The fault theory was proved, although the exact location of the central fault was still in doubt.

Well 14A

Now that the fault theory has been proved, water injectors should be located on the west side of Fault M and as far north as possible. Well 14A was chosen as the next location for an injector in the northern part of the western fault block.

Directional drilling difficulties resulted in the well drifting off target by about 260 ft (79.3 m) eastward. This was not critical, since the well target was supposed to be about 500 ft (150 m) west of Fault M. In fact, Well 14A penetrated the reservoir in the eastern

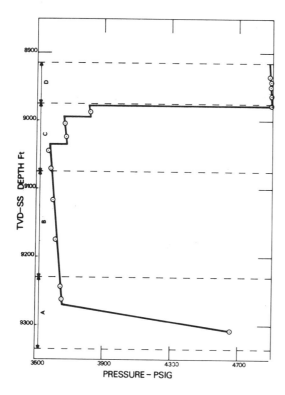

Fig. 11 — RFT pressures, Well 13A.

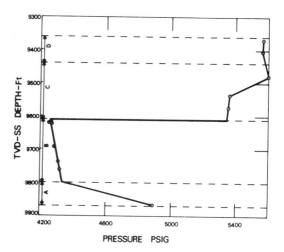

Fig. 12 — RFT pressures, Well 211/19a-7.

fault block on the downthrown side of Fault M since the sand tops were 250 ft (76 m) deeper than predicted, which was consistent with elevations in the eastern fault block. Logs and RFT were run and the well was sidetracked 820 ft (250 m) to the west. Sand elevations in the sidetracked hole showed the well to have penetrated the Brent sand on the upthrown side of Fault M. This was confirmed by a seismic survey run later in 1979 that showed that the earlier seismic lines were located incorrectly; the error was about 500 ft (152.4 m). The 1979 survey showed clearly that the original Well 14A had penetrated the Brent sand just east of the fault. The directional drilling control problems could have been due to the well's proximity to the fault plane. Pressure measurements proved conclusively that the original and sidetracked holes were in different pressure regimes. The RFT in the original hole showed high pressure, greater than 5,300 psig (36 542 kPa), in all sands. A pressure buildup survey in Sand B in the sidetracked hole showed low pressure of about 3,800 psig (26 200 kPa).

The well was completed as a Sand B injector. It has excellent sand quality and a water injection capacity of 50,000 B/D (7950 m³/d).

Wells 15A Through 22A

No surprises were encountered in these wells except for Well 16A, which was aimed for the eastern fault block but ended in the west. This showed that although we had a general idea where the central fault may be, we still could not pinpoint it accurately.

The sand quality has been excellent, indicating that poor sand quality in Well 12A was a local phenomenon. Water injection in the western fault block has increased reservoir pressure from 3,600 psig (24 821 kPa) in June 1978 to 4,300 psig (29 647 kPa) in June 1980. The aim is to operate the reservoir at approximately 5,000 psig (34 475 kPa). The increase in reservoir pressure has resulted in a direct increase in oil production capacity.

Appraisal Well 211/19a-7

Drilling high-deviation wells from the platform into areas of reservoir uncertainty is risky, since it involves the possibility of sidetracking the platform development well if the drill encountered the wrong pressure regime. The sidetracking could be even more complicated if intermediate 9⅝-in. (24.4-cm) casing already had been run. A vertical appraisal well could reduce this risk significantly. Thistle field economic analysis showed that it is worthwhile to drill a vertical appraisal well, Well 211/19a-7, to find whether Fault P (Fig. 8) is sealing. The results would allow the platform Well 17A to be located correctly as a water injection well at the southernmost end of Thistle field.

The RFT pressures in Well 211/19a-7 were low (Fig. 12) and proved that Fault P was not a barrier to flow. Sand B pressure was nearly identical to Well 10A. Because pressures of Sands C and D are lower than original, there is a small amount of vertical communication between all the sands in the southern fault block.

Fault Pattern, Third Version

A seismic survey was run in May 1979 to help identify the faults better. Although the data quality was better than the 1974 survey and the positioning was significantly more accurate, there was no clear indication of the existence of a major fault in the central part of Thistle field. The faulting pattern shown in Fig. 10 is the third version, based on data from 22 development wells. The series of faults in the center of the field is the only continuous trend that could be established from the seismic data. Work is still continuing on different reprocessing techniques to see whether a better definition of faults is possible. The reason for low Sand D RFT pressures in Well 11A still is not understood fully. The accepted explanation is that Well 11A had crossed the central barrier fault. Sand D, therefore, is in the western fault block and its low initial pressure is a result of Sand D oil production from Well 09A.

Pressure Measurements

Three sources of pressure measurements were used in Thistle field to monitor reservoir behavior, wellhead flowing pressures, bottomhole pressure buildup surveys, and RFT pressures. Wellhead flowing pressures were measured twice a day with dead-weight gauges and were used for day-to-day monitoring of individual well performance. The wellhead flowing pressures, taken while a well was tested, were used for history matching. First, they were converted to bottomhole flowing pressures by Orkiszewski's method. Second, the bottomhole flowing pressures were converted to grid-block pressures by Peaceman's method. Both these methods were checked by subsurface pressure surveys, which gave flowing and bottomhole pressures and permeability and skin values.

Orkiszewski's method was accurate to ±50 psi (345 kPa) for wells flowing at up to 20,000 BOPD (3180 m³/d oil). Peaceman's correction was especially useful for high-permeability wells. For wells with average permeabilities less than 100 md, more regular pressure buildup surveys are needed to ensure that changes in rate have not altered the value of skin.

The RFT tool has been a very valuable and reasonably trouble-free tool. It has been run in Thistle field wells with average deviation as high as 63°. The pressure information it gave was invaluable. Without this pressure information, the task of understanding the geometry of the Thistle reservoir would have been much more difficult, especially concerning small leaks between the various sands. It would have been difficult to prove that Sand A was only in partial communication with Sand B. It would have been impossible to find the various pressure subunits in Sand C.

Conclusions

1. A flexible approach to development planning is essential to optimize field productivity and recoverable reserves.

2. The reservoir engineers analyzing Thistle field performance found that a continuously updated, mathematical reservoir model was an extremely useful tool in identifying, understanding, and correcting deviations from the expected reservoir performance.

3. The Thistle field is divided into three fault blocks: western, eastern, and southern. Although these faults are not fully sealing, the fault blocks should be treated as separate units for practical reservoir development.

4. The drilling of vertical appraisal wells may be an economical method of collecting data and may save considerable platform drilling rig time, especially in areas of uncertain reservoir description.

5 The RFT tool gave invaluable pressure data.

6. Wellhead flowing pressures converted to grid-block pressures were found very reliable for pressure matching purposes.

Acknowledgments

I acknowledge the vital contribution of members of BNOC Thistle Reservoir Management Team to development of the Thistle reservoir. I thank BNOC and the other Thistle field unit participants for allowing publication of confidential data.

References

1. Nadir, F.T. and Hay, J.T.C.: "Geological and Reservoir Modelling of the Thistle Field," paper EUR 88 presented at the European Offshore Petroleum Conference and Exhibition, London, Oct. 24-27, 1978.
2. Hallett, D.: "Refinement of Geological Model of Thistle Field," paper presented at the 1980 Inst. of Petroleum Conference on Petroleum Geology of Continental Shelf of North West Europe, London, May 1980.
3. Orkiszewski, J.: "Predicting Two-Phase Pressure Drops in Vertical Pipe," *J. Pet. Tech.* (June 1967) 829-838.
4. Peaceman, D.W.: "Interpretation of Well-Block Pressures in Numerical Reservoir Simulation," *Soc. Pet. Eng. J.* (June 1978) 183-194.

SI Metric Conversion Factors

$$\text{bbl} \times 1.589\ 873\ \text{E}-01 = \text{m}^3$$
$$\text{ft} \times 3.048^* \quad \text{E}-01 = \text{m}$$
$$\text{mile} \times 1.609\ 344^*\text{E}+00 = \text{km}$$
$$\text{psi} \times 6.894\ 757\ \text{E}+00 = \text{kPa}$$

*Conversion factor is exact.

JPT

Original manuscript received in Society of Petroleum Engineers office Aug. 4, 1980. Paper accepted for publication Aug. 14, 1981. Revised manuscript received Aug. 14, 1981. Paper (SPE 10547, EUR 165) first presented at the European Offshore Petroleum Conference and Exhibition, held in London, Oct. 21-24, 1980.

Reprinted by permission of the American Institute of Mining, Metallurgical, and Petroleum Engineers. Published in *Petroleum Transactions*, AIME, V. 186, pp. 39-48.

CAPILLARY PRESSURES – THEIR MEASUREMENT USING MERCURY AND THE CALCULATION OF PERMEABILITY THEREFROM

W. R. PURCELL, JUNIOR MEMBER AIME, SHELL OIL COMPANY, HOUSTON, TEXAS

ABSTRACT

An apparatus is described whereby capillary pressure curves for porous media may be determined by a technique that involves forcing mercury under pressure into the evacuated pores of solids. The data so obtained are compared with capillary pressure curves determined by the porous diaphragm method, and the advantages of the mercury injection method are stated.

Based upon a simplified working hypothesis, an equation is derived to show the relationship of the permeability of a porous medium to its porosity and capillary pressure curve, and experimental data are presented to support its validity.

A procedure is outlined whereby an estimate of the permeability of drill cuttings may be made with sufficient acuracy to meet most engineering requirements.

INTRODUCTION

The nature of capillary pressures and the role they play in reservoir behavior have been lucidly discussed by Leverett[1], Hassler, Brunner, and Deahl[2], and others. As a result of these publications the value of determining capillary pressure curves for cores has come to be generally recognized within the oil industry. While considerable attention has been directed toward the subject in an effort to provide a reliable method of estimating percentages of connate water,[3,4,5] it has been recognized that capillary pressure data may prove of value in other equally important applications.

This paper describes a method and procedure for determining capillary pressure curves for porous media wherein mercury is forced under pressure into the evacuated pores of the solids. The pressure-volume relationships ob-

tained are reasonably similar to capillary pressure curves determined by the generally accepted porous diaphragm method. The advantages of the method lie in the rapidity with which the experimental data can be obtained and in the fact that small, irregularly shaped samples, e.g., drill cuttings, can be handled in the same manner as larger pieces of regular shape such as cores or permeability plugs.

Based upon a simplified working hypothesis, a theoretical equation will be derived which relates the capillary pressure curve to the porosity and permeability of a porous solid, and experimental data will be presented to support its validity. This relationship applied to capillary pressure data obtained for drill cuttings by the procedure described provides a means for predicting the permeability of drill cuttings.

METHODS FOR DETERMINING CAPILLARY PRESSURES

Several techniques have so far been employed in determining capillary pressure curves and these fall into two principal categories:

(1) Liquid is removed from, or imbibed by, the core through the medium of a high displacement pressure porous diaphragm[3,4,5,6].

(2) Liquid is removed from the core which is subjected to high centrifugal forces in a centrifuge[4,6].

There are, however, certain limitations inherent in both methods.

The greatest capillary pressure which can be observed by method (1), above, is determined by the maximum displacement pressure procurable in a permeable diaphragm which at the present time appears to be less than 100 psi. An even more serious limitation of the diaphragm method is imposed by the fact that several days may be required to reach saturation equilibrium at a given pressure; hence, the time re-

quired to obtain a well-defined curve may be measured in terms of weeks. Furthermore, to date, no suitable technique for handling relatively small, irregularly shaped pieces of rock, such as drill cuttings, has been reported and, therefore, measurements must be made, in general, on cores, or portions thereof.

The centrifuge method offers the distinct advantage over the porous diaphragm method of arriving at saturation equilibrium in a relatively short time by virtue of the elimination of the transfer medium for the liquid. The calculation of capillary pressures from centrifuge speeds is somewhat tedious[6], however, and the equipment required is fairly elaborate. While there exists the possibility that this method might be adaptable to the determination of the capillary pressures of cuttings, this particular ramification has not been investigated, as far as is known.

In view of the limitations of the two principal methods for determining capillary pressures, the apparatus described in the following sections has been devised in order that difficulties previously encountered might be circumvented.

MERCURY INJECTION METHOD FOR DETERMINING CAPILLARY PRESSURES
Theory

The methods described above for determining capillary pressures are characterized by the fact that one of the fluids present within the pore spaces of the solid is a liquid which "wets" the solid, i.e., the contact angle which the liquid forms against the solid is less than 90° as measured through that phase.

For these "wetting" liquids the action of surface forces is such that the fluid spontaneously fills the voids within the solid. These forces likewise oppose the withdrawal of the fluid from the pores of the solid.

Manuscript received at office of the Branch September 1, 1948. Paper presented at Branch Fall Meeting, Dallas, Texas, Oct. 4-6, 1948.
[1] References are given at the end of the paper.

There is, however, a second type of system which may be considered in the study of capillary pressures. This system involves the porous solid and a single "non-wetting" fluid (mercury) which forms a contact angle of greater than 90° against the solid. In this case the action of the surface forces involved opposes the entrance of the liquid into the solid and pressure must be applied to the liquid to cause penetration of the pores of the solid.

This type of system has been employed by Drake and Ritter[7] in studying the pore size distribution of catalysts, and the apparatus to be described below has been developed in order that similar techniques might be applied to materials exhibiting pore sizes of the order of that found in naturally occurring rock formations.

Apparatus and Procedure

An apparatus suitable for determining capillary pressures of porous media is shown in Fig. 1. The essential components of the apparatus are a mercury displacement pump A, a sample holder B, both shown in detail, and a manifold system C, shown schematically, wherein the gas pressure may be varied from small absolute values (high vacuum) to about 2000 psi, gauge.

The mercury pump consists of a piston-cylinder arrangement, the former being moved by means of an accurately machined screw, the pitch of which is such that one turn of the driving mechanism moves the piston through a distance sufficient to displace one cubic centimeter. The volume of liquid displaced from the pump is determined by successive readings of the scale D, and vernier E which is attached to the hub of the hand wheel.

The sample holder consists of two parts, both of which carry a lucite window, G, of frusto-conical shape which is cemented into the body of the holder and held rigidly in place by bushings. The displacement pump is connected to the sample holder and manifold by means of diametral conduits through the two lucite plugs. Reference marks, H, are incorporated in these conduits at about the midway point of the lucite windows and may be viewed through the openings in the supporting bushings.

The manifold is connected, as shown, to both a vacuum system and a high pressure (2000 psi) nitrogen bottle. To this manifold are also connected a manometer and pressure gauges suitable for measuring gas pressures ranging from a few millimeters of mercury, absolute, to 2000 psi, gauge.

In operation, one or more plugs drilled from a core, or a number of drill cuttings, which have been extracted and dried, are placed in the cavity, F, of the sample holder. The top portion of the sample holder is positioned and the two parts brought together by a make-up nut. A suitable gasket makes the seal pressure tight.

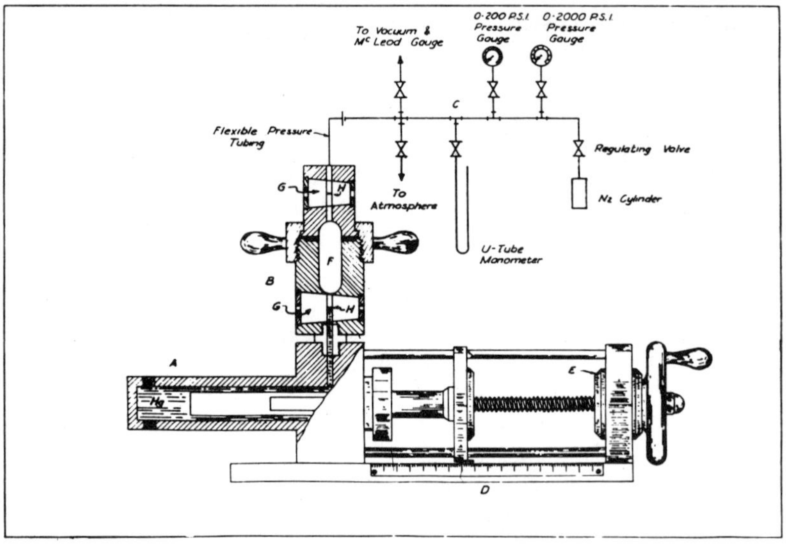

With the mercury level somewhat below the reference line of the lower lucite window, a vacuum is drawn on the system until an absolute pressure of 0.005 mm. of mercury, or less, is registered by the McLeod gauge. The mercury level is then accurately positioned at the lower reference mark by advancing the piston of the displacement pump. The scales attached to the volumetric pump are set at zero following which the piston is further advanced until the mercury meniscus reaches the reference mark in the top lucite window. At this point a scale reading is made which indicates the amount of mercury required to fill the cell with the sample in place. This quantity is subtracted from the known volume of the sample holder (between the reference marks) to provide a measure of the bulk volume of the sample under test.

The vacuum pump is isolated from the manifold and gas admitted to the system in increments, thereby increasing the pressure on the mercury surrounding the sample. The entrance of mercury into the pores of the core or cuttings is indicated by a recession of the mercury-gas interface from the upper reference line, and the degree of penetration is determined by advancing the displacement pump piston until the mercury meniscus returns to this reference mark.

The procedure of alternately building up the pressure to cause recession of the mercury meniscus and advancing the pump piston to return the meniscus to the reference mark, thereby determining the amount of mercury injected into the porous solid under various pressures, is repeated until the pressure of the nitrogen cylinder is reached.

A pressure-volume correction curve is established for the apparatus by carrying out a run as described above without a sample in the holder. The volume readings obtained when testing cores or cuttings are corrected by subtracting amounts as determined by this blank run at corresponding pressures.

The pressure on the mercury entering the sample is taken as the pressure in the gas phase plus a hydrostatic head due to the weight of the column of

FIG. 1 — APPARATUS FOR DETERMINING MERCURY CAPILLARY PRESSURES

mercury between the upper reference line and the midway point of the sample. This hydrostatic head may be determined by direct measurement.

Inasmuch as saturation equilibrium is reached very rapidly at any particular pressure, an entire curve may be determined in from 30 to 60 minutes. Temperature fluctuations of the system are ordinarily not sufficiently great during this time to require corrections for thermal expansion or contraction.

Experimental Results

In Figs. 2 to 8, inclusive, (Pages 44 and 45), mercury capillary pressure curves determined in the manner just described are compared with curves determined with water* and air using the porous diaphragm method. The air permeabilities and total porosities of the samples are shown on the graphs.

In comparing these curves it must be recalled that the magnitudes of the capillary pressures are proportional to the product of the surface tension of the liquid being used and the cosine of its angle of contact against the solid. While the surface tensions of the liquids involved can be measured with fair accuracy, the uncertainty in values of the contact angles makes it difficult to predict in advance the exact ratio which should exist between mercury and water/air capillary pressures at corresponding saturations. As a first approximation, however, the following values can be assumed:

Surface tension of water: 70 dynes per cm.

Surface tension of mercury: 480 dynes per cm.[7]

Contact angle of water against the solid: 0°

Contact angle of mercury against the solid: 140°[7]

The required ratio is then,

$$\frac{\text{(Mercury Capillary Pres.)}}{\text{(Water/Air Capillary Pres.)}}_{\substack{\text{corresponding}\\\text{saturations}}} =$$

$$\frac{-(480)\ (\cos 140°)}{(70)\ (\cos 0°)} \cong 5.$$

It will be seen that the curves of Figs. 2 to 8 have been plotted on scales such that this approximate ratio of 5 to 1

* The water used in all tests referred to in this paper contained 5 per cent by weight of sodium chloride.

is taken into account and hence they may be easily compared qualitatively by visual observation.

As evidenced by these curves, which are typical of those so far obtained, relatively close agreement has been found between mercury and water/air capillary pressure curves for the various types of formations studied and over the range of permeabilities and porosities encountered.

Fig. 9 (P.45) is included to show a capillary pressure curve over the entire saturation range (from 0% to 100%). With the apparatus just described it is possible to measure mercury capillary pressures of the order of 2000 psi. which corresponds approximately to a water/air capillary pressure of 400 psi. The apparatus could be adapted for measurements at higher pressure if this were deemed advisable.

The advantages of the mercury injection method over those previously used are that an entire curve of as many as twenty to thirty points may be determined in about an hour's time and small irregularly shaped pieces can be handled in the same manner as larger portions of regular shape. In addition, the range of capillary pressures that can be observed is considerably greater than for the porous diaphragm method.

CALCULATION OF PERMEABILITY FROM CAPILLARY PRESSURE CURVES

In reservoir analysis and production practice, the importance of that property of rock formations which is referred to as permeability has long been recognized by the exploitation engineer. The determination of air permeability has, for some time, been a routine core analysis test, but in order to obtain the required measurements for this determination, it is first necessary to procure a sample of regular shape and of appreciable dimensions. To accomplish this end, the expensive operation of coring is generally resorted to.

In the course of drilling, cuttings are usually available which, although too small to be suitable for permeability measurements, provide satisfactory samples with which to determine other important characteristics of the rock

formation from which they were cut. An apparatus has been described in the preceding sections which enables the measurement of capillary pressure curves for cuttings. An equation will now be developed which provides a means for calculating the permeability of a porous medium from capillary pressure data. This, in turn, of course, makes possible the estimation of the permeability of drill cuttings.

Theory

The rate of flow Q/t, of a fluid of viscosity, μ, through a single cylindrical tube or capillary of length, L, and internal radius, R, is given by Poiseuille's equation:

$$\frac{Q}{t} = \frac{\pi R^4 P}{8\mu L}, \qquad \ldots \ldots \ldots (1)$$

wherein P is the pressure drop across the tube. Since the volume, V, of this capillary is $\pi R^2 L$, equation (1) may be written as,

$$\frac{Q}{t} = \frac{V R^2 P}{8\mu L^2}, \qquad \ldots \ldots \ldots (2)$$

The capillary pressure for this single tube, is given by the Pressure of Displacement Equation,

$$P_c = \frac{2\sigma \cos \theta}{R}. \qquad \ldots \ldots \ldots (3)$$

where P_c, the capillary pressure, is the minimum pressure required to displace a wetting liquid ($\theta < 90°$) from or inject a non-wetting liquid ($\theta > 90°$) into a capillary of radius R when the surface or interfacial tension at the interface is σ and the angle of contact which this interface forms with the solid of the capillary is θ.

Equation (3) indicates that capillary pressure is inversely proportional to pore radius and hence may be used as a measure of capillary size. Substituting equation (3) in (2) we have,

$$\frac{Q}{t} = \frac{(\sigma \cos \theta)^2 V P}{2\mu L^2 (P_c)^2} \qquad \ldots \ldots (4)$$

Consider now a system composed of a large number, N, of parallel, cylindrical capillaries of equal length but random radii, each tube being identical in all respects except internal area to the capillary under discussion above. The total rate of flow $(Q/t)_s$ through this system must be equivalent to the sum of the contributions made by each of the N single tubes or capillaries.

227

The flow through each individual tube is given by equation (4); the total flow, therefore, may be represented as follows:

$$(Q/t)_* = \frac{(\sigma \cos \theta)^2 P}{2\mu L^2} \sum_{i=1}^{N} \frac{V_i}{(P_c)_i^2} \cdot \quad (5)$$

On the other hand, the rate of flow $(Q/t)_*$ through this same system of capillaries is also given by Darcy's Law:

$$(Q/t)_* = \frac{K A P}{\mu L}, \quad \ldots \ldots (6)$$

where K is the permeability of the system, L is the length of the tubes, A is the total cross-sectional area of the system, and P is the pressure differential causing flow.

By equating the right-hand members of the equations (5) and (6) the following is obtained:

$$K = \frac{(\sigma \cos \theta)^2}{2AL} \sum_{i=1}^{N} \frac{V_i}{(P_c)_i^2} \ldots \ldots (7)$$

To simplify equation (7) the volume, V_i, of each capillary may be expressed as a percentage, S_i, of the total void volume, V_T, of the system, i.e.,

$$\frac{V_i}{V_T} \times 100 = S_i \quad .$$

Furthermore, since AL is the bulk volume of the system we may introduce the per cent porosity, f, where

$$f = \frac{V_T}{AL} \times 100 \quad ,$$

and equation (7) becomes

$$K = \frac{(\sigma \cos \theta)^2 f}{2 \times 10^4} \sum_{i=1}^{N} \frac{S_i}{(P_c)_i^2} \ldots \ldots (8)$$

Equation (8) relates the permeability of a system of parallel cylindrical capillaries of equal lengths, but various radii, to the porosity of the system and to the capillary pressures and volumes of its component parts.

Equation (8) is derived for a porous medium composed of non-interconnected capillaries of circular cross-section and equal length. Certainly no such system obtains in naturally occurring rock formations. The path by which a fluid may travel through a rock

is circuitous, for the pore spaces within these materials are usually interconnected to a greater or lesser degree. Furthermore, this path is neither uniform nor circular in cross-section. It is necessary, therefore, to modify equation (8) by introducing a so-called lithology factor, F, to account for differences between the flow in the hypothetical porous medium for which equation (8) was derived and that in naturally occurring rocks. Equation (8) becomes

$$K = \frac{F(\sigma \cos \theta)^2 f}{2 \times 10^4} \sum_{i=1}^{N} \frac{S_i}{(P_c)_i^2} \quad . \quad (9)$$

The amount of variation in the factor F (to be discussed below) for samples of equivalent and of different lithology will determine the practical utility of equation (9).

The evaluation of the quantity

$$\sum_{i=1}^{N} \frac{S_i}{(P_c)_i^2}$$

can best be presented by reference to Fig. 10, which shows a typical capillary pressure curve for a naturally occurring porous media such as a reservoir sandstone.

Consider the change in saturation, $\Delta\rho$, that occurs when the pressure is increased from $(P_c)_1$ to $(P_c)_2$. This change in saturation is a result of liquid either entering (non-wetting liquid) or receding from (wetting liquid) all pores having capillary pressures lying between $(P_c)_1$ and $(P_c)_2$. All pores in this interval may be treated as if they exhibited some intermediate average capillary pressure, $(P_c)_{av.}$, where,

$$(P_c)_1 \: (P_c)_{av.} \: \forall \: (P_c)_2 \quad .$$

If a number, r, of such intervals are chosen, and if n_j is the number of pores in the jth interval, then

$$\sum_{i=1}^{n_j} \frac{S_i}{(P_c)_i^2} \cong \left[\frac{\Delta\rho}{(P_c)_{av.}^2} \right]_j \quad .$$

Likewise,

$$\sum_{i=1}^{N} \frac{S_i}{(P_c)_1^2} = \sum_{j=1}^{r} \sum_{i=1}^{n_j} \frac{S_i}{(P_c)_1^2} \cong$$

$$\sum_{j=1}^{r} \left[\frac{\Delta\rho}{(P_c)_{av.}^2} \right]_j \: ,$$

and

$$\lim_{\substack{\Delta\rho \to 0 \\ r \to \infty}} \sum_{j=1}^{r} \left[\frac{\Delta\rho}{(P_c)_{av.}^2} \right]_j = \int_{\rho=0}^{\rho=100} \frac{d\rho}{(P_c)^2}$$

$$= \sum_{i=1}^{N} \frac{S_i}{(P_c)_i^2}$$

It is seen, therefore, that the quantity

$$\sum_{i=1}^{N} \frac{S_i}{(P_c)_i^2} \quad \text{is equal to}$$

the integral of the reciprocal of the square of the capillary pressure expressed as a function of per cent liquid saturation. Such a function is shown in Fig. 10 (P.45). This integral may be determined by planimetering or, more readily, by applying Simpson's Rule.

In the calculation of permeability by by equation (9) from mercury capillary pressure data a surface tension of 480 dynes per cm. and a contact angle of $140°$[7] are assumed. Equation (9) then reduces to

$$K = 0.66Ff \int_{\rho=0}^{\rho=100} \frac{d\rho}{(P_c)^2}, \quad (9a)$$

where K is the permeability in millidarcies, f is per cent porosity, ρ is per cent of total pore space occupied by the liquid, and P_c the capillary pressure expressed in atmospheres.

Experimental Results

In an attempt to determine experimentally the utility of equation (9a), mercury capillary pressure curves have been obtained for numerous rock samples of two types of formations, the Upper Wilcox of Eocene age and the Paluxy of Cretaceous age. Samples of such size were used that their air permeabilities could be measured directly and these observed permeabilities were employed in determining the magnitude of, and the variation in, the factor F of equation (9a). The data obtained

TABLE 1

Observed Values of F

Sample No.	Factor F, Eq. (9a) Required to Make Calculated and Observed Permeabilities Identical	Permeability Calculated from Eq. (9a) Using an Average F of 0.216	Observed Air Permeability (md)
Upper Wilcox			
1	0.085	3.04	1.2
2⟨1⟩	0.122	21.2	12.0
3	0.168	17.3	13.4
4	0.149	53.5	36.9
5	0.200	61.9	57.4
6	0.165	91.6	70.3
7	0.257	92.3	110
8	0.256	97.5	116
9	0.191	163	144
10	0.107	680	336
11 (Fig. 6)	0.216	430	430
12	0.273	348	439
13	0.276	388	496
14⟨1⟩	0.185	902	772
15	0.282	816	1070
16	0.363	865	1459
Paluxy			
17	—	0.003	<0.1
18	—	0.10	<0.1
19	0.182	42.2	35.7
20	0.158	54.9	40.2
21	0.231	172	184
22	0.276	183	235
23	0.215	308	307
24	0.163	422	320
25	0.284	383	506
26	0.272	502	634
27 (Fig. 5)	0.338	734	1150
	Av. 0.216		

⟨1⟩ "Cuttings"

with typical samples are recorded in Table I.

It will be seen in Table I that the factors F for the Upper Wilcox and Paluxy sands are of the same general order of magnitude. However, there is some indication that this factor may vary with permeability from a minimum of about 0.1 for samples of low permeability to about 0.4 for samples of high permeability.*

It may be possible, with the further accumulation of data, to establish values of the factor F for various perme-

ability ranges. Likewise, it may be found that the factor F will be different for samples of widely divergent lithology** since the two formations studied in detail and reported here do not vary greatly in lithology. The Upper Wilcox is in general a poorly sorted sandstone while the Paluxy is a well sorted sandstone. The establishment of more precise values of F, however, will serve merely to increase the precision of the method for estimating permeabilities from capillary pressure data. For the moment, with the information at hand, quite satisfactory accuracy may be obtained by merely taking an arithmetical average of the factors F over the entire permeability range.

Table I gives observed air permeabilities together with permeabilities as calculated from equation (9a) using an average value of the factor F of 0.216. These values have been plotted in Fig. 11.

In Table I, samples 2 and 14 are designated as "cuttings". These samples were prepared by crushing cores of known air permeability to obtain a number of small pieces in approximately the size and shape of drill cuttings. Mercury capillary pressure curves were determined for these sets of "cuttings" in exactly the same manner as employed for the larger single pieces. It will be noted in Fig. 11 that the data for these "cuttings" fit the average line as well as the data for the cores.

The procedure for estimating the permeability of drill cuttings (or any porous medium, regardless of size or shape) consists of determining the capillary pressure curve by the method described above, calculating a permeability from the curve by means of equation (9) using a predetermined value of F, and then reading the corresponding air permeability from a plot such as that shown in Fig. 11. The spread of points of Fig. 11 is such as to indicate that the permeability so determined will be sufficiently accurate for most engineering requirements.

SUMMARY AND CONCLUSIONS

A method of determining capillary pressures for porous media wherein mercury is forced under pressure into the pores of the evacuated solid has been found to yield results which are reasonably similar to those obtained using the porous diaphragm technique. This method offers the following advantages over previously decribed procedures:

1. An entire capillary pressure curve consisting of as many as 20 to 30 points can be determined in a matter of hours rather than weeks.

2. Small, irregularly shaped pieces, such as drill cuttings, can be handled in exactly the same manner as larger, regularly shaped samples such as cores or permeability plugs.

3. The range of capillary pressures which can be observed is 5 to 10 times that of conventional methods.

---

* It may be shown both theoretically and experimentally[8] that for close-packed spheres of uniform size the length of the path by which a fluid may pass through such a system is $\pi/2$ times as great as the length of the pack. Since the length of the path enters into the denominator of equation (6) as a squared term, it can be shown that the factor F for this pack is $(2/\pi)^2$ or about 0.4. The difference between this value and those actually observed (0.1 to 0.4) may be due in part to inaccuracies in the assumptions of values for σ and θ, but is probably chiefly the result of the difference in pore structure between that of the ideal pack of uniform spheres and that of naturally occurring rock structures. It is interesting to note, however, that the observed values for F of natural rocks are of the same order of magnitude as, but somewhat less than, that which can be calculated for an ideal "sand" pack.

** While F values have been determined extensively for only the Upper Wilcox and Paluxy sandstones, it is interesting to note that the following values are obtained from the data presented in Figs. 2, 3, 4, 7, and 8.

Fig. No.	Formation	Factor, F	Air Permeability (Md.)
2	Frio Sand	0.073	23
3	Frio Sand	0.271	170
4	Frio Sand	0.276	950
7	San Andres Lime	0.133	35
8	San Andres Lime	0.150	43

Although these values are in line with those reported in Table I for samples of comparable permeability, additional data must be obtained before it can be concluded that all four types of formations exhibit similar F factors.

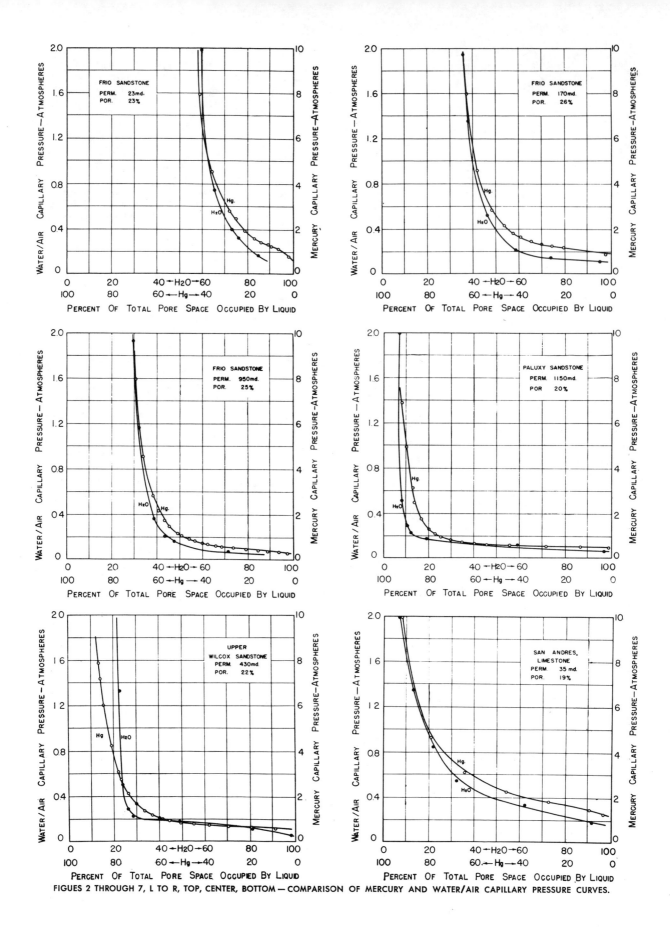

FIGUES 2 THROUGH 7, L TO R, TOP, CENTER, BOTTOM — COMPARISON OF MERCURY AND WATER/AIR CAPILLARY PRESSURE CURVES.

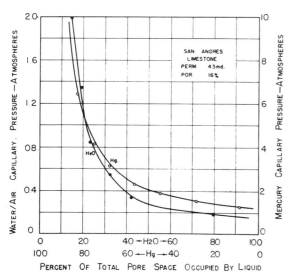

FIG. 8 — COMPARISON OF MERCURY AND WATER/AIR
CAPILLARY PRESSURE CURVES

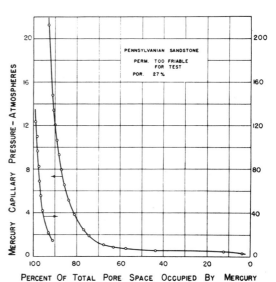

FIG. 9 — MERCURY CAPILLARY PRESSURE CURVE OVER
THE ENTIRE SATURATION RANGE

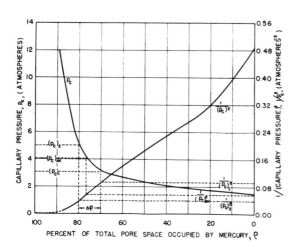

FIG. 10 — PERCENT SATURATION AS A FUNCTION OF
CAPILLARY PRESSURE AND RECIPROCAL OF SQUARE
CAPILLARY PRESSURE

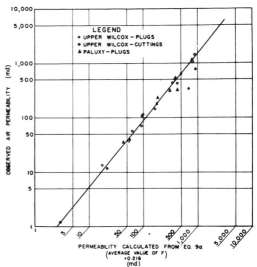

FIG. 11 — OBSERVED AIR PERMEABILITY AS A FUNCTION
OF PERMEABILITY CALCULATED FROM EQUATION 9A.

An equation has been derived which indicates theoretically the relationship which should exist between the permeability of a porous medium and its capillary pressure curve. Experimental data have been presented to show that this equation provides a fairly reliable method of calculating permeability from capillary pressure data.

A combination of the method for measuring the capillary pressure curve of drill cuttings and the equation relating permeability to the capillary pressure curve so determined makes possible the estimation of the permeability of those cuttings.

ACKNOWLEDGMENT

The author wishes to express his appreciation for the aid and cooperation of Mr. G. E. Archie in the formulation of the problem, for the assistance rendered by Messrs. M. A. Westbrook and C. F. Blankenhorn in designing and assembling the apparatus, and for the many valuable suggestions offered by Mr. J. P. Murphy during the preparation of the manuscript.

REFERENCES

1. M. C. Leverett: Capillary Behavior in Porous Solids. *Trans.* AIME 142, 152, (1941).

2. G. L. Hassler, E. Brunner, and T. J. Deahl: The Role of Capillarity in Oil Production. *Trans.* AIME 155, 155, (1944).

3. O. F. Thornton and D. L. Marshall: Estimating Interstitial Water by the Capillary Pressure Method. *Trans.* AIME 170, 69, (1947).

4. J. J. McCullough, F. W. Albaugh, and P. H. Jones: Determination of Interstitial-Water Content of Oil and Gas Sand by Laboratory Tests of Core Samples: *Drilling and Production Practice* (A.P.I.) 180 (1944).

5. W. A. Bruce and H. J. Welge: Restored-State Method for Determination of Oil in Place and Connate Water. *Oil and Gas Journal*, **46**, 223, (1947).

6. G. L. Hassler and E. Brunner: Measurement of Capillary Pressures in Small Core Samples. *Trans.* AIME **160**, 114, (1945).

7. R. L. Drake and L. C. Ritter: Pore-Size Distribution in Porous Materials. *Ind. Eng. Chem.*, An. Ed. **17**, 782, (1945), and Macropore-Size Distributions in Some Typical Porous Substances. *Ibid* **17**, 787, (1945).

8. F. E. Bartell and H. J. Osterhof: Pore Size of Carbon and Silica Membranes. *J. Phys. Chem.* **32**, 1553, (1928).

DISCUSSION

☆

By Walter Rose, Gulf Research and Development Company, Pittsburgh, Pa.

It is appropriate certainly to compliment Mr. Purcell on the interesting manner in which he has called attention to another laboratory method for measuring capillary pressure phenomena, and to another useful application for capillary pressure data. It appears that conventional methods for measurement have not been utilized fully to serve the purposes of routine laboratory core analysis because of the instrumentation and associated difficulties which have been encountered; and also it appears that the heretofore emphasized application for capillary pressure data to evaluate initial fluid phase distributions in virgin petroleum reservoirs can be discredited to some extent due to theoretical uncertainties. However, it is not to be expected that the results of Mr. Purcell's work immediately provide solutions for all the problems related to the evaluation and application of capillary pressure phenomena. This is because the data obtained by the mercury penetration method is not equivalent entirely to that obtained by conventional procedures, as perhaps implied by Purcell, in ways now to be discussed. Elsewhere it will be shown, however, that the data obtained by the mercury penetration method indeed lead to the other useful results suggested by Purcell.

It is to be assumed that when a difference in pressure is established at the interfaces of contact between immiscible fluids saturating the interstices of porous media, this capillary pressure (as an explicit function of fluid saturation and an implicit function of fluid distribution) is in effect a measure of the interstitial pore widths containing these interfaces of contact, assuming conditions of static equilibrium obtain. From this standpoint then it is apparent that Purcell's method and conventional methods of capillary pressure measurement all yield potentially the same kinds of data. To obtain the exact equivalence of these data by the various methods, however, requires that consideration be given to the well-known hysteretic possibilities, that the configurational character of the interstitial spaces not be variably affected by the method of measurement employed, and that the dynamic mechanism of wetting and non-wetting fluid flow (as static equilibruim is approached) be unrelated to the method of measurement employed. Presumably, the first requirement has been satisfied by Purcell's experimentation for he attempts comparison of data obtained only by the capillary drainage mechanism (i.e. where the invading non-wetting phase permits only drainage and no intermediate imbibition of the wetting phase) so that hysteretic uncertainties need not be considered. Regarding the second requirement, it is felt that only in the case of "inert" porous media will the character of the interstitial spaces be unrelated to the physico-chemical properties of the saturating fluids. As an example, the presence of interstitial surface clay coating in certain types of naturally occurring porous media will certainly result in a different pore structure when mercury and rarified gas are substituted for oil and brine as the saturating fluids, due to the dependence of the clay swelling effects on the nature of the fluids contacting the clays. Thus, even in conventional capillary pressure experimentation it has been recognized that use of reservoir-like fluids is to be preferred so that the data will truly reflect on the pore character representative of the natural reservoir condition. This refinement evidently cannot be employed in the use of Purcell's method. However, it is the inevitable disre-

gard for the third requirement as above stated which most seriously limits the possibility of attaining equivalence between the results of Purcell's method and conventional methods, but which nonetheless leads to the potential usefulness for Purcell type experimentation as described in his paper. For it can be anticipated on theoretical grounds, and indeed it can be established from an examination of Purcell's data (c.f. Purcell's Figures 2 through 8), that "irreducible" minimum wetting phase saturations (reflecting on zero wetting phase relative permeability due to the attainment of pendular configurations) will not be a result of the mercury penetration method, since this procedure involves a compression rather than a flow of the wetting phase as desaturation occurs. It is evident then that the data reported by Purcell reflect principally on a complete distribution of pore radii, as originally advanced by Ritter and Drake (c.f. Reference 7 of Purcell's paper), providing information not directly derivable from conventional capillary pressure curves and leading to a method for approximating permeabilities and lithology characteristics of porous media. However, it must be emphasized that exact equivalence between conventional capillary pressure data and that obtained by the mercury penetration method is not to be expected as a common result.

The interesting feature about Purcell's equation for permeability is that it can be arrived at by several processes, leading to a more complete interpretation of the lithology factor, F. For instance, in a recent paper (not yet generally available)* an expression for permeability is given as:

$$k = \frac{\left[\lim_{\rho_w \to 1} j(\rho_w)\right]^2 \left[\sigma \cos\theta\right]^2 f}{P_D^2 \times 10^2}$$

Eq. (1)

where ρ_w is the fractional wetting phase saturation, $j(\rho_w)$ is Leverett's capillary pressure function, P_D is the displacement pressure as commonly defined, and where the other notation follows that as employed by Purcell. In this cited reference it is shown that the limiting (minimum) value for the capil-

———
*Walter Rose and W. A. Bruce, "Evaluation of Capillary Character in Petroleum Reservoir Rock", *Jnl. of Petr. Tech.*, in press, (1949).

lary pressure function can be identified with the "rock" textural constant, t, appearing in the Kozeny equation for streamline fluid flow through porous media, as:

$$\lim_{\rho_w \to 1} j(\rho_w) = \left(\frac{1}{t}\right)^{1/2} \qquad \text{Eq. (2)}$$

Now, it is observed that the constant, t, as used in Eq. (2) refers to its maximum value, which has been established as being greater than 5 (and sometimes, for instance, as great as 500) as related to the various types of unconsolidated and consolidated porous media encountered in natural reservoir rock. This implies that Purcell's lithology factor generally will be less than 0.4, as indeed predicted and observed, for it is seen on combination of Eq. (1) above and Purcell's Eq. (9) or (9a) that:

$$\frac{F}{2} > \left(\lim_{\rho_w \to 1} j(\rho_w) \right)^2, \qquad \text{Eq. (3)}$$

since:

$$\int_0^1 \frac{d\rho}{P_c^2} < \frac{1}{P_D^2}. \qquad \text{Eq. (4)}$$

Moreover, it can be shown from these equations that the lithology factor, F, is defined fundamentally as:

$$F = \frac{2}{\displaystyle\int_0^1 \frac{d\rho}{j(\rho_w)^2}} = \frac{2}{t_{min}}. \qquad \text{Eq. (5)}$$

The above considerations imply that in texturally similar porous media the lithology factor, F, and the limiting value of the j function will be related

to each other by some constant factor characteristic of the media, and that in texturally dissimilar media a rough proportionality will still obtain between these parameters. This is clearly shown by Figure 1, constructed from Purcell's data where F has been plotted versus $\lim_{\rho_w \to 1} j(\rho_w)$. It would appear, therefore, that Eq. (1) above has the same sort of validity and usefulness as Purcell's Eq. (9a), and its use is to be preferred since from the experimentation standpoint it is easier to evaluate. For only the initial segment of the capillary pressure curve need be measured (actually, only the displacement pressure need be measured) to solve Eq. (1) and to provide values reflecting on lithology or permeability characters of porous media equivalent to those obtained by Purcell. In this connection, a plot of F versus the logarithm of the corresponding permeability value is also given in Figure 1 and the linearity obtained by this plot suggests that in the classification of texturally dissimilar porous media a knowledge of permeability alone suffices sometimes to predict qualitatively lithological differences.

Figure 2 shows the results of some experimentation conducted recently to verify that the mercury penetration method of capillary pressure measurement would yield data equivalent to that obtained by conventional methods for inert porous media. Thus, Curve A of Figure 2 shows the capillary retention replot of the original Ritter and

Drake data characterizing pyrex fritted glass media. Curve B is a similar plot of data obtained on other fritted glass media by the conventional capillary pressure technique, and the close conformance between these two curves can be taken as evidence that both methods have provided a measurement of the same phenomena, at least at the saturation levels where the wetting phase is continuous throughout the interstitial spaces (i.e. at saturations greater than those obtaining when the wetting phase configuration is pendular). These results are offered to support further the validity of the experimentation described by Purcell, for it is regarded that Purcell's technique for laboratory

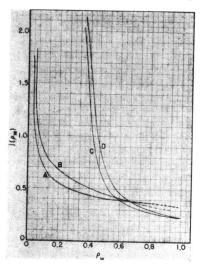

FIGURE 2
CAPILLARY RETENTION CURVES

CURVE A IS A PLOT OF THE RITTER AND DRAKE DATA.

CURVE B IS A PLOT OF DATA CONVENTIONALLY OBTAINED ON POROUS MEDIA TEXTURALLY SIMILAR TO THAT USED BY RITTER AND DRAKE.

CURVE C IS A REPLOT OF PURCELL'S FIG. 3.

CURVE D IS A REPLOT OF PURCELL'S FIG. 4.

measurement must have high order accuracy and precision. This is shown also by the replot of the Frio core data as Curves C and D in Figure 2 from Purcell's Figures 3 and 4, since essentially the same capillary retention curve has been obtained for these two cores of widely different permeability. For when capillary pressure data yield the same capillary retention plot for different porous media, a condition of textural similarity between these media

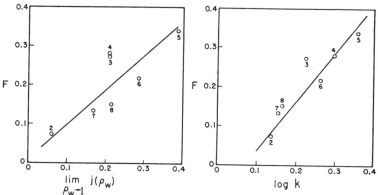

FIGURE 1

CORRELATION BETWEEN LITHOLOGY FACTOR, F, AND THE LIMITING VALUE FOR THE CAPILLARY PRESSURE FUNCTION, j(ρ_w), (C.F. LEFT DIAGRAM), OR THE LOGARITHM OF PERMEABILITY, LOG K, (C.F. RIGHT DIAGRAM). NUMBERS ASSOCIATE DATA POINTS WITH PURCELL'S FIGURE NUMBERS.

is implied, as already was established in the case of the Frio cores by the observed approximate equivalence of the measured lithology factor.*

To conclude, it is apparent that Mr. Purcell, in applying his method for capillary pressure evaluation, has accomplished probably as much as will be forthcoming from experimentation of this sort. That is, he has developed suitably a rapid and precise method to make valid measurements of capillary pressure on small rock fragments reflecting on the permeability and the lithological character of porous media. Such data undoubtedly will be found to be useful for correlation purposes in the interpretation of the results of other core analysis experimentation. However, it is noted that for some purposes at least conventional experimentation still will be required, and this should not be neglected because of the features of simplicity associated with the mercury penetration type experimentation, especially because it appears that Mr. Purcell's objections to use of conventional methods can be obviated somewhat through suitable instrumentation improvements. For instance, use of cellophane type capillary barriers permits measurements of capillary pressure in excess of 300 p.s.i., which is well above the values required to produce "irreducible" minimum saturations of the wetting phase in petroleum reservoir rock of practical interest. Also, it is noted that the equilibrium values of saturation can be obtained by extrapolation of rate of displacement data, thereby decreasing the time required for conventional laboratory experimentation. Finally, it is noted that the recently proposed "multi-core" procedure for capillary pressure evaluation (c.f. Rose and Bruce paper, loc. cit.) has a potential use for the study of small core fragments similar to that characterizing Purcell's method of evaluation. In this connection, however, it must be realized that no method presently described in the literature will provide satisfactory capillary pressure data in those instances where the core fragments are so small that their total bulk surface area approaches the total interstitial surface area, thereby imposing a lower limit on these possibilities.

The consideration presented in this discussion have been directed to bear on the significance of the results of Mr. Purcell's excellent paper. It is to be recognized, however, that Mr. Purcell's paper, independent of these considerations, is a most valuable contribution to the literature, and the author is to be congratulated for his developments which contribute so significantly to our knowledge of capillary pressure phenomena.

Author's reply to Mr. Walter Rose:

It is indeed gratifying that Mr. Rose, through his discussion, has provided a link between the theoretical approach to the problem which is exemplified by his recent paper entitled "Theoretical Generalizations Leading to the Evaluation of Relative Permeability"† and the experimental approach of the author. It is believed that through a close intermeshing of these two general types of investigation our knowledge of capillary phenomena should increase substantially in the future.

Mr. Rose has very aptly pointed out certain conditions which must be satisfied if exact equivalence between mercury capillary pressures and those obtained by other means is to be obtained; likewise he has indicated that all of these conditions are not necessarily fulfilled. It would seem appropriate, therefore, to state that the author does not intend to imply, as indicated by Mr. Rose, that the mercury penetration method is exactly equivalent to conventional procedures but instead has chosen to show experimentally (as evidenced by Figures 2 to 8, inc.) that for the various types of formations studied and over the range of permeabilities and porosities encountered a reasonable similarity exists between mercury and water/air capillary pressures. In the paper no conclusions are drawn from the comparison tests other than the one of similarity between the two types of curves and this same conclusion has been obtained by Mr. Rose for his fritted glass plates. Furthermore, it should be pointed out that only one application of capillary pressure data, namely that of estimating permeability, is discussed.

Mr. Rose's confirmation of equation 9 by the derivation outlined in his discussion adds materially to the subject. The calculation of permeability from but a single point of the capillary pressure curve, namely the displacement pressure, should perhaps be used with caution, although it must be admitted indications are that the results so obtained would compare favorably with those obtained by an integration of the entire curve. In this connection, it might be pointed out that an accurate determination of the displacement pressure is exceedingly difficult by means other than mercury penetration.

It is stated in the discussion that difficulties usually encountered in conventional methods of determining capillary pressures, can be alleviated by suitable instrumentation. While it is apparent that the range of pressures can be extended through the use of cellophane membranes, this may be accomplished, perhaps, only through the aggravation of a second difficulty, namely that of the length of the time involved. Yuster** has recently stated that the extrapolation method of estimating equilibrium saturations from rates of displacement data gave accurate results in only about 25 per cent of the tests for which it was employed; hence, it may be concluded that this method of reducing the time involved in conventional tests should be used with reservation. It is to be hoped that the paper of Rose and Bruce referred to in the discussion will soon be made generally available for it is believed that any method of determining capillary pressure curves for cuttings will find immediate application.   ★ ★ ★

---

* Note, from Curves C and D of Figure 2 the validity of the definition presented as Eq.(5) above is readily verified.

† *Jnl. of Petr. Tech.*, in press (1949).

** Producer's Monthly, December 1948, p. 24.

FRACTURES

The American Association of Petroleum Geologists Memoir
16, *Stratigraphic oil and gas fields,* edited by Robert E. King,
copyright 1972, pp. 82-106

Reservoirs in Fractured Rock[1]

DAVID W. STEARNS and **MELVIN FRIEDMAN**

Center for Tectonophysics, Texas A&M University, College Station, Texas 77843

Abstract In recent years three developments which have evolved more or less independently, when related, may be of value to the petroleum industry. First is the recognition, through normal oil field development, that fractures are significant to both reservoir capacity and performance. Second is the fact that controlled laboratory experiments have produced, in increasing quality and quantity, empirical data on rupture in sedimentary rocks. These data have been segregated to demonstrate the individual control on rupture of several important parameters: rock type, depth of burial, pore pressure, and temperature. The third development consists of the discovery of new methods to recognize, evaluate, use, and, in some cases, see fractures in the subsurface. This discussion of these three developments may help geologists and engineers to find new approaches to exploration and exploitation of fractured reservoirs. Reservoir and production engineers presently make the greatest use of fracture data, but geologists should find this information useful in exploration for oil and gas trapped in subsurface fractures. Except in the search for extensions to proved fracture reservoirs, there is in the literature a paucity of clear-cut examples of the use of fracture porosity data in advance of drilling. For this reason, several speculative exploration methods discussed herein implement mapping of fracture facies as well as stratigraphic facies.

INTRODUCTION

Scope

In the past 15 to 20 years the petroleum industry has shown an increasing interest in fractured reservoirs. During this same period, laboratory results have provided an empirical evaluation of the factors that influence fracturing in rocks, and advances in methods for studying fractures in the subsurface have been numerous. Nevertheless, few fractured reservoirs have been discovered as a result of exploration specifically for such traps. The succeeding review of pertinent data on rupture in rocks and of selected methods for studying fractures in reservoirs may lead to exploration specifically for

fractured reservoirs or, at least, to their recognition early in the development of a field. Knowledge of the fracture control in a given reservoir not only aids in the primary recovery of hydrocarbons, but also guides in the design of secondary recovery programs.

Basic Definitions and Concepts

A material is said to rupture when it loses cohesion along a more or less planar surface, separates into discrete parts, and is no longer able to support a stress difference. For most materials, this definition is unambiguous, but, for sedimentary rocks in the earth's crust, the concept is not quite so clear. Scale and boundary conditions become important factors. For example, a single rock layer interbedded with several other lithologic types may contain many "fractures"; however, it cannot be separated from its geologic environment, and the entire interbedded sequence may not be "fractured" by our definition. Without the constraints by the surrounding beds, the fractured layer would not be able to support a stress difference; but, as a part of the entire layered mass which has not lost cohesion, it still may be able to carry a part of the load and, indeed, even to rupture further. It is essential in the consideration of fractured reservoirs to differentiate between the behavior of the individual fractured layer and the bulk behavior of the entire sedimentary section.

A discussion of fractures and stratigraphic traps in a single volume might be questioned, because the common cause of fracturing is tectonism. Although fractures are associated with structure, a fractured reservoir might be considered as a form of stratigraphic trap—because lithology controls, to a large extent, the formation of fractures, and a stratigraphic trap is ". . . a type of trap which results from a variation in lithology of the reservoir rock . . . ." (AGI, 1957). Since most rocks are fractured in the subsurface, fractures influence the productivity of nearly all reservoirs.

Fractures play a multiple role in reservoir alteration. They can change the porosity or the permeability, or both. If the fractures are filled with secondary minerals, they may inhibit fluid

[1] Manuscript received, December 19, 1969.

We especially thank John Handin for his thoughtful reading and constructive criticism of the manuscript and Robert Berg for his many helpful suggestions.

The fracture-orienting device described in this paper was designed while we were employed by the Exploration and Production Research Laboratories of Shell Development Company, Houston, Texas. We are grateful to Shell Development Company for their permission to publish information on this device. We also wish to thank Shell Development for the use of permeability data from cores.

235

flow. However, even in rocks of low matrix porosity, fractures may increase the pore volume so that hydrocarbons can be recovered profitably. In other rocks the fractures may not add substantially to the total pore volume, but they may connect previously isolated spaces. Moreover, fractures actually may contribute to reservoir performance by both increasing the porosity and connecting isolated matrix porosity to the well bore.

Good general reviews of fractured reservoirs and their economic significance were given by Hubbert and Willis (1955), Smekhov (1963), and Drummond (1964). The typical fractured reservoir is (1) developed in brittle rock with low intergranular porosity; (2) characterized by high permeability (*e.g.*, up to 35 darcys; Regan, 1953) and low bulk porosity (*e.g.*, <6 percent; Hubbert and Willis, 1955); (3) described in terms of two distinct elements, fractures and intact rock, each of which has its own characteristic porosity and permeability; and (4) recognized initially by loss of drilling fluid, data derived by use of certain logging techniques, production many times that expected from intergranular porosity and permeability, pressure interference with offset wells as much as 50 mi (80 km) distant (O'Brien, 1953), erratic productivity from different wells in the field, and the general enhancement of production by artificial stimulation. In the past, fractured reservoirs usually have been discovered by accident.

Significant differences exist among fractured reservoirs. 1. The pores of the host rock may or may not contain hydrocarbons (Walters, 1953; Doleschall *et al.*, 1967). 2. The reservoir potential may or may not be indicated by open-hole drill-stem tests (Drummond, 1964; Harp, 1966). 3. Under certain conditions the fluids segregate into zones of high oil and high gas saturation that result in the increase of gas-oil ratios early in the depletion history (Pirson, 1953). 4. For some reservoirs, it is possible, on the basis of buildup and/or drawdown plots (Warren and Root, 1963), to distinguish between fractured reservoirs and those involving only intergranular porosity; for others it is not (Odeh, 1964). 5. The fractures creating the reservoir may be genetically related to local faults and folds (Braunstein, 1953; O'Brien, 1953; Daniel, 1954; Kafka and Kirkbride, 1960; Ogle, 1961; Park, 1961; Pohly, 1962; Ells, 1962; Malenfer and Tillous, 1963; Martin, 1963; Pirson, 1967), may be part of a regional system (Hunter and Young, 1953; Walters,

1953; Wilkinson, 1953; Elkins, 1953; Durham, 1962; Pampe, 1963; Ward, 1965), or they may be related to both (Regan, 1953; Daniel, 1954).

DISCUSSION OF FRACTURE PROCESS
Fracture Phenomena

There are two types of fractures. Shear fractures involve movement parallel with the plane, but there is no perpendicular movement. In extension fractures, the walls move apart. No adequate theory explains either the formation or the propagation of either type of fracture. However, the geometric relations of the fractures with respect to the causative stress state have been well established empirically in the laboratory. For any triaxial stress state[2] there are two potential shear-fracture orientations and one potential extension-fracture orientation. The two shear-fracture planes form with a dihedral angle of about 60°. The axis of greatest principal stress bisects this acute angle; the axis of least principal stress bisects the obtuse angle; and the intersection of the two planes is parallel with the intermediate principal stress axis. The extension fracture is parallel with the plane containing the greatest and intermediate principal stress axes and normal to the least principal stress axis.

The geometry of shear fractures is in accord with the Coulomb (1776) criterion, which states that shear fracturing is controlled not only by the initial shear strength of the material but also by its internal friction. Coulomb suggested that shear fractures would occur, not along planes of maximum shear stress (45° to σ_1), but, because of internal friction, at some angle of less than 45°. The departure from 45° is a function of internal friction and is a property of the material. Specifically, shear fractures develop at $\pm 45° \mp \phi/2$, where ϕ is the angle of internal friction. Though this criterion is nearly 200 years old, its significance has been appreciated by geologists only recently. Experimental work (Handin, 1966) and field work (Stearns, 1967) substantiate that, even in the absence of good mechanistic theory, the Coulomb criterion can be used to predict the geometric relation of shear fractures to the axes of the three principal stresses in the rock at the

[2] In this paper the greatest principal stress axis will be indicated by σ_1, the intermediate principal stress axis by σ_2, and the least principal stress axis by σ_3. Therefore, a triaxial stress state is one in which $\sigma_1 > \sigma_2 > \sigma_3$ compressive stresses are considered positive.

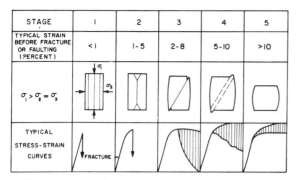

STAGE	1	2	3	4	5
TYPICAL STRAIN BEFORE FRACTURE OR FAULTING (PERCENT)	< 1	1-5	2-8	5-10	>10
$\sigma_1 > \sigma_2 = \sigma_3$					
TYPICAL STRESS-STRAIN CURVES					

FIG. 1—Generalized spectrum of deformation characteristics of rocks as measured and observed in laboratory (after Griggs and Handin, 1960).

time of rupture. The position of the extension fracture, though not included in the Coulomb criterion, can be predicted from the results of this same experimental and field work.

Terminology

Several useful descriptive adjectives can be used to modify the term "fracture." In this paper the term *conjugate* implies fractures that are related by a common origin and that form under a single state of stress. A complete conjugate pattern contains a left-lateral shear, a right-lateral shear, and an extension fracture. Because conjugate fractures arise from the same state of stress, the whole pattern of the three potential fractures relative to one another can be determined at any point where the orientation of only one of the fractures is known. Conjugate shear fractures form in most rock materials in such a way that the angle between them is about 60°.

The term *orthogonal* will imply only that fractures are normal to one another. Because of the restriction to a single stress state placed on conjugate fractures, orthogonal fractures cannot also be conjugate. This is not to say that orthogonal fractures cannot have a common geologic origin; but, because they are normal to one another, whether shear or extension fractures, they cannot be caused from a single stress state, as far as is now known.

Regional fractures are those that pervade a wide area, such as the Colorado Plateau, and apparently are unrelated to local structures. Conversely, structure-related fractures are directly associated with an individual structural feature. These fractures also may be found throughout vast areas, but only where similar local structural features also pervade that area.

Fracture number refers to the average number of parallel fractures in a given set per linear

distance measured in a direction normal to the fracture plane. It is a measure of relative frequency of fracturing. A volumetric or areal measure which includes the surface areas of the fractures would be more meaningful, but ordinarily it cannot be made in naturally deformed rocks. It is assumed, therefore, that the relative number of fractures also reflects the relative surface area of these same fractures. The denominator of the fracture number is 100 ft; thus, a fracture number of 100 implies that there is about one fracture per foot in a given set.

Factors Known to Contribute to Fracture Development

Though not restricted to the brittle domain, rupture is everywhere involved in the deformation of brittle rocks. It follows that those parameters that tend to increase rock ductility tend also to decrease the dependency on rupture. Figure 1 is a summary of the deformation characteristics of rock as measured and observed in the laboratory. Each of the drawings represents a generalized final configuration in longitudinal cross section of an originally intact right cylinder of rock material that has been compressed parallel with the cylinder axis. Ductility increases from left to right, as can be seen from the typical stress-strain curves at the bottom of the illustration or from the average permanent longitudinal strain indicated at the top of each of the columns. Very brittle behavior is characterized by a nearly linear stress-strain response up to the rupture point (stage 1). In most cases only extension fractures are observed. This generalization applies only to macroscopic fractures in standard triaxial tests. The sequence of formation between shear and extension fracturing in general is unresolved. If the stress-strain curve departs from linearity prior to rupture, the observable deformation may be caused by shear fracturing that forms a wedge and terminates in an extension fracture (stage 2). The small wedges give way to failure along discrete throughgoing shear fractures (stage 3). At this stage, some observable permanent distortion of the cylinder wall can be seen, and the specimen is said to have "barreled." Under even more ductile conditions (stage 4), the major offset is along a shear zone rather than along a discrete fracture. This zone contains numerous small fractures, all of which can be of the shear type or mixtures of both shear and extension fractures. Very ductile rocks are characterized by large permanent strain without evidence of macroscopic fracture (stage 5).

Extensive laboratory studies of the mechanical properties of common sedimentary rocks have been published (Handin and Hager, 1957; 1958; Handin *et al.*, 1963; Handin, 1966). Though precise constitutive relations are unknown for rock materials because of insufficient data on time effects, these studies have provided an empirical basis for the single effects of several important parameters. The factors that affect rock ductility and, therefore, rupture are: rock type, temperature, effective confining pressure,[3] and strain rate. Thus, there are several ways that the spectrum of behavior shown in Figure 1 can be achieved. Increasing effective confining pressure or temperature or decreasing strain rate tends to increase ductility. Just how much the ductility increases depends on rock type. Quartzites and dolomites, for example, never become as ductile as stage 5 specimens under environmental conditions encountered in sedimentary basins. Limestones, however, may behave like specimens in stages 1 and 2 at or near the surface, but at 25,000 ft (7,620 m) of burial their behavior may be more similar to stage 5 specimens. Contrasted with quartzite, halite is ductile even at atmospheric conditions. Sandstones, depending on the degree of cementation, behave in a fashion intermediate between those of limestones and dolomites. Because of their variable composition and fabric, shales show the widest range of behavior for any given set of conditions. These statements exclude flow by large-scale cataclasis, which can, of course, lead to grossly ductile behavior but which is accomplished in part by rupture in constrained beds.

The change in ductility of common sedimentary rock types is plotted in Figure 2 against depth of burial, using normal gradients of overburden pressure, pore pressure, and temperature. On the basis of laboratory work, very little difference in ductility of the common sedimentary rock types at near-surface conditions would be expected, but, under several thousand feet of burial, large differences should exist.

In natural rock deformation, overburden, pore pressure, and temperature are not independent variables. In order to compare natural rock deformation under the same environmental conditions, one can restrict his observations to narrow stratigraphic intervals and similar structural positions where the effective overburden pressure and temperature would be about

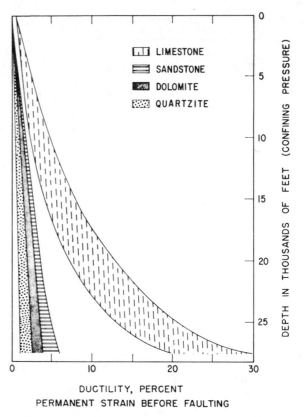

FIG. 2—Ductilities of water-saturated rocks as a function of depth. Effects of confining (overburden) pressure, temperature, and normal formation (pore) pressure are included (after Handin, *et al.*, 1963).

the same. The lithologic effect then can be isolated and studied as a single variable. The results of such a study are shown in Figure 3. The fracture number is computed for many stations in each of the lithologic types and is averaged. These stations are at the same structural position, and they cover a stratigraphic interval that is very small compared to the total depth of burial at the time of deformation. The vertical arrows above quartzite and dolomite indicate that at some stations the rock is too shattered for meaningful measurements. The number above each column indicates the average for that lithologic type exclusive of stations where there is shattering. From these data it can be concluded that, though absolute ductilities (Fig. 2) may differ in nature, primarily because of lower strain rates, the *relative* ductilities of the common sedimentary rock types as measured in the laboratory are validly applicable to natural deformations.

These data become very significant in prospecting for fractured reservoirs. Figure 4A is a thin section made from an experimentally de-

[3] Effective confining pressure is the external hydrostatic confining pressure (analogous to total overburden pressure) less the internal hydrostatic pore pressure (analogous to formation pressure).

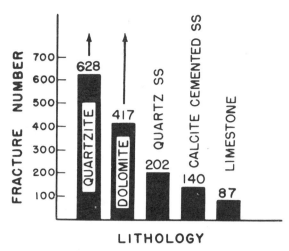

FIG. 3—Average fracture number for several common rock types naturally deformed in the same physical environment (after Stearns, 1967).

formed composite specimen of limestone and dolomite. A standard half-inch right cylinder of limestone was center-bored, and a smaller, tight-fitting right cylinder of dolomite was inserted into the bore. This composite sample then was deformed under 5 kbar confining pressure in an extension test (σ_3 parallel with cylinder axis). The contrasting behavior between the limestone and dolomite is evident in Figure 4A. The limestone deformed by intra-

granular flow mechanisms whereas the dolomite deformed by fracture. Figure 4B shows the same behavioral relation between limestone and dolomite except that they were deformed in a natural environment. It is noteworthy that the vast majority of fractured reservoirs in the Permian basin are in Ellenburger dolomites, not the limestones. From the foregoing data, it is evident that knowledge of the mechanical properties of all common sedimentary rocks, not just limestone and dolomite, should be used actively in prospecting for or developing fractured reservoirs.

NATURAL FRACTURE SYSTEMS
General

Fracture systems that are both pervasive and consistently oriented throughout a large volume of rock can be subdivided into two major classes: regional orthogonal fractures and structure-related fractures. Without further explanation this categorization can be ambiguous. Vast areas commonly may contain one or more fracture patterns, each of which is composed of two very regular and continuous fracture sets that are normal to one another. Their restored orientations are independent of local structure, and these fractures are commonly well developed, even in flat-lying beds. Because all fracturing results from a stress difference, the ori-

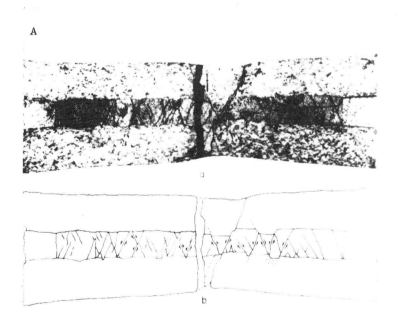

FIG. 4—**A.** Thin section of experimentally deformed composite specimen of limestone (upper and lower units) and dolomite (middle unit; after Griggs and Handin, 1960). **B.** Sequence of naturally deformed limestone (upper and lower units) and dolomite (middle unit). Rock type in either **A** or **B** can be recognized on basis of fracture number.

239

gin of these fractures must be regarded as a problem of structural geology. However, these fractures are excluded from the structure-related class because they are not the result of stress states associated with specific geologic structures. Rather, they are somehow involved in the structural development of an entire region and are classified, therefore, as regional fractures. Structure-related fractures, however, are associated with specific features like folds or faults. Contributing to the ambiguity of this classification is the fact that these structure-related patterns can pervade entire regions where similar parallel structural features also are common. However, any local change in the structural trend changes the trend of the restored orientation of structure-related fractures, whereas restored regional fractures would be unaffected. The structure-related fracturing is a deformation mechanism within a specific structure; it can be homogeneous only where the local structures are also homogeneous throughout the region.

Regional Orthogonal Patterns

Many areas like the Colorado Plateau, Uinta basin, and Piceance Creek basin are characterized by one or more sets of regional orthogonal fractures. In the Uinta basin, for example, a ubiquitous set of fractures strikes nearly north-south. A second set strikes east-west. Both sets are found throughout the basin, are independent of local structures, and are thus regarded as regional orthogonal fractures. They are rotated in places by later folding but, where the beds return to an undisturbed position, the fractures return to their monotonous regional orientations. Throughout the basin, the intersection of the two sets remains normal to the bedding, and any change in strike is due to bed rotation by later folding.

Speculative origins of regional orthogonal fractures vary; they have been discussed by several writers (Blanchet, 1957; Price, 1959; Hodgson, 1961). Low-level cyclic stress differences like those due to lunar tides, which could cause fatigue failure in rocks, are appealing (Blanchet, 1957), but unlikely, sources for regional orthogonal fractures. In some areas, regional fractures are not developed at all, even though the lithologic types are similar to those of other localities where such fractures are found. The lunar tides, of course, affect the entire surface of the earth. Furthermore, two orthogonal patterns (four fractures) which are commonly found in the same area cut rocks of Tertiary age. If lunar tides were the cause, the existence of two patterns would imply a drastic change in the tidal direction.

Hodgson (1961) proposed that regional fractures are inherited, *i.e.,* the pattern is propagated upward into younger beds as they become lithified. This explanation is made doubtful by field evidence from at least one area. On the Uncompahgre Plateau, a regional orthogonal fracture pattern is present in the lower and upper beds of the Jurassic System but is absent in the intervening layers. Here the Entrada Sandstone is a calcite-cemented sandstone about 125 ft (40 m) thick. It is completely free of fractures. Indeed, local inhabitants refer to it as "slick rock." The underlying Kayenta and the overlying Summerville sandstones both contain the same regional orthogonal fracture patterns. That the fractures in the Summerville could be inherited from the underlying beds through a completely unfractured layer 125 ft thick seems most implausible.

Price (1959) gave theoretically, but not always geologically, sound arguments for the formation of regional orthogonal fractures. His mathematical arguments for an origin by uplift are logical and fit certain geologic facts— namely, the common association with plateau-type uplift. Price's arguments, however, do not explain the occurrence of two superposed patterns (four fractures).

Even though none of the explanations for regional orthogonal fractures is satisfactory, several empirical statements can be made concerning their occurrence. These fractures are commonly continuous as a single break or a narrow zone for long distances. On the Colorado Plateau, for example, patterns that are continuous for several miles can be seen on air photographs. If such a fracture were intersected by a borehole, it could greatly affect fluid communication to the well.

Vertical continuity depends on the nature of the layered sequence. These fractures rarely terminate vertically within a bed, but continue vertically through several beds. They generally, but not everywhere, terminate at the contact with a shale unit. In many areas the fractures transect thick units of shale and sandstone with no refraction at the lithologic interfaces. Outcrop conditions usually limit measurements of total vertical continuity, but fractures that cut 400–500 ft (120–150 m) of section are not uncommon. Thus, holes deviated through a producing interval to intersect several regional orthogonal fractures should improve well performance. Furthermore, the orientations of regional orthogonal fracture systems are extremely consistent, and, where they are known from core or outcrop studies, the best direction for deviation can be established.

Fractures Associated with Faults

Fractures associated with faults are generally assignable to the same stress state that caused the fault. The shear fractures are miniatures of the fault. It is not surprising, therefore, that their orientations, as well as that of the associated extension fracture, can be predicted from knowledge of the attitude of the fault. The converse is true also: knowledge of the fracture orientations reveals the strike of the fault as well as its sense of shear.

The orientations of fractures strictly associated with faults change with the fault attitude. This statement holds for both dip and strike. What cannot be predicted for any fault is the relative development of the three potential fractures or the width of the fractured zone parallel with the fault plane. There is no way to estimate the number of fractures that will be associated with a given fault. The total displacement on a large fracture can be accomplished by cumulative slip on several smaller parallel fractures. The amount of throw on a fault lends no insight either. Some large faults have narrow zones of fracturing whereas many small faults have wide zones of small-scale rupture. Furthermore, the shear fracture parallel with the fault is not everywhere well developed; the conjugate shear fracture is commonly more conspicuous.

However, the most likely direction of fluid communication that results from fractures associated with faults can be determined. The strikes of all three potential fractures, as well as their intersection, parallel the faults, so the direction of lateral communication can be predicted. The proper deflection of a hole to intersect the greatest number of fractures depends on the attitude of the fault and the relative development of the associated fractures. For low-dipping faults (of the order of 30°), no deflection would theoretically increase the likelihood of intersection. For vertical faults, deflection either away from or toward the fault in a direction normal to its strike would increase chances of intersection. For normal faults, knowledge of the relative development of the two shear fractures is required. If the fracture that parallels the fault is better developed, deflection toward the fault in the downthrown block and

away from the fault in the upthrown block is most likely. If the fracture conjugate to the fault is better developed, drilling in the reverse directions in the two blocks has the best chance of intersecting the greater number of fractures. Core or outcrop analyses are the best methods upon which to base a decision.

Unlike shear fractures associated with folding, many that are associated with faulting display small visible offset. Movements along the shear fractures can act either to increase or decrease the porosity and permeability. If the movement causes rigid-body rotation of discrete blocks along the fractures, permeability and porosity in the fault zone are increased. Because fracture surfaces are rarely perfectly planar, the displacement can cause a poorer "fit" between adjacent sides of the fracture and thus increase fluid flow along the fracture. If there is also fragmentation into chip-sized material along the fracture, these chips can prop any openings. Decrease of fluid flow, both across and along the fractures, may result from mylonitization, crushing, or smearing—especially if the reservoir is a dirty sandstone. One example is the North dome of Kettleman Hills, where shear fractures are associated with normal faults. In the Etchegoin Formation, shear fractures show very little offset, yet they contain zones of accumulated clay and clay-sized material. This fracture-zone material is so resistant to weathering that the fracture zones stand out in relief relative to the unfractured rock. These fractures would most probably be permeability barriers, because the displacements are less than the formation thickness. If an area is crossed by more than one fault trend, a series of isolated production blocks can result. In these more complicated cases of multiple faulting, the procedure of fracture analysis is essentially the same as for a single trend; nevertheless, additional effort should be made to determine not only the several ages of faulting but also ages relative to oil migration. The nature of the fracture zone associated with any particular fault is best determined from outcrop studies, but considerable information also can be gathered from cores (Friedman, 1969).

A second complication arises where folding is associated with the faulting. All the fracture patterns found on folds then have to be considered along with those associated strictly with the fault. If the only folding is slight drag along the fault, it may have little influence on the pattern. However, there may be a rotation problem. If the fractures associated with the fault-

241

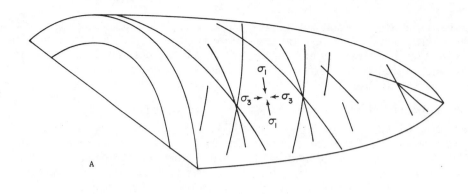

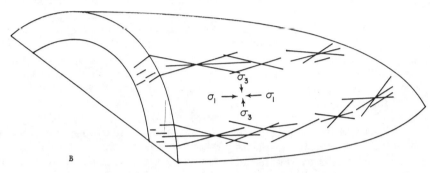

FIG. 5—Schematic illustration of most common fractures associated with a fold. **A.** Pattern 1. **B.** Pattern 2. Both patterns maintain consistent relation to bedding, but not to fold trend.

ing formed early, before much drag had occurred, they would have to be rotated later.

The mechanical interpretations of fractures associated with faults are not a subject of this paper. However, fracturing is best understood in light of the mechanical properties of the rocks (Handin and Hager, 1957), the mechanics of faulting (Anderson, 1951; Hubbert, 1951; Hafner, 1951; Sanford, 1959), and field studies of fractures (Stearns, 1967).

Fractures Associated with Folds

Fractures associated with faults are related to the same stress state that caused the fault. This relation is not true of fractures associated with folds, because analyses indicate that any particular volume of rock can undergo several different stress states through the folding history.

Five major fracture patterns have been found to be associated with folds (Stearns, 1964; 1967). Though all the patterns are significant to an understanding of the folding process, only two have fracture numbers high enough to warrant consideration here. The first pattern consists of two conjugate shear fractures and an extension fracture which indicate that the intermediate principal stress axis (σ_2) is normal to bedding, the greatest principal stress and least principal stress axes (σ_1, σ_3) are in the plane of bedding, and σ_1 is the dip direction (Figs. 5A, 6). On the flank of the anticline the extension fractures of this pattern are in the *ac* geometric plane. This pattern represents shortening in the dip direction, elongation in the strike direction, and no change normal to bedding. The second pattern also consists of two conjugate shear fractures and an extension fracture, thus indicating that σ_2 is still normal to bedding and σ_1 and σ_3 are still in the bedding plane—but σ_1 is parallel with strike and σ_3 is the dip direction (Figs. 5B, 7). On the anticlinal flank the extension fracture of this pattern parallels the *bc* geometric plane. The geometry of pattern 2 represents elongation parallel with dip, shortening parallel with strike, and no change normal to bedding.

That these fractures are the result of the folding process itself and not of the regional stresses that initiated folding is best seen from

FIG. 6—Shear fractures of pattern 1 on flank of fold. Bisector of acute angle between fractures defines dip direction of bed.

the consistent relation of the fractures to bedding orientation. Both patterns maintain their relations to bedding even on the noses of folds, where attitudes depart widely from the average trends of the folds. Figure 8 is a photograph of bedding taken partway around such a nose. The strike of the bed is approximately 45° to the bedding strike in Figures 6 and 7, and to the anticlinal axial trend. The fractures are the shears of pattern 1. Notice that the bisector of the acute angle between the shear fractures remains in the dip direction, and the intersection remains normal to bedding despite the difference between the strike of the bed and the fold axis. A consistent relation is equally true for pattern 2. Therefore, all six fractures are related to the local rotation axis, not to the average trend of the fold axis. The same conclusion is supported by studies of folds that are sinuous along strike. A slight change in strike marks a slight change in the geographic orientation of the fractures, but not in their relation to bedding.

The shear fractures of these two patterns rarely show visible offset. They are designated as a shear or extension type solely on the basis of the geometry of the assemblage, but the consistent and ubiquitous evidence, from both the laboratory (Handin and Hager, 1957; Handin, 1966) and the field (Sheldon, 1912; Melton, 1929; Cloos, 1936; Friedman, 1964; Stearns, 1964; 1967; 1969), cannot be denied. The relative ages of the two patterns cannot be determined absolutely, but pattern 1 probably begins to form earlier than pattern 2. This relation is based on the observation that pattern 1 commonly is present on low-dipping folds without much cross-sectional curvature, even though pattern 2 is absent. However, development of

the two patterns must overlap in time, because a single fracture belonging to either pattern in many places terminates several fractures from the other pattern (Fig. 9). Though the two patterns represent different stress states, they can occur in the same bed (Fig. 9). Pattern 2 represents elongation normal to the anticlinal trend, and as such it probably does not develop until the folding progresses far enough to exceed the ability of the rocks to deform elastically. Fractures of pattern 2 can be explained by large-scale cataclastic flow of the folded sequence in bulk (Stearns, 1969).

The morphologies of the fractures of the two patterns are somewhat different. Many of the fractures of pattern 1 are continuous as single breaks or as zones of parallel subfractures across an entire structure (Fig. 10). Ground inspection of such lineations reveals that they are large fractures with a single orientation, and not an assemblage of all three orientations. These fractures are continuous over large distances laterally and also through several hundred feet of section. Although these fractures are commonly enormous, they occur in all sizes down to the scale of ruptured sand grains. Their homogeneous orientation on all scales is remarkable. Fabric diagrams of pattern 1 shear fractures look the same whether they represent fractures seen in quartz grains or lineations from air photographs. The larger of these fractures could affect fluid communication over vast distances.

The fractures of pattern 2 never attain the very large sizes of pattern 1. They range up to several tens of feet long, but most are only a

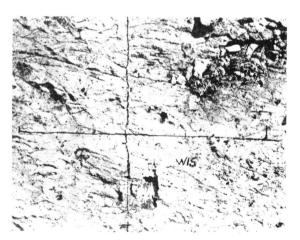

FIG. 7—Fractures of pattern 2 on flank of fold. Horizontal line (W15) is parallel with bedding strike and is 18 in. long. Bisector of acute angle between shear fractures parallels bedding strike.

243

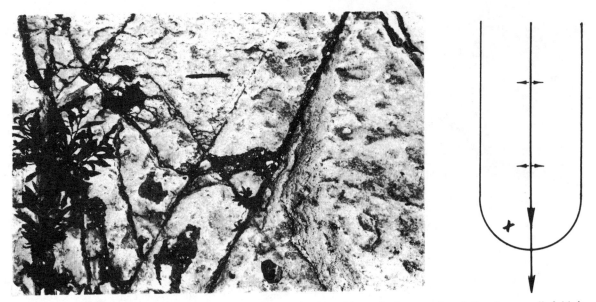

FIG. 8—Shear fractures of pattern 1 on nose of anticline (position of photograph relative to overall fold is indicated on sketch). Pen parallels bedding strike, which is at angle of about 45° to anticlinal trend.

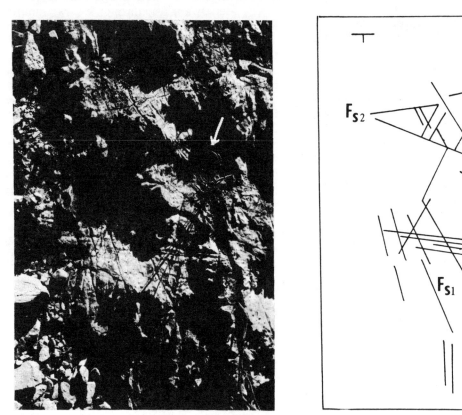

FIG. 9—Photograph and sketch of fractures of both patterns 1 and 2 in same bed. Shear fractures, pattern 1 (F_{s1}); extension fractures, pattern 1 (F_{e1}); shear fractures, pattern 2 (F_{s2}); extension fractures, pattern 2 (F_{e2}).

FIG. 10—View of anticline showing shear-fracture zones of pattern 1 (dark lineations form acute angle in dip direction) and zones of pattern 2 (lineations of trees are parallel with anticlinal trend). Pattern 2 zones are made up of three fracture sets—two shear fractures and an extension fracture.

few inches or a few feet long. Nevertheless, long lineations parallel with the fold trend (the trend of the extension fracture, pattern 2) commonly are found on folds (Fig. 10). Close ground inspection reveals that the lineations usually consist of zones which contain all three fractures in close association. These are measurable in inches or at most in feet, and it is only the trend of the zone that is really continuous parallel with the fold strike, not the individual fractures. The interplay of the fracture sets is more pronounced than that in pattern 1. For example, an extension fracture a few inches long may terminate at one end in a short left-lateral shear and at the other in a short right-lateral shear. The shear fractures in turn may terminate in other extension fractures or perhaps in their conjugate fractures. The overall effect is a lacy pattern in the bedding (Fig. 7). The vertical continuity of these zones can be large also. The most notable exception to the common presence of small fractures is that in a few places a single, isolated extension fracture of this set will be both horizontally and vertically continuous over several hundred feet. Though the continuity of single fractures is greater in general for pattern 1, the fractures of pattern 2 may be more effective fluid passages. There are no data to support this idea, but it is plausible. Because the fractures of pattern 2 are smaller, local rotations of the small blocks

within the zone are likely. This rotation would result in a more open system than that of a single fracture or of elongate parallel slivers. The dense fractures of pattern 2 also may reflect considerable elongation locally that results in dilation (decrease of bulk density) along the fracture zone. Because of the large size and isolation of the fractures in pattern 1, there are three possible directions of communication in any well that intersects them. However, for wells intersecting pattern 2 fracture zones, communication is favored in one direction— parallel with the structural trend.

Factors that Affect Number of Fractures

All other factors being constant, fracture number is most affected by rock type (Figs. 2, 3). This is true of both patterns 1 and 2, but only if fracturing took place at sufficient depth for the effective confining pressure to be significant to the behavior of rocks. Figure 2 shows that under near-surface conditions all rocks are about equally brittle. The divergence in ductility is the result of effective confining pressure, and only under sufficient overburden is lithology an important factor. Exact depth control cannot be specified because of lack of knowledge of the effect of time. However, the laboratory work of Heard (1962) indicates that reduction of strain rates even seven orders of magnitude has little effect on the brittle-ductile transitions at the relatively low temperatures encountered in oil wells.

A second factor that affects fracture number is bedding thickness. It is known from observation that, for given rock type, structural position, and depth of burial, a thin bed will contain more fractures than a thick bed under the same conditions (Harris *et al.*, 1960). A possible explanation is that fracture spacing depends in part on instabilities developed in the extended multilayered medium. Such instabilities depend on bedding thickness as well as contrasts in physical properties of beds (G. M. Sowers, personal commun., 1969).

Thickness and lithology affect fracture numbers in both patterns 1 and 2. However, structural position is also significant to pattern 2, because here fracturing is an active mechanism of cataclastic flow. The intensity of fracturing is greatest where the rate of change of dip is greatest. This parameter should not be confused with the steepness of dip. Steep dips can result simply from rigid-body rotations between hinge points, and strain need not be large. However, changes in dip over short lateral distances can result in large local strains (elongation or shortening). The elongation can be accomplished by fracturing (pattern 2) along zones parallel with the fold hinge. In these zones, fracture numbers as high as 15,000 have been recorded (Stearns, 1969). In some hinges where the dip change is very great over short distances (60° or more in a few hundred feet laterally), the rock units, including limestones, can be completely shattered by unordered fracturing. The lack of order may result from numerous local stress concentrations developed at fracture intersections.

Influence of Fracture on Reservoir Porosity and Permeability

Estimate of Fracture Porosity

Fracture width, area, spacing, surface roughness, and filling are the primary factors governing fracture porosity. Reported values of fracture porosity are low, from <0.05 percent (Snow, 1968) to 6 percent (Regan and Hughes, 1949, p. 47). These low values are not surprising if one realizes that a porosity of less than 2 percent is provided by unfilled fractures 0.04 in. (1 mm) wide developed around the six faces of 1 cu ft of rock. The ability to estimate fracture porosity is, however, essential to reservoir evaluation. Several approaches to this problem have been reported, and a few examples are discussed below.

Elkins (1953, p. 181–182) measured fracture widths in cores of Spraberry Sandstone and estimated the spacing of vertical fractures from their observed frequency in 3.5-in.-diameter (8.9 cm) cores from five wells and from pressure-buildup data. He found fracture widths up to 0.013 in. (0.33 mm), but the average was 0.002 in. (0.051 mm); he concluded that spacing ranged from a few inches up to a few feet. From these data he calculated a fracture volume of 110 bbl per acre-foot.

A geometric approach applicable to folded rocks was developed by Murray (1968, p. 60). On the assumption that extension fractures form in the outer layers of curved beds, he showed that the fracture porosity varies directly as the product of bed thickness and curvature, *i.e.*,

$$\phi f = \frac{T}{2R} = 1/2T \frac{d^2z}{dx^2}, \tag{1}$$

where T is thickness of bed, R is the radius of curvature, and d^2z/dx^2 is the curvature (dz/dx is dip). This approach is particularly attractive because the calculation is independent of fracture spacing, although spacing is considered in deriving the equation. Moreover, R can be determined from dip information because it is the reciprocal of d^2z/dx^2. Murray applied this analysis to the Sanish pool, McKenzie County, North Dakota, and demonstrated a good coincidence between areas of maximum curvature and areas of best productivity. Using maximum and minimum values of T and R, respectively, he calculated a fracture porosity of 0.05 percent.

Snow (1968) described a method for calculating fracture porosity, openings, and spacings from permeabilities measured in drillholes. The method applies only to rocks with negligible intergranular porosity; it is not valid if solution enlargements, porous interbeds, breccia zones, or fracture spacing less than about a third of the separation between straddle packers is present. Results from studies of selected dam sites indicate maximum near-surface fracture porosities of 0.05 percent. These decrease by an order of magnitude each 200 ft (60 m) within the depth of usual dam-site exploration. Fracture widths are typically 0.004 in. (0.101 mm) at 50 ft (15 m) to 0.002 in. (0.051 mm) at 200 ft (60 m). Average fracture spacing is 4 ft (1 m) near the surface and increases to 14 ft (4 m) at depths of 300 ft (90 m).

Fracture porosity also can be estimated from conventional electric logs where certain conditions obtain and the porosity of the intact reservoir rock is known (Pirson, 1967; Pirson et al., 1967). A fracture-intensity index, FII, is related to fracture porosity through total porosity, ϕt, and the porosity of the unfractured reservoir rock, ϕb, as follows:

$$FII = \frac{\phi t - \phi b}{1 - \phi b}. \tag{2}$$

Pirson et al. (1967, p. 6) demonstrated that FII is calculated from relations between the resistivities of the mud filtrate, the fracture system filled with mud filtrate, the intact rock, the formation water, the invaded zone (from the short-normal curve), and the zone of 100 percent water saturation (from the long-normal or induction curves). The fracture porosity then can be determined if ϕb is known. Applying this method to a section of Austin Chalk from a producing well in the Salt Flat–Tenney Creek field, Caldwell County, Texas, they calculated a fracture porosity of 0.2 percent.

Estimates of Fracture Permeability

Although the absolute porosity provided by fractures is low, the effective porosity is high because the available fracture-void volume is connected. It follows that the influence of fracturing on reservoir permeability is most important. Pertinent analyses of fluid flow through parallel plates (or fractures) were given by Lamb (1932), Muskat (1937), Huitt (1956), and Parsons (1966), among others. The equation for volumetric flow rate between two smooth parallel plates (after Lamb, 1932) is the common point of departure as follows:

$$q = \frac{-g_c b W_f^3}{12\mu} \frac{dp}{dL}, \tag{3}$$

where:

q = volume of flow per unit length
W_f = width of fracture
b = depth of fracture
dp/dL = pressure gradient in direction of flow
g_c = a conversion factor
μ = absolute viscosity of fluid.

This equation is approximately valid for viscous, laminar flow for variable values of W_f, provided that the gradient dW_f/dL is small; it applies even if both bounding surfaces are curved (Huitt, 1956, p. 259). Huitt considered the influences of surface roughness and turbulent flow and concluded that, if flow in fractures is laminar, surface roughness has no appreciable effect on the resistance to flow. Surface rough-

ness becomes a significant factor where the flow is turbulent. From consideration of production rates from oil wells, however, both Huitt and Parsons regarded turbulent flow in fractured reservoirs as unlikely.

Baker (1955, p. 384) illustrated the importance to permeability of fracture width, using the basic relations of equation (3). For example, a single fracture 0.01 in. (0.25 mm) wide has the equivalent permeability of 454 ft (188 m) of unfractured rock with a uniform permeability of 10 md; an 0.05-in-wide (1.27 mm) fracture is equivalent to 568 ft (173 m) of rock with a permeability of 1,000 md.

Parsons (1968) modified equation (3) so as to include the permeability of the total fracture-intact rock system in which vertical fractures occur in sets of specified spacing and orientations relative to the overall pressure gradient. He expressed the total permeability of the fracture-rock system as:

$$k_{fr} = k_r + \frac{W a^3 \cos^2 \alpha}{12A} + \frac{W b^3 \cos^2 \beta}{12B} + \cdots, \tag{4}$$

where

k_{fr} is the permeability of the total system
k_r is the permeability of the intact rock
Wa, Wb are the widths of fractures in sets a, b, etc.
α, β are angles between fracture sets a, b, respectively, and overall pressure gradient
A, B are the fracture spacing (perpendicular distance between fractures for sets a, b, etc.).

Further insight into the role of fractures in reservoir permeability has been gained through study of electrical analog models. McGuire and Sikora (1960) showed that the width of artificial fractures is much more important than their length in affecting communication between natural fractures. However, little is gained by widening isolated short fractures because the unfractured formation beyond their ends controls the rate of flow. Huskey and Crawford (1967) kept their simulated fractures at constant width but varied fracture shape, orientation, density, and total length. They found that fracture density correlated closely with production capacity and that the influence of fracture shape was small. Fractures oriented parallel with the isopotential lines [$\alpha = 90°$, equation (4)] do not increase the effective permeability.

In summary, the existing knowledge of fluid flow in fractures provides at least a first approximation of measured flow rates and, what is important here, an appraisal of those attributes of fractures that are truly significant to the problem. The permeability is directly proportional to the cube of the fracture width, inversely proportional to the spacing between fractures, and dependent upon the fracture orientation relative to the direction of the pressure gradient. The permeability decreases with fracture filling and roughness of the fracture surfaces (where the flow is turbulent). It follows that the permeability of a fracture system can be expected to be greatest where the reservoir bed contains wide, closely spaced, smooth fractures oriented parallel with the fluid pressure gradient.

Estimates of fracture permeability range from a few millidarcys to many darcys. Methods used to estimate fracture permeability include pressure-buildup studies, interpretations of volumes of oil produced relative to production from unfractured reservoirs of known permeability, and certain geometric relations. Elkins (1953, p. 184–186) used pressure-buildup data to calculate the average effective permeability in the Spraberry Sandstone reservoir (16 md). This permeability is the same order of magnitude as that deduced by Dyes and Johnston (1953) from a similar analysis of the Spraberry trend. Elkins tested his result by calculating the fracture widths and spacings that would yield this permeability. He found that the 16-md permeability would be provided by fractures .0011 in. wide (0.28 mm), spaced 4 in. (10 cm) apart, or by .0015-in.-wide (0.38 mm) fractures 10 in. (25 cm) apart, or by .002-in.-wide (0.51 mm) fractures 24 in. (61 cm) apart. These predicted widths compare favorably with the average of 0.002 in. actually measured in cores. Some other analyses pertinent to pressure-buildup data are by Miller *et al.* (1950) and Adams *et al.* (1968).

Permeability estimates commonly are made from the volume of oil produced. Regan (1953, p. 213) estimated a maximum permeability of 35 darcys and an average of 10–15 darcys from production rates for the chert zone early in the history of the Santa Maria Valley field, California. He did this by comparing the production rates from the fractured chert with those from "oil sands" in various other fields.

Murray (1968, p. 60–61), reasoning from equation (3), demonstrated that, in folded beds with extension fractures normal to bedding and parallel with the fold axis, the permeability and the porosity (see preceding discussion) are functions of bed thickness and curvature, as follows:

$$K = \frac{A^1}{48T^2}\left(T\frac{d^2z}{dx^2}\right)^3, \qquad (5)$$

where T is the bed thickness, d^2z/dx^2 is the curvature, and A is the average area per fracture perpendicular to the direction of flow. After conversion to millidarcys and evaluation of A in terms of an assumed fracture spacing of 6 in. (15 cm), equation (5) reduces to:

$$K = 4.9 \times 10^{11}\left(T\frac{d^2z}{dx^2}\right)^3. \qquad (6)$$

The following examples of permeability values were given by Murray (1968, Table 1, p. 61):

$T = 5$ ft		$T = 10$ ft	
$\frac{d^2z}{dx^2}(\times 10^{-5})$	K (md)	$\frac{d^2z}{dx^2}(\times 10^{-5})$	K (md)
1	0.06	1	0.49
2	0.49	2	3.92
4	3.92	4	31.20
6	13.20	6	106.00

He demonstrated that zones of high productivity in the Sanish pool coincide with those in which the curvature exceeds a certain minimum value. This fact would seem to verify his approach; however, an independent evaluation of permeability was not given.

As stated, fractures can either enhance or restrict fluid flow. McCaleb and Wayhan (1969) reported an interesting case in which both restriction and increase in fluid communication as a result of fractures are found in the same reservoir. At the Elk Basin field, fractures form a pressure-communication barrier in the "A" zone which causes an abrupt pressure drop of 2,000 psi across the fractured area. However, they reported that in other areas the "D" zone has a sufficiently high fluid movement to be assignable as "a fracture-permeability reservoir" (McCaleb and Wayhan, 1969, p. 2111). Emmett *et al.* (1969, p. 3) discussed fractures in the Little Buffalo Basin field (Tensleep reservoir) that are so open that ". . . upon completion the well produced 100-percent water even though analysis of the matrix rock indicates it should be oil productive." McCaleb and Willingham (1967, p. 2126), in their study of the Cottonwood Creek field, stated: ". . . where developed, the fracture systems probably offer the primary means available for fluid movement in the reservoir."

DETECTION OF FRACTURES

Early recognition of a fractured reservoir will influence the location and number of subsequent development wells, and, therefore, is of major economic significance. Some of the methods used to detect and study fractures *in situ* are listed below in the approximate order that they would be used.

1. Loss of circulating fluids during drilling is widely recognized as a positive indication that a fractured or cavernous formation has been penetrated. (Very narrow fractures that can greatly increase the effective permeability may be missed, however.) The axiom is (Daniel, 1954): field wells in which a large volume of mud is lost almost without exception become the major producers if they produce at all.

2. Fractures in cores provide direct information on fracture development if natural fractures can be distinguished from those induced by coring, and if fracture spacing and orientation permit adequate sampling in cores (Friedman, 1969, p. 369–375). For some rocks, poor core recovery suggests intensely fractured zones. At least, porosity and permeability data for unfractured samples of the reservoir rock are obtained from the cores. These measurements are preferably made at the pressures of the reservoir (Fatt and Davis, 1952; Fatt, 1953, 1958).

3. Well-test analyses provide at least two types of data from which the influence of fractures can be evaluated: production rates and pressure-buildup interference. Perhaps the most common clue to a fractured reservoir is a flow rate many times that to be expected from the porosity and permeability data of the unfractured reservoir rock.

Use of pressure-buildup data was reviewed by Odeh (1964), Matthews and Russell (1967), and Adams *et al.* (1968), among others. Pressure-buildup data consist basically of well-bore pressure versus shut-in time (Horner, 1951; Black, 1956; Gray, 1962). The shape and slope of the observed curve or line are compared with those expected from various models of the reservoir. For example, Warren and Root (1963) analyzed such data in order to determine both intergranular porosity and fracture or vugular porosity on the assumption that the former contributes significantly to the pore volumes but negligibly to the flow capacity. Adams *et al.* (1968, p. 1193), using 2-week pressure-buildup tests run every 2 to 5 months for several years, recognized sharp downward bends in the pressure-time curve. They stated that ". . . although many factors—faults, stratification, heterogeneity, fracturing, etc.—are known to cause a pressure buildup curve to

bend upwards, very few effects other than a permeability increase or other improvements in rock flow character are known to cause a decrease in the straight line slope." They interpreted the increase in effective permeability to be caused by a fracture (or other discontinuity) at some distance from the well bore. The time of the buildup-curve bend can be used to estimate the radial distance to the fracture. This approach is one of the few techniques capable of detecting the influence of fractures away from the borehole, and it is widely used.

4. Certain well-logging techniques may be helpful in recognizing and locating fractures (Pickett and Reynolds, 1969). Pirson (1963, 1967), Nechai and Mel'nikov (1963), and Pirson *et al.* (1967) discussed the use of conventional logs to evaluate fracture porosity and thus to detect fractures (see equation 2). Sonic and sonic-amplitude logs have been used to locate fractures, but their interpretations are complicated by many factors (see review by Morris *et al.*, 1964).

5. Several downhole direct and indirect viewing systems, including downhole photographic and television cameras, packer impressions, and "acoustical picture" reconstructions, have been used to detect fractures on the borehole wall. All these devices include a system for obtaining the azimuthal orientation of the record. Direct optical systems employing 16 mm and 35 mm multiframe borehole cameras have been described by Dempsey and Hickey (1958), Jensen and Ray (1965), and Mullins (1966). Briggs (1964) described a television system for immediate viewing of the borehole wall. Each of these visual systems requires a transparent environment and light source. Each is limited in depth by ambient temperatures. Packer-impression techniques require the inflation of an impressionable tubing against the walls of the borehole (Fraser and Pettitt, 1962; Anderson and Stahl, 1967). When the tubing is partly deflated and withdrawn from the borehole, the trace of a fracture can be observed where the tubing had been injected into the opening.

The Borehole Televiewer described by Zemanek *et al.* (1969) uses an entirely new concept to provide a picture of the reservoir rock. A piezoelectric transducer probes the borehole wall with acoustic energy. The amount of energy reflected back to the transducer is a function of the properties of the rock. In general, irregularities reduce the amplitude of the reflected signal. Return signals are observed on an oscilloscope and photographed for permanent record. A flux-gate magnetometer senses the earth's magnetic field and provides the means for determining the orientation of the log. As with the camera techniques, features on the cylindrical borehole are observed only in two dimensions (Fig. 11). A typical Borehole Televiewer log (Fig. 12) clearly shows fractures and small vugs. The tool now can be used at temperatures up to 300°F and in a borehole filled with any homogeneous, gas-free liquid like fresh water, saturated brine, crude oil, or drilling muds. All these techniques give information on fracture development directly adjacent to the borehole. When used in combination with well-test analyses, they can provide the confirmation of fracturing necessary for complete reservoir evaluation.

6. One common characteristic of fractured reservoirs is that the productivity of wells is greatly increased by artificial stimulation (Elkins, 1953; Walters, 1953; Hunter and Young, 1953; Daniel, 1954; Baker, 1955; Hassebroeck and Waters, 1964; and Shearrow, 1968). For example, in the Spraberry Sandstone of West Texas, initial potentials of wells range up to 1,000 bbl/day after fracture treatment as compared to estimated capacities of 5–10 bbl/day if oil had to flow into the borehole through the siltstone itself.

Acidizing of fractured carbonate reservoirs is done primarily to increase the width of natural and artificial fractures and the size of the borehole. Increasing the diameter of the borehole enhances the probability of intersecting natural fractures. In addition, Baker (1955, p. 384–387) showed that for oil with 50 lb-ft³ density and 1 cp viscosity, channeled to a 6-in. (15 cm) diameter borehole through a horizontal fracture 0.04 in. (1.00 mm) wide with a radius of 660 ft (200 m), most of the pressure drop occurs within a foot of the hole. An acid treatment to increase the borehole diameter to 9 in. (23 cm) would reduce the total pressure drop and increase the flow capacity significantly.

7. Other techniques for fracture detection become useful where several wells penetrate the same reservoir. All are based on evidence of fluid communication between wells which can be explained only by the presence of fractures. Pressure interference is perhaps the most common of these, and it generally is cited as clear-cut evidence of a fractured reservoir (Elkins, 1953, 1960; O'Brien, 1953; Baker, 1955). In the Agha Jari field, Iran, pressures in two ob-

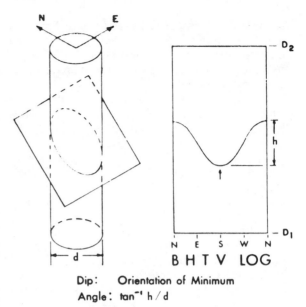

Dip: Orientation of Minimum
Angle: $\tan^{-1} h/d$

BHTV LOG

FIG. 11—Isometric sketch of fracture or bedding plane intersecting borehole at moderate dip angle, and corresponding Borehole Televiewer log (after Zemanek et al., 1969, Fig. 7).

servation wells, 11.5 mi (18.4 km) apart, fell at nearly the same rate within a lag of only a few pounds per square inch (Baker, 1955, p. 386). Similar evidence of communication over distances of 40–50 mi (64–80 km) have been reported (O'Brien, 1953; Baker, 1955). Elkins (1960) used pressure-interference data to determine the effective fracture orientation in parts of the Spraberry field. Furthermore, from the observation that all development wells within a given area in the Spraberry field exhibit reduced initial reservoir pressures (i.e., they all drain the same interconnected system of fractures), he concluded that the number of such wells could have been reduced and their spacing increased.

Other methods in this category include monitoring water-breakthrough times in waterflood tests and gas-injection tracer tests (Barfield et al., 1959, and Elkins, 1960). McCaleb and Willingham (1969) described the Permian Phosphoria reservoir, Cottonwood Creek field, Wyoming, in which the true nature of the reservoir (fracture plus intergranular porosity) was discovered only after poor gas- and water-injection performances were noted. The rapid channeling of injected phases stimulated a review of available geologic, engineering, and production data, which resulted in recognition of the dual nature of the reservoir. Data on communication times and directions relative to

FIG. 12—Borehole Televiewer log in fractured West Texas Precambrian formation. Well bore was filled with drilling mud (after Zemanek *et al.,* 1969, Fig. 9).

surrounding producing wells can be used to calculate effective permeabilities.

TECHNIQUES THAT AID IN SUBSURFACE-FRACTURE INTERPRETATIONS

Fractures in Cores

Studies of cores containing natural fractures can provide statistical information on fracture spacing, width, and orientation. In addition, the effective directional permeability due to small fractures can be measured in places, and indirect methods for fracture detection can be calibrated against the observed occurrence. These studies require, however, that (1) natural fractures are distinguishable from those artificially induced by coring, (2) biases inherent in sampling fractures in cores are considered, and (3) core orientation is known. Care must be exercised from the time the core is permanently removed from the barrel to record whatever information is available on core orientation and to recover and preserve maximum, intact lengths of core. Fractures are best observed if full cores are washed and slabbed along several cuts parallel with the core axis. Very fine frac-

tures can be detected by X-ray techniques or by use of fluorescent-dye penetrants (Gardner and Pincus, 1968).

There commonly is a problem in distinguishing natural fractures from those induced by coring, from partings along bedding planes, or from separations that might have occurred during handling of the cores. A workable classification is as follows (Friedman, 1969, p. 369): *unequivocal* natural fractures are partly or completely filled with gouge or vein material or are unfilled but parallel with nearby filled fractures (including microfractures in individual grains as observed in thin section); *very probable* natural fractures are those with slickensided surfaces and fractures parallel with them; *probable* natural fractures have clean fresh surfaces and are accompanied by parallel incipient fractures (or microfractures) that, in turn, are parallel with unequivocal fractures in the same or adjacent length of core. Criteria to recognize fractures induced by coring are not so straightforward. We have observed cores broken repeatedly by fresh clean fractures that within short distances curve from subparallel to sub-

251

perpendicular to the core axis. These may result from twisting or bending of the rock cylinder during coring. The fractures that produce disked cores are also probably artificially induced. These fractures are evenly and closely spaced and are typically oriented nearly perpendicular to the core axis. The disking may be caused by coring rock in which the *in situ* differential stress is high (Obert and Stephenson, 1965). Disking thus may suggest the possibility of potential casing collapse or formation failure during the production of the well. Though disking may be triggered by the coring, it may occur also along incipient unequivocal natural fractures (Friedman, 1969) or parting planes. We have found that certain phenomena like fracturing, which perfectly parallel the axis of the core or the development of plumose or concentric markings on fracture surfaces, are not reliable criteria for distinguishing between natural and artificial fractures.

Two factors introduce bias in sampling macrofractures (*i.e.*, fractures visible without optical magnification and measurable with hand tools; see below). These are the spacing of the fractures relative to the diameter of the core and the orientation of the fractures relative to the hole deviation (core axis). If, for example, the natural macrofractures are actually vertical and are spaced 1 ft (30 cm) apart, they obviously are poorly sampled by even a 6-in. (15 cm) core bit in a vertical hole. The sampling improves, however, as the angle between the fracture and the core axis increases (Friedman, 1969, p. 374). It is not generally realized that, if a perfectly uniform three-dimensional array of fractures (*i.e.*, fractures developed equally in all possible orientations) is cored, the frequency distribution of observed angles (θ) between the core axis and the normal to the fracture surface (reduced to a two-dimensional plot) is Gaussian, having a peak at $\theta = 45°$. This Gaussian distribution exists because the distribution of angles between fracture surfaces and core axis (after Bloss, 1957) is:

$$P = 100 \int_{\theta_1}^{\theta_2} \sin \theta d\theta = 100 (\cos \theta_1 - \cos \theta_2), \quad (7)$$

where P is the probability and $\theta_1 - \theta_2$ is the cell width in degrees; and the preferential sampling of fractures by a core, which increases as the sine of the angle between the core axis and the fracture plane, is:

$$P' = P_\theta \sin (90° - \theta), \quad (8)$$

where P' is the final probability and P_θ is the probability at a given θ from equation (7).

Friedman (1969, p. 374) used the relation in equation (8) to weight observed frequencies of different fracture sets in cores from the Saticoy field, California, in order to determine their order of relative abundance. The sampling biases do not influence the study of microfractures from individual sand grains, because the core dimensions are very large compared to the scale of the grains. A sufficient number of microfractures in any possible orientation can be measured in three mutually perpendicular thin sections to give a statistically sound three-dimensional sample. It is also important to realize that poor core recovery from a given interval may imply the presence of a highly fractured zone.

The absolute orientation of fractures in cores can be obtained only when the *in situ* orientation of the core itself is known. First the core must be oriented, then the fractures in the oriented core must be accurately measured. Modern oriented coring devices satisfy the first requirement, although most cores are not oriented during routine operations. Cores not oriented *in situ* can be oriented later from the traces of bedding on the surfaces and from dipmeter and hole-deviation data. Cores taken from holes deviated in excess of 10° from the vertical in thin-bedded, horizontal rocks can be oriented from the elliptical trace of bedding on the deviated core. Even where oriented cores are available, it is a cumbersome task to hold the core in its *in situ* orientation during measurement of the orientations of fractures. Moreover, the fractures are visible only at the surface of the opaque core. In order to alleviate these difficulties, we designed a fracture-orienting device which permits both the orientation in space of a core and the accurate measurement of the strike and dip of fractures or other planar or linear features that transect the core. The method is to (1) insert the core into a suitably sized plastic cylinder, (2) draw onto the surface of the cylinder the traces of bedding and usable fracture planes (Fig. 13A, B), (3) orient the cylinder (now representative of the core) as it was *in situ* from knowledge of hole deviation and strike and dip of bedding (Fig. 14), and (4) measure the attitudes of the fracture traces. A plastic cylinder is used rather than the core itself, because the tracings of planar features clearly define easily measured planes when viewed through the transparent cylinder, and because of the mechanical difficulties involved in supporting and rotating the heavy, oddly shaped core. Cores taken from a

252

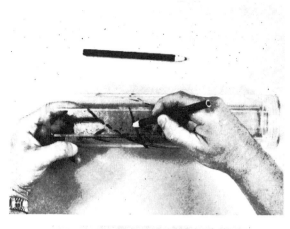

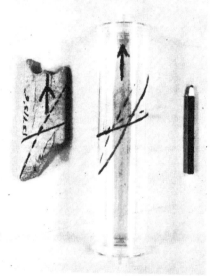

FIG. 13—Photographs of core and plastic-cylinder facsimile: *upper*—Core is held in plastic cylinder of similar diameter while traces of bedding and fracture are marked on surface of cylinder; *lower*—Traces of planar features are clearly visible through plastic cylinder.

deviated well represent the most complicated case for which the apparatus was designed. The apparatus can be used also for cores taken from vertical wells or for cores which previously have been oriented.

The fracture-orienting device was used by Friedman (1969) in the analysis of fractures in cores from the Saticoy field, Ventura County, California. The study shows that both the macro- and microfracture patterns are reasonably homogeneous along the length of the field and that these fracture patterns, as well as small changes in the relative abundance of microfractures, can be used to predict the orientations and relative positions of the two major

faults in the field. The producing sandstones are the lower Santa Barbara and the upper Pico (upper Pliocene); they are not regarded as fractured reservoirs, but both macro- and microfractures do influence productivity. For example, the fractures introduce a marked directional permeability. The permeability across the array of *unequivocal* natural fractures (corehole 2, Fig. 15) is 10 md, that parallel with the fractures and containing the gouge material (corehole 3, Fig. 15) is 40 md, and that parallel with the fractures and not in the gouge (corehole 1, Fig. 15) is 140 md. This differential permeability probably exists in much of both reservoirs because there is a statistically homogeneous fracture pattern developed along the length of the Saticoy field. Emmett *et al.* (1969) also noted a directional effect by fractures on permeability. They reported a 10–30 percent reduction in permeability normal to fractures as compared to permeability parallel with the fractures in the Little Buffalo Basin field.

Fracture Prediction as an Exploration Method

In the introduction we noted that most fractured reservoirs are "discovered" after the initial discovery. We know of no clear-cut example in the published literature of methods of exploration for fractured reservoirs. Therefore, this section contains suggestions as to what could be done rather than methods which have been proved.

Sufficient laboratory data are available to allow construction of fracture-facies maps. The first requisite of such a construction is the ability to establish the burial history at the time of hydrocarbon migration. From burial data, limits can be placed on the effective confining pressure and temperature throughout a basin. Knowing the conditions under which the fracturing had to take place enables one to establish the lithologic types most likely to contain a high density of fractures. This approach can be augmented profitably if precise outcrop studies have been made in the exploration area. Environmental parameters then can be integrated even more closely into the mapping, and absolute rather than relative fracture numbers can be determined for certain rock types or lithofacies. The effect of bed thickness likewise can be evaluated, and those assemblages of rocks that would best serve as fractured reservoirs can be mapped just as any other facies element. Furthermore, if the fractures are structure-related,

FIG. 14—Photograph of fracture-orienting device showing plastic cylinder oriented as corresponding core *in situ*. Borehole deviation is 24° to the S22°W, read at indexes (A) and (B), respectively. Strike of planar features determined by sighting along "strike bar" (C) and read at index (D). Dip of planar feature read at (E).

certain overlay maps may outline targets more exactly. For example, areas of anticipated exceptionally high fracture density, like trends of high rate of change of dip, can be plotted on the fracture-facies map. Overlay maps that include the orientations of various types of fractures (pattern 1 or 2, or fractures associated with faults) may pinpoint specific locations within the potential fracture facies where intersecting fracture zones would occur. Updip limits of fracture facies can be mapped with the same accuracy as for any other facies mapping.

An obvious example of the fracture-facies approach would involve mapping the distribution of dolomitic or siliceous facies relative to limestone or shale facies. If the brittle-ductile facies contact crosses a structure so that the ductile facies is updip from the brittle facies, dry holes in the structurally high areas do not preclude potential discoveries in the downdip brittle facies.

Specific testing of physical properties of lithologic types in a given area also may aid in exploration. If, early in the exploitation of an oil province, a given lithofacies is recognized as a fractured reservoir, then the question arises

as to what other facies may also contain fractured reservoirs. The Sooner Trend on the northeast flank of the Anadarko basin in Oklahoma is a case in point (Durham, 1962; Ward, 1965). Brittle, cherty members of the thick Mississippian carbonate section yield high initial potentials. The cherty members grade laterally into siltstone and shelf carbonate facies.

One approach to the development of such an area would be to restrict drilling to the proved fractured, cherty facies. This strategy would require, however, an accurate knowledge of the distribution of that facies, and other fractured reservoirs might be missed. A second approach would be to investigate experimentally whether other associated facies could be potentially fractured before attempting any stepout drilling. Assume, for example, that facies A forms a fractured reservoir and we wish to determine if facies B and C might also be fractured. Assume also that all three facies occur in the same structural framework and have the same deformation history. Typical specimens of the three facies could be fractured experimentally under properly simulated physical conditions reconstructed from the geologic setting. These

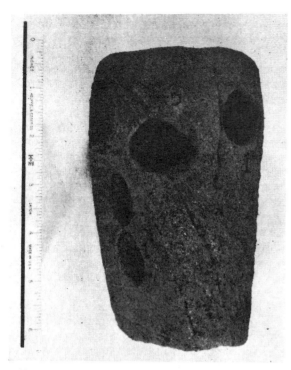

FIG. 15—Photograph of core from Saticoy No. 14, 9,400–9,420 ft, Saticoy field, Ventura County, California, showing location of perm-plugs (1–3) discussed in text. Permeability data were furnished by Shell Development Co.

tests could be repeated at the same temperatures and effective pressures but at different strain rates. If all three facies should prove to be brittle and strain-rate insensitive, but should fracture at different magnitudes of differential stress, certain conclusions could be drawn. If facies A has a higher fracture strength than do facies B and C, it can be concluded that B and C are potential fracture reservoirs. No such prediction could be made for any facies stronger than A. Should a facies fail by ductile flow at a yield stress lower than the fracture strength of facies A, the reservoir quality of that lithofacies would depend on the survival of intergranular porosity during the flow process.

Surface Measurements of Fractures

The purpose of measuring fractures in the field is to establish the orientations and morphologic characteristics of the fracture sets in a given region. In the absence of core data, these surface measurements should be projected into the subsurface. The extrapolation may well be uncertain, but even a qualitative prediction of subsurface conditions is better than none at all.

Obviously, every single fracture cannot be measured or even seen in the field. A major difficulty, therefore, as with fractures in cores, is one of sampling. If the data are to be projected eventually into the subsurface, the collection of these data should be aimed at a good statistical sampling of the following: (1) separation of all fractures into sets and determination of the orientations of these sets; (2) measurements of fracture numbers for each set including variations with bed thickness, lithology, and structural position; (3) establishment of the horizontal and vertical continuities of all sets; and (4) observations of morphologic differences between the sets.

Erosion is a large deterrent to sampling. Highly developed fracture sets or systems commonly accelerate both mechanical and chemical weathering processes, which may result in removal of the fractured rock or in burial under erosional debris. Weathering along fracture zones frequently develops erosional gullies. Debris collects at the bottoms of depressions and obscures the fractures from view. If an area contains parallel gullies, their bottoms should be examined carefully. Of course, the mere existence of parallel drainage systems does not necessarily prove a fracture-controlled origin.

Erosional control by fracture creates still another difficulty of good sampling, namely, that of outcrop boundary orientations. Rock ledges usually spall parallel with major fracture systems. Eventually, ridges develop parallel with these fractures so that the maximum lateral distance of outcrop exposes the least number of fractures. In order to compare the number of fractures between fracture sets, it must be determined that the outcrop examined did not favor the exposure of one particular fracture set. If the outcrop pattern is linear, it is advisable to record its direction. The use of air photographs simplifies this problem. If the preferred development of some particular orientation is suspected but there are too few measurements to be certain, outcrops that favor the suspected fracture set should be sought.

In outcrop studies, the following data should be collected at each station: (1) attitude of bedding, (2) average attitude and variation in attitude of each fracture set, (3) fracture number for each fracture set, (4) lithology, (5) complete description of fracture morphology by set, and (6) an oriented sample (necessary for correlation of the microfracture and macrofracture fabrics).

Fracture number can be recorded as either absolute or relative. An individual worker can develop his own subjective system of relative fracture number such as high, medium, or low, based upon his own impression at each outcrop; arbitrary but consistent limits should be set on the relative values. With experience, this system becomes reliable, but the relative measurements of several workers should not be intermixed. The easiest way to make absolute measures of fracture number is to mark an outcrop with a tape in any convenient direction, independent of the orientation of the fracture sets for which the fracture number is to be determined. By knowing the orientation of the fracture set and measuring the azimuth and plunge of the tape as it lies on the outcrop, one can calculate the angle between the line of the tape and the normal to the given fracture plane. This angle also can be measured from a stereographic net. The actual fracture number is:

$$F_n = \frac{F_t}{L} \cos \alpha \times 100 \qquad (9)$$

where F_t is the number of fractures in a given set which intersect the tape, L is the taped length, and α is the angle between the line of the tape and the normal to the fractures. The meaningful fracture number for a given lithology is found by averaging results from several similar stations. However, in analyzing the data one must make sure that there is no systematic change in, for example, structural position.

Before treating the bulk data, one should rotate the bedding strikes to a common strike line to determine the true nature of the fracture patterns. To make a proper interpretation, special attention should be given to stations where the entire pattern—two shear fractures and one extension fracture—occurs in a single outcrop, and these observations should strongly influence the interpretation of the bulk data. If data are gathered from many stations where the fractures all belong to patterns 1 and 2, the fractures will tend to group around the anticlinal dip and strike directions. Even though each grouping spreads ±30° from these directions, one may be tempted to interpret this as a single pattern of orthogonal fractures. This error can be avoided by referring to those stations where the total pattern is recorded in a single outcrop. If the beds at most of the stations dip more than about 5°, they should be rotated to a horizontal position before one makes the final interpretation. This is particularly true of fractures that strike at low angles to the bedding but dip at high angles to the bed.

Measurements from Aerial Photographs

The use of aerial photographs in fracture studies has advantages and disadvantages. The decision to use aerial photographs is, therefore, dictated by the final use of the data derived from them. One advantage is that strong fracture trends can be determined for large areas quickly and inexpensively. A disadvantage is that fractures are three-dimensional features, but any single surface trend is only a two dimensional representation and may represent several different fracture sets. Furthermore, no data concerning fracture morphology can be derived from air photos. If the desired end result involves only the surface trends of fractures—e.g., in secondary recovery programs—aerial photography can be a useful tool (Alpay, 1969). Aerial photographs should not be used without extensive ground control if the desired end result requires knowledge of: (1) fracture morphology, (2) small-scale fracture densities, (3) relative development of specific orientations of a conjugate pattern, or (4) determination of total fracture geometries.

Probably the most successful application of aerial photographs to fracture problems is a combined field and photo study. Initially, the major trends, changes in trend, and outcrop areas can be determined by measurements on the photomosaics. Specific locations for ground stations can be selected to sample properly the aerial control. Ground measurements can be made at these stations (as described in a preceding section), and any changes in attitude, relative development, morphology, etc., can be integrated over the photographs from the surrounding ground control. This method not only saves time over studies that are totally ground-controlled, but it also makes fracture studies possible in areas with limited accessibility (because of topography or the attitude of landowners). However, even with this method caution must be exercised to avoid selection of anomalous ground stations which would have significant bearing in the analysis.

All complete surface studies should include examination of aerial photographs, because some fracture systems are too large to be observed on the ground. For example, the individual fractures of pattern 2 that make up the lineation in Figure 10 can be measured on the ground, but the fact that they form continuous zones can be seen only on the air photos.

SUMMARY AND CONCLUSIONS

A basis for the prediction of the relative development and orientations of fractures in an unknown province is provided by an understanding of the interactions between three primary factors:

1. The physical environment at the time of fracturing, *i.e.*, effective confining pressure, temperature, and strain rate. These three parameters strongly affect the behavior of rock material.

2. The magnitudes and orientations of the three principal stresses in the rock body at the time of fracture. The relative stress differences control the locations of the fractures, and the orientation of the stress field determines the potential fracture orientations.

3. The nature of the sedimentary layer, including the degree of induration at the time of fracturing and the relative thickness of the rock units. Thickness and lithology determine which rocks in a mixed sequence are more likely to be fractured.

Precise predictions of fracture spacing, areal extent, width, types (shear or extension), and exact locations within a rock body are not possible because the processes of fracturing and the compositions of the deformed body are so complex. It is possible to specify, at least approximately, the history of burial of a rock unit and something of its present structural geometry. From the former, the maximum temperature and overburden pressure affecting the rock can be determined. Qualitative predictions of the expected fracture development and orientation can be made from knowledge of the general geology, the known associations of fractures with structures, and the relations of fracturing to rock type, thickness, and structural position. Outcrop studies can be made initially and projected into the subsurface. As more seismic and well data become available, the prediction will become more quantitative. Though fracturing is a complicated process, laboratory and field studies provide as good a basis for estimating fracture development and trends as is available for many other geologic phenomena which are fearlessly predicted during the exploration of an area.

The recent literature reflects our growing ability to use fracture analysis as a productive geologic tool. The number of papers that pertain to laboratory techniques, field studies, and theory is increasing. The general relations of fracturing to lithology and geologic structure were summarized by Stearns (1967). Harris *et al.* (1960) described fracture development in the Big Horn basin, and this can be extended to any Rocky Mountain intermontane basin. Investigations like those of Braunstein (1953), McCaleb and Willingham (1967), Murray (1968), Martin (1963), and Friedman (1969) illustrate how fracture problems can be treated in specific oil fields. McCaleb and Willingham demonstrated the importance of structural residual mapping and trend-surface analyses to increase production in fractured reservoirs. Murray applied quantitative methods to producing fields, and Friedman and Martin used subsurface data in a fracture analysis. The theoretical work of Hafner (1951) and Sanford (1959) showed the expected stress fields associated with certain types of faulting; from these the anticipated fracture patterns can be calculated. Other theoretical studies of folding—*i.e.*, Biot (1961) and Dieterich and Carter (1969)—consider not only how the stress trajectories change with time in the folding process, but also the delineation of the zones of highest strain where fracturing should be most intense. All these papers and many others have contributed to the knowledge of the three primary factors and to field application. There is now a firm basis from which to work, and fractured reservoirs should be found in the future—not by accident, but by intent.

REFERENCES

Adams, A. R., Ramey, H. J., Jr., and Burgess, R. J., 1968, Gas-well testing in a fractured carbonate reservoir: Jour. Petroleum Technology, v. 20, p. 1187–1194.

Alpay, A. O., 1969, Application of aerial photographic interpretation to the study of reservoir natural fracture systems: Soc. Petroleum Engineers, Ann. Fall Mtg., October, SPE No. 2567, 11 p.

American Geological Institute (J. V. Howell, chm.), 1957, Glossary of geology and related sciences: Natl. Research Council Pub. 501, 325 p.

Anderson, E. M., 1951, Dynamics of faulting and dyke formation, with applications to Britain, 2d ed.: Edinburgh, Oliver and Boyd, 206 p.

Anderson, T. O., and Stahl, E. J., 1967, A study of induced fracturing using an instrumental approach: Jour. Petroleum Technology, v. 19, p. 261–267.

Baker, W. J., 1955, Flow in fissured formations: 5th World Petroleum Cong. Proc., Sec. II/E, p. 379–393.

Barfield, E. C., Jordan, J. K., and Moore, W. D., 1959, An analysis of large scale flooding in the fractured Spraberry trend area reservoir: Jour. Petroleum Technology, v. 11, p. 15–19.

Biot, M. A., 1961, Theory of folding of stratified viscoelastic media and its implications in tectonics and orogenesis: Geol. Soc. America Bull., v. 72, p. 1595–1620.

Black, W. M., 1956, A review of drill-stem testing techniques and analysis: Jour. Petroleum Technology, v. 8, p. 21–30.

Blanchet, P. H., 1957, Development of fracture analysis as exploration method: Am. Assoc. Petroleum Geologists Bull., v. 41, p. 1748–1759.

Bloss, D. F., 1957, Anisotropy of fracture in quartz: Am. Jour. Sci., v. 225, p. 214–226.

Braunstein, Jules, 1953, Fracture-controlled production in Gilbertown field, Alabama: Am. Assoc. Petroleum Geologists Bull., v. 37, p. 245–249.

Briggs, R. O., 1964, Development of a downhole television camera: 5th Ann. SPWLA Logging Symposium, Midland, Texas, Houston, Texas, May; Soc. Prof. Well Log Analysts.

Cloos, Hans, 1936, Einführung in die Geologie: Berlin, Gebrüder Borntraeger, 503 p.

Coulomb, C. A. de, 1776, Sur une application des règles maximis et minimis à quelques problèmes de statique, relatif à l'architecture: Acad. Sci. Paris Mém. Math. Phys., v. 7, p. 343–382.

Daniel, E. J., 1954, Fractured reservoirs of Middle East: Am. Assoc. Petroleum Geologists Bull., v. 38, p. 774–815.

Dempsey, J. C., and Hickey, J. R., 1958, Use of a bore-hole camera for visual inspection of hydraulically-induced fractures: Producers Monthly, June, p. 18–21.

Dieterich, J. H., and Carter, N. L., 1969, Stress-history of folding: Am. Jour. Sci., v. 267, p. 129–154.

Doleschall, S., et al., 1967, The examination of the functional mechanism in the Nagylengyel-type fractured limestone reservoirs: Banyaszati Lapok, v. 100, p. 268–275.

Drummond, J. M., 1964, An appraisal of fracture porosity: Bull. Canadian Petroleum Geology, v. 12, p. 226–245.

Durham, C. A., Jr., 1962, Petroleum geology of Southeast Lincoln oil field, Kingfisher County, Oklahoma: Shale Shaker, v. 13, p. 2–10.

Dyes, A. B., and Johnston, O. C., 1953, Spraberry permeability from build-up curve analyses: AIME Trans., v. 198, p. 135–138.

Elkins, L. F., 1953, Reservoir performance and well spacing, Spraberry trend area field of West Texas: AIME Trans., v. 198, p. 177–196.

—— 1960, Determination of fracture orientation from pressure interference: AIME Trans., v. 219, p. 301–304.

Ells, G. D., 1962, Structures associated with the Albion-Scipio oil field trend: Michigan Geol. Survey Div., 86 p.

Emmett, W. R., Beaver, K. W., and McCaleb, J. A., 1969, Little Buffalo basin, Wyoming, Tensleep heterogeneity—its influence on infill drilling and secondary recovery: Soc. Petroleum Engineers Ann. Fall Mtg., October, SPE No. 2643, 8 p.

Fatt, I., 1953, The effect of overburden pressure on relative permeability: Jour. Petroleum Techology, v. 5, Tech. Note 194, October, p. 15–16.

—— 1958, Pore volume compressibilities of sandstone reservoir rock: Jour. Petroleum Technology, v. 10, Tech. Note 2004, Mar., p. 64–66.

—— and Davis, D. H., 1952, Reduction in permeability with overburden pressure: Jour. Petroleum Technology, v. 4, Tech. Note 147, Dec., p. 16.

Fraser, C. D., and Pettitt, B. E., 1962, Results of a field test to determine the type and orientation of a hydraulically induced formation fracture: Jour. Petroleum Technology, v. 14, p. 463–466.

Friedman, M., 1964, Petrofabric techniques for the determination of principal stress directions in rocks, p. 451–552 in State of stress in the earth's crust: New York, Elsevier, 732 p.

—— 1969, Structural analysis of fractures in cores from Saticoy field, Ventura County, California: Am. Assoc. Petroleum Geologists Bull., v. 53, p. 367–398.

Gardner, R. D., and Pincus, H. J., 1968, Fluorescent dye penetrants applied to rock fractures: Internat.

Jour. Rock Mechanics and Mining Sci., v. 5, p. 155–158.

Gray, K. E., 1962, How to plot pressure build-up curves: World Oil, v. 154, p. 82–91.

Griggs, D. T., and Handin, J. W., 1960, Observations of fracture and a hypothesis of earthquakes: Geol. Soc. America Mem. 79, p. 347–364.

Hafner, W., 1951, Stress distribution and faulting: Geol. Soc. America Bull., v. 62, p. 373–398.

Handin, J. W., 1966, Strength and ductility, Sec. 11 in Handbook of physical constants—revised edition: Geol. Soc. America Mem. 97, p. 223–289.

—— and Hager, R. V., 1957, Experimental deformation of sedimentary rocks under confining pressure—tests at room temperature on dry samples: Am. Assoc. Petroleum Geologists Bull., v. 41, p. 1–50.

—— and —— 1958, Experimental deformation of sedimentary rocks under confining pressure—tests at high temperature: Am. Assoc. Petroleum Geologists Bull., v. 42, p. 2892–2934.

—— et al., 1963, Experimental deformation of sedimentary rocks under confining pressure: Pore pressure tests: Am. Assoc. Petroleum Geologists Bull., v. 47, p. 717–755.

Harp, L. J., 1966, Do not overlook fractured zones: World Oil, v. 162, p. 119–123.

Harris, J. F., Taylor, G. L., and Walper, J. L., 1960, Relation of deformational fractures in sedimentary rocks to regional and local structure: Am. Assoc. Petroleum Geologists Bull., v. 44, p. 1853–1873.

Hassebroek, W. E., and Waters, A. B., 1964, Advancements through 15 years of fracturing: Jour. Petroleum Technology, v. 4, p. 760–764.

Heard, H. C., 1962, The effect of large changes in strain rate in the experimental deformation of rocks: PhD thesis, Univ. California, Los Angeles.

Hodgson, R. A., 1961, Regional study of jointing in Comb Ridge–Navajo Mountain area, Arizona and Utah: Am. Assoc. Petroleum Geologists Bull., v. 45, p. 1–38.

Horner, D. R., 1951, Pressure build-up in wells: 3d World Petroleum Cong. Proc., Sec. II.

Hubbert, M. K., 1951, Mechanical basis for certain familiar geologic structures: Geol. Soc. America Bull., v. 62, p. 355–372.

—— and Willis, D. G., 1955, Important fractured reservoirs in the United States: 4th World Petroleum Cong. Proc., Sec. I/A/1, p. 58–81.

Huitt, J. L., 1956, Fluid flow in simulated fractures: Am. Inst. Chem. Engineers Jour., v. 2, p. 259–264.

Hunter, C. D., and Young, D. M., 1953, Relationship of natural gas occurrence and production in eastern Kentucky (Big Sandy gas field) to joints and fractures in Devonian bituminous shale: Am. Assoc. Petroleum Geologists Bull., v. 37, p. 282–299.

Husky, W. I., and Crawford, P. B., 1967, Performance of petroleum reservoirs containing vertical fractures in the matrix: Soc. Petroleum Engineers Jour., v. 7, p. 221–228.

Jenson, O. F., Jr., and Ray, William, 1965, Photographic evaluation of water wells: Log Analyst, March, p. 15–26.

Kafka, F. T., and Kirkbride, R. K., 1960, The Ragusa oil field, in Excursion in Sicily, May 27–30, Rome Petroleum Explor. Soc. Libya: p. 61–85.

Lamb, H., 1932, Hydrodynamics, 6th ed.: New York, Dover Publications, p. 581–582.

Malenfer, J., and Tillous, A., 1963, Study of the Hassi Messaoud field stratigraphy, structural aspect, study of detail of the reservoir: Inst. Franç. Pétrole Rev., v. 18, p. 851–867.

Martin, G. H., 1963, Petrofabric studies may find fracture porosity reservoirs: World Oil, v. 156, p. 52–54.

Matthews, C. S., and Russell, D. G., 1967, Pressure buildup and flow tests in wells: Soc. Petroleum Engineers Monograph Ser., no. 1.

McCaleb, J. A., and Wayhan, D. A., 1969, Geologic reservoir analysis, Mississippian Madison Formation, Elk Basin field, Wyoming-Montana: Am. Assoc. Petroleum Geologists Bull., v. 53, p. 2094–2113.

—— and Willingham, R. W., 1967, Influence of geologic heterogeneities on secondary recovery from Permian Phosphoria reservoir, Cottonwood Creek field, Wyoming: Am. Assoc. Petroleum Geologists Bull., v. 51, p. 2122–2132.

McGuire, W. J., and Sikora, V. J., 1960, The effect of vertical fractures on well productivity: AIME Trans., v. 219, p. 401–403.

Melton, F. A., 1929, A reconnaissance of the joint systems in the Ouachita Mountains and Central Plains of Oklahoma: Jour. Geology, v. 37, p. 733–738.

Miller, C. C., Dyes, A. B., and Hutchinson, C. A., Jr., 1950, The estimation of permeability and reservoir pressure from bottom hole pressure build-up characteristics: AIME Trans., v. 189, p. 91–104.

Morris, R. L., Grine, D. R., and Arkfeld, T. E., 1964, Using compressional and shear acoustic amplitudes for the location of fractures: Jour. Petroleum Technology, v. 4, p. 623–632.

Mullins, J. E., 1966, New tool takes photos in oil and mud-filled wells: World Oil, v. 163, p. 91–94.

Murray, G. H., Jr., 1968, Quantitative fracture study—Sanish pool, McKenzie County, North Dakota: Am. Assoc. Petroleum Geologists Bull., v. 52, p. 57–65.

Muskat, M., 1937, The flow of homogeneous fluids through porous media: New York, McGraw-Hill Book Co.,

Nechai, A. M., and Mel'nikov, D. A., 1963, A study of reservoir properties of strata in Northeast Caucasus areas by geophysical means: Vses. Nauchno-Issled. Inst. Geofiz. Metodov Razved. Trudy, p. 44–54 (in Russian).

Obert, L., and Stephenson, D. E., 1965, Stress conditions under which core discing occurs: Soc. Mining Engineers Trans., v. 232, p. 227–235.

O'Brien, C. A. E., 1953, Discussion of fractured reservoir subjects: Am. Assoc. Petroleum Geologists Bull., v. 37, p. 325–326.

Odeh, A. S., 1964, Unsteady-state behavior of naturally fractured reservoirs: Soc. Petroleum Engineers of AIME Proc., 39th Ann. Mtg., Houston, SPE Preprint No. 966, 6 p.

Ogle, B. A., 1961, Prospecting for commercial fractured "shale" reservoirs, Rocky Mountains: Am. Assoc. Petroleum Geologists Bull., v. 45, p. 407.

Pampe, C. F., 1963, Fort Trinidad's 25 sq. mile area promises big reserves: Oil and Gas Jour., v. 16, no. 27, p. 162–166.

Park, W. H., 1961, North Tejon oil field: California Div. Oil and Gas, California Oil Fields—Summ. Operations, v. 47, p. 13–22.

Parsons, R. W., 1966, Permeability of idealized fractured rock: Soc. Petroleum Engineers Jour., v. 6, p. 126–136.

Pickett, G. R., and Reynolds, E. B., 1969, Evaluation of fractured reservoirs: Soc. Petroleum Engineers Jour., v. 9, March, p. 28–38.

Pirson, S. J., 1953, Performance of fractured oil reservoirs: Am. Assoc. Petroleum Geologists Bull., v. 37, p. 232–244.

—— 1963, Handbook of well log analysis: Englewood Cliffs, Prentice-Hall, Inc., p. 303–314.

—— 1967, How to map fracture development from well logs: World Oil, v. 164, p. 106, 108, 113–114.

—— J. P. Trunz, Jr., and Gomez N., P., 1967, Fracture intensity mapping from well logs and from structure maps: 8th Ann. SPWLA Logging Symposium Trans., Denver; Houston, Texas, Soc. Prof. Well Log Analysts, p. B1–B23.

Pohly, R. A., 1962, Gravity work may aid search for Trenton fracture zones: World Oil, v. 154, p. 85–88.

Price, N. J., 1959, Mechanics of jointing in rocks: Geol. Mag., v. 96, p. 149–167.

Regan, L. J., 1953, Fractured shale reservoirs of California: Am. Assoc. Petroleum Geologists Bull., v. 37, p. 201–216.

—— and Hughes, A. W., 1949, Fractured reservoirs of Santa Maria district, California: Am. Assoc. Petroleum Geologists Bull., v. 33, p. 32–51.

Sanford, A. R., 1959, Analytical and experimental study of simple geologic structures: Geol. Soc. America Bull., v. 70, p. 19–51.

Shearrow, G. G., 1968, The story of Ohio's southeastern sleeper: Oil and Gas Jour., v. 66, p. 210–212.

Sheldon, Pearl, 1912, Observations and experiments on joint planes: Jour. Geology, v. 20, p. 54–79.

Smekhov, E. M., 1963, Fractured oil and gas reservoir problem and present status of its study: 2d U.N. Develop. Petroleum Resources, Asia and Far East Symposium Proc., v. 1, p. 476–481.

Snow, D. T., 1968, Rock fracture spacing openings and porosities: Am. Soc. Civil Engineers Proc., Jour. Soil Mechanics and Found. Div., v. 94, no. 1, p. 73–91.

Stearns, D. W., 1964, Macrofracture patterns on Teton anticline, northwest Montana (abs.): Am. Geophys. Union Trans., v. 45, p. 107–108.

—— 1967, Certain aspects of fracture in naturally deformed rocks, p. 97–118 in R. E. Riecker, ed., NSF advanced science seminar in rock mechanics: Bedford, Massachusetts, Air Force Cambridge Research Lab. Spec. Rept.

—— 1969, Fracture as a mechanism of flow in naturally deformed layered rocks: Conference on Research in Tectonics Proc.; Canada Geol. Survey Paper 68–52, p. 79–96.

Walters, R. F., 1953, Oil production from fractured Pre-Cambrian basement rocks in central Kansas: Am. Assoc. Petroleum Geologists Bull., v. 37, p. 300–313.

Ward, L. O., 1965, Mississippian Osage, northwest Oklahoma platform: Am. Assoc. Petroleum Geologists Bull., v. 49, p. 1562–1563.

Warren, J. E., and Root, P. J., 1963, The behavior of naturally fractured reservoirs: Soc. Petroleum Engineers Jour., v. 3, p. 245–255.

Wilkinson, W. M., 1953, Fracturing in Spraberry reservoir, W. Texas: Am. Assoc. Petroleum Geologists Bull., v. 37, p. 250–265.

Zemanek, J., et al., 1969, The Borehole Televiewer—a new logging concept for fracture location and other types of borehole inspection: Jour. Petroleum Technology, v. 246, p. 762–774.

Published as Chapter 1 in *Geologic Aspects of Naturally Fractured Reservoirs.*

1
Geologic Aspects

Present-day fractured reservoirs have been accidentally discovered when looking for some other type of reservoir. McNaughton and Garb

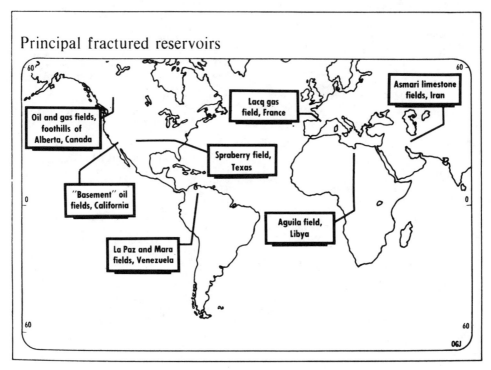

Fig. 1-1 Locations of principal oil and gas fields producing from fractured rocks. (After McNaughton & Garb)

estimate that ultimate recovery from currently producing fractured reservoirs will surpass 40 billion stock tank barrels of oil (STBO). In spite of this very attractive figure, fractured reservoirs had not received the attention they deserved until recently.

Fig. 1-1 shows the location of some important fractured reservoirs in the world. McNaughton and Garb indicate that recovery from fractured limestones in Iran and Iraq (1) will probably be over 30 billion STBO. Gas recovery from Lacq field in France (2) will probably be over 8 trillion cu. ft. Recovery from fractured igneous and metamorphic rock and fractured limestones in Venezuela (4) will probably be over 1.5 billion STBO. Other important fractured reservoirs, such as Aguila field of Libya (3), the Spraberry trend of Texas (5), the oil and gas fields in the foothills of Alberta, Canada (6), and the "basement" oil fields of California (7) are also shown in Fig. 1-1.

Requirements for Hydrocarbon Accumulation

In general, a petroleum reservoir consists of source rock, reservoir rock, seal rock, trap, and fluid content.

Source rock, or source environment, is believed to be responsible for the origin of petroleum. Most geologists believe that the origin of petroleum is organic, related mainly to vegetables which were altered by pressure, temperature, and bacteria. It is difficult to prove that oil actually came from a definite source. However, it is believed that the source rock is usually near the hydrocarbon reservoir, i.e., that petroleum was formed within that particular area. Source rock is difficult to identify because it usually contains no visible hydrocarbons. Snider indicates that the main source rock is shale, followed by limestone.

Reservoir rock is provided by porous and permeable beds. Precise determination of porosity is important for accurate calculations of hydrocarbon-in-place. Permeability is an important parameter in calculating flow capacities.

Igneous, sedimentary, or metamorphic rocks can make an acceptable reservoir rock. However, most of the world's hydrocarbon accumulations occur in sandstones and carbonate rocks.

Seal rock confines hydrocarbons in the reservoir rock because of its extremely low level of permeability. Usually, seals have some plasticity, which allows them to deform rather than fracture during earth crust movements. The most important seal is shale, followed by carbonate rocks and evaporites.

A *trap* is formed by impervious material which surrounds the reservoir rock above a certain level. The trap holds the hydrocarbons in the reservoir. Traps are formed by a variety of structural and stratigraphic features. Landes provides a simplified classification of oil and gas traps:

I. Structural traps
   a. dry synclines
   b. anticlines
   c. salt-cored structures
   d. hydrodynamic
   e. fault
II. Stratigraphic traps
   a. Varying permeability caused by sedimentation.
   b. Varying permeability caused by ground water.
   c. Varying permeability caused by truncation and sealing.

The combination of structural and stratigraphic features is not uncommon.

Fluid content is the water and hydrocarbon that occupy the porous beds.

Porosity

Porosity represents the void space in a rock. It can be quantified by dividing the void space by the bulk volume of the rock. In general, porosity can be classified as primary and secondary.

Primary porosity

Primary porosity is established when the sediment is first deposited. Thus, it is an inherent, original characteristic of the rock.

For example, a sandstone rock usually has primary porosity. The value of primary porosity depends on many factors, including its arrangement and distribution, cementation and degree of interconnection among the voids.

Therefore, it is necessary to distinguish between total primary porosity and effective porosity. Total primary porosity is the ratio between the total primary void spaces and the bulk volume of the rock. Effective primary porosity is the ratio between the interconnected void space and the bulk volume of the rock. The commercial point of view is interested in effective porosity.

Graton and Fraser have evaluated porosity of the systematic packing of spheres (Fig. 1-2). For example, in the cubic or wide-packed system, porosity can be evaluated by first considering the volume of the spheres (or sand-grain volume).

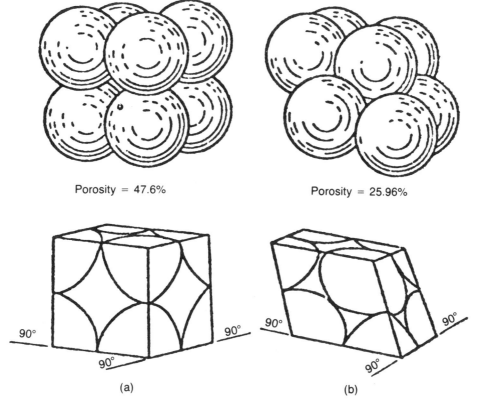

Porosity = 47.6%   Porosity = 25.96%

(a)   (b)

Fig. 1-2 Unit cells and groups of uniform spheres for cubic and rhombohedral packing. (a) Cubic or wide-packed; (b) rhombohedral, or close-packed. (After Fraser & Graton)

$$Sphere\ volume = \frac{4\pi r^3}{3} \qquad (1\text{-}1)$$

where r is the radius of the sphere. The unit cell presented in the lower part of Fig. 1-2 is a cube with each side equal to 2r, or:

$$Bulk\ volume = (2r)^3 = 8\ r^3 \qquad (1\text{-}2)$$

By definition, porosity (ϕ) is equal to the void space divided by the bulk volume of the rock, or:

$$\phi = \frac{void\ space}{bulk\ volume} = \frac{bulk\ volume - sphere\ volume}{bulk\ volume} \qquad (1\text{-}3)$$

For cubic packing:

$$\phi = \frac{8\ r^3 - (4\pi r^3/3)}{8\ r^3} = 0.476\ or\ 47.6\% \qquad (1\text{-}4)$$

263

Notice in Eq. 1-4 that porosity for the cubic or wide-packed system is only a function of packing and is independent of the radius of the sphere as these radii cancel out.

For the rhombohedral or closed-packed system, Graton and Fraser found a porosity of 25.96%. In general, primary porosities are lower than the above theoretical values due to cementation, irregularity of grain size, and shaliness. For example, the porosity of a clean, average sandstone is about 20%.

Secondary Porosity

Also known as induced porosity, secondary porosity is the result of geologic processes after the deposition of sedimentary rock and has no direct relation to the form of the sedimentary particles. Most reservoirs with secondary porosity are either limestones or dolomites.

In general, secondary porosity is due to solution, recrystallization and dolomitization, and fractures and joints.

Secondary porosity by solution can be generated by percolating acid waters which dissolve mostly limestones and dolomites, thus improving their porosities.

Dolomitization improves porosity of carbonates. The equation describing dolomitization can be written as:

$$\text{limestone} \qquad \qquad \text{dolomite}$$
$$2\ CaCO_3 + Mg\ CL_2 \longrightarrow CaMg(CO_3)_2 + CaCL_2$$

Stage	1	2	3	4	5
Typical strain before fracture or faulting (%)	<1	1-5	2-8	5-10	>10
$\sigma_1 > \sigma_2 = \sigma_3$					
Typical stress-strain curves					

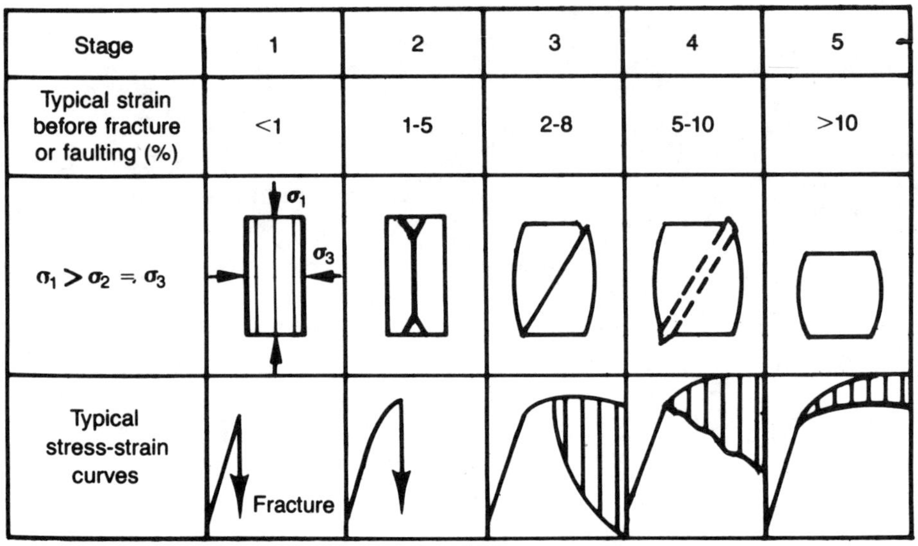

Fig. 1-3 Generalized spectrum of deformation characteristics of rocks as measured and observed in laboratory. (After Griggs & Handin)

264

Porosity is improved because the transformation from calcite (limestone) to dolomite results in a volume shrinkage which creates voids in the rocks (Hohlt).

Fractures and joints are usually formed in rocks which are brittle.

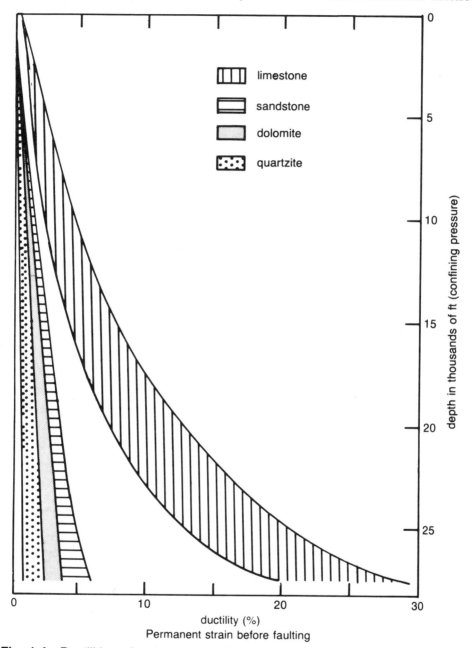

Fig. 1-4 Ductilities of water-saturated rocks as a function of depth. Effects of confining (overburden) pressure, temperature, and normal formation (pore) pressure are included. (After Handin et al.)

265

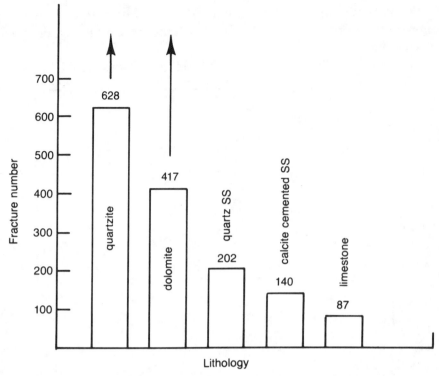

Fig. 1-5 Average fracture number for several common rock types naturally deformed in the same physical environment. (After Stearns)

After analyzing the deformation characteristics of rocks in the laboratory, Griggs and Handin found that rock with very low ductility was characterized by a nearly linear stress-strain relationship, as shown in stage 1 of Fig. 1-3. As ductility is increased (stages 2 to 4), the stress-strain curve departs from linearity; however, some fractures are still formed. For very ductile rocks (stage 5), there is a large, permanent strain and a lack of fracturing.

Handin ascertained that ductility is affected by rock type, temperature, and net overburden. Fig. 1-4 shows how ductilities vary as a function of depth for various lithologies. Quartzite is less ductile (more brittle), followed by dolomite.

These findings were corroborated by Stearns who measured the relative frequency of fractures in various lithologies. He found the highest degree of fracturing was present in quartzite, followed by dolomite (Fig. 1-5). With Friedman, Stearns has done an excellent, comprehensive study of ductility.

Evaluating the value of fracture porosity with a satisfactory level of certainty is difficult even with whole core analysis, because cores usually break along the natural fracture planes. However, fracture systems do provide important storage capacity in some reservoirs.

Important hydrocarbon reservoirs are found in fractured cherts, shales, limestones, dolomites, siltstone, sandstone, igneous, and metamorphic rocks.

Fracture widths are usually very small, varying from paper-thin to 6 mm and more. The other two dimensions are also variable.

Fracture generation is attributed to three main causes:

- Diastrophism as in the case of folding and faulting. The faulting tends to generate cracks along the line of fault, which in turn produce a zone of dilatance. The dilatancy effect is probably responsible for a large part of the migration and accumulation of petroleum in fracture reservoirs.
- Deep erosion of the overburden that permits the upper parts to expand, uplift, and fracture through planes of weakness.
- Volume shrinkage, as in the case of shales that lose water, cooling of igneous rocks, and dessication of sedimentary rocks.

Storage Capacities

Developing naturally fractured reservoirs has led to numerous economic failures. Initial high oil rates led engineers to overestimate production forecasts of wells. Reservoir engineers usually assume two things: (1) the fractures have a negligible storage capacity and are only channels of high permeability that allow fluids to flow; and (2) the matrix has an important storage capacity, but a very small permeability.

The first assumption has led to many fiascos in developing naturally fractured reservoirs. In fact, many reservoirs that produce at high initial rates decline drastically after a short period of time. This occurs because the producible oil has been stored in the fracture system. Consequently, it is important to estimate oil-in-place reasonably within the fracture system.

The second assumption must be considered carefully. If the permeability of the matrix is very low, then the oil bleed-off from the matrix into the fractures will be very slow and only the oil originally within the fractures will be produced in a reasonable span of time. If the matrix has a reasonable permeability, then the storage capacity of the matrix becomes of paramount importance.

It is important to visualize that the storage capacity of naturally fractured reservoirs varies extensively, depending on the degree of fracturing in the formation and the value of primary porosity. The greater the value of primary porosity, the greater the success possibilities of naturally fractured reservoirs.

The storage capacity in the matrix porosity of Fig. 1-6A is large compared with the storage capacity in the fractures. For the schematic of

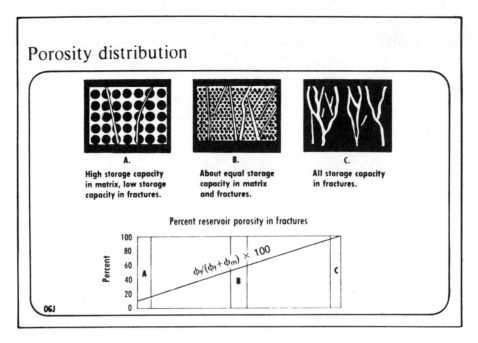

Fig. 1-6 Schematic sketches showing porosity distribution in fractured reservoir rocks. (After McNaughton & Garb)

Fig. 1-6A, 10% of the total porosity is made out of fractures. In this case the fractures may create problems during drilling operations, such as mud losses, blowouts, and fires.

The schematic of Fig. 1-6B shows a rock with about the same storage capacity in fracture and matrix porosities. In this case, the reservoir matrix is tight and the fractures provide avenues of nearly infinite permeability. According to McNaughton and Garb, this ideal combination of porosities has facilitated production of over 100 million STBO from individual wells in Iran.

Fig. 1-6C shows the schematic of a rock where the matrix porosity is zero and all the storage capacity is due to fractures. Reservoirs of this type are generally characterized by initially high production rates that decline to uneconomic limits in a short period of time.

However, there are exceptions. For example, the Edison and Mountain View fields in the San Joaquin Valley of California and the El Segundo, Wilminton, and Playa de Rey fields in the Los Angeles Basin produced above 15,000 bo/d from fractured pre-Cretaceous basement schist. The storage in the basement rock of the La Paz-Mara oil fields in western Venezuela is in the fracture system. This field produced over 80,000 bo/d from the basement reservoir. Matrix porosity contributes very little, if at all, to the overall reservoir capacity of the Osage and

268

Meramec limestones in the Eastern Anadarko basin. In these limestones essentially all the oil is within the fracture system.

In summary, there is probably enough evidence to banish the generalized assumption that the storage capacity of a fractured system is negligible compared to the storage capacity of the matrix.

Permeability

Permeability is a property of the porous medium and is a measure of the capacity of the medium to transmit fluids. Reservoirs can have primary and secondary permeability. Primary permeability is also referred to as matrix permeability. Secondary permeability can be either by fractures or solution vugs.

Matrix permeability

Matrix permeability can be evaluated with the use of Darcy's law:

$$v = - \frac{k}{\mu} \times \frac{d\,p}{d\,l} \qquad (1\text{-}5)$$

where:
- v = apparent flow velocity, cm/sec
- μ = viscosity of the flowing fluid, cp
- dp/dl = pressure gradient in the flow direction, atm/cm
- k = permeability of the rock, darcys

Darcy's law applies under the following conditions: (1) steady-state flow, (2) horizontal and linear flow, (3) laminar flow, (4) isothermal conditions, (5) constant viscosity, and (6) pore space 100% saturated with flowing fluid.

For the case of linear, incompressible fluid flow, permeability can be calculated from the equation:

$$k = v\,\frac{\mu\,L}{\Delta p} = \frac{q\,\mu\,L}{A\,\Delta p} \qquad (1\text{-}5a)$$

where:
- q = flow rate, cm³/sec
- A = area, cm²

Since the darcy unit is too large for most cases, permeability is usually expressed as one thousandth of a darcy, or a millidarcy (md). Darcy's law in petroleum field units is expressed as:

$$q = \frac{0.001127\,k\,A\,\Delta\,p}{\mu\,L} \qquad (1\text{-}6)$$

where:

$$q = flow\ rate,\ b/d$$
$$k = permeability,\ md$$
$$A = area,\ sq\ ft$$
$$\Delta p = pressure\ differential,\ psi$$
$$\mu = viscosity,\ cp$$
$$L = distance,\ ft$$

Solution vug permeability

In some carbonate reservoirs, the percolation of acid waters can improve porosity and permeability by dissolution of the matrix. Poiseuille's law for capillary flow and Darcy's law for flow of liquids in permeable beds can be combined to estimate permeability in solution channels. Craft and Hawkins made a comprehensive study of this problem, modeled in the following discussion.

Assume a capillary tube with the following characteristics:

$$L = length\ of\ capillary,\ cm$$
$$r = inside\ radius,\ cm$$
$$A = area,\ cm^2$$

A fluid of viscosity μ in poises is flowing in laminar or viscous flow under a pressure drop equal to p_1-p_2 dynes/sq cm. As the fluid wets the walls of the capillary, the velocity at the walls is considered to be zero and the velocity at the center of the tube is a maximum. The viscous force can be expressed by:

$$F = \mu A\ \frac{dv}{dx} \tag{1-7}$$

where dv/dx is cm/sec/cm. The phase area of a capillary equals $2\pi rL$. Consequently, for a cylinder the previous equation can be written as:

$$F = \mu(2\pi rL)\ \frac{dv}{dr} \tag{1-8}$$

If the fluid is not accelerating, the driving forces plus the viscous forces must equal zero. The driving force is equal to the pressure differential (p_2-p_1) times the cross-sectional area πr^2. Consequently,

$$\pi r^2\ (p_1 - p_2) + \mu\ (2\pi rL)\ \frac{dv}{dr} = 0 \tag{1-9}$$

and

$$dv = \frac{-\ (p_1 - p_2)rdr}{2\mu L} \tag{1-10}$$

Integrating,

$$\int dv = \frac{-(p_1 - p_2)}{2\mu L} \int r dr \qquad (1\text{-}11)$$

$$v = \frac{-(p_1 - p_2)r^2}{4\mu L} + C \qquad (1\text{-}12)$$

The integration constant (C) is evaluated from the previous equation by taking $v = 0$ at $r = r_o$, or

$$0 = \frac{-(p_1 - p_2)r_o^2}{4\mu L} + C \qquad (1\text{-}13)$$

Consequently,

$$C = \frac{(p_1 - p_2)r_o^2}{4\mu L} \qquad (1\text{-}14)$$

Inserting Eq. 1-14 into Eq. 1-12 leads to

$$v = \frac{(r_o^2 - r^2)(p_1 - p_2)}{4\mu L}$$

The previous equation indicates that the liquid velocity in a capillary varies parabolically and reaches a maximum at the center of the tube. Velocity at the walls is zero.

The flow rate (dq) through an element (dr) is $dq = vdA$, where the area dA equals $2\pi r dr$, or

$$q = \int_0^q dq = \int_0^{r_o} vdA \qquad (1\text{-}15)$$

and,

$$q = \int_0^{r_o} \frac{(r_o^2 - r^2)(p_1 - p_2)}{4\mu L} 2\pi r \, dr \qquad (1\text{-}16)$$

The integration leads to Poiseuille's law for viscous flow of liquids through capillary tubes:

$$q = \frac{\pi r_o^4 (p_1 - p_2)}{8\mu L} \qquad (1\text{-}17)$$

Darcy's law for steady state linear flow of incompressible fluids can be written as:

$$q = \frac{9.86 \times 10^{-9} kA (p_1 - p_2)}{\mu L} \qquad (1\text{-}18)$$

where:

$A = $ area available to flow $= \pi r^2$, sq cm
$k = $ permeability, darcys

Combination of Eq. 1-17 and 18 leads to

$$k = \frac{A r_o^2 (p_1 - p_2)}{8\mu L} \times \frac{\mu L}{9.86 \times 10^{-9} A (p_1 - p_2)}$$

and,

$$k = 12.5 \times 10^6 r_o^2 \; darcys \tag{1-19}$$

where r_o is in centimeters.

If the inside radius (r_o) is in inches rather than centimeters, the permeability is given by:

$$k = 80 \times 10^6 r_o^2 = 20 \times 10^6 D^2 \; darcys \tag{1-20}$$

where D is the capillary diameter in inches. Therefore, the permeability of a circular opening 0.0001 in. in diameter is 0.20 darcys or 200 md. The permeability of a circular opening 0.01 in. in diameter is 2,000 darcys or 2,000,000 md.

Average permeability from a vug-matrix system can be obtained from the relationships:

$$k_{av} = \frac{k_v \; N \; \pi r^2 + k_b (A - N\pi r^2)}{A} \tag{1-21}$$

where:

k_v = solution vug permeability, darcys
N = number of solution channels per section
A = cross-sectional area, in.^{2}
k_b = matrix permeability, darcys
r = solution channel radius, in.

Example 1-1. Consider a cube of limestone rock, 1 ft on each side with a matrix permeability of 1 md. The rock contains 20 solution channels, each one having a diameter of 0.05 in. What is the average permeability of the system if the channels lie in the direction of flow? What percentage of the fluid is stored in the matrix and what percentage in solution channels if the matrix porosity is 2%?

The permeability of each solution channel is obtained from the equation:

$$k_v = 20 \times 10^6 \times 0.05^2 = 50,000 \; darcys$$

The average permeability of the system is obtained from the equation:

$$k_{av} = \frac{(50,000 \times 20 \times \pi \times .025^2) + 0.001 (144 - 20 \times \pi \times .025^2)}{144}$$

$$k_{av} = \frac{1963.50 + 0.14}{144} = 13.64 \; darcys$$

The pore volume of the 20 solution channels is obtained from:

$$\pi r^2 \, N \, L = \pi \times 0.025^2 \times 20 \times 12 = 0.47 \; in.^3/ft^3$$

The pore volume of matrix rock is obtained from:

$$\phi_b \, (V_{rock} - V_{channels}) = 0.02 \, (12^3 - 0.47) = 34.55 \; in.^3/ft^3$$

The percentage of fluids stored in the matrix is:

$$\frac{34.55 - 0.47}{34.55} = 98.6\%$$

And the percentage of fluids stored in the solution channels is:

$$100 - 98.6 = 1.4\%$$

Example 1-2. Calculate the flow rate through a cube, 1 ft on each side with a matrix permeability of 1 md (Amyx). If there is a solution channel 0.05 in. in diameter, calculate the flow rate through the solution channel and the combined flow rate.

Darcy's law (Eq. 1-6) can determine the flow rate in the matrix, or

$$q = 1.127 \; \frac{kA \, \Delta p}{\mu L} = 1.127 \times \frac{0.001 \times 1 \times 1}{1 \times 1} = 0.001127 \; b/d$$

where: $k = 50,000 \; darcys/solution \; channel$

$$A = \frac{\pi \, 0.05^2}{4 \, (144)} = 1.3635 \times 10^{-5} \; ft^2/solution \; channel$$

$$q = \frac{1.127 \times 50000 \times 1.3635 \times 10^{-5} \times 1}{1 \times 1} = 0.768422 \; b/d$$

Consequently, the combined rate is $0.001127 + 0.768422 = 0.769549$ b/d. The importance of the solution channel can be better appreciated in terms of percentage increase in flow rate which for this case is 68,183%.

Fracture permeability

The presence of unhealed fractures greatly increases the permeability of a rock. It is possible to estimate fracture permeability and flow rates through fractures by following a development similar to the one presented for solution vugs.

Assume a fracture with a width equal to w_o, a length equal to L, and a lateral extent of the fracture equal to h. For this system the cross-sectional area open to flow is equal to $w_o h$. The driving force on the fracture is the pressure differential (p_1-p_2) acting on the area w h, or (p_1-p_2) w h dynes. The viscous forces are given by:

$$F = \mu A \frac{dv}{dw} \qquad (1\text{-}22)$$

where A is the area equal to h L. If the liquid is not accelerating, the driving force plus the viscous force must equal zero, or:

$$(p_1 - p_2)\, wh + \mu hL\, \frac{dv}{dw} = 0 \qquad (1\text{-}23)$$

separating variables and integrating,

$$(p_1 - p_2)\, \textstyle\int wdw = -\, \mu L \textstyle\int dv \qquad (1\text{-}24)$$

and,

$$(p_1 - p_2)\, \frac{w^2}{2} = -\, \mu L\, v + C \qquad (1\text{-}25)$$

The integration constant may be evaluated at $v = 0$ and $w = w_o/2$, or

$$(p_1 - p_2)\, \frac{(w_o/2)^2}{2} = C \qquad (1\text{-}26)$$

and,

$$(p_1 - p_2)\, \frac{w_o{}^2}{8} = C \qquad (1\text{-}27)$$

Inserting Eq. 1-27 into Eq. 1-25 leads to:

$$(p_1 - p_2)\, \frac{w^2}{2} = -\, \mu L\, v + (p_1 - p_2)\, \frac{w_o{}^2}{8} \qquad (1\text{-}28)$$

and

$$(p_1 - p_2)\, \left(\frac{w^2}{2} - \frac{w_o{}^2}{8}\right) = -\mu L\, v \qquad (1\text{-}29)$$

Consequently,

$$\frac{(p_1 - p_2)}{\mu L}\, \left(\frac{w_o{}^2}{8} - \frac{w^2}{2}\right) = v \qquad (1\text{-}30)$$

The flow rate (dq) through an element (dw) equals vdA, where the area (dA) is given by 2hdw. Consequently,

$$q = \textstyle\int_o^q dq = \textstyle\int_o^{w_o} vdA \qquad (1\text{-}31)$$

274

and,

$$q = \int_o^{w_o} \frac{(p_1 - p_2)}{\mu L} \left(\frac{w_o^2}{8} - \frac{w^2}{2} \right) 2hdw \qquad (1\text{-}32)$$

Integrating,

$$q = \frac{w_o A (p_1 - p_2)}{12\mu L} \qquad (1\text{-}33)$$

The previous equation can be combined with Darcy's law (Eq. 1-18) to obtain a relationship for permeability as follows:

$$k = \frac{w_o^2 A (p_1 - p_2)}{12 \mu L} \times \frac{\mu L}{9.86 \times 10^{-9} A (p_1 - p_2)} \qquad (1\text{-}34)$$

and,

$$k = 8.45 \times 10^6 w_o^2 \; darcys \qquad (1\text{-}35)$$

where w_o is in centimeters. If the fracture width (w_o) is in inches rather than in centimeters, the permeability is given by:

$$k = 54 \times 10^6 w_o^2 \; darcys \qquad (1\text{-}36)$$

Consequently, the permeability of a fracture 0.01 in. thick would be 5,400 darcys or 5,400,000 md. These extremely high values of permeability clearly indicate the importance of fractures on production of tight reservoirs which otherwise would be noncommercial.

Example 1-3. Calculate the average permeability of a rock which contains three fractures, each one 0.01 in. wide. Dimensions of the rock are 1 ft × 1 ft × 1 ft. Matrix permeability is 1 millidarcy.

$$K_{av} = \frac{0.001 \, [144 - 3 \, (12 \times 0.01)] + [(54 \times 10^6 \times 0.01^2) \, 3 \, (12 \times 0.01)]}{144}$$

$$K_{av} = 13.51 \; darcys = 13,510 \; md$$

Migration and Accumulation

One reasonable explanation for petroleum migration and accumulation in fractured reservoirs is provided by the theory of dilatancy. The principle of this theory is explained with the use of Fig. 1-7, as in the case of earthquakes. Fig. 1-7A shows a fault under tectonic stresses. In Fig. 1-7B the stresses have built up sufficiently to fracture the rock.

Then, fluids start moving into the dilatant zone, due to the vacuum produced by the fractures. In Fig. 1-7C the fluids have already filled the

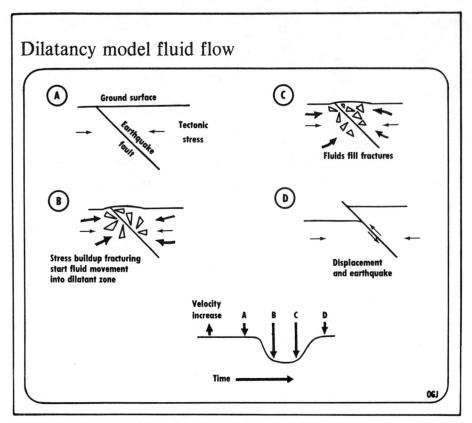

Fig. 1-7 Fluid flow according to dilatancy model. (After Kanamori)

fractures. In Fig. 1-7D a displacement and earthquake occurs. As certain seismic velocities decrease in stages B and C, it is possible to predict the occurrence of an earthquake within reasonable time limits.

McNaughton and Garb have analyzed the possibility of the same sequence of events for the migration and accumulation of petroleum in naturally fractured reservoirs. Upon the breaking of brittle rock by tectonic stresses, oil, water, or gas migrate toward the zone of dilatancy due to the vacuum produced by the fractures. The geological requirement for this hydrocarbon migration is a source rock contiguous to the brittle rock.

Fig. 1-8 depicts the evolution of a basement reservoir according to the theory of dilatancy. In stage A, the basement metamorphic rocks are fractured, generating pressure gradients across the unconformity. The vacuum produced by the fractures causes oil and water to move into the fractures. In stage B, continued fracturing caused by pulsatory tectonic movements establishes cell-like, sizable oil reservoirs below an impermeable capping. This capping may be formed by deposition of calcite in fractures of the metamorphic rock.

276

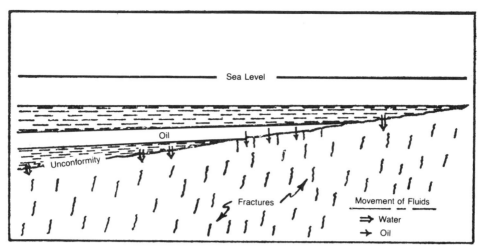

Stage A. Fracturing of metamorphic rocks establishes pressure gradients across the unconformity. Oil, gas, and water move into the dilated basement complex.

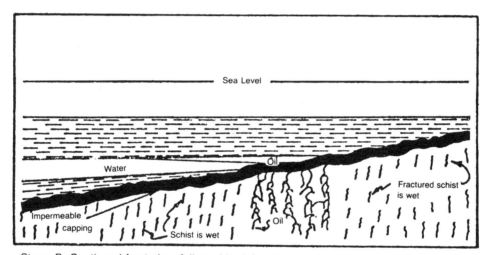

Stage B. Continued fracturing, followed by influent seepage and cementation, has established discrete, cell-like reservoirs below impermeable capping. Fractured schists bordering the reservoirs are wet.

Fig. 1-8 Schematic sketches showing evolution of basement oil pool according to dilatancy hypotheses. (After McNaughton)

Another theory for explaining the migration and accumulation of petroleum in fractured rocks is depicted in Fig. 1-9. According to this theory the fractures were formed before the generation of petroleum.

Fig. 1-9A shows how deep erosion of the overburden permitted uplift of the rock, generating fractures in the uplifted "hill." In Fig. 1-9B coarse rock debris accumulated in the lower slopes of the hill. The hill became deeply buried with continuous subsidence and sedimentation.

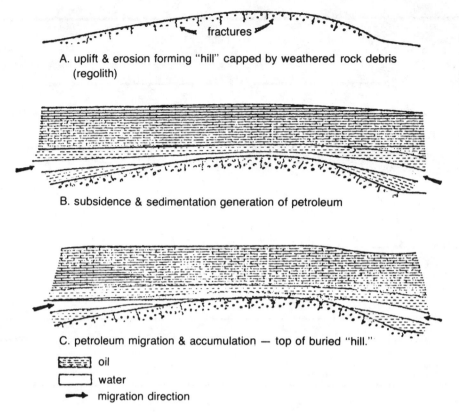

A. uplift & erosion forming "hill" capped by weathered rock debris (regolith)

B. subsidence & sedimentation generation of petroleum

C. petroleum migration & accumulation — top of buried "hill."

oil
water
→ migration direction

Fig. 1-9 Possible geologic evolution of petroleum accumulation in fractures formed before time of petroleum migration. (After McNaughton & Garb)

In Fig. 1-9C petroleum is generated and migrates up the crest of the hill, indicated by the arrow. This petroleum is trapped in both the sediments and the fractured basement.

Another theory indicates that in some reservoirs the oil may enter the reservoir by upward migration along fractures from some deeper bed. This appears to be the case of the Ain Zalah field in Iraq (Daniel).

Direct Sources of Information

Core analysis, oriented cores, drill cuttings, and downhole cameras provide direct sources for evaluating fractured reservoirs.

Core analysis

Whole core analysis is an important tool for direct examination of fractures. However, it is important to distinguish whether fractures are

278

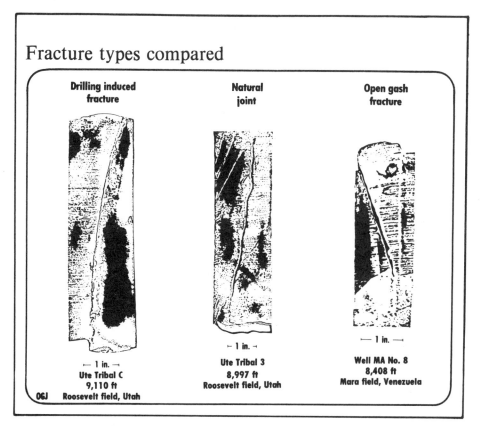

Fracture types compared

Drilling induced fracture

Natural joint

Open gash fracture

— 1 in. —
Ute Tribal C
9,110 ft
Roosevelt field, Utah

— 1 in. —
Ute Tribal 3
8,997 ft
Roosevelt field, Utah

— 1 in. —
Well MA No. 8
8,408 ft
Mara field, Venezuela

OGJ

Fig. 1-10 Comparison of drilling induced fracture, natural joint, and open-gash fracture. (After Sangree)

natural or artificially induced. Sangree suggests various criteria for differentiating natural from artificially induced fractures in cores.

The fracture is probably natural if:

1. Cementation is observed along the fracture surface. (Be careful that crystals on fracture surface are not halite deposited by evaporation of core fluids on other materials deposited from drilling fluids). In general, any fracture surface which appears to be a fresh break (i.e., unweathered and free of mineralization) should not be counted as a natural fracture unless there is some special supporting evidence.

2. Fracture is enclosed in core. One end (penetrating) or both ends (enclosed) of the fracture occur in the core.

3. Parallel sets of fractures are observed in a single core.

4. Slickensides (friction grooves) are observed on fracture. Unfortunately, drilling-induced slickensides are not uncommon, particularly in semiplastic shales or marls drilled at shallow depths. This criterion should be used with care.

The fracture is probably artificially induced if:

1. An uncemented vertical fracture angles in abruptly from the core edge in the down hole direction (Fig. 1-10). Watch out for this type—it is most probably induced during drilling or pulling cores. Drilling-induced fractures commonly split the core into equal halves, often with a slight rotation about the core axis.

2. Fractures are conchoidal or very irregular. Natural joints tend to be relatively plane. An exception occurs in highly porous, coarse-textured rocks where natural fracture surfaces may be quite irregular.

Whole core analysis provides a valuable tool when the core does not break along the plane of fracture. A method developed by Kelton for analyzing fractured and vugular limestones reveals that fractures and vugs may provide an important part of the storage capacity in reservoirs with double-porosity systems. Table 1-1 shows a comparison of matrix and whole core data in the Ellenbunger formation, Fullerton field, for four groups of samples.

Group 1 consisted of 14 samples with limited fracture development, while Group 2 provided 13 samples. Group 3 used 15 samples with more development of intergranular porosity. Group 4 consisted of four samples which had intergranular and/or highly vugular porosity. Notice from Table 1-1 the importance that secondary porosity can have on storage capacity. Also remarkable are the differences between matrix and whole core (fractures plus vugs plus matrix) permeabilities.

Locke and Bliss (1950) presented a technique which allows the direct determination of matrix and fracture porosity. This technique has been used successfully by Pirson to evaluate two-porosity systems. The

Table 1-1
Matrix vs. Whole Core Data, Ellenburger, Fullerton Field
(After Kelton)

Group	1	2	3	4
Gas, % bulk	0.08	0.91	0.63	3.72
Oil, % bulk	0.06	0.06	0.88	1.84
Water, % bulk	2.07	1.65	1.66	3.27
% pore	94	63	52	39
Matrix porosity, % bulk	1.98	1.58	2.56	7.92
% total pore	90	60	81	94
Total porosity, % bulk	2.21	2.62	3.17	8.40
k_{max}, md	10	409	23	94
$k_{90°}$, md	0.6	1.2	10	38
Matrix k, md	0.3	0.2	0.3	3.7
Partitioning coefficient, v	0.1	0.39	0.19	0.05

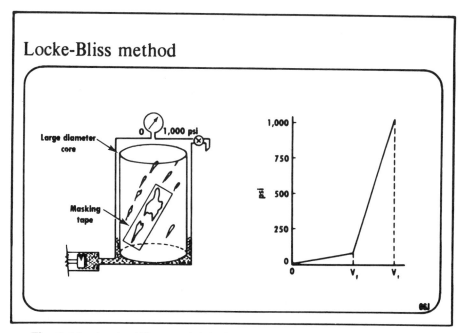

Fig. 1-11 Locke-Bliss method of porosity partitioning. (After Pirson)

method consists essentially of masking the fractures (and/or vugs) with adhesive tape before submerging the core into a water pressuring chamber. Water is injected at controlled volumes and the pressures are recorded. The result is shown in Fig. 1-11.

Between O and V_f the water invades the coarse porosity; consequently, no drastic pressure increase is noticeable. When the coarse porosity is water-saturated, the water starts penetrating the matrix porosity and the pressure increase is more pronounced. The total pore volume of the core is the total water injected V_t.

The breaking point in Fig. 1-11 represents the partitioning coefficient, which has been defined by Pirson as:

$$v = \frac{V_f}{V_t} = \frac{V_f}{V_f + V_b\, \phi_b} \qquad (1\text{-}37)$$

where:

$V_f = fracture\ volume$
$V_b = matrix\ volume$
$\phi_b = matrix\ porosity,\ fraction$

Oriented cores. These are useful for direct analysis of naturally-fractured rocks. The most important information extracted from oriented cores is the fracture orientation in the subsurface.

The technique consists essentially of placing the fractured core in the laboratory at its reservoir position. This permits direct determination of fracture dip.

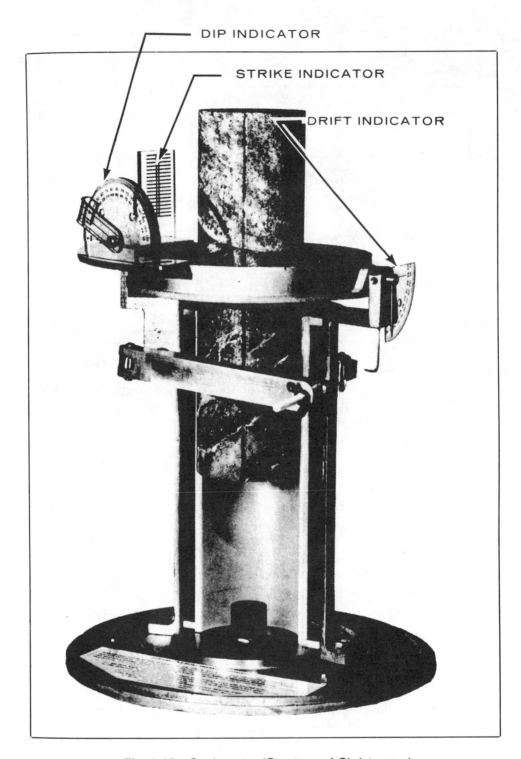

DIP INDICATOR

STRIKE INDICATOR

DRIFT INDICATOR

Fig. 1-12 Goniometer (Courtesy of Christensen).

Fig. 1-12 shows a goniometer which helps calculate dip direction, dip angle, and strike. With this information, core plugs taken from the parent core can be oriented.

Drill cuttings. These can detect natural fractures in only a few instances. However, natural fractures may not be preserved in cuttings due to breakage along fractures. Consequently, the reservoir might be naturally fractured even if the cuttings do not show any fractures.

Downhole cameras. Cameras can obtain direct information regarding bed boundaries, faulting, fractures, hole size, and hole shape (Fons). They also provide valuable information during fishing operations.

A compass may be added below the camera to take photographs which help determine borehole deviation from the vertical axis and orientation of the fractures intersecting the well bore. In this sense, downhole cameras and oriented cores provide similar information.

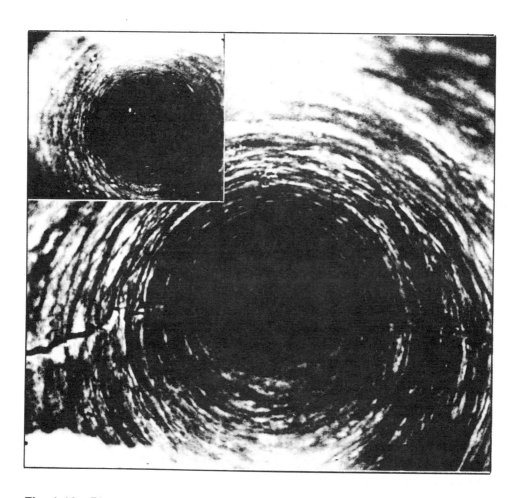

Fig. 1-13 Photograph of fracture obtained with a borehole camera. (After Fons)

The method applies equally to dry or gas-filled wells. One of the cameras uses 16-mm rolled film and can take 1,000 pictures in a single trip. After each photo is taken, the film automatically rolls to the next position.

Operations are restricted to temperatures below 200°F. and pressures below 4,000 psi. The entire tool is 4½ in. in diameter and 4 ft long. The usual problems of photography, like clean lenses and focus depth, are associated with this type of technique. Fig. 1-13 is a photograph of a fracture obtained with a downhole camera.

Indirect Sources of Information

Indirect sources of information for evaluating naturally fractured reservoirs include drilling history, log analysis, well testing, inflatable packers, and production history.

Drilling history provides valuable information regarding mud losses and penetration rates. Mud losses might be associated with natural fractures, vugs, underground caverns, or induced fractures. Penetration rates can be increased considerably while drilling all types of secondary porosity.

Log analysis is a powerful tool for detecting and evaluating naturally fractured reservoirs. There are logs which, in some cases, are run specifically to locate fractures. For example, sonic amplitude logs, variable intensity logs, the borehole televiewer, and dipmeter logs have been used successfully in detecting fractures. Conventional logs can be used in some instances for quantitative analysis of fractured reservoirs.

Well testing is also a powerful tool for evaluating fractured reservoirs. Much progress has been made on the subject of pressure analysis of fracture media. Most pressure time curves of odd shapes can now be analyzed by straightforward analytical procedures and/or numerical techniques. In some instances, it is possible to evaluate parameters such as fracture, matrix, and combined permeability and/or porosity from well testing.

Inflatable packers obtain impressions of the borehole on the pliable material of the packer. Fraser and Pettit reported the results of a field test using inflatable packers to determine the type and orientation of the fractures. Care must be exercised when working with these devices, since the packer can be accidentally over-inflated and blown out in enlarged boreholes.

Fig. 1-14 shows the photo of a borehole impression obtained with an inflatable packer. The impression of the fractures is recorded by a special rubber element when the packer is inflated to a certain pressure and maintained under these conditions for a certain length of time.

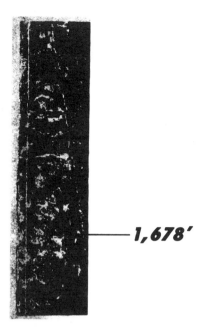

—*1,678'*

Fig. 1-14 Impression: top of fracture at borehole (Well 2). (After Anderson)

Production history provides qualitative information regarding the presence of natural fractures. If, for example, the matrix permeability of a sandstone reservoir is less than 0.1 md, its porosity is 5%, and well rates exceed 1,000 bo/d, then it is clear that the well performance is a function of fracture permeability. In some cases, early water or gas breakthroughs in secondary recovery projects indicate the presence of fractures and their directions. This type of situation must be avoided in practice by using all possible means to determine fractured trends before a secondary recovery project is started.

Mapping Fractured Trends

Aerial photography has been used successfully in some areas to map fracture trends. Alpay has reported a field application of aerial photography in eight reservoirs of West Texas. He found that, in general, a good match was obtained between predominant fractured trends determined from aerial photography and subsurface trends determined from reservoir performance.

The rationale supporting aerial photography indicates that surface fracture trends conserve essentially the same direction with depth in tectonically undisturbed areas. Consequently, surface-fractured trends

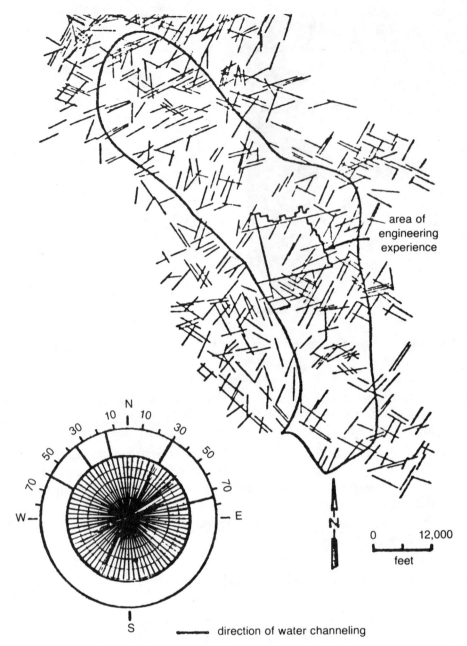

area of
engineering
experience

direction of water channeling

Fig. 1-15 Surface fracture traces in Field 4. (After Alpay)

can provide valuable information with respect to potential fracture direction in underground hydrocarbon reservoirs. One advantage of aerial photography is that large areas (and consequently large fractured trends) can be evaluated quickly and rather inexpensively.

Fig. 1-15 shows surface fracture traces determined from aerial

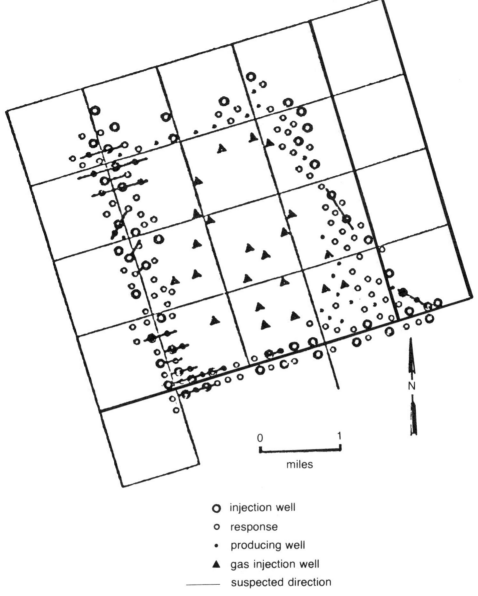

O injection well

o response

· producing well

▲ gas injection well

——— suspected direction

Fig. 1-16A Waterflood experience to date in Field 4. San Andres Reservoir. (After Alpay)

photography in West Texas. Also shown is the area where engineering experience is available, and a rosette which summarizes the surface fracture traces on a field-wide basis. The interior of the rosette indicates three prominent directions of the surface fractures: (1) a N25–35E with a mode at N30E, (2) a N60–75E with a mode at N65E, and (3) a N55–75E with twin modes at N60W and N70W.

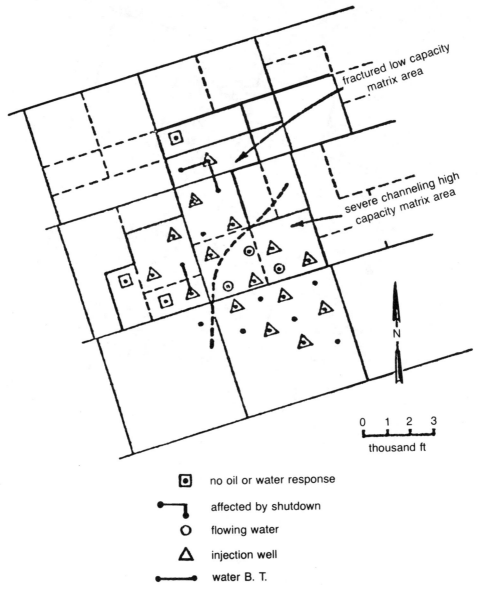

fractured low capacity matrix area

severe channeling high capacity matrix area

N

| 0 | 1 | 2 | 3 |

thousand ft

◙   no oil or water response

•┐   affected by shutdown

○   flowing water

△   injection well

•——•   water B. T.

Fig. 1-16B   Waterflood experience in Field 4, Upper Clearfork Reservoir. (After Alpay)

The exterior part of the rosette indicates the preferential direction of water channeling in the area of engineering experience. The agreement between direction determined from the photo study and direction determined from water channeling (Fig. 1-16A and B) is good.

Mapping fractured trends from aerial photography must be done by experts, since fences, power lines, roads, pipelines, etc., may be mis-

288

taken for natural fractured trends. To avoid these possible mistakes, it is better to find areas which have no unnatural features.

If, in addition to the delineation of fractured trends, it is necessary to know fracture morphology, small-scale fracture densities, and determination of total fracture geometries, Stearns and Friedman recommend that extensive ground control be used in conjunction with the aerial photography interpretation.

Fracture intensity indices derived from conventional well logs can map fracture trends in some cases. Pirson indicates that fracture porosity is possibly the factor which better measures quantitatively the intensity of deformational shattering of brittle rocks. According to this rationale, the value of the fracture intensity (FII) increases as a fault is approached. With the value of FII, it is possible to estimate lateral distance to a fault.

Remote sensing imagery appears to be a promising tool for delineating fractured trends. Rabshevsky indicates that the most important step in using remote sensing imagery is interpreting and analyzing data. The interpreter must become familiar with image processing, especially when using LANDSAT imagery. In the image processing and interpretation stages, he can use his ingenuity. Other steps, such as mission planning and image acquisition, are beyond the interpreter's control.

Fractured Reservoirs

Naturally fractured reservoirs are found in sandstones, carbonates, shales, cherts, siltstones and basement rocks. The percentage of the total porosity made from fractures ranges from very small to 100%.

Fractured sandstones

The Spraberry field of West Texas is an example of a fractured sandstone reservoir. The main producing structure in the Spraberry is a fracture permeability trap on a homoclinal fold which is made of alternate layers of sands, siltstones, shales and limestones. These layers were deposited in the deep Midland basin.

Fig. 1-17 shows a generalized stratigraphic section of the area. The oil is stored primarily in the sandstone matrix, and paper-thin fractures provide the channels to conduct the oil to the well bore.

Another naturally fractured sandstone reservoir is found in the Altamont trend, Uinta basin, Utah. Production from this reservoir comes from fine-grained sandstones of Tertiary age and fractured rocks consisting of sandstone, carbonate, and highly calcareous shale. Initial production rates of over 1,000 bo/d are not uncommon in this reservoir of low

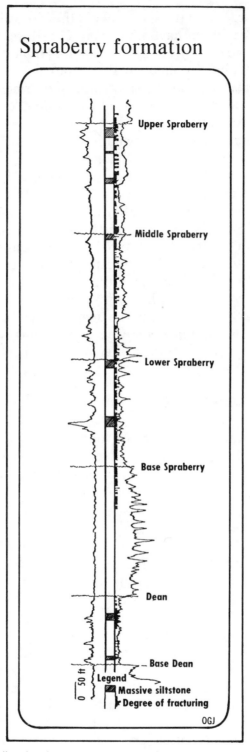

Fig. 1-17 Generalized columnar section of Spraberry formation. Diagonal shading, massive siltstone; solid black, degree of fracturing. (After Wilkinson)

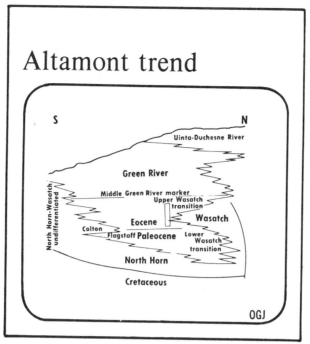

Fig. 1-18 Abnormally pressured Altamont trend productive section lies in the transition between Green River source rocks and Wasatch red bed facies. (After Baker & Lucas)

porosity ($\phi \simeq 3 - 7\%$) and low permeability ($k < 0.01$ md). Fig. 1-18 shows a stratigraphic section of the area. Practical experience and detailed economic evaluations indicate that 640 acres is the optimum spacing for the Altamont trend. Closer spacing proved uneconomical.

Finn studied the presence of fractures in the Oriskany sandstone in Pennsylvania and New York. These fractures have been responsible for larger flow rates than expected in the area.

Fractured carbonates

Carbonates are limestones, dolomites, and the intermediate rocks between the two.

Daniel provides an excellent description of three fractured reservoirs of the Middle east: Ain Zalah and Kirkuk in Iraq, and Dukham in Qatar.

The Ain Zalah reservoir is extremely tight and has very low porosity. However, due to the presence of fractures, it can produce at high rates during limited periods. Fig. 1-19 shows a generalized stratigraphic section of the field, a structure-contour map, and a structural cross section.

Daniel found that the oil possibly entered the "present reservoir by

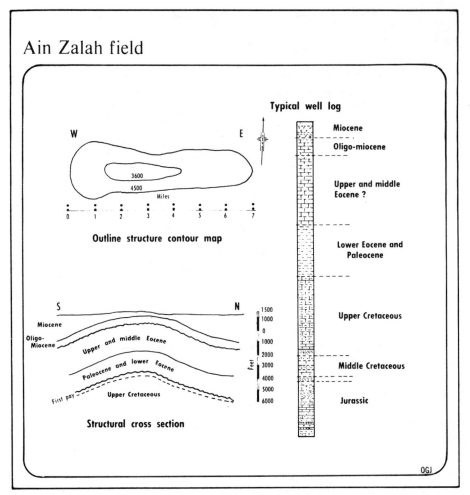

Fig. 1-19 Ain Zalah oil field. Diagrammatic structure-contour outline and stratigraphic section. Length of field about 7 miles; thickness of stratigraphic
column about 8,330 ft. (After Daniel)

upward migration along fractures from some deeper zone, perhaps the
middle Cretaceous or the Jurassic." Due to the high degree of fracturing
of the formation, he indicated that drainage from the first and second pay
at Ain Zalah could probably be achieved with two or three wells.

The Kirkuk field limestone reservoir has a higher average porosity
and varied permeability which depends on stratigraphy. Fig. 1-20 shows
structural maps of the Khurmala, Avanah, and Baba domes.

According to Daniel, the fractures at this 61-mile structure are so
closed that only a few wells located at the base of the highest dome (Baba)
would be enough to drain the entire reservoir and a 2-mile spacing could
give adequate drainage (spacing of approximately of 1,280 acres).

The Dukhan field has limestones which are moderate to highly

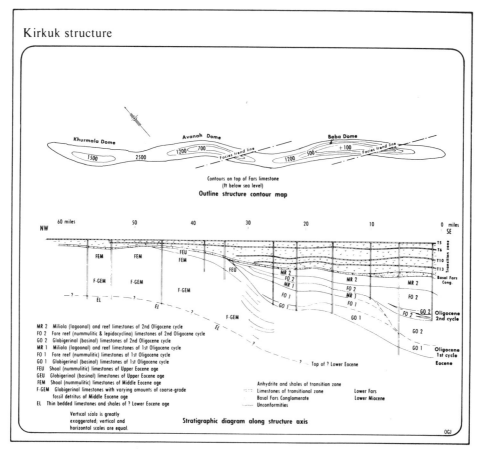

Khurmala Dome

Avanah Dome

Baba Dome

Contours on top of Fars limestone
(ft below sea level)
Outline structure contour map

MR 2   Miliola (lagoonal) and reef limestones of 2nd Oligocene cycle
FO 2   Fore reef (nummulitic & lepidocyclina) limestones of 2nd Oligocene cycle
GO 2   Globigerinal (basinal) limestones of 2nd Oligocene cycle
MR 1   Miliola (lagoonal) and reef limestones of 1st Oligocene cycle
FO 1   Fore reef (nummulitic) limestones of 1st Oligocene cycle
GO 1   Globigerinal (basinal) limestones of 1st Oligocene cycle
FEU   Shoal (nummulitic) limestones of Upper Eocene age
GEU   Globigerinal (basinal) limestones of Upper Eocene age
FEM   Shoal (nummulitic) limestones of Middle Eocene age
F-GEM Globigerinal limestones with varying amounts of coarse-grade
         fossil detritus of Middle Eocene age
EL   Thin bedded limestones and shales of ? Lower Eocene age

Anhydrite and shales of transition zone
Limestones of transitional zone   Lower Fars
Basal Fars Conglomerate   Lower Miocene
Unconformities

Vertical scale is greatly
exaggerated; vertical and
horizontal scales are equal.

Stratigraphic diagram along structure axis

Fig. 1-20   Kirkuk oil field. Stratigraphic diagram along the axis of the Kirkuk structure; structure-contour map. (After Daniel)

porous and permeable. The degree of fracturing is lower than at Ain Zalah and Kirkuk; consequently, the appropriate drainage of the reservoir requires a closer spacing.

Fig. 1-21 shows a structural map, structural cross section, and a typical lithological log of the Dukhan oil field. The length of the anticline is about 31 miles and maximum well depth is 6,600 ft.

The Savanna Creek gas field in the Canadian Rocky Mountains has a 900-ft carbonate (Mississippian Livingstone formation) which is essentially impermeable and produces through fractures. Wells have produced from zero to over 57 MMcfd. Matrix porosity is very low—about 4%. However, it is increased considerably by open fractures and brecciation.

Fig. 1-22 is a structural map of La Paz-Mara fields in Venezuela. The intergranular porosity of the Cretaceous limestone reservoir (Colon-Cogollo formations) is 2-3%. Permeability is smaller than 0.1 md.

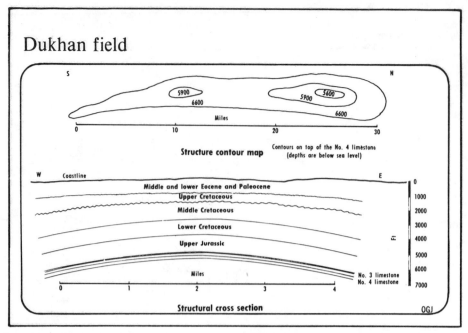

Fig. 1-21   Dukhan oil field. Diagrammatic structure-contour outline and cross section. (After Daniel)

However, these fields produced at 250,000 bo/d during 1951 when they were described by Smith. This production was clearly related to fracture permeability of the formation.

Lowenstam found that the Marine reservoir of Illinois is characterized by a matrix limestone porosity, enlarged by a system of fractures. Lowenstam believed that fractures connected the discontinuous producing streaks with each other and with the main reef core.

Boyd indicated that fractures contributed significantly to the storage capacity of the Silurian dolomite in the Howell gas field in Michigan. In fact, he noted gas reserves exceeding those calculated by conventional methods and concluded that the excess gas was stored in fracture networks.

Braunstein reported oil production from the Selma fractured chalk (Navarro and Taylor age) in Gilbertown field, Alabama. He indicated that no matrix porosity was found in the chalk, and that fracture porosity was associated with the zone of fault. The fracture porosity provided a secondary trap for oil which migrated from the lower Eutaw sands.

Production from the Selma chalk comes only from wells located in the down-thrown side near the fault. Fig. 1-23 shows a schematic cross section of Gilbertown field. The dry well encountered a complete section of the Selma chalk, but was far away from the fault and, consequently,

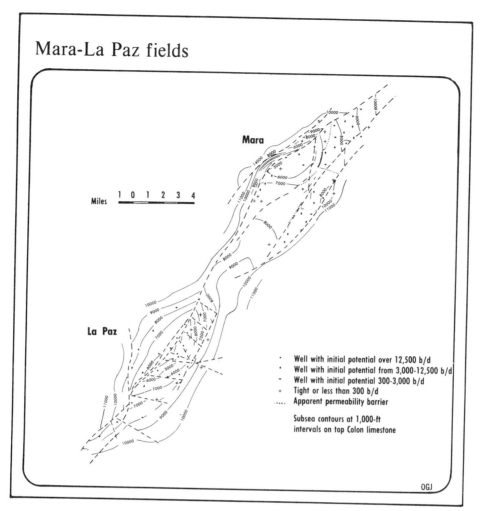

Fig. 1-22 Structure contours on Mara-La Paz Fields in Venezuela. (After Smith)

did not penetrate any fractures. The middle well near the fault produced oil from the fracture Selma chalk. Finally, the well to the right of the fault produced oil from the Eutaw sand, but not from the Selma chalk.

Hanna and Pirson have also reported fracture porosity associated with faults in the Taylor marl and Austin chalk.

The Asmari limestone of Iran is characterized by an extensive fracture system. Lane estimated that 80% of the recoverable oil was stored in the matrix and 20% in the fractures.

The Tamaulipas limestone in Mexico has produced oil of 12.5° API at rates as high as 30,000 bo/d from a single well. The fracture permeability varies so much that wells only 200 ft apart from a producing well are dry.

Excellent rates have been obtained from the Reforma (dolomitic

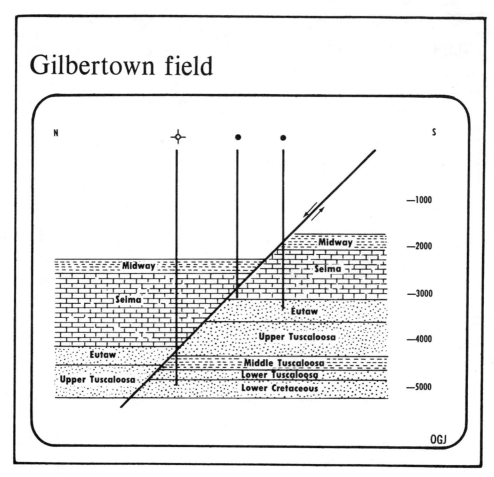

Fig. 1-23 Diagrammatic north-south section across Gilbertown field. (After Braunstein)

limestone) giant Cretaceous reservoirs in Mexico (Fig. 1-24). Delgado describes the reservoir as characterized by dolomitic limestone of low primary porosity and very good secondary permeability, due to the presence of natural fractures and caverns.

Fractured shales

Fractured shales have produced gas since the early 1900's along the western margin of the Appalachian Basin. The amount of gas within these shales has been estimated at 460 quadrillion scf. Since shales do not have effective porosity, production is achieved only through fracture networks (Hunter). Stimulation is usually necessary to obtain commercial production.

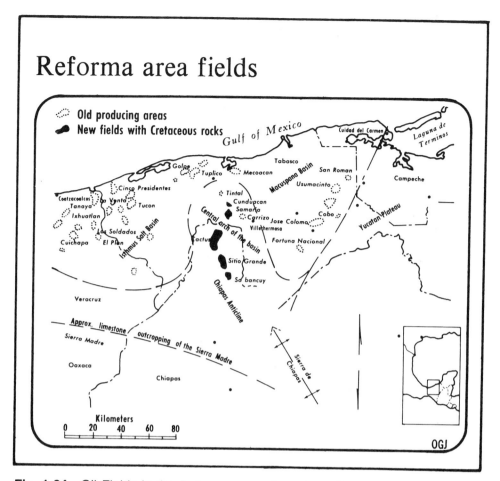

Fig. 1-24 Oil Fields in the Reforma area of southern Mexico. (After Delgado & Loreto)

Devonian shales are present in 26 states of the United States, six provinces and territories of Canada, along the U.S.-Mexican border, South America, Africa, and Europe. Gas production is also obtained from Cambrian shales in the St. Lawrence lowlands of Quebec.

In some cases, oil production has been obtained from shales. Results presented by Peterson indicate that high-grade oil has been produced from the Mancos fracture shale, Rangely field, Colorado. These fractures are associated with the axial bending and arching of the Rangely anticline.

Fractured chert

Regan studied oil production from fracture shale and chert reservoirs in the Santa Maria Coastal District and the San Joaquin Valley in

297

California. The rock is made of fractured chert and siliceous shale of upper Miocene age. Production ranged from 200 to 1,000 bo/d. Therefore, fracture shales have become very important with new economic incentives.

Fractured basement reservoirs

In some cases oil production can be obtained from fractured basement rock. Smith (1956) reported that in 1953 a well drilled in a basement of igneous and metamorphic fractured rocks in the La Paz-Mara fields of Venezuela produced 3,900 bo/d from a depth of 8,889 ft. By 1955 the fields were producing 80,000 bo/d from 29 wells drilled in the fractured basement rocks. Since there is essentially no matrix porosity in basement rocks, it can be inferred that all oil produced from a basement reservoir is stored within the fractures.

Walters researched oil production from fractured pre-Cambrian basement rocks in central Kansas. Sixteen wells in the Orth field, Rice County, Kansas, had produced over one million barrels of oil from fractured pre-Cambrian basement rock by 1953. Oil production had also been obtained from four wells in the Kraft-Prusa field and from one well each in the Beaver, Bloomer, Eveleigh, and Trapp fields in Barton County.

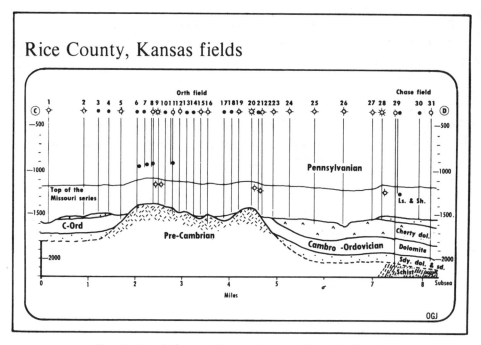

Fig. 1-25 Orth and Chase fields. (After Walters)

298

In these reservoirs, no oil was found when the basement rocks were structurally lower. Only when the Pennsylvanian beds were resting directly on the pre-Cambrian was oil found in the basement rocks.

Fig. 1-25 shows a NW-SE cross section through the Orth and Chase fields. Wells 6, 10, 11-B, 12, 13, 16, 20, and 21 produce oil from fractured pre-Cambrian quartzite.

Eggleston did a comprehensive summary of oil production from fractured basement rocks in California. Fifteen thousand bo/d were being produced from fractured basement rocks by 1948. This represented about 1.5% of the total California production at the time (918,000 bo/d). Production from all kinds of fractured reservoirs was 55,000 bo/d, or about 6% of the total state production.

The importance of these figures lies in the fact that oil production from basement rocks had been considered nearly impossible up to that time. Production from basement rocks in California came mainly from Edison, Santa Maria Valley, and Wilmington. Smaller production came from El Segundo and Playa del Rey fields (Eggleston).

Exploring for Naturally Fractured Reservoirs

Most naturally fractured reservoirs have been discovered by accident. Possible exceptions are some fields of Venezuela, Iran, and Iraq, where these types of reservoirs have been sought.

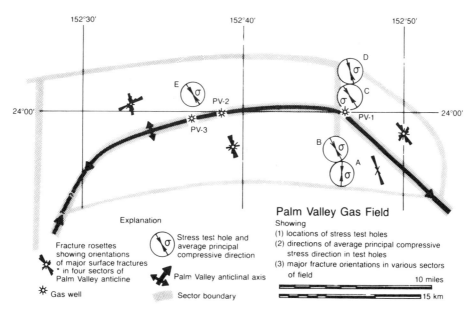

Fig. 1-26 Palm Valley gas field—surface fracture orientations and residual stress orientations. (After McNaughton & Garb)

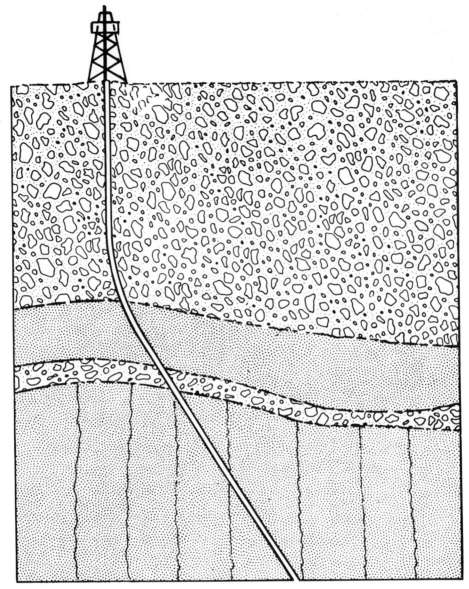

Fig. 1-27 Artist's conception of "slant-hole" development of the Palm Valley gas field. (After McNaughton & Garb)

McNaughton and Garb related exploration methods to the concept of dilatancy, i.e., to the search for large bodies of brittle rock which are either underlain or overlain by petroleum source rocks.

Modern seismology may provide a valuable tool for detecting fractured reservoirs. In fact, acoustic velocity should decrease in open, unhealed fractured rocks. A decrease in acoustic velocity associated

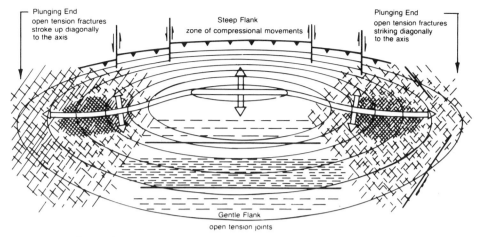

Fig. 1-28 Open fractures and fracture pattern in asymmetrical anticline. (After Martin)

with brittle fractured rocks has been established in the Amadeus basin of central Australia.

Mapping fractured trends facilitates searching for areas of more intense shattering. The combination of photogeology, subsurface information, and orientation of residual stresses are significant tools for discovering fractured reservoirs.

Fig. 1-26 shows surface fracture orientation and residual stress orientation of the fractured Palm Valley gas fields in Australia (Magellan Petroleum Corporation). Fracture rosettes based on photogeology indicate that northwesterly trends dominate in the south, southeast, and northeast of the field. These trends are corroborated by residual stress analysis in test holes. Note that fractured trends in a reservoir tend to be parallel to the principal horizontal stress measured in surface outcrops.

The finding, with respect to fracture direction, led McNaughton to recommend drilling holes deviated toward the northeast and southwest to intercept the northwesterly fractured trends (Fig. 1-27).

Martin shows how petrofabric studies may help find fracture-porosity reservoirs. Fig. 1-28 is a schematic top view of an asymmetrical anticline. Open tension joints occur in the plunging ends and on the gentle flank. The steep flank is a zone of compressional movements, where joints are often brecciated and cemented by secondary mineralization.

Contrary to popular belief, the crestal area of an asymmetrical anticline contains little fracture porosity. The best wells of Kirkuk field in Iraq, La Paz field in Venezuela, and Raman field in Turkey are located over the gentler flanks. Crestal wells are rare or less prolific than the ones in the plunging ends.

Important consideration in this model is given to faulting. The best possibility of finding open fractures is in the upthrown block because of tension, while fractures in the downthrown block may be closed due to compression.

REFERENCES

Aguilera, Roberto. "Log Analysis of Gas-Bearing Fractured Shales in the Saint Lawrence Lowlands of Quebec." SPE 7445 presented at the Annual Meeting of SPE of AIME, Houston (October 1–4, 1978).

Alpay, O.A. "Application of Aerial Photographic Interpretation to the Study of Reservoir Natural Fracture Systems." SPE 2567 presented at the 44th Annual Meeting of the SPE of AIME, Denver (September 28–October 1, 1969).

Amyx, J.W., D.M. Bass, and R.L. Whiting. *Petroleum Reservoir Engineering—Physical Properties*, McGraw-Hill Book Company (1960), 538.

Anderson, T.O., and E.J. Stahl. "A Study of Induced Fracturing Using an Instrumental Approach." *Journal of Petroleum Technology* (February 1967).

Atkinson, B., and D. Johnston. "Core Analysis of Fractured Dolomite in the Permian Basin." *Transactions.* AIME (1949), 128–132.

Baker, D.A., and P.T. Lucas.: "Strap Trap Production May Cover 280 Plus Square Miles." *World Oil* (April 1972). 65–68.

Beach, J.H. "Geology of Edison Oil Field, Kern County, California." *Structure of Typical American Oil Fields.* AAPG (1948), 58–85.

Beiers, R.J. "Quebec Lowlands: Overview and Hydrocarbon Potential." Proceeding of the Seventh Appalachian Petroleum Geology Symposium, Morgantown, West Virginia (March 1–4, 1976), 142.

Braunstein, J. "Fracture-Controlled Production in Gilbertown Field, Alabama." *Bulletin.* AAPG (February 1953), 245–249.

Craft, B.C., and M.F. Hawkins. *Applied Petroleum Reservoir Engineering.* Prentice-Hall Inc. (1962).

Daniel, E.J. "Fractured Reservoirs of Middle East." *Bulletin.* AAPG (May 1954), 774–815.

Delgado, O.R., and E.G. Loreto. "Reforma's Cretaceous Reservoirs: An Engineering Challange." *Petroleum Engineer* (December 1975), 56–66.

Dempsey, J.C., and J.R. Hickey. "Use of Borehole Camera for Visual Inspection of Hydraulically Induced Fractures." *Producers Monthly* (April 1958), 18–21.

Eggleston, W.S. "Summary of Oil Production from Fractured Rock Reservoirs in California." *Bulletin.* AAPG (July 1948), 1352–1355.

Elkins, L.F. "Reservoir Performance and Well Spacing, Spraberry Trend Area Field of West Texas." *Transactions*. AIME (1953) 301–304.

Finn, F.H. "Geology and Occurrence of Natural Gas in Oriskany Sandstone in Pennsylvania and New York." *Bulletin*. AAPG (March 1949), 303–335.

Fons, L.C. "Downhole Camera Helps Solve Production Problems." *World Oil* (1960), 150–152.

Fraser, C.D., and C.D. Pettit. "Results of a Field Test to Determine the Type and Orientation of a Hydraulically Induced Formation Fracture." *Journal of Petroleum Technology* (1961), 463–466.

Fraser, H.J., and L.C. Graton. "Systematic Packing of Spheres—With Particular Relation to Porosity and Permeability." *Journal of Geology* (November-December 1935), 785–909.

Griggs, D.T. and J.W. Handin. "Observations of Fracture and a Hypothesis of Earthquakes." Geol. Soc. America, Mem 79 (1960), 347–364.

Handin, J.W. "Experimental Deformation of Sedimentary Rocks Under Confining Pressure: Pore Pressure Test." *Bulletin*. AAPG (1963), 717–755.

Hanna, M.A. "Fracture Porosity in Gulf Coast," *Bulletin*. AAPG (February 1953), 266–281.

Hohlt, R.B. "The Nature of Origin of Limestone Porosity." *Colorado School of Mines Quarterly* (1948) No. 4.

Hunter, C.D., and D.M. Young. "Relationship of Natural Gas Occurrence and Production in Eastern Kentucky (Big Sandy Gas Field) to Joints and Fractures in Devonian Bituminous Shale." *Bulletin*. AAPG, No. 2. (February 1953), 282–299.

Kanamori, Kiroo: "Earthquake Prediction." California Institute of Technology (1974).

Kelton, F.C. "Analysis of Fractured Limestone Cores," *Transactions*. AIME, 189 (1950) 225–234.

Landes, K.K. *Petroleum Geology*. 2 ed. John Wiley and Sons Inc. (1959).

Lane, H.W. "Oil Production in Iran." *Oil and Gas Journal* (August 18, 1949), 128.

Levorsen, A.I. *Geology of Petroleum*. 2 ed. W.H. Freeman and Company (1967).

Locke, L.C., and J.E. Bliss. "Core Analysis Technique for Limestone and Dolomite." *World Oil* (September 1950), 204.

Lowenstan, H.A. "Marine Pool, Madison County, Illinois, Silurian Reef Producer." *Illinois Geological Survey* 131 (1942).

McNaughton, D.A. "Dilatancy in Migration and Accumulation of Oil in Methamorphic Rocks." *Bulletin*. AAPG No. 2 (February 1953) 217–231.

McNaughton, D.A., and F.A. Garb. "Finding and Evaluating Petroleum Accumulations in Fractured Reservoir Rock." *Exploration and Economics of the Petroleum Industry*, Vol. 13. Matthew Bender & Company Inc. (1975).

Martin, G.H. "Petrofabric Studies May Find Fracture-Porosity Reservoirs." *World Oil* (February 1, 1963), 52–54.

Mead, W.J. "The Geologic Role of Dilatancy." *Journal of Geology* (1925), 685–698.

Muir, J.M. "Limestone Reservoir Rocks in the Mexican Oil Fields." *Problems of Petroleum Geology*. AAPG (1934), 382.

Peterson, V.E. "Fracture Production from Mancos Shale, Rangely Field, Rio Blanco County, Colorado." *Bulletin*. AAPG 39 (April 1955), 532.

Pickett, G.R., and E.B. Reynolds. "Evaluation of Fractured Reservoirs." *Society of Petroleum Engineering Journal* (March 1969), 28.

Pirson, S.J. "How to Map Fracture Development From Well Logs." *World Oil* (March 1967), 106–114.

Pirson, S.J. "Petrophysical Interpretation of Formation Tester Pressure Buildup Records." *Transactions*. SPWLA (May 17–18, 1962).

Porter, L.E. "El Segundo Oil Field, California." *Transactions* 127. AIME (1943), 451.

Provo, L.J. "Upper Devonian Black Shale—Worldwide Distribution and What It Means." Proceeding of the Seventh Appalachian Petroleum Geology Symposium, Morgantown, West Virginia (March 1–4, 1976).

Rabshevsky, G.A. "Optical Processing of Remote Sensing Imagery." Proceeding of the Seventh Appalachian Petroleum Geology Symposium, Morgantown, West Virginia (March 1–4, 1976), 100.

"Recommended Practice for Determining Permeability of Porous Media." American Petroleum Institute (September 1952).

Regan, L.J. "Fractured Shale Reservoirs of California." *Bulletin* 37. AAPG (February 1953), 201–216.

Sangree, J.B. "What You Should Know to Analyze Core Fractures." *World Oil* (April 1969), 69–72.

Scott, J.C., W.J. Hennessey, and R.S. Lamon. "Savanna Creek Gas Field, Alberta." *Bulletin of Canadian Minerologists and Metallurgists* 533 (1958), 270–278.

Smith, J.E. "The Cretaceous Limestone-Producing Areas of the Mara and Maracaibo District—Venezuela." *Proceedings of Third World Petroleum Congress* (1951), 56–71.

Smith, J.E. "Basement Reservoir of La Paz-Mara Oil Fields, Western Venezuela." *Bulletin* 40, AAPG (February 1956), 380–385.

Snider, L.C. "Current Ideas Regarding Source Beds for Petroleum." *Problems of Petroleum Geology*. AAPG (1934), 51–66.

Stearns, D.W. "Certain Aspects of Fracture in Naturally Deformed Rocks." NSF Advanced Science Seminar in Rock Mechanics: Bedford, Massachussetts. R.E. Riccker, ed. (1967), 97–118.

Stearns, D.W., and M. Friedman. "Reservoirs in Fractured Rock." *AAPG Memoir* (1972), 82–106.

Walters, R.F. "Oil Production from Fractured Pre-Cambrian Basement Rocks in Central Kansas." *Bulletin* 37. AAPG (February 1953), 300–313.

Wilkinson, W.M. "Fracturing in Spraberry Reservoir, West Texas." *Bulletin* 37. AAPG (February 1953), 250–265.

Reprinted with permission. From: The Rocky Mountain
Association of Geologists' 1977 Symposium, *Exploration
frontiers of the central and southern Rockies,* copyright 1977,
pp. 95-101.

INTERFORMATIONAL CONTROL OF REGIONAL FRACTURE ORIENTATIONS

by

Ronald A. Nelson[1] and David W. Stearns[2]

ABSTRACT

Regional fractures are exceptionally well developed in the Lake Powell area of Utah and Arizona. It is concluded that the changes in strike from formation to formation of the regional fractures in this area are due to large scale primary sedimentary structures within the rocks that make up each formation. These sedimentary features can create mechanical anisotropies within a formation which control the orientation of subsequent regional fractures when subjected to regional tectonic forces.

Two sedimentary structures which create mechanical differences in ultimate strength, strain, and secant modulus are cross-bedding in eolian sediments and preferred bedding-plane grain-orientations in fluvial sediments. Laboratory measurements indicate that anisotropies in the order of 50-100% of the differential stress, longitudinal strain, and secant modulus at fracture are due to these sedimentary features.

Emperical correlations indicate that regional fractures tend to form perpendicular to the paleo-wind direction in eolian sediments and parallel to the paleo-current direction in fluvial sediments.

INTRODUCTION

Natural rock systems contain numerous fractures that arise from many different causes. The distribution and orientation of these fractures often influence or determine the major holding capacity and flow capabilities of subsurface reservoirs. The aim of most field and industry-related fracture analyses is to determine the geometry and genesis of naturally occurring and induced fracture networks. Most fractures occurring in outcrops are related to deformation associated with local or nearby geologic structures such as folds, faults, or domes. These are termed "tectonic fractures" (Stearns, 1971). Many fractures, however, are not related to structure and have a consistent orientation over large areas. These are termed "regional fractures" and are developed over large areas of the earth's surface with relatively little change in orientation; they show no evidence of offset; and they are always perpendicular to major bedding surfaces (Price, 1959; Stearns, 1971).

The major change in orientation of these regional fractures often occurs interformationally through the stratigraphic section. It is this interformational shift in regional fracture orientation that is the subject of this study.

The area of study covers 13,000 km² (5000 mi²) immediately surrounding Lake Powell Reservoir which is a 300 km (190 mi) long, man-made reservoir on the Colorado River in the Colorado Plateau region of southeastern Utah and northeastern Arizona (Fig. 1).

The geology of this area is typical of the entire Colorado Plateau which is characterized by widely spaced structures, 80 km (50 mi), with relatively small vertical displacement. The stratigraphic section of interest to this study extends from the base of the Permian Cedar Mesa Sandstone to the top of the Jurassic Entrada Sandstone (Fig. 2). These formations consist perdominantly of alternating fluvial and eolian clastics. Regional dip is less than 10° to the southwest. Superposed on the regional dip are wide-spread folds of varying amplitudes.

PREVIOUS WORK

Several authors have reported fracture orientation studies including all or part of the Lake Powell Region. Kelley and Clinton (1960) present a fracture-orientation map for the entire Colorado Plateau based on the air-photo coverage available at that time. They divide the Plateau into a number of areas of homogeneous fracture orientation. The immediate Lake Powell Area is divided into the Kaiparowits Basin, Echo Cliffs Uplift, Henry Mountains Basin, Monument Warp, White Canyon Slope, and Piute Fold Belt (see Fig. 1).

Hodgson (1961) reported on fracture orientations south of the San Juan arm of Lake Powell between Navajo Mountain and Comb Ridge. He was interested in the origin and classification

---

[1] Amoco Research Center, Tulsa, Oklahoma
[2] Department of Geology, Texas A&M University, College Station, Texas

HENRY
HENRY
CENTERS
BASIN
HITE
CATARACT CANYON
MONTICELLO
ESCALANTE
CIRCLE
CLIFFS
UPLIFT
KAIPAROWITS
SYNCLINE
KAIBAB
HALLS CANYON
CANYON
SLOPE
UPWARP
BLANDING
BASIN
HOLE IN THE ROCK
BASIN
WHITE
SAN JUAN
RIVER
BLUFF
MEXICAN HAT
NAVAJO MTN
PIUTE
UTAH
ARIZONA
MONUMENT
NAVAJO CREEK
CLIFFS
LEES FERRY
FOLD
BELT
UTAH
ECHO
KAYENTA
N
0
2
4 Miles

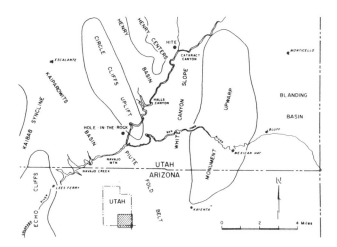

Fig. 1 — Generalized location map of the Lake Powell Area showing the location of the major structural features.

of all fractures present, as well as their spatial distribution. Ziony (1966) restudied a portion of the area previously described by Hodgson (1961) with primary interest in the origin of regional fractures.

ORIGIN OF REGIONAL FRACTURES

The origin of regional fractures has long been an enigma because they defy obvious mechanical explanation. Recall that they are developed over large areas with relatively little change in orientation; they show no evidence of offset; and they are always perpendicular to major bedding surfaces. Regional fractures are commonly developed in orthogonal sets (Price, 1959; Stearns, 1968a and b; Gay, 1973); therefore, their derivation from a single stress field is difficult to explain. Yet the fracture orientations and spacings are consistent over areas as large as a quarter of the Colorado Plateau, which would indicate that the causative stress field was large in area and extremely homogeneous with respect to direction and intensity.

Regional fractures are very well developed in the entire Colorado Plateau Province; and, while the Lake Powell area accordingly has well developed regional fractures, they are unusual because the fractures do not occur in typical orthogonal patterns at any one outcrop. Only one dominant regional fracture orientation is developed in any single formation. On area-wide formational composite diagrams, however, a hint of orthogonality is displayed.

Figures 3 and 4 present the total fracture data from the study area with all those fractures attributable to local tectonic events such as folding and faulting omitted (the N15°W orientation has not been associated with any local structure, but it is found in abundance only in the local area surrounding Glen Canyon Dam and is, therefore, disregarded). Figure 3 presents the fracture data for each formation individually. Figure 4a is a composite of all ground-based regional fracture measurements while Figure 4b is a composite of all air-photo measurements. It is important to note that the strike of the regional fractures changes from formation to formation. Because of this change, any theory of the origin of regional fractures must support either the creation of separate stress fields in each stratigraphic unit or account for the possibility that the different fracture orientations arise from strength anisotropies developed within the stratigraphic units.

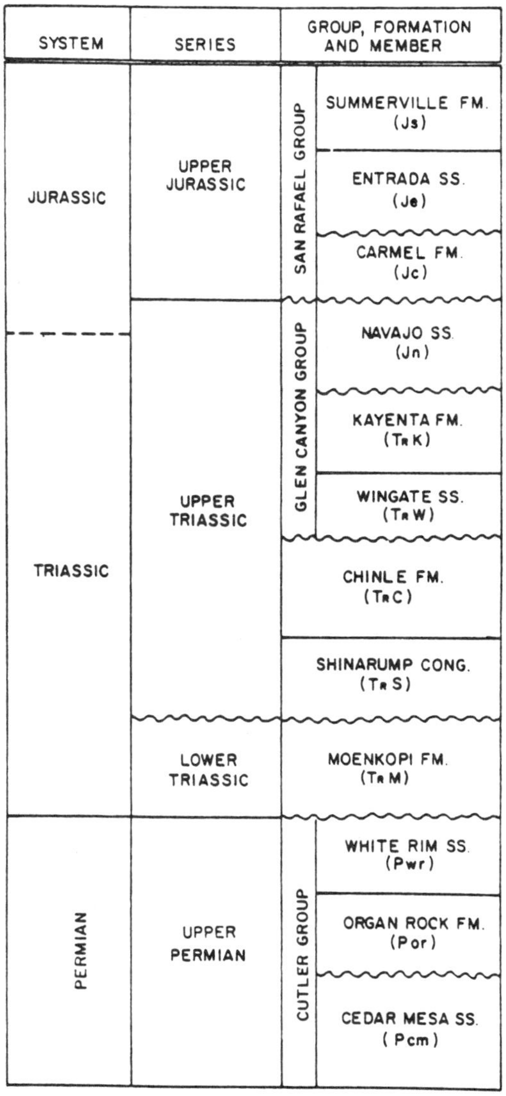

SYSTEM
SERIES
GROUP, FORMATION AND MEMBER
JURASSIC
UPPER JURASSIC
SAN RAFAEL GROUP
SUMMERVILLE FM. (Js)
ENTRADA SS. (Je)
CARMEL FM. (Jc)
TRIASSIC
UPPER TRIASSIC
GLEN CANYON GROUP
NAVAJO SS. (Jn)
KAYENTA FM. (TRK)
WINGATE SS. (TRW)
CHINLE FM. (TRC)
SHINARUMP CONG. (TRS)
LOWER TRIASSIC
MOENKOPI FM. (TRM)
PERMIAN
UPPER PERMIAN
CUTLER GROUP
WHITE RIM SS. (Pwr)
ORGAN ROCK FM. (Por)
CEDAR MESA SS. (Pcm)

Fig. 2 — Generalized stratigraphic section of the rocks outcropping in contact with Lake Powell (after Gregory, 1917).

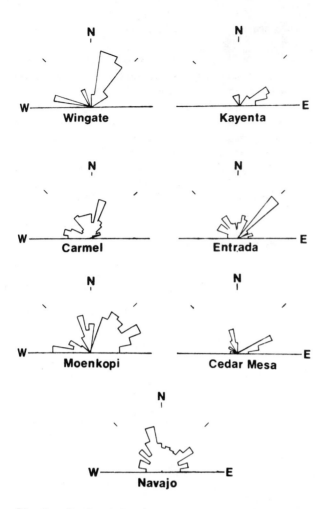

Fig. 3 — Regional fracture orientation composites for formations in the Lake Powell area. The scales for each diagram are different. This figure represents a total of about 2500 measurements.

indicate that no slip could have occurred along the fracture plane, at least at the time of fracturing (Hodgson, 1961).

LOADING DIRECTION AND AGE

Because regional fractures are extension fractures, the maximum principle stress at the time of fracturing was parallel to the fracture plane. Because the regional fractures cut the Jurassic Entrada Sandstone and are rotated on the flanks of Laramide folds, stresses must have been active between Late Jurassic and Early Tertiary time. Because of the large scale of the fractures, these stresses could have been related to either pre-folding Laramide deformation or possibly the activation of the Rocky Mountain geosyncline in Cretaceous time. No large-scale horizontal deformation is known to have occurred within this area during either of these time periods. Therefore, this study will investigate vertical maximum principal compressive stresses as the cause of regional fracturing within this area.

CONTROL OF REGIONAL FRACTURE ORIENTATION

The fact that regional fractures have different orientations from formation to formation presents a problem with regard to

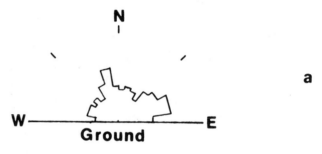

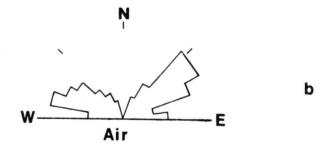

Fig. 4 — Total regional fracture orientation composites for all formations. (a) derived from 2500 ground measurements; (b) derived from 2000 air-photo measurements.

In order to discuss the mechanics of regional fracturing within this area, it is necessary to determine if the fractures are shear or extension fractures. Due to the lack of conjugate fractures and the observation that regional fractures are often developed in orthogonal sets, it is concluded that the regional fractures within the area of this study are extension fractures (Nelson, 1975). As such, the maximum principal stress direction (σ_1) was parallel to the fracture plane at failure. This conclusion is not inconsistent with field observations in that the regional fracture planes observed in the field show no evidence of offset or slippage parallel to the trace, while known tectonic shear fractures do show offset and gouge development. Further evidence that the regional fractures are of the extension type is found on the fracture surface where plumose structures

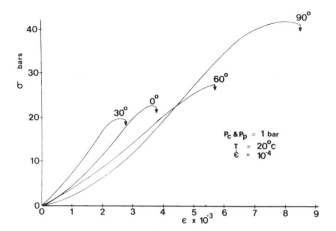

Fig. 5a — Stress (σ) - Strain (ϵ) diagram for triaxial compression tests on Navajo Sandstone. Curves indicate various angles between cross-bedding and load axis.

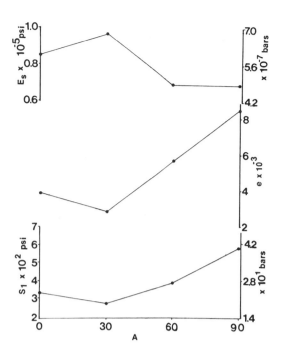

Fig. 5b — Secant modulus (E_s), longitudinal strain (e), and axial stress (s_1) at failure for Navajo Sandstone versus the angle (A in degrees) between cross-bedding and load axis.

their genesis. These changes could be due to an anisotropic state of stress in the layered sequence, strength variations within the units themselves, or a combination of the two.

Because these fractures are of wide extent, it must be concluded that the loading conditions causing the deformation were homogeneous over a large area. Given homogeneous loading conditions and the assumption that fractures result from an anisotropic stress field, then welded contacts would seem to be precluded. This is because it would be impossible to shift the principal stress directions across bedding planes or formational boundaries which is necessary to produce extension fractures with discretely different orientations. If, on the other hand, the contacts were unwelded, small shifts in the orientations of the principal stresses could have occurred due to large differences of elastic moduli across formational boundaries. Such large moduli differences, however, do not exist in this stratigraphic section.

The most obvious candidate for cause of interformational strength anisotropies in the Mesozoic clastic rocks of this area is cross-bedding. The Cedar Mesa, Wingate, Navajo, and Entrada Formations are all eolian sandstones with inherent cross-bedding assuming statistical orientations determined by the prevailing paleo-wind directions at the time of deposition.

Two types of strength anisotropy could arise from sedimentary bedding, including cross-bedding. Strength anisotropy can occur within or across bedding planes. Due to the sphericity of wind blown sands, preferred fabric orientation within the planes of eolian cross-beds of the Navajo Sandstone that would produce an anisotropy within bedding planes is rare. Anisotropy does exist, however, across the plane of the cross-bed.

Testing of these rocks has shown that cross-bedding has a pronounced effect on their mechanical properties. Compressive and tensile strengths, longitudinal strain (e_1) and secant modulus (E_s) at fracture are all a function of the orientation of cross-bedding with respect to loading direction. Data on these anisotropies are found in USBR (1957), Robinson (1970), Dunn and others (1973), and Figure 5 of this report.

In general, for the Navajo Sandstone, strength and elastic modulus at fracture increase, and longitudinal strain at fracture decreases as the angle between cross-beddings and the load axis inclines from 60° to 90°. According to Potter and Pettijohn (1963), the overall strike of a sand dune is perpendicular to the wind direction at the time of formation. The inclination of cross-beds with respect to true bedding is a maximum (about 26°) in the direction parallel to the wind direction, and it approaches zero in the direction perpendicular to the wind direction. This preferential orientation of dips of cross-beds, coupled with the fact that the physical properties of the subsequently formed sandstone are controlled by the inclination of cross-bedding with respect to loading, can create a substantial strength anisotropy in the rock mass. The orientation of the bulk anisotropic effect is related to the average paleo-wind direction at the time of deposition of a sandstone. The average paleo-wind direction for several formations was calculated by

Table 1

Paleo-wind direction data
(Poole, 1962)

Formation	Average wind direction	Number of localities	Standard deviation (degrees)
Navajo	41°W	28	28
Wingate	57°W	23	22
Cedar Mesa	47°W	5	15
Entrada	63°E	2	11
upper sand	68°W	3	17
Moab tongue	E-W	2	13

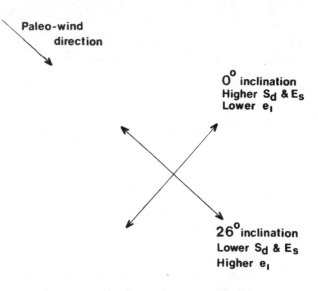

Fig. 6 — General mechanical anisotropy developed due to cross-bedding in the horizontal plane and its relation to the paleo-wind direction.

Poole (1962) from statistical treatment of measurements of cross-bedding. These data are presented in Table 1.

These figures represent the overall maximum and minimum cross-bed dip directions for each formation throughout the entire area. They should also be a clue to existent bulk strength anisotropies.

Anisotropies developed with respect to paleo-wind direction are shown in Figure 6. According to the data in Figure 6, vertical extension fracturing would form more easily perpendicular to the paleo-wind direction (parallel to the long axis of a dune). The relation between this direction for each eolian formation and the regional fracture orientations is shown in Figure 7. In the Wingate and Navajo Sandstone diagrams, the correlations are very high. There is a good correlation for the Cedar Mesa Sandstone, but virtually none for the Entrada Sandstone. It is not surprising, however, that the poorest correlations are found for the two formations for which measurements of paleo-wind-directions are the fewest (Cedar Mesa, 5 stations; Entrada, 7 stations). In formations with sufficient data, there is a surprisingly good correlation between Triassic and Jurassic wind directions and regional fracture orientations within a 13,000 km² (5000 mi²) area.

This geometric relation between cross-bed and fracture directions can be found in the data of Hodgson (1961) and from specific outcrop measurements of fracture and cross-bed orientations near Glen Canyon Dam, Figure 8, (Nelson, 1975). Picard (1966, 1969) has described the identical geometric relationship between vertical shrinkage cracks (extension fractures) and ripple marks in both siltstone and limestone.

Comparisons of current directions by Poole (1961) for two of the fluvial formations in the Lake Powell area with the orientations of their regional fractures show a different correlation. Regional fractures tend to parallel the current direction in the Moenkopi and Kayenta Formations.

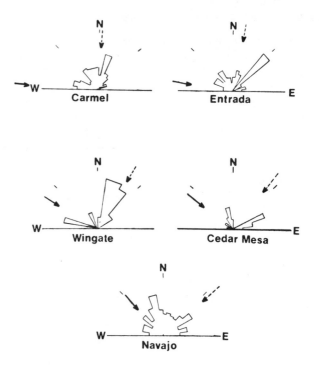

Fig. 7 — Relationship between paleo-wind directions and regional fracture orientations by formations. Solid arrow indicates the averaged paleo-wind direction, dashed arrow indicates the perpendicular to the paleo-wind direction.

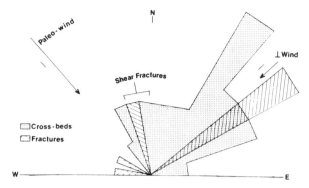

Fig. 8 — Relationship between fractures, paleo-wind direction, and cross-bedding in the Navajo Sandstone for the area just east of Glen Canyon Dam.

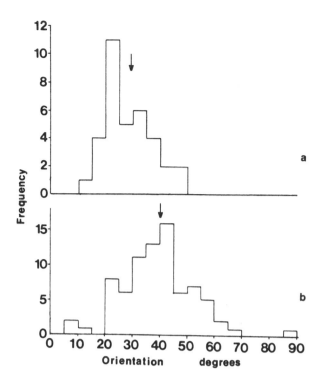

Fig. 9 — (a) Frequency versus angle between arbitrary mark and fractures, measured to the right of the mark. (b) Frequency versus angle between arbitrary mark and long grain axes and hemetite stringers, measured to the right of the mark.

Cross-bed orientations in fluvial sediments still indicate current directions, but they are often more variable areally than in eolian sediments. This is due to the fact that current flow within a channel can frequently change strike. In fluvial sediments, there is no appreciable change in compressive strength between the 60° and 90° orientations of cross-bed with respect to load axis and no consistent trend in e_1 and E_S over the same interval, Dunn and others (1973). This indicates that anisotropy across the fluvial cross-beds is low compared to that of the eolian sediments. This is why the regional fractures do not orient themselves perpendicular to fluvial sediment-transport directions as in the eolian sandstones. They do, however, orient themselves approximately parallel to the transport direction. This may imply a substantial anisotropy within the bedding plane. Potter and Pettijohn (1963) show that, in general, sediments exhibit a preferred orientation of the long axes of sand grains in the direction parallel to fluid transport. This orientation is not evident in eolian sands because the individual grains tend to be spherical. Figure 9a is a plot of extension fracture orientations as measured from an arbitrarily chosen direction in point-load tests on unoriented samples of Kayenta Sandstone, Friedman and Logan (1970), Nelson (1975). The specimens used were 1 by 2 cm disks cut and loaded perpendicular to the bedding plane. The data show a pronounced preferred fracture direction. Thin sections were made of several of these samples, and grain orientations were measured. The preferred fracture direction is within 11° of the long-grain fabric (Fig. 9).

The control of regional fracture orientations by large-scale strength anisotropies within the formations also relates to the development of only single sets of regional fractures in the Lake Powell area. Regional fractures usually exist in two mutually perpendicular orientations in most areas. Price (1959), page 160, describes ways in which this pattern can be developed.

"The moment fracture takes place, the least principle stress changes, the tensile stresses are released and are replaced by a compressive stress ... The former intermediate principle stress thus becomes the least principle stress. Further uplift will cause a second set of 'Tension' Joints to develop; the two sets forming an orthogonal system. If the lateral stresses were initially almost exactly equal, the formation of the sets of joints will be practically simultaneous."

Price (1959, 1966, 1974) suggests almost equal stresses in the plane of isotropic bedding are necessary for simultaneous fracturing. However, if a preferred weak direction existed within the bed, only the first fracture which forms would be developed. The subsequent shift of principle stresses in the bedding plane would not create a high enough tensile stress to cause fracture in the stronger direction. Thus, the rocks in the Lake Powell area with high strength anisotropies could contain only a single fracture orientation instead of the classic orthotonal pattern.

311

CONCLUSIONS

We conclude, therefore, that regional anisotropies can result from large scale sedimentary structures and that these anisotropies can mechanically control the orientations of regional fractures ($\pm 20°$ in this study). Regional fractures tend to be parallel to the area-wide paleo-current directions in fluvial sandstones and perpendicular to the area-wide paleo-wind direction in eolian sandstones.

The most noteworthy papers concerning regional fractures in the Western United States are Kelley and Clinton (1960) and Hodgson (1961). Their studies have treated regional fracturing from a geological point of view and contributed a great deal to our understanding of the existance of such fractures. Modern studies of geologic deformations should, however, consider rocks as distinct mechanical units which possess unique strengths, heterogeneities, and anisotropies. When regional fractures are studied with a knowledge of the physics of fracture, many of the observations which have defied explanation become understandable. Such an observation is the shift in regional fracture orientation through the stratigraphic section. Analysis and interpretation of existing experimental data on the strength of rocks permit a quantitative account of this shift. A descriptive, qualitative approach does not lead to such an understanding.

REFERENCES

Dunn, D. E., La Fountain, L. T., and Jackson, R. E., 1973, Porosity dependence and mechanism of brittle fracture in sandstones: Jour. Geophysical Res., v. 78, p. 2403-2417.

Friedman, M., and J. M. Logan, 1970, The influence of residual elastic strain on the orientation of experimental fractures in three quartzose sandstones: J. Geophys. Res., v. 75, p. 387-405.

Gay, S. P., 1973, Pervasive orthogonal fracturing in the earth's continental crust: Salt Lake City, Utah, Amer. Stero Map Co., Tech Publication No. 2, 121 p.

Hodgson, R. A., 1961, Regional study of jointing in Comb Ridge-Navajo Mountain area, Arizona and Utah: Amer. Assoc. Petrol. Geol. Bull., v. 45, p. 1-38.

Kelley, V. C. and Clinton, N. J., 1960, Fracture systems and tectonic elements of the Colorado Plateau: Univ. of New Mexico Publ. in Geology 6, 104 p.

Nelson, R. A., 1975, Fracture permeability in porous reservoirs: An experimental and field approach: Ph.D. Dissertation, Texas A&M University, 171 p.

Picard, M. D., 1966, Oriented linear-shrinkage cracks in Green River Formation (Eocene), Raven Ridge area, Uinta basin, Utah: Jour. Sed. Petrology, v. 36, no. 4, pp. 1050-1057.

_____, 1969, Oriented linear-shrinkage cracks in Alcova Limestone Member (Triassic), Southeastern Wyoming: Contrib. Geol., v. 8, no. 1, pp. 1-7.

Poole, R., 1961, Current directions in Triassic rocks of the Colorado Plateau: U.S. Geol. Survey Prof. Paper 424C, 139-141 p.

_____, 1962, Wind directions in Late Paleozoic to Middle Mesozoic time on the Colorado Plateau: U.S. Geol. Survey Prof. Paper 450D, 147-151 p.

Potter, P. E. and Pettijohn, R. J., 1963, Paleocurrents and basin analysis: New York, Academic Press, Inc.

Price, N. J., 1959, Mechanics of Jointing in Rocks: Geol. Mag., v. XCVI, no. 2, pp. 149-167.

_____, 1966, Fault and Joint Development in Brittle and Semi-Brittle Rock: London, Pergamon Press, 176 p.

_____, 1974, The development of stress systems and fracture patterns in undeformed sediments: Proc. Third Cong. Intern. Rock Mech., TA487-496 p.

Robinson, E. S., 1970, Mechanical disintegration of the Navajo Sandstone in Zion Canyon, Utah: Geol. Soc. America Bull., v. 81, p. 2799-2806.

Stearns, D. W., 1968a, Fracture as a mechanism of flow in naturally deformed layered rock, in Baer, A. J. and Norris, D. K. (eds.) Kinkbands and brittle deformation: Geol. Survey Canada Paper 68-52, p. 79-95.

_____, 1968b, Certain aspects of fracture in naturally deformed rocks, in National Science Foundation Advanced Science Seminar in Rock Mechanics, A. E. Riecker (ed.), Special Report: Air Force Cambridge Research Laboratories, Bedford, Mass., AD 669375 I, p. 97-118.

_____, 1971. Mechanisms of drape folding in the Wyoming Province: Wyoming Geol. Assoc. 23rd Annual Field Conference Guidebook, p. 125-143.

U.S. Bureau of Reclamation, Department of the Interior, 1957, Technical data for the use of the consultants, Glen Canyon Dam, Colorado River Storage Project, prepared by the office of the Assist. Commissioner and Chief Engineer.

Ziony, J. I., 1966, Analysis of systematic jointing in part of the Monument Upwarp, Southeastern Utah: Ph.D. Dissertation, University of California at Los Angeles, 152 p.

Reprinted by permission of the University of Chicago Press.
Published in the *Journal of Geology*, V. 33, No. 7
(October-November 1925), pp. 685-698.

THE GEOLOGIC ROLE OF DILATANCY

WARREN J. MEAD

University of Wisconsin

ABSTRACT

Dilatancy, the expansion of granular masses when deformed due to the rearrangement of the grains, was described and illustrated with many interesting experiments by Osborn Reynolds, an English physicist. In this paper the term is used in a broader sense to include all volume increase due to deformation. The consequences of dilatancy by deformation have far-reaching geologic significance. It is suggested that dilatancy is important in inducing faulting and jointing in unconsolidated sediments, in the movement of fluids—oil, water, and gas—contained in rocks, in initiating magmas and in certain of the processes accompanying intrusion and crystallization of magmas. The factors controlling the manner of deformation of unconsolidated granular masses are applied by analogy to a consideration of the manner of deformation of solid rocks by fracture and by rock flowage.

Dilatancy. The property of granular masses of expanding in bulk with change of shape. It is due to the increase of space between the individually rigid particles as they change their relative positions.—*Century Dictionary.*

This property of granular aggregates was described by Osborn Reynolds,[1] whose interest was apparently confined to its application in his mechanistic conception of the ether as applied to the development of a theory of gravitation. Dilatancy is a property of many earth materials, and is fundamentally related to the manner of response of these materials to deformation. It is the purpose of this paper to indicate the possible role of dilatancy in certain geologic processes. The term will be used in a somewhat broader sense, to include all volume increases due to deformation.

DILATANCY IN DEFORMATION OF INCOHERENT GRANULAR MASSES

Hard, spherical grains, such as shot, shaken down in a container, tend to arrange in a condition of maximum-density packing. If the grains are spherical and of uniform size, each grain is in contact with twelve neighboring grains. This arrangement has a minimum of voids, 25.9 per cent. It is obviously impossible to change the shape without increasing the volume of this aggregate (assuming that the grains are not deformed), as any differential movement between the grains involves a change in the system of packing, which of necessity requires increase of voids and consequently of volume. This is illustrated in Figure 1, by a cross-section of spherical grains in close packing at the left, and the increase in volume occasioned by deformation of this mass at the right.

A mass made up of spherical grains of uniform size serves best for illustrative purposes. The property of dilatancy, however, is not confined to masses of spherical grains, but is common to all granular aggregates, regardless of angularity of grains or degree of assortment.[2]

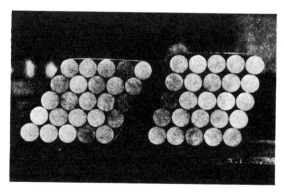

FIG. 1.—Illustrating the increase in voids and consequent increase in volume resulting from deformation of a mass of close-packed spherical grains. Cylinders of wood held by rubber bands and strips of cardboard serve to represent spherical grains in cross-section. This serves as a two-dimensional representation of dilatancy.

---

[1] Osborn Reynolds, *Scientific Papers,* Vol. II (1901), p. 217.

[2] When particles are so small that the adsorbed liquid or gaseous layers on their surfaces are thick in proportion to their radii, movement between the particles apparently takes place through deformation of the envelopes. This is illustrated by the behavior of aerated, fine dry powders, which simulate fluid in their behavior, and by damp, fine-grained clay.

Quoting Reynolds, "The most striking evidence of dilatancy is obtained from the fact that since dilatant material cannot change its shape without increase in volume, by preventing change of volume all change of shape is prevented."[3] This fact is beautifully illustrated by one of Reynolds' experiments. A toy rubber balloon, filled with sand and containing just sufficient water to saturate the sand after it has been shaken down into a condition of dense packing, resists deformation to an astonishing degree, because sand so contained, when deformed, must increase in volume, and requires more water than is available to fill the increased proportion of voids. The result is that the fluid pressure within the container is greatly reduced and atmospheric pressure forces the rubber very tightly about the mass of sand, preventing dilatation, increasing intergranular friction, and thereby inducing rigidity. The tensile strength of the water may also be a factor in the rigidity of the mass. When deformed (although Reynolds, not being interested in this phase of the experiment, makes no mention of it) it fails along definite shear plans, rather than by plastic yielding of the entire mass.

The same phenomenon can be illustrated by a modification of Reynolds' experiment. If a rubber container, such as a toy balloon, is filled with sand and the air is pumped out, atmospheric pressure forces the rubber tightly against the contained sand, and the mass becomes exceedingly rigid. An 8-inch rubber balloon filled with dry sand and exhausted by an ordinary laboratory aspirator becomes as rigid as a solid rock, and gives out a musical tone, as does a stoneware jar, when struck with the knuckles. Rigidity is due to the containing pressures, but probably also to removal of air cushions from between the grains, both factors increasing intergranular friction. This evacuated, sand-filled, rubber sack behaves as a brittle solid, which fails by shear when deformed under pressure. A thin-walled cylinder filled with sand (with a small proportion of flowers of sulphur to serve as a binder), closed at both ends, and evacuated, fails along a definite shear plane under longitudinal pressure. By heating the deformed cylinder to the melting-point of sulphur, the mass of sand is solidified. Such a mass is illustrated with the container removed in Figure 3. Granular masses free to dilate deform by flow; when dilatation is prevented or restricted, deformation causes fracture. Sand, either wet or dry, under relatively slight containing pressures, behaves essentially as a solid (as dilatation is restricted) and, when deformed, fails largely by fracture along shear planes. This manner of failure requires a minimum increase in volume and involves dilatation only in the shear zone. Plastic deformation of close-packed grains involves the entire mass deformed, and causes a much greater volume increase than that required by failure along definite shear planes.

The mechanical properties of fine, dry sand under no other containing pressures than its own weight are interesting in connection with the phenomena of dilatancy, and afford a means of simple experiments quite instructive in connection with problems of the mechanics of rock deformation. If sand, fine enough to pass through an 80-mesh screen, is poured out on two overlapping sheets of paper so that the pile of sand is partly on one sheet and partly on the other, and is leveled off to a thickness of approximately 2 cm., a graben fault may be produced by pulling the upper sheet of paper away from the lower in a direction at right angles to its covered edge. See Figure 4. If, with the sand again prepared as above described, the upper sheet of paper is moved *toward* its covered edge, a wedge of

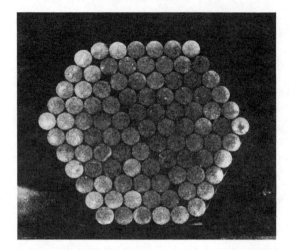

FIG. 2.—Wooden cylinders in triangular packing are held together by several very strong rubber bands. Increase in area is restricted by the containing pressure of the bands. One-sided pressure results in failure by shear with increase in area as shown. If increase in area is not prevented, the "grains" are all free to move relative to one another and single fractures as shown above do not develop.

FIG. 3.—Failure by shear of dry sand held in a cylindrical rubber container and made rigid by pumping out the air. The mass was deformed by longitudinal compression.

[3] Osborn Reynolds, *Nature,* Vol. XXXIII, p. 30.

sand will be faulted upward in a horst. See Figure 5. If the upper sheet of paper is rotated about a point at its edge beneath the sand, a combination of graben and horst is produced. See Figure 6. If the sand is leveled out, as previously described, in a layer 2 cm. thick, and a small block is thrust slowly into the mass by sliding it along the paper, the sand displaced in front of the block will rise and override the undisturbed sand in a series of overthrust faults,[4] resembling in a rather striking way the well-known structure of the Scottish Highlands. In this case the easiest relief is upward, and movement is accomplished by faulting in this direction. If the sand is piled into a rather long, narrow ridge, as in Figure 8, and the block is thrust longitudinally into the ridge of sand, the direction of easiest relief is lateral, and the sand faults along nearly vertical planes at an angle to the direction of the movement.

The geologic significance of the previously described properties of sand and allied granular aggregates is that even under very moderate load they tend to fail by fracture rather than by plastic flow, and that joints and faults in unconsolidated sediments are to be expected. The faulting and jointing in fine-grained glacial material so commonly observed (and so frequently marveled at) exemplify this property of granular aggregates. It seems probable that at least part of the jointing in sandstones and other sedimentary rocks may have had its origin in the deformation of these rocks prior to consolidation. The ''grain'' (a capacity to split along parallel surfaces across the stratification), so common in sandstones, may be the result of more open packing of the grains along planes of precementation dilatation.

EFFECT OF DILATANCY ON THE MOVEMENT OF FLUIDS CONTAINED IN SEDIMENTS

The changes in the manner of packing of grains, with consequent changes in voids, involves a movement of any fluids which may be contained in the voids. An experiment illustrating this may be made by filling a thin-walled rubber cylinder (a section of automobile tire tubing) with sand, tightly closing the ends with large corks, and filling the voids in the sand with water to the point of complete saturation. If a glass tube is inserted through a hole in one of the corks and bent at right angles to a vertical position and partly filled with colored water it serves as an indicator of the change in volume of the sand. It is found that this sand-filled cylinder cannot be bent or squeezed or deformed in any manner without lowering the level of the water in the vertical tube. This suggests that there must be a movement of contained fluids in sediments toward regions of deformation, as these are regions of dilatation, due to both rearrangement of grains and to rock fracture. It is believed that this phenomenon should not be neglected as one of the processes controlling the movement of oil, water, and gas, and as one of the reasons for the movement of oil toward anticlines and monoclines.

It has been shown that if the amount of fluid in the granular aggregate is only sufficient to fill the voids in the condition of maximum-density packing, the mass fails largely by fracture, and not by plastic deformation. If fluid is available slightly in excess of this amount, the aggregate is easily deformed plastically up to the point where the

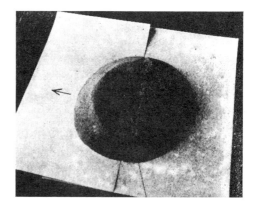

Fig. 4.—Graben fault in fine, dry sand, developed by moving the upper sheet of paper in the direction of the arrow.

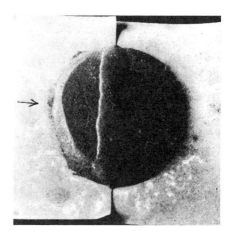

Fig. 5.—Faults producing a horst in fine, dry sand, developed by moving the upper sheet of paper in the direction of the arrow.

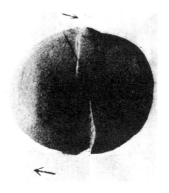

Fig. 6.—Combination of graben and horst produced by a rotational movement of one of the overlapping sheets of paper, as indicated by the arrows.

---

[4] R.T. Chamberlin and F.P. Shepard have described repeated overthrust faults in unconsolidated sand (*Journal of Geology,* Vol. XXXI (1923), pp. 511-12). Faults in unconsolidated sand and clay were produced by thrusting a pressure block horizontally into the mass. The nature of the faulting was tested by means of straws stuck into the sand and by the action of the moving sand on smoked glass plates inserted in the mass at right angles to the pressure block.

increased voids absorb the available fluid. At this point the mass becomes rigid. This is well illustrated by a toy balloon filled with sand and water, with the latter somewhat in excess of the amount required to saturate the sand in a condition of dense packing. The balloon so filled is soft and easily deformed *up to a certain point*. If squeezed in the hand, it suddenly becomes rigid when the volume of voids and the volume of water become equal. If more water is added, a condition is reached where the balloon is soft and easily deformed to any extent without becoming rigid. There is enough water present to fill the voids when the sand is in the most open arrangement possible. In this condition deformation does not involve an increase in volume of the balloon, and the mass changes its shape by plastic flow and offers very little resistance to deformation. This state of affairs is illustrated in nature by quicksand, which contains sufficient water to permit the most open arrangement of the grains—usually found in a situation where an upward current of water keeps the sand in open packing. Quicksand becomes very firm when this excess of water is drained away.

The mechanics of certain types of landslides, of mud flows, of failure of earth dams, involve in a large measure the factors of dilatancy. A comparatively small change in the water content of a mass of earth or sand changes it from a stable, comparatively rigid condition to a relatively fluid condition. Initial movement of the mass is an important factor, as it causes dilatation, which increases the voids, draws in water, and causes the mass, on passing the critical point, to move as a fluid.

DILATANCY IN THE DEFORMATION OF SOLID ROCKS

It is convenient to refer to the hard grains as the solid phase and to the material between the grains as the fluid phase. Using these terms, then, the experimental work seems to demonstrate two general principles. (1)When the fluid phase is sufficient only to fill the voids with the grains in a condition of maximum-density packing, deformation of the mass requires increase in volume. (2)When the available fluid phase is sufficient to fill the voids with the grains arranged in minimum-density packing, the mass may be deformed to any extent without increase in volume.

The writer believes that this conception can safely and profitably be carried over to a consideration of the mechanics of deformation of the solid rocks. Rocks in general may be regarded as granular aggregates. To the extent that they are porous, the pores represent the volume of a fluid phase, but the amount of fluid phase is, with the exception of a few special cases, too small to play much part in determining the manner of deformation. A sandstone, if cemented sufficiently to merit the term, has less porosity than sand. The small proportion of a fluid phase causes all solid rocks at or near the surface to yield to deformation by fracture, with increase in volume. This fracturing of rocks clearly involves dilatation, and the net volume of the fractured mass has been increased by the total volume of the openings produced.

That the deformation of brittle materials involves an increase in volume *prior* to their failure has been generally recognized. Bucher[5] has discussed this matter in a consideration of the mechanical interpretation of joints, and quotes Chwolson, who gives a formula con-

[5] Walter H. Bucher, *Journal of Geology,* Vol. XXIX, p. 1.

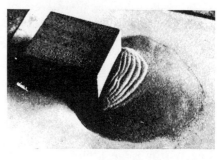

Fig. 7a.—Repeated overthrust faults in fine, dry sand, developed by shoving the wooden box horizontally along the paper into the flat mass of sand. The direction of easiest relief is upward.

Fig. 7b.—Cross-section of faulted mass developed as shown in Figure 7a. Stratification is produced by sifting alternate layers of sand of different colors. The sand is cemented after deformation by heating the mass to a temperature sufficient to melt a small proportion of sulphur which had been incorporated in the mass.

Fig. 8.—The sand was piled in a narrow ridge and the box thrust longitudinally into the mass. The direction of easiest relief is lateral and the sand faults along nearly vertical planes inclined to the direction of the movement.

316

necting the modulus of volume increase with Young's modulus and Poisson's ratio. He also quotes the work of Kahlbaum and Seidler, and of Lea and Thomas, as giving experimental evidence for increase in volume accompanying deformation under one-sided compression. That this increase in volume occasioned by deformative stresses imparts a greatly increased rigidity to the rock when under great containing pressures appears very probable. Bucher offers no explanation of how this increase in volume is accomplished, whether by change in the physical nature of the material itself, or by the development of voids.

When rocks are deformed under certain conditions involving high confining pressures and a proper rate of application of deforming stresses, they yield to deformation by plastic flow with the development of schistose textures characteristic of rock flowage. This manner of yielding to deformation does not involve a general fracturing of the rock, and probably does not require increase in volume. If the analogy of the requirements for plastic deformation of an unconsolidated granular mass be carried over to the case of rock flowage of a solid rock, it is necessary to conceive of the latter as consisting of a solid phase of hard grains and of a fluid phase surrounding these grains of a sufficient amount to permit the movement of the hard grains without occasioning dilatation of the mass by their interference. The solid phase is represented by the harder, more resistant minerals. The fluid phase is represented by those constituents of the rock which are relatively mobile, as evidenced by their rearrangement to schistose structures through processes of crystallization and recrystallization. This involves a complete atomic or molecular rearrangement of that portion of the rock.

Solid rocks in general, then, may be considered as granular aggregates consisting of a solid phase and a *potentially* fluid phase, which is caused to function as a fluid phase under certain conditions of composition, pressure, temperature, and rate of deformation. To make this clear let us revert for the time being to a modification of the simple concept of a granular mass consisting of sand and water. In this case a mass of sand, well shaken down into close packing with the voids filled with asphaltum instead of water, is used. The sand grains constitute a solid phase, the asphaltum a potentially fluid phase. A block of this asphalt-cemented, closely packed sand cannot fail except by fracture, even if deformation is applied very slowly, because the amount of fluid phase is not sufficient to permit plastic deformation. If, however, a mixture of sand and asphaltum be taken, with sufficient asphaltum to permit freedom among the sand grains equivalent to that in cubic or minimum-density packing, the resulting aggregate is such that under slow deformation at ordinary temperatures, or under more rapid deformation at higher temperatures (under which the asphaltum would soften), it would yield by plastic deformation of the entire mass without increase in volume. The asphaltum, or potentially fluid phase, would function as a fluid. If, on the other hand, the aggregate is deformed rapidly, the asphalt would not be able to function as a fluid, and the mass would fail by fracture, with an increase in volume.

This analogy of the asphaltum-sand aggregate may be applied to the flowage or fracture of rocks. Under conditions of rock flowage that portion of the rock which undergoes molecular or atomic rearrangement constitutes the fluid phase. If the rate of deformation under the existing conditions of temperature and pressure is too great to permit this rearrangement, the rock fails by fracture; and if the rate of deformation is slow enough to permit this rearrangement, failure is by plastic flow, provided the solid phase is sufficiently dispersed to prevent dilatation. If the amount of the fluid phase is intermediate, the deformation will be by combination of plastic flow and

317

fracture, the grains of the solid phase being fractured and granulated through their mutual interference, the fractures being filled by movement of the fluid phase into the openings; the net volume change is determined by the degree to which the increase of volume due to fracturing of the solid grains is balanced by the decrease in volume due to development of new compounds of higher density in the fluid phase.

The zones of schist so commonly found marking planes of movement in massive igneous rocks differ in composition from the massive rock by having more combined water and less lime and soda. Mineralogically they are characterized by an increased amount of platey minerals, micas, and chlorite. That these schists may have been developed along zones previously weathered or hydrothermally altered is a possible explanation for the change in chemical composition. It is equally probably that the increase in water and loss of other constituents has in some instances been contemporaneous with the deformation of the rock along this zone. Because the development of a schist from a strong massive rock presupposes deep burial with intense rock pressures and deformative stresses of great magnitude, it has been difficult to reconcile the introduction of water, and the escape of constituents known to be lost, with the condition of great pressure, which on first thought would appear to tend toward the squeezing out of any water present and the prevention of any ingress of fluids. If it be assumed that deformation has been too rapid to permit the potentially fluid phase of the rock to function, there has been increase in volume as a result of deformation within the elastic limit, and as a result of crushing and fracturing. This dilated zone is a region of low *fluid* pressures, and any available fluids will move to fill these openings. Under the conditions of temperature and pressure they will have highly reactive and solvent properties, and metamorphic changes with the development of hydrous minerals would ensue. Continued and perhaps slower movement along this zone of yielding finds a rock partly adapted by composition to rock flowage, and any excess water, with a high content of dissolved mineral matter, is squeezed out.

It is interesting to speculate on the possible relationship of dilatancy to the manner of deformation of rocks at depth. The fact that yielding by fracture involves increase in volume, while rock flowage does not, suggests that the increased prominence of flowage with increase in depth is due to the fact that flowage does not require the lifting of the equivalent of the gravitational load, and therefore may not require stress differentials as great as would be demanded for rock fracture.

DILATANCY AND ORIGIN OF MAGMAS

If the conclusion be accepted that deformation of a rock by flowage requires that the rate of deformation be slow enough to permit the potentially fluid phase of the rock to function as a fluid, and that deformation at a more rapid rate must of necessity produce failure by fracture, it follows that great and comparatively rapid deformations of the earth's crust may extend far below the surface and well into the zone normally characterized by rock flowage. If this is at a depth where the rocks are at a temperature above their melting-point but are kept solid by pressure, the result of fracture-dilatation would be immediate liquifaction of the rock in that zone to an extent measured by the increase in volume. This fluid rock migrating by way of the fracture zone to regions of lower pressure would remain fluid and contain sufficient excess heat to fuse a certain amount of rock in its path. The presence and movement of this fluid material would considerably upset the dynamic stability of the whole, and result in

318

the development of magma and magmatic activities of greater or lesser extent, depending on the magnitude of the original deformation.

DILATANCY IN MAGMATIC PROCESSES

An igneous magma in the process of solidification can be considered as a granular mass made up of a solid phase (the mineral crystals) and a fluid phase. The mass has the properties of a fluid until solidification has proceeded to a point where the solid phase is so abundant that the grains are no longer free to move, but interfere with one another when the mass is deformed. In this condition the mass can no longer behave as a fluid, and is incapable of being injected into openings in the surrounding rock due to the great rigidity imparted to it by deformative stresses. It follows from this that plastic flow of igneous rock must occur before cooling has developed the solid phase sufficiently to make the mass potentially dilatant. After this potentially dilatant condition of the magma has been reached, it seems reasonable to suppose that any openings or fissures in the surrounding rock will be filled, not by movement of the magma as a whole, but by the flowing out of its fluid phase under the fluid pressure of the liquid magma from the interior of the mass. This fluid phase would differ in composition from the solid phase, and consequently from the magma as a whole, by being in general more acid and containing a larger amount of mineralizers, in short, pegmatitic or aplitic in nature.

The potentially dilatant mass itself would also be capable of fracture under rapid deformation, and these fractures would be filled likewise by the fluid phase, resulting in thin dike-like occurrences of pegmatitic or aplitic material having vaguely defined, blending boundaries of the type so commonly observed in igneous masses.

Zones of deformation not resulting in fracture of the partly solidified magma would conceivably produce zones of more open packing of the solid grains, resulting in a greater proportion of the fluid phase and a consequent difference in composition of these zones of movement. If these zones of deformation or shear were closely spaced and parallel, as is quite reasonable, it is conceivable that the result would be a banding of the rock in the manner of certain of the primary gneisses.

Protoclastic structure may be developed by the interference of the solid particles during deformation of a potentially dilatant magma.

The movement of the fluid phase from the hotter portions of the magma would result in a reheating of the zone into which the hotter fluid magma migrated, with consequent interruption in the normal growth of a solid phase, cause re-solution, and the development of zonal growths of changed composition.

SUMMARY

Incoherent, granular masses, such as sand, in a condition approaching maximum-density packing, are dilated by deformation. In a condition of open packing they deform without dilatation. Prevention of free dilatation by enclosing pressures induces failure by fracture or shear when the mass is deformed, and with the development of joints and faults along thin zones of dilatation.

Deformation of a potentially dilatant mass causes *decrease* in pressure of the fluid portion, and therefore fluids in rocks—water, oil, gas—move toward regions of dilatation.

The mechanics of response to deformation of incoherent granular masses is applied by analogy to solid rocks by conceiving of them as having a solid and a *potentially* fluid phase. When the latter functions as a fluid, the rock yields to deformation by flowing, otherwise by fracture.

Dilatation occasioned by deeply penetrating zones of fracture initiates magmas. Cooling magmas become potentially dilatant when the solid phase develops beyond a certain proportion. The fluid phase alone is then mobile and forms dikes and veins in the surrounding rock and in the granular mass itself. The flow of the fluid phase into cooler parts causes reheating of those parts.

Reprinted by permission of the University of Chicago Press.
Published in the *Journal of Geology*, V. 48, No. 8
(November-December 1940), pp. 1007-1021.

FOLDING, ROCK FLOWAGE, AND FOLIATE STRUCTURES

W. J. MEAD
Massachusetts Institute of Technology

ABSTRACT

Variations in opinions as to the usefulness of cleavage in the interpretation of folded structures is due in part to lack of discrimination among the several varieties of foliate structures and the assumption by some observers that all cleavage is flow cleavage. Five varieties of foliate structures—bedding fissility, bedding foliation, fracture cleavage, flow cleavage, and shear cleavage—are described and discussed from the standpoint of physical characteristics and genesis. It is emphasized that compressional folding necessarily involves plastic deformation. Such plastic response is of two kinds, intergranular and interatomic. The folding of an incompletely cemented sedimentary series may occasion only intergranular plastic deformation and grain-crushing, without development of cleavage, until the limits of this type of plastic response are reached by consolidation. If folding continues beyond the limits of intergranular plasticity, interatomic plasticity functions, with development of flow cleavage. Dynamic metamorphism reduces interatomic plasticity, and continued deformation may develop shear failures in the form of thrust faults or shear cleavage.

INTRODUCTION

In a region of folded rocks the problem of the structural geologist is frequently the building of a reasonable three-dimensional picture of folded structures from the fragmental evidence afforded by scattered outcrops, drill holes, test pits, and mine workings. Dip, strike, and lithology of beds alone are inadequate because they do not afford unique solutions. Determinations of "top or bottom" of beds are of great value because they make possible unique solutions of many situations, and it is now generally recognized by structural geologists that every reasonable effort should be made to ascertain the stratigraphic sequence of beds. Unfortunately, sedimentation features useful as indicators, such as cross bedding, ripple marks, and grain-size gradation, are not always available and have a perverse way of being absent when most vitally needed.

Irving, Van Hise, Leith, and other students of folded pre-Cambrian rocks have made the important contribution that the mechanics of folding may develop certain secondary structures—fracture cleavage, flow cleavage, and drag folds—which enable the observer to infer correctly the stratigraphic sequence of the beds and the inclination and pitch of the folds. The application of these principles has been of inestimable value to structural geologists and has made possible the correct solution of countless structural problems not otherwise soluble.

In more recent years, under the leadership of Sander, Schmidt, and their students, much attention has been directed to the study of the orientation of the nonplaty minerals as a means of inferring the mechanical history of deformed rocks. The application of the principles and techniques of petrotectonics requires the use of a microscope and the universal stage. Foliation, or cleavage, and its relation to stratification are available for immediate study and interpretation in the field.

It is the observation of the writer that much confusion exists among geologists as to the usefulness of cleavage in the interpretation of the structures of folded rocks. Unfortunately, it is frequently a matter of belief or disbelief, faith or doubt, instead of judgment based on discriminating observation and an understanding of the factors and principles involved. Much of the confusion is occasioned by failure to realize that there are several types of cleavage or foliation, developed in a variety of manners. Flow cleavage and fracture cleavage may develop incidental to, and as a consequence of, folding and may, therefore, be of reliable usefulness in the interpretation of the structures of folded rocks. However, fracture cleavage or flow cleavage also may be developed in zones of shear deformation quite unrelated to folding. Other types of foliation to be described later—types which are neither flow cleavage nor fracture cleavage and which may occur or be developed in folded rocks but not as a consequence of folding—serve further to confuse the structural geologist and call for most careful discrimination in the field.

In this paper it is proposed to review in as simple and elementary a manner as possible the several varieties of cleavage or foliate structures and their causes and the relationships of folding and cleavage, in the hope that such a review may to some extent clarify the situation and may stimulate further discussion and investigation.

In the following discussion the term "foliate structures" is used in a broad sense to include the textural or structural properties of certain rocks which permit them to be cleaved or parted along approximately parallel surfaces or lines. This property has a wide range of appearances and may be developed in the rocks in various manners. In the following paragraphs an attempt is made to describe and differentiate the principal varieties.

Bedding fissility.—This is the capacity to part parallel to the stratification in many of the finer-grained sedimentary rocks. This property is attributed to compositional and grain-size variation between layers but may be and probably is in many cases due to the low-angle attitudes of platy and elongate rains. It seems reasonable to presume that bedding fissility is accentuated during consolidation by the rotation of platy and elongate particles toward horizontal positions.

Bedding foliation.—In deeply buried, thin-bedded sediments possessing an initial bedding fissility, it seems reasonable that metamorphism induced by heat, pressure, and the permeation of mineralizing solutions, possibly from magmatic sources, would accentuate the bedding by growth and enlargement of the primary platy and elongate constituents, developing a gneissic or a schistlike rock without the operation of plastic deformation or shear usually assumed to be necessary in the development of schists. It is not intended to imply that all flat-lying schists are developed in this manner or that they have not undergone shear or flow. It is unreasonable to assume that all bedding foliation has escaped subsequent deformation. In some instances the development of this type of foliation may well be an accompaniment of flow or creep due to load, described by R. A. Daly[1] as load cleavage.

Fracture cleavage.[2]—This type of foliate structure is a phenomenon of rock fracture and consists of closely spaced, parallel, shear joints or ruptures usually confined to a weaker layer between stronger layers. It is frequently observed in the limbs of moderate folds where

[1] "Metamorphism and Its Phases," *Bull. Geol. Soc. Amer.*, Vol. XXVIII (1917), p. 375.

[2] C. K. Leith, "Rock Cleavage," *U.S. Geol. Surv. Bull. 239* (1905), pp. 119 ff.

it has developed as a consequence of interlayer shear incidental to folding. Fracture cleavage is in no way dependent on the orientation of constituent minerals, and the plates between the shear ruptures are themselves not cleavable. This type of cleavage is commonly but not necessarily an accompaniment of folding. It is also found in fault gouges and shear zones and serves as an admirable indicator of the direction of shear displacement.

Flow cleavage.—This type of cleavage has been variously described as "flow cleavage,"[3] "slaty cleavage,"[4] and "axial-plane cleavage."[5] It is due to the parallel orientation of platy and elongate minerals and is the result of plastic deformation by interatomic rearrangement of constituents. It is a consequence of dynamic metamorphism. It is homogeneously or evenly developed in the rock and not spaced into parallel surfaces as is the case of shear cleavage. Its orientation is perpendicular to the short axis of strain, which is equivalent to saying that it is perpendicular to the direction of the resultant of compressional stresses. It may likewise be described as developing in the directions of greater elongations of the plastically deformed mass. Its relationship to strain places it approximately parallel to the axial plane of folds. The simple relationship between flow cleavage and strain makes it reliably useful in the interpretation of folds. It is generally but not exclusively a consequence of folding.

Flow cleavage may be accentuated and rendered coarser grained by subsequent static metamorphism in a manner similar to the development of bedding foliation described in the foregoing section.

Shear cleavage.—This type of cleavage consists of roughly parallel, closely spaced surfaces of shear displacement on which platy minerals may have developed and into which they may have been dragged. The spacing of shear surfaces varies from a millimeter or less to such wide spacing that the term "cleavage" becomes no longer applicable, and the rock is best described as having undergone minor parallel faulting. When the spacing is unusually close, shear cleavage may simulate and be easily confused with flow cleavage.

[3] *Ibid.*, pp. 23 ff.   [4] *Ibid.*, p. 15.

[5] H. W. Fairbairn, "Notes on Mechanics of Rock Foliation," *Jour. Geol.*, Vol. XLIII (1935), p. 591.

Shear cleavage may be developed in rocks having earlier flow cleavage or bedding foliation. It may have any angular relationship to these earlier foliate structures but is most likely either to parallel earlier foliation or to cross it at relatively large angles. Small angular differences between shear cleavage and an earlier foliation are not to be expected, because the earlier foliation affords planes of weakness which prevent the formation of shear failures at angles close to them. Closely spaced shear cleavage, when it crosses earlier flow cleavage, may nearly or completely obliterate the latter. When it is parallel to an earlier flow cleavage, it may be difficult or impossible to recognize its presence.

Shear cleavage is a phenomenon of rock flowage rather than of fracture, but it is not a consequence of homogeneous plastic yielding of the rock. Deformation is accomplished by shear along more or less closely spaced surfaces, but the mineral rearrangement along these surfaces is largely by interatomic adjustments without the development of fracture ruptures in the rock. Brittle layers and brittle minerals may be locally fractured.

This type of cleavage is oriented approximately $45°$ from the axis of the compressive resultant of force. Each shear surface may be considered as a minute thrust fault. Since two directions of shear approximately perpendicular to each other are always possible, and since one may develop to the exclusion of the other, the inferring of the direction of the causal stresses is not a simple problem. Shear cleavage, although commonly found in folded rocks, is not a consequence of folding. This statement will be amplified in a later section of this paper.

FOLDING

Definitions are difficult, dangerous, but necessary. For the purposes of this discussion folds are defined as undulations formed by the flexing of layered rocks.[6]

There is a wide range in the degree or intensity of folding, from very gentle undulations which may or may not be original or depositional to extremes of folding in which the degree of deformation

---

[6] It is not intended to imply that other types of folds such as shear folds, contorted masses of marble, rock salt, or ice, or the folded structures formed by differential loading or squeezing of muds or clay, are not properly called "folds."

and structural disturbance is intense. The present discussion is not concerned with the more gentle or moderate phases of folding and is intended to apply to deformation which might be described as more-than-moderate folding.

Very gentle flexing of layered rocks may conceivably occur, with negligible plastic deformation, by minute fracturing, crushing, and opening of joints. In folding that is more-than-moderate, plastic deformation of at least some of the layers is imperative. It may be stated as a fundamentally important principle that folding requires the plastic behavior of part or all of the layers involved—that plasticity is demanded in proportion to the degree of folding.

PLASTIC DEFORMATION (ROCK FLOW)

The term "plastic deformation" is here used broadly to include homogeneous change in shape, regardless of the internal mechanism by which it is accomplished. Plastic deformation involves the differential, positional shifting of constituent materials.

In fine-grained soft rocks (clays, muds, compaction shales, marls, etc.), positional shifting is largely intergranular, and these materials may be said to possess intergranular plasticity. This type of plasticity may involve a certain amount of grain-crushing. It does not develop cleavage. It may develop crenulations and minor distortions, including drag folds, in thinly bedded sediments. In special cases, not believed to be common or important, the mica plates in extremely micaceous muds may be oriented along flow directions, producing some degree of foliation.

In compact, dense, well-cemented rocks, intergranular plasticity is minimized and positional shifting of constituents is largely by interatomic rearrangement by physical or chemical means. This type of plasticity functions by recrystallization, neo-crystallization, and gliding, with a minor amount of granular rotation and shifting, and may be designated as interatomic plasticity. The results of this type of plastic deformation are generally described as dynamic metamorphism.

Depending on the composition of the rock, interatomic plastic deformation may or may not develop cleavage. The cleavage of dynamically metamorphosed rock is dependent on the parallel orientation of elongate or platy minerals, largely the latter. Rocks

326

such as limestone, marble, quartzite, and rock salt may be deformed plastically without the development of cleavage. These rocks are not compositionally capable of producing mineral types essential to cleavage.

The shales and related rock types—arkose, graywacke, impure limestones, and impure quartzites are compositionally capable of developing micas, chlorite, and other platy and elongate minerals under conditions of dynamic metamorphism. Leith[7] developed the now generally recognized principle that in flow cleavage the orientation of the platy and elongate minerals is determined and controlled by the orientation of strain in the plastically deformed rock, and that the cleavage due to the parallel arrangement of the platy and elongate minerals developed by dynamic metamorphism is perpendicular to the shortest axis of strain. This cleavage may be variously described, therefore, as being perpendicular to the axis of compression or being parallel to the elongations of the rock mass.

Intergranular plasticity is obviously dependent on freedom of intergranular movement and is limited by intergranular friction. Adsorbed water films and interstitial moisture contribute to the intergranular plasticity of clays and other fine-grained rocks, and this type of plasticity is diminished by consolidation. In so far as plastic deformation squeezes out water, reduces voids, and consolidates the rock, the material becomes less intergranularly plastic.

Interatomic plasticity also has its modifications and limitations. Pure marble, pure quartzite, and rock salt, when plastically deformed, respond by intracrystal gliding and recrystallization. In rocks of the shale group interatomic plasticity functions largely by the development of the platy and elongate minerals. In such rocks interatomic plasticity is to a degree determined by the compositional ability to produce these new minerals. As dynamic metamorphism proceeds, this ability of the rock is reduced and its interatomic plasticity is correspondingly diminished, except in rocks having a sufficient content of carbonates to dominate the situation.

RELATION OF PLASTICITY AND CLEAVAGE TO DEFORMATION

With the foregoing summary of the various types of foliate structures and with emphasis on the fact that plastic deformation, which

[7] *Op. cit.*, p. 112.

may be accomplished either by intergranular rearrangement or by interatomic reorganization, is an essential accompaniment of folding, we may turn to a consideration of the series of events involved in the compressional deformation of a thick accumulation of sediments.

Fine-grained, argillaceous rocks, which may be called "shale" under a broad usage of the term, may be presumed to predominate in any great series of sediments. It has been estimated by F. W. Clarke[8] and others that sediments of the shale type account for approximately 80 per cent of all sediments. The behavior of a great accumulation of sediments during the earlier stages of compressional deformation may therefore be assumed to be largely controlled by the behavior of shale. The mechanically stronger and stiffer layers of sandstone and carbonate rocks commonly referred to as the more competent layers, of course, play an important part in determining the type and characteristics of the folds developed.

Under load of superimposed sediments the consolidation of clays and muds into shale proceeds with astonishing slowness. This is due to the fact that consolidation can be accomplished only by the squeezing-out of interstitial water, and this loss of water is inhibited by the low permeability of the material itself and by the fact that in these fine-grained materials, containing much matter of colloidal size, a large proportion of the water is rather firmly held on the grain surfaces by adsorption. The retention of water in shales is illustrated in the following specific cases.

A large number of moisture determinations on core-boring samples of Fort Union shale of Tertiary age in North Dakota showed an average moisture content equivalent to 40 per cent of the volume of the shale. An average of a great number of moisture determinations on core-boring samples from the Bear Paw shale of Cretaceous age at the Fort Peck Dam site in eastern Montana showed an average content equivalent to 30 per cent of the volume of the shale. Similar moisture determinations on a large number of core samples of shale of Pennsylvanian age from the Possum Kingdom Dam site on the Brazos River in Texas gave an average moisture content equivalent to 18 per cent of the volume of the shale. Some years ago the writer had occasion to examine in detail a 500-foot section

[8] "Data of Geochemistry," *U.S. Geol. Surv. Bull. 770* (1924), p. 34.

of Pennsylvanian shales underlying the La Salle limestone in Illinois. No moisture determinations were made, but slaking tests showed that the entire shale section was compaction shale[9] which had undergone very little, if any, intergranular cementation and presumably possessed intergranular plasticity.

The conclusion seems inescapable that folding, developed in the earlier stages of compressional deformation of a thick mass of sediments, is accomplished by intergranular plasticity and consequently without the development of cleavage. Fine-grained sediments may be folded to an extreme degree without the development of secondary cleavage where conditions have not produced sufficient consolidation to prevent intergranular plasticity. Closely folded shales devoid of flow cleavage are not uncommon. Folded shales of this type are excellently exposed in the Claremont formation along the "Tunnel Road" east of Oakland, California.

Pressure and deformation involved in the folding of a mass of shale are very effective in promoting consolidation. As folding proceeds, with reduction of intergranular plasticity by consolidation, a condition is eventually reached where further plastic deformation requires the functioning of interatomic plasticity, and this stage marks the initiation of the development of flow cleavage.

The amount of folding that is possible before the limits of intergranular plasticity are reached depends on a variety of factors, among which are fineness of grain, which influences permeability and, consequently, degree of consolidation under load, and thickness of shale formations, because thick formations consolidate by compaction less readily, owing to the longer travel distance for the escaping water through the impervious material. Thin layers of shale between sandstone beds consolidate more easily because of the opportunity for escape of water.

After the limit of intergranular plasticity is reached, continued deformation demands interatomic plasticity, and folding proceeds with the development of flow cleavage essentially parallel to the axial planes of the folds. It is not unreasonable to expect to find flow cleavage developed in thin shale layers between layers of sandstone and absent in thicker shale layers of the same series, because the

[9] W. J. Mead, "Geology of Dam Sites," *Civil Engineering* (1937), p. 392.

329

limits of intergranular plasticity may be reached earlier in the more easily consolidated thin layers, as explained above.

It should be emphasized that the development of flow cleavage does not necessarily begin in the early stages of folding of a thick sedimentary series. Flow cleavage is not initiated until the limits of intergranular plasticity are approached, which means that it may not make its appearance until the later stages of folding and thus escapes the distortion and change in orientation that would have been necessary had it developed in the early stages of folding.

It is not inconceivable that a series of sedimentary beds may be thoroughly lithified by consolidation and cementation prior to deformation. Such rocks would have little intergranular plasticity, and folding would call upon interatomic plasticity from the start, resulting in the initiation of flow cleavage in the early stages of folding.

As folding proceeds by interatomic plasticity and the development of flow cleavage, interatomic plasticity is progressively diminished. Dynamic metamorphism of shales involves changes in composition, both chemical and mineralogical, and develops a slate or a schist less interatomically plastic than the parent-rock. Just as a limit of intergranular plasticity is reached by consolidation, a limit of interatomic plasticity is approached by metamorphism.

When compressional stresses continue after a thorough dynamic metamorphism of the shale has been accomplished, the rock, less willing to deform homogeneously in the manner of a plastic material, may respond by yielding along surfaces of shear failure. This may result in the development of thrust faults or, depending on conditions of restraint and manner of easiest relief from stress, deformation may be accomplished on multiple surfaces of shear dislocation. If conditions are those of rock flowage, these are tight shears accomplished by interatomic rearrangement and are not shear fractures. The phenomenon is one of rock flowage rather than of rock fracture but differs from flow cleavage by being confined to separate surfaces or thin zones of shear dislocation.

If the rate of deformation exceeds the time requirements for interatomic rearrangement, shear fractures necessarily result, and fracture cleavage as distinguished from shear cleavage is developed.

In a mass of rock which has been deformed in this manner, it is to be expected that certain lithological types will undergo tight-shear foliation (shear cleavage), while other types, less amenable to flowage phenomena, exhibit shear-fracture foliation (fracture cleavage). There is little importance in distinguishing between these two varieties of cleavage, because they have the same relationship to causal stresses and accomplish the same deformational results.

This type of deformational response which is a consequence of the reduction of interatomic plasticity is not a continuation of folding. The folded mass of rock has lost to a considerable extent its layered nature necessary to folding, and the continuance of deformation by multiple shearing cannot be regarded as a continuation of flexing. The folds themselves are deformed, not accentuated.

Masses of layered sediments which have been folded to a degree to develop thorough dynamic metamorphism and flow cleavage may later be subjected to deformative stresses operating in directions different from those which caused the original folding. If this deformation occurs at relatively shallow depth or too rapidly to permit flow phenomena, faults and fracture patterns are developed. If the deformation occurs under sufficient load and at a rate slow enough to permit flowage phenomena, it appears probable that such deformation would be accomplished by shear rather than by folding. Thrust faults might result, but, depending on load and restraint, shear cleavage might well develop as a means of accomplishing the required deformation of the mass. Folding does not seem probable because the mass has lost its layered condition and because interatomic plasticity has been so greatly reduced by the earlier dynamic metamorphism.

INTERPRETATION OF FOLIATE STRUCTURES

Intelligent utilization of cleavage in the interpretation of the structures of folded rocks requires correct classification of the type or types of foliate structures observed. Much of the existing confusion in the utilization of cleavage as a guide to structural interpretation has been due to the incorrect assumption that all cleavage is flow cleavage. The statement that flow cleavage develops essentially parallel to the axial planes of folds has been doubted and discredited by geologists who have observed cleavage following around the crests and troughs of folds. The writer has heard many discussions, in which one observer insists that all cleavage is axial-plane cleavage and does not follow the bedding around the convolutions of folds, another insisting that the first man is wrong because cleavage follows the beds in folded rocks. One has seen flow cleavage (axial-plane cleavage); the other had observed bedding foliation which had been folded or developed subsequent to folding.

No hard and fast specifications can be written for the identification of foliate structures. The problem perhaps falls more into the category of an art than a science. Experience and familiarity with a wide variety of occurrences is of value. A geologist whose experience has been confined to the folded rocks of the Lake Superior region, where flow cleavage uncomplicated by bedding foliation or shear cleavage is so successfully applicable to the structural problems, may well experience some astonishment in other regions where bedding foliation and shear cleavage complicate the structural picture. A student of structural geology, initially confronted with the foliation complexities of the folded rocks of New England, may doubt the practical usefulness of cleavage and the statements in the literature that "cleavage develops parallel to the axial planes of folds."

The outstanding characteristic of flow cleavage is that it develops homogeneously through a given bed. Since it is a product of plastic deformation, the entire bed has been uniformly affected. Every cubic millimeter has taken part, and a homogeneous development of cleavage has resulted. This is particularly true in the fine-grained rocks. Flow cleavage may easily be confused with bedding foliation, and the best means of identification is to find flow cleavage in angular relationship to stratification.

Shear cleavage, in contrast to flow cleavage, is not homogeneously distributed through the mass but is spaced into parallel surfaces, each one of which is a surface of shear failure with some degree of shear displacement, however minute. When shear cleavage consists of extremely closely spaced surfaces, it may simulate flow cleavage, and identification may be difficult or impossible in the hand specimen. In the field, however, the true nature of the shear cleavage is usually discernible in neighboring beds or on tracing it into areas in the vicinity where it is less closely spaced.

The angular relationship between shear cleavage and the effective resultant of causal compressional stress is not a simple one in contrast to the simple perpendicular relationship of flow cleavage to the least axis of strain. Shear cleavage is a manifestation of shear failure and presumably develops at angles somewhat less than 45° to the direction of compression.

There seems to be a reasonable analogy between the orientation of thrust faults and shear cleavage in their angular relationship to causal stresses. When the direction of easiest relief is upward, thrust faults strike at right angles to the direction of compression and dip either way, depending on whether we consider them as overthrusts or underthrusts. Similarly, where easiest relief is upward, shear cleavage strikes at right angles to the direction of compression and may dip either way from that strike. Wherever two sets of shear failures are possible, one set generally develops to the exclusion of the other, as either set is capable of providing the necessary deformational response. Shear cleavage, dipping in opposite directions in separate portions of a single region of deformation, may therefore have been developed by the same deformation.

Where the direction of easiest relief is lateral (horizontal), steep-angle thrust faults are developed with horizontal displacement striking somewhat less than 45° either to the right or to the left of the direction of compressive force. Continuing the analogy between shear cleavage and compressional faults, we may infer that where the direction of easiest relief is lateral, shear cleavage with vertical dip may develop, striking somewhat less than 45° either to the right or to the left of the direction of compression and that locally within the deformed mass either of these two directions of strike may develop to the exclusion of the other.

It should not be inferred from the two preceding paragraphs that the direction of easiest relief is necessarily either vertical or laterally horizontal. Where the direction of easiest relief is inclined, the dip of shear cleavage should be correspondingly intermediate between 45° and 90°, with a shift in the strike directions called for by the geometry of the situation.

Since each surface in shear cleavage is actually a minute thrust fault, it is frequently possible to observe the direction of displacement, which aids materially in determining the approximate direction of the axis of compression. Drag-lineation, minute slickenslides, or scoring are also useful in indicating the line of displacement.

Shear cleavage developed by rotational deformation in a weaker layer between two or more competent layers should have the same relationship to causal stresses as does fracture cleavage. The writer has seen instances of fracture cleavage in the field with development of platy minerals between the cleavage plates, strongly suggesting that the cleavage should have been interpreted as shear cleavage rather than fracture cleavage.

In regions which have had a complex history of deep-seated deformation, it is reasonable to expect a discouraging complexity of shear cleavages, usually superimposed with varying angular relationships on an earlier flow cleavage.

Bedding foliation is an accentuation of primary stratification by metamorphic mineralizing processes involving grain enlargement, recrystallization, and replacement, with implied compositional changes both mineralogical and chemical. There seems to be no reason why it should not develop prior to folding, during folding, or subsequent to folding. Its development depends upon temperature, pressure, and the availability of mineralizing solutions. The development of bedding foliation may reasonably be looked upon as an early stage of granitization. Mineralogically, it may differ from flow cleavage and shear cleavage in that albitization, the development of garnets, staurolite, etc., may be an accompaniment. However, since similar mineralization may be superimposed on flow cleavage or shear cleavage, its presence may not be considered as diagnostic.

In fine-grained, homogeneously thin-bedded rocks, bedding foliation may simulate flow cleavage in appearance. In general, the spaced or layered nature of the foliation and its faithful parallelism with stratification are the best criteria for distinguishing it from flow cleavage.

Folding may occur subsequent to the development of bedding foliation and produce a folded schist in which the schistosity follows faithfully around the crests and troughs of the flexures. Such folded schists have been incorrectly pointed out as a refutation of the usefulness and reliability of cleavage as an aid to the interpretation of folded structures. Most folded schists possess only bedding foliation that follows the original stratification. The writer has seen no instances in which flow cleavage developed in the folding of schists, but such cases may, and probably do, exist.

It is of interest to note that the Z or Gothic type of folding is frequently found in folded schists. An explanation of this type of folding may lie in the fact that the metamorphism which developed the foliation reduced the interatomic plasticity of the rock, and a type of folding calling for a minimum of rock flowage is the one to develop. The mechanical requirements for folds formed with a minimum of plastic deformation in laminated materials having abundant interlayer shear facilities are sharply angular crests and troughs with a minimum of curvature in the limbs.

PRODUCTION GEOLOGY

and Metallurgy. Published in the *Journal of Canadian Petroleum Technology,* part 1, V. 5, No. 4 (October-December 1966), pp. 153-164; part 2, V. 7, No. 3 (July-September 1968), pp. 128-144.

Interpretation, Averaging and Use of the Basic Geological- Engineering Data

D. HAVLENA*

(17th Annual Technical Meeting, The Petroleum Society of C.I.M., Edmonton, May, 1966)

ABSTRACT

The purpose of this paper is to emphasize that the taking, assembling and averaging of the basic data for further use in an analysis of a reservoir requires a profound knowledge of many of the specialized disciplines of modern geology and petroleum engineering.

The various aspects of "Static Data" as presented in this paper, analyze that problem and specifically discuss the following major subheadings: (A) Porosity, (B) Permeability, (C) Facies Mapping, (D) Connate Water, (E) Relative Permeabilities and Capillary Pressures, and (F) Method of Least Squares.

This paper is essentially a general and philosophical discussion of the necessity for obtaining data quality and of the need for an integrated approach to be applied in a detailed analysis of any reservoir.

INTRODUCTION AND OBJECTIVES

The correct taking, interpretation and usage of reservoir data are among the most important tasks of the petroleum engineer, and particularly of the reservoir engineer and geologist. On the quality, and to a much lesser degree on the quantity, of this information depends the success as well as the accuracy of all the subsequent geological and engineering studies and finally of the economic evaluations. Despite this paramount importance, very little is found on this subject either in petroleum literature or in university curricula. There are exceptions—e.g., the normal types of pressure build-up curves and the determination of the areal and volumetric average pressures, without explaining when to use which and why. At the present stage of knowledge of the science of reservoir physics, it appears that the analytical methods of solution (equations) are normally, from a mathematical and physical standpoint, much more exact than the basic data and particularly the "averages which are fed" into them. In layman's terms, the reservoir analyst often performs an exact multiplication of doubtfully reliable numbers. This lack of interest in the accuracy and representativeness of the basic data and in the methods of averaging them is usually explained by the simplicity of this problem, which can be solved by "common engineering sense;" this explanation is correct if it is backed by a thorough and uniform theoretical as well as practical, understanding of this complex problem.

During the last fifteen years, many scientists, be they mathematicians, physicists, geologists, engineers, programmers, etc., advanced tremendously the understanding of reservoir mechanics and put it on firm theoretical, scientific foundations. Unfortunately, nearly nothing has been done in securing more representative data; most of the methods presently applied in obtaining, interpreting and averaging the basic data are still the same as those used when reservoir engineering was in its infancy and when an experienced reservoir analyst had only to compute the initial oil in place by volumetric methods.

In this paper, an attempt is made to show that the taking and assembling of some of the "GIVEN" is a real challenge and not a mundane task of obtaining some answer for any price. Great discretion and judgment has to be exercised in working with the basic data in order to obtain the most probable and physically logical answer to a specific reservoir problem. Because of its idiosyncratic nature, it is impossible to prepare a "universal cookbook" which satisfies all conditions: each reservoir and even

each well presents a specific problem to be considered separately when planning, taking, interpreting and finally using the surface and subsurface data.

It is hoped that, by this "scratching of the surface," the industry, and particularly its technical personnel, will realize the magnitude and importance of proper and systematic data collecting, of their interpretation and of the averaging problems.

All the systematic data collecting programs have to be formulated with the following chief concerns in mind:

(a) To obtain an optimum coverage of data and testing among the reservoirs and wells at reasonable frequency, having due regard to the cost of obtaining such data.

(b) To achieve a consistency in procedures, as much as is practical, so that the data represent the actual reservoir conditions and are thus directly comparable to each other.

(c) To understand that quality is always better than quantity.

(d) To consider at all times the purpose of taking the data and their final usage; unnecessary data collecting constitutes economic waste.

(e) To obtain a maximum utility from the available data with a minimum of assumptions, and particularly eliminating any "cooking."

Reservoir data may be divided into two groups; i.e., Static Data and Dynamic Data.

The *static data* represent direct measurements of some static property of the reservoir or its fluids; for example, the porosity, which numerically defines the amount of void space, the permeability, as measured on samples in the lab, and which gives a value on the transmissibility (usually to air), the connate water saturation, temperature, chemical composition, etc.

The *dynamic data* refer more to the level of energy and the potential and also to forces contained or induced into the reservoir; for example, potentials, pressures, PVT's, flood fronts, permeabilities evaluated from pressure build-ups, interference tests, etc.

As is evident, some types of data will always be static, others always dynamic, and a few might be, depending on the nature of the measurements and/or on their final usage, in one or the other category; for example, the permeability or skin factor. It should be clearly understood, however, that this division into static and dynamic data is rather arbitrary and refers more to a group name than to some rigid classification which would have a unique physical meaning.

This paper deals only with the static data. The dynamic data will be dealt with in a later paper.

STATIC DATA

Because we are primarily interested in the quantitative evaluations that will define the rock volume, porosity and water saturation of a potential reservoir, we tend to think of such a reservoir as a static, homogenous, interconnected single rock volume. It is the intent of this dissertation to suggest that we should have a concept of a geologic unit that is porous and heterogeneously homogenous, and consistently changing with time. As information is coming in continuously, the three-dimensional picture of the reservoir in the mind of a production geologist and of a reservoir engineer should represent a flowing process which might change with each new piece of information.

Within a single geological unit, we can then expect that the horizontal and vertical differences are important, as well as the

Senior Staff Engineer, Hudson's Bay Oil and Gas Company Ltd., Calgary, Alta.

fact that the beds lying above and below the particular reservoir may have lateral changes that can affect the reservoir itself; fractures may cause leakage after some pressure drawdown, or the cap-rock facies may change laterally into a porous and permeable material on an erosional subcrop surface. In some cases, the porosity cut-off, if applicable at all, should be considered a variable with time in order to correlate and calibrate the core analysis and logs with the results of material balance solutions and of other dynamic information, such as, for example, well-to-well interference tests, decline curves (23), etc. The geologist and engineer must be made aware that their concept of the rock facies must change if they are looking for such detectable differences. However, should we assume, naturally after having a proof for such an assumption, that the rock facies is more or less consistent within a statistical error, then the horizontal and vertical changes would be ignored or assumed to be insignificant. Each year's production-pressure performance might change our concept of the reservoir rock volume, any possible cut-offs, the radius of drainage and the effect that the surrounding environment is having on the particular accumulation of hydrocarbons.

The important reservoir rock characteristics that we attempt most often to understand with a statistical definition of a reservoir are the porosity, the gross or net rock volume that is effectively being drained, the *in-situ* connate water saturation, the gas-oil, oil-water or gas-water transition zones, the interwell relative permeability to gas, oil or water, the grain size distribution, the relative interwell heterogenity and the capacity distribution. Thus, in order to develop a meaningfully useful, most probable interpretation of the static-dynamic configuration of a reservoir, an "integrated approach" is required. The term integrated refers to the necessity for a progressive reservoir analyst to apply and to catalyze the knowledge of an engineer, geologist, physicist, statistician, stratigrapher and several other specialists employed by a modern oil company.

Porosity

As few reservoirs are cored completely, and normally all wells are logged, the logs are generally used to fill in that missing knowledge. Porosity values derived from core analysis are used as a standard, and it is usually implicitly assumed that the quantitative changes in rock porosity and permeability that occur from the time that the core is removed from the reservoir until the measurements are made are insignificant. It is thus accepted that the lab data express quantitatively the *in-situ* conditions (3).

One method of correlating the core analysis porosity values with the porosity-type logs, such as the sonic, formation density, neutron logs and the microlog, can be accomplished as follows:

1.—Convert the core analysis porosity values into a *running average* that is consistent with the interval of investigation of the petrophysical log to calibrated; i.e.:

Petrophysical Log:	*Core Analysis*:
1' Sonic	1' Running Average
3' Sonic	3' Running Average
Neutron	2' — 3' — 4' Depending on type of detector
Formation Density	2' — 3'

2.—Adjust the formation depth of the core analysis to correspond with the depth of the petrophysical porosity log. A drilling time log has sometimes been used to justify the depth correlation of a core - faster drilling often suggests higher porosity.

3.—Either (a) plot the respective core porosity as a function of the corresponding petrophysical reading and draw the best correlation curve, *or* (b) apply a forced fit technique, i.e., arrange the measured core porosity as well as the log values in ascending order for both of these parameters (see Table I).

The sets of values shown in the table are plotted, and the best correlation determined by inspection or by the method of least squares *(Figures 1a and 1b)*. If a sufficient number of correlatable data points are available, standard statistical methods (11, 14, 20, 21, 22), eventually including standard deviations and confidence interval evaluations, should be applied. The gamma curve is often used as a qualitative control to restrict (select) the

TABLE I			
METHOD (A): NATURAL (Fig. 1A)		**METHOD (B): FORCED FIT (Fig. 1B)**	
Core Porosity %	1.0 ft. Sonic Log Reading μ sec.	Core Porosity %	1.0 ft. Sonic Log Reading μ sec.
9.5	55.0	7.9	55.0
11.4	58.0	8.5	55.5
10.1	58.4	8.8	57.0
10.2	57.7	9.5	57.7
11.5	61.0	10.1	58.0
10.3	61.6	10.2	58.4
8.8	59.0	10.3	59.0
7.9	57.0	11.4	61.0
8.5	55.5	11.5	61.6

values which are used in a correlation. Where there is a uniform level of radioactivity through a formation or stratigraphic interval, those points that lie at radioactive levels higher than a preassigned random statistical variation would be considered to be subject to elimination. If a wide scattering occurs, the geologist and stratigrapher should determine if such variations are not caused by lumping together data of different lithological character. Thus, the reservoir analyst must always be aware of any geological, lithological and stratigraphic factors, such as lateral lithological and facies changes, or, in layered limestone or dolomite beds, of the fact that a correlation determined for one porous interval might be changed even in the same geologic bed if the percentage of dolomite changes.

In fields which cover a large area, say one or more townships, the analyst should investigate the core porosity-log correlations of each cored well or group of wells in order to detect any vertical variations and also any well-to-well or area-to-area variations; in some cases, a particular rock-type variation may be detected by correlating log characteristics of the individual submembers.

Logging devices, other than the micro-devices, average the rock and the fluid properties and reflect some kind of a porosity averaged over the interval of investigation. Porosity alone is not a reliable index for distinguishing between pay and non-pay zones. The micro-devices provide detail, i.e., they do not average, and in addition to the responses of the "pinpoint" effective porosity recording they also indicate the presence of a filter cake build-up; filter cake build-up therefore suggests a permeability, which is the other necessary property of net pay.

Well and logging conditions and the desired type of geological engineering information determine which combination of logs should be run. Although log estimates of pay and porosity values are routinely made, such estimates are evaluated from information recorded by a device which is usually averaging over several feet. Log evaluations are often quite reliable (21), and this is particularly true when dealing with gross values; however, the net values obtained from logs should always be considered qualitative until corroboration is obtained by independent means. Particular attention should be given to the contact logs in determining the gross pay section; those portions in the gross thickness which, on the basis of resistivity, radioactivity (lithology), transit time, neutron count, the absence of mud filter cake, etc., appear to be nonpay should be eliminated.

In cases where well conditions permit, and the desire for information warrants the costs, a core of the hydrocarbon-bearing zone is recovered for inspection and laboratory testing. It is not usual to have a core on every well in a pool, and frequently, at best, only a portion of the zone has been cored, the remainder having been drilled through. In all cases, the laboratory tests reveal the behaviour of the core under the conditions set in the laboratory, and these conditions seldom duplicate *in-situ* conditions in the reservoir.

In routine analyses, those sections of core which are not obviously tight are inspected and a few selected samples are often subjectively assumed to represent the whole. These samples (often only small plugs) are then extracted and dried, thus

337

removing the fluid content of the rock, possibly also some solid fines and the water of crystallization. Measurements of porosity are ordinarily conducted in a Boyles' Law porosimeter with air. A combination of factors results in a porosity value which often inclines toward optimism (3), some of the obvious causes being the removal of solids from the core, the removal of crystallization water from the core and the effect of overburden pressure on porosity; eventually, also, alterations of the *in-situ* rock properties during the coring operations must be considered. In carbonates, furthermore, a clear distinction should always be made between the effective and the total porosity value measurements.

Permeability

To understand the averaging of the permeability data entering into engineering calculations, some familiarity with the process of deposition is mandatory.

Petroleum reservoirs may be considered to be composed of aggregates of sedimentary particles of like or similar origin which have been brought together individually under the action of a driving force and which may or may not have been sorted into a size order. The argillaceous content, if any, of the rock decreases with the degree of sorting, and there is a functional form to the particle size distribution (1, 2). The amount of clay and the particle size at a point within the reservoir is a function of these properties at an adjacent point.

The framework of particles may be cemented together or chemically altered by ion replacement at deposition or subsequently, thus adding a secondary porosity system to that pore order which existed at deposition. There may be a sharp demarcation line between pay and non-pay or a gradual transition reflecting the rate of change or deposition from point to point—laterally if deposited at the same geological time, or vertically if deposited at different geological times.

The deposition process, which has been oversimplified above, may be subject to periodic forces, allowing muds, or non-pay material to be interbedded within the reservoir deposit. The presence of organic deposits generally precludes this type of interbedding, as a widespread change in conditions of deposition usually kills the organic growth. Frequently, periodic changes may be quite localized, resulting in the transgressing and regressing of a single organic bed to yield two or more vertically separated beds.

It is the task of the reservoir analyst, using an integrated approach, to define that portion of the reservoir rock which contains the producible, active, moveable hydrocarbons and, based upon laboratory tests on a few samples of the reservoir rock and their fluid contents, to evaluate the amounts and economic rates of fluid recovery from the entire reservoir as a function of real time. It should be noted that data in the interwell area are always incomplete and, as a result, the reservoir parameters are always interpretative and inter- or extrapolated. Unfortunately, it is less well understood that the quantitative data available on the individual wells are usually also incomplete and always highly subjective.

The permeability measurement in routine analysis is usually an air value which is therefore subject to a slippage or Klinkenberg (4) correction and possibly even to a turbulence correction (5). Air permeability in sands is usually greater than the "unknown true" permeability to hydrocarbons by a factor related to the absolute permeability level. At a permeability range of 100 mdcs or less, the air permeability may be two to many times as large as the "unknown true" value (6) which governs the movement of reservoir fluids; i.e., oil, gas, water.

Limestones and dolomites usually show little correlation between the amount of correction and the permeability level. This might be expected because of the lack of a Kozeny-type (7) relationship between porosity and permeability. According to Kozeny, for a sand of a given porosity level within fixed limits, the permeability may often be predicted reasonably well. For a limestone or dolomite, any eventual relationship between porosity and permeability is much weaker or usually non-existent, so that, for a given porosity level, the air permeability as reported in the laboratory may be almost anything.

A very permeable section, such as may occur in limestone or dolomite as fractures, fissures or vug may be beyond the evaluation capacity of the test equipment; in any case, such sections are not amenable to representative measurements.

The transverse permeability (8) *(Figure 2)* of core is commercially measured using a geometical shape factor. The permeability at 90 degrees to the first measurement is subsequently taken, with the high value reported as K_{max} and the low value recorded as $K\ 90°$. No other effort is usually made to orient the core beyond the above-mentioned doubtfully reliable and always highly subjective procedure.

The permeability of a reservoir may often be considered to consist of two systems: a matrix system and a vug or fracture (fissure) system. Permeability in the matrix system is a vector, and as such it consists of two components separated by an angle. If the azimuth angle of the K_{max} system is approximately the same for each plug, the permeability is said to be oriented. It should be clear that the direction of flow and the direction of maximum pressure gradient need not coincide (9).

The vertical permeability is generally small for stratified beds, but it may be essentially the same as the horizontal permeability

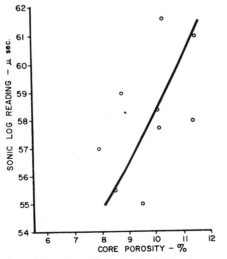

Figure 1a—Method (a)—Natural

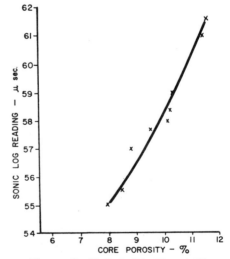

Figure 1b—Method (b)—Forced Fit

Figure 1.—Correlation of the Sonic Log Reading with the Core Analysis Porosities.

338

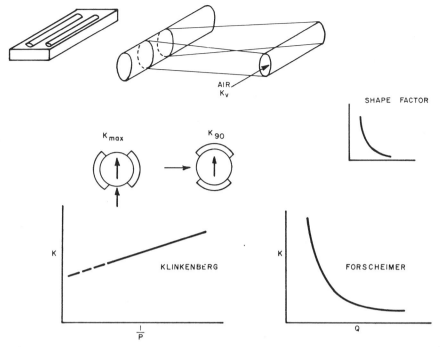

Figure 2.—Permeability Measurements.

for unstratified, homogenous reservoirs such as uniform sands, and limestone and dolomite bodies.

If coring were more or less complete on all wells in a given area, the core analysis results may be regarded as an unbiased statistical sample of the entire reservoir. The arrangement of the permeability values into an order might sometimes allow a fit between porosity and permeability which should result in a curve joining the modes of all the permeability distributions at the different porosity levels or conversely. In such a case, a three-dimensional diagram of porosity, permeability and relative frequency will result (10) *(Figure 3)*. In a general case, the individual well models of this type will not have the same appearance as the total field model. Individual wells or beds in individual wells will represent only a specific volume fraction of the total field model. Sometimes, however, a good proportion of the individual wells will have a model rather similar to the total field model, thus indicating that some common denominator applies more or less uniformly to all of the sedimentation.

Presuming a knowledge of the continuity or discontinuity between the pay sections in adjoining wells, several numerical techniques are available which will allow the computation in three dimensions of steady state and unsteady state flow problems (12). In most cases, however, these sophisticated methods require large electronic computers; for simplification purposes, therefore, the problem is commonly reduced to two dimensions to provide only an areal coverage (13), and such two-dimensional results are subsequently modified by a vertical sweep (14). Briefly, these methods consist, in essence, of defining a permeability and also a porosity map, indicating the production history of the individual wells on the wells of the model, and noting the corresponding pressures. These computer-calculated pressures are then compared with the pressures actually measured in the field and, depending on the nature and magnitude of the discrepancies, the model is adjusted. By modifications of the maps of permeability and porosity an adequate agreement is finally obtained between the model-calculated and the field-measured pressures.

The practising engineer, if without access to a large computer, will commonly determine an areal sweep efficiency from published correlations (15) and then apply one of the vertical sweep approximations (14).

Statistical methods of core analyses are often applied in determining whether bedding correlations exist and whether oriented or directional permeability prevails within the reservoir (9).

In the absence of a three-dimensional technique for the treatment of the permeability and porosity distribution, it is generally possible to arrange the reservoir into horizontal units which best fit the bedding and lithology and then to sum the resultant predictions (15). For waterfloods having approximately a unit mobility ratio, it is of little consequence if the units are physically separated as beds; for mobility ratios much different from unity, the beds must be physically separated for the purpose of predictions and then numerically combined if the predictions are expected to be realistic (15).

Looking at a single stratigraphic unit which is a fairly uniform bed not containing any major correlatable impermeable streaks, it will be noted that there is a variation in porosity and permeability at each well which often follows some type of distribution law *(Figure 4a and 4b)*; porosity and permeability will often change in the direction of thinning and toward the limits of the bed.

If variations of core data within an individual well represent a random distribution for that stratigraphic bed, and if it can be assumed that the same statistical distribution extends into the reservoir, it has been shown that the best representation of the permeability variations for that bed is the geometric mean permeability (16) *(Figure 5)*.

A series type of variation of bed permeabilities toward the limits of the reservoir will produce a harmonic mean permeability distribution as the best representation for that reservoir (17). This would also be the best representation of one bed if it had a

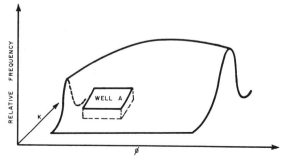

Figure 3.—φ - k - Relative Frequency.

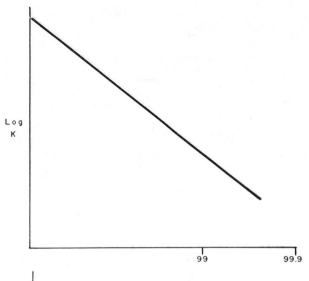

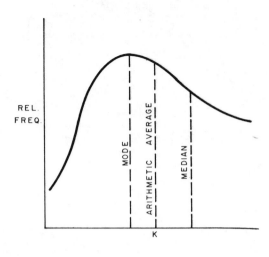

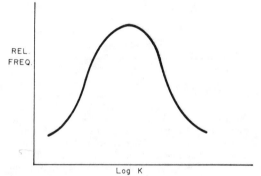

dependent upon what has already happened; i.e., dependent on adjacent values.

$$\text{mode} = \sqrt[n]{k_1 \times k_2 \times k_n}$$

Figure 4a.—LOG NORMAL.

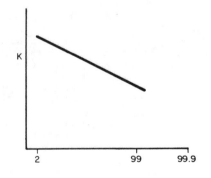

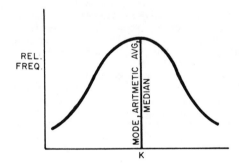

event completely un-related to any other event—random.

$$\text{mode} = \frac{\sum\limits_{i}^{n} k_i}{n}$$

Figure 4b.—NORMAL.

Definitions:

1. *Average:* A single number typifying or representing a set of numbers of which it is a function. It is a single number computed such that it is not less than the smallest nor larger than the largest of a set of numbers.

> *Arithmetic Average:* or arithmetic mean, is the average of the sum of n numbers divided by their number.
>
> *Harmonic Average:* is the reciprocal of the arithmetic average of reciprocals of the numbers.
>
> *Geometric Average:* or geometric mean, is the n-th root of the product of a set of n positive numbers.

2. *Median:* is the middle measurement when the items are arranged in order of size; or, if there is no middle one, then it is the average of the two middle ones.

3. *Mode:* is the most frequent value of a set of measurements.

constant porosity level and if the permeability varied in a series fashion *(Figure 5)*.

If porosity and permeability values remain constant throughout the pool, the best value for such a reservoir is the arithmetic average permeability; unfortunately, such an idealized simple case is very rare *(Figure 5)*.

Normally, the geometric mean value will give the most satisfactory correlation for any one well; this could be expected, because the permeability-capacity evaluated from pressure build-ups or draw-downs measures rock properties away from the wellbore but not at too great a distance from the well. Therefore, a gradual decrease in permeability or in the harmonic mean effect should not be too significant.

This general discussion on the arrangement of porosities and permeabilities in a reservoir has a direct bearing on the averaging of special core study data.

In a pool study, if only some ten to twenty plugs were evaluated, it will often be found that there is no correlation between the results of flow tests and the porosity and/or permeability of the analyzed cores and perhaps not even a relationship with the lithology, if changes in lithology of the individual samples are not too severe. In such a case, it becomes obvious that the reservoir can best be represented by a single "average" curve and that deviations from this curve are probably only statistical change occurrences. The best representation for a random distribution of permeabilities then becomes the geometric mean of all the plug values (16). As small samples are subject to large statistical accidental errors, usually at least some forty to fifty samples are required to obtain even moderately reasonable quantitative results.

Occasionally, there will be a relationship between porosity or permeability and the capillary pressure curve, but this is generally not the case in limestone or dolomite and should not be taken for granted unless at least some twenty plugs indicate that such a relationship exists. In that event, if the reservoir can logically be broken into units of varying porosity and permeability, the capillary pressure of each bed of a given porosity should be evaluated separately by using the proper curve. Where no correlation appears, the average curve, defined as the geometric mean of all of the curves, may be used. When in doubt in representing the permeability stratification, it is useful to plot a cumulative frequency curve on probability paper in order to determine whether the distribution is normal or log-normal. If the distribution is normal, the mode, median and average will coincide; if log-normal, the mode, average and median will not coincide. Most permeability distributions are log-normal, and the geometric mean therefore most nearly represents where the majority of the data lie, which is at the mode *(Figure 4a)*.

Log-normal distributions result from measurements (data) in which a point within the reservoir depends upon results measured at an adjacent point not completely disassociated from it (2). A normal distribution might be expected in a batch process where an entire body of sediment were moved *en masse* to deposition in random lumps of various sized particles; a normal distribution is not expected in a particle-by-particle growth process in which the character of each particle is dependent to some degree on what is happening at a neighbouring point *(Figure 4b)*.

Facies Mapping

As one of the main purposes of this paper is to stress the importance of an integrated approach in the analysis of a reservoir, it appears to be appropriate to mention, at least superficially, the modern applications of mathematical, graphical and computer methods to handle geological and stratigraphic data. Such an approach may be called *Stratigraphic Facies Mapping*. It shows significant trends and interpretations of lithological characters; for example, lateral variations of composition, lateral variations of vertical arrangements, the probability of a bed occurring, etc.

Some of the techniques of facies mapping are *Percentage Mapping* and *Ratio Mapping*. The former presents the relationship between two variables *(Figure 6a)*; the latter maps one variable versus the sum of two *(Figure 6b)*. Other examples are: the *Percentage Triangle*, which allows the mapping of three varia-

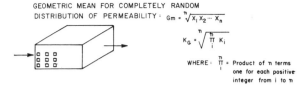

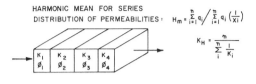

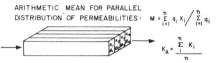

Figure 5.—Permeability Arrangement.

bles *(Figure 6c)*; the *Entropy Function*, which expresses the content of the components in a mixture (*Relative Entropy* investigates the degree of mixing); and the *D- Function*, which classifies data plotted in triangulate coordinates into seven groups, according to which component predominates, and assigns mixing indices to each group.

In a general case, it is not sufficient to evaluate and report only an element and/or component within a stratigraphic unit, and therefore "Vertical Variability" techniques must be applied. It is usually required to know where and how many discrete units occur; i.e., to evaluate quantitatively the vertical arrangement of their occurrence. Such a quantitative vertical evaluation of discrete units within a specific stratigraphic interval can be conveniently made by calculating and mapping their respective centers of gravity *(Figure 6d)*. As it is usually also desirable to evaluate quantitatively the spread of the units whose vertical distribution is being studied, standard statistical methods, as for example variance and standard deviation, become useful tools. For example, standard deviation defines quantitatively the spread relative to the center of gravity discussed above. Combined center-of-gravity and standard-deviation maps can then be constructed to show the variations within the stratigraphic unit studied.

Other methods found useful in modern geological-stratigraphic-engineering studies include *Trend Surface Analysis* and *Factor Analysis*.

The principle of trend surface analysis is to divide observed data into "regional" and "residual" data. The purpose is to find, to map and eventually to investigate the evidence of local structures or significant data anomalies which may be obscured by variations in regional trends. The regional trend is assumed to be represented by smooth data; therefore, a plane, quadratic or higher order surface is developed which most closely fits the observed data. The residuals, which are the positive and negative variations of observed data from the calculated smoothed surface, are then examined for significance.

Factor analysis is basically a computer method of sorting data. The method identifies variables which increase or decrease together or which vary inversely with each other. It is a type of statistical analysis (multiple correlation analysis) which can be used to investigate, for example, the quantitative observations made by a geologist or a stratigrapher in describing a series of rock samples. These observations are sorted, arranged and analyzed to determine whether the variations represented can be accounted for adequately by a number of basic categories (facies) smaller than that with which each observation started. Thus, data composed of a large number of variables in the first instance may be explained in terms of a smaller number of reference variables.

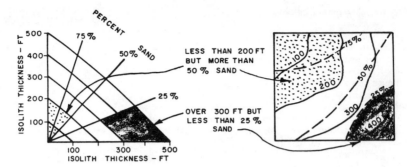

Figure 6a.

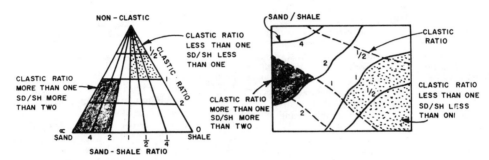

Figure 6b.

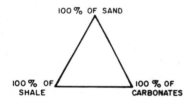

Fine	Lagunal
Medium	Limestone
Coarse	Dolomite
Rare	Anhydrite
Common	Non-Clastic
Abundant	Shale
Reefal	Sandstone
Basinal	

Figure 6c.—Examples of Other Groups.

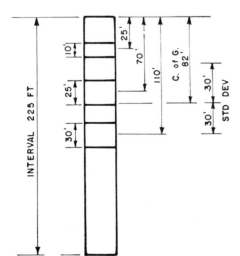

Center of gravity equals sum of product of distance times thickness divided by sum of thicknesses.

$$\text{C. of G.} = \frac{d_i\,t_i}{t_i}$$

$$
\begin{array}{rcl}
25 \times 10 & = & 250 \\
70 \times 25 & = & 1750 \\
110 \times 30 & = & 3300 \\
\hline
65 & = & 5300
\end{array}
$$

$$\text{C. of G.} = \frac{5300}{65} = 82 \text{ ft.}$$

Relative C. of G. $= \frac{82}{225} \times 100$ or 36% of the distance down from the top of the interval.

Figure 6d.

342

The above procedures of mathematical, graphical and computer applications of the integrated analysis of a reservoir are derived from notes on a facies mapping course offered by Mr. John Rivette, a Calgary-based consulting geologist.

Connate Water

The water content at a specific point within the reservoir is related to the structural position of that point, to the pore geometry and to the wettability of the rock (17). As the wettability of the rock at reservoir conditions is difficult to determine by laboratory measurements of the wettability of a core (18), due to possible changes in conditions which affect the characteristics of rock, it is usually most acceptable to establish the connate water saturation by direct measurements. Because of the usual indisputability of direct measurements, the preferred method of determining the connate water saturation of rock is thus through the analysis of cores cut with non-water-base muds (21). Unfortunately, it is too often forgotten that even if coring with oil-base mud, the amount of connate water subsequently determined is greatly influenced by the general drilling and core retrieving and storing practices.

When results are not available from non-water-base cores, the next preferred method of estimation of the interstitial water content is the capillary pressure or ''restored-state'' method. This method is developed on the theory that the reservoir rock which has lain for geological ages at equilibrium will have ''drained'' its water saturation to an irreducible value and will show a definite distribution arrangement in accordance with its capillary pressure characteristics.

The curve of capillary pressure (related to the height above level of 100 per cent water saturation) versus water saturation is determined in the laboratory on a water-saturated core sample using either air, oil or mercury as a displacing fluid.

The transition zone between oil and water is usually short, perhaps a few feet or so, although there are instances of transition zones much more than 100 feet thick, depending upon the pore size distribution of the rock and the characteristics of wetting and non-wetting fluids.

Numerous checks of interstitial water saturation determined by capillary pressure versus oil-base core results have been made and often found to be satisfactory (19). The capillary pressure method is therefore commonly used today. Capillary pressure data are only useful in connate water calculations if the reservoir rock is water wet. If the reservoir rock is only partially water wet or if it is oil wet, the capillary method would not apply. For an oil-wet system, the connate water saturation is much lower than it is for a water-wet system, because the irreducible minimum saturation is displaced from next to the rock and then becomes an insular saturation.

If samples of capillary pressure curves are available, so that a reasonable study of the capillary characteristics of the reservoir rock can be made, it is possible to determine whether or not the reservoir rock is oil wet by comparing its irreducible minimum water saturation with the oil-base core and log interpretation clues. Statistical tests can show whether any given difference is systematically significant (20.)

Interstitial water contents may also be computed from electric logs with some degree of success, provided that the logging program was rigidly controlled (21). Other methods using tracers have been tried, but with only fair results (19).

For a particular formation, the interstitial water content often shows a rough relation to permeability, with the lower permeability rocks having more interstitial water; unfortunately, however, there is no correlation as yet known between formations in this respect. Formations with a higher clay and silt content generally have higher interstitial water saturations than the cleaner ones.

In Western Canada and in some oil regions of the United States, it has been found that correlations of connate water saturations with the effective porosity (17, 22) often give better and more consistent results than a relationship with permeability, particularly because the differences between the means, modes and orientation of samples do not need to be accounted for. A discussion of such a correlation and of methods of averaging the connate water saturations of a reservoir has been comprehensively discussed by Buckles (22).

It must be strongly stressed that, in order to use and average any data on connate water saturation, each rock type must be lithologically identified and considered as a separate unit. In that way, the systematic variances often noticeable when lumping together several rock types can be reduced and a physically meaningful correlation of S_w with either ϕ and/or K can then be obtained.

Averaging Relative Permeability Ratios and Capillary Pressure Data

It is assumed that the relative permeability and capillary pressure curves are primarily functions of the pore size distribution. In order to average several curves for any of these two variables, a normalization procedure for the saturation axis is required.

Relative Permeability: Because of the usually wide range of scattering of ($k_{non-w.}/k_w$) values, the relative permeability ratios are commonly plotted on the log scale of semilog paper. The central portion of such a curve is usually linear, and may then be represented by the expression:

$$(k_{non-w.}/k_w) = a\, e^{-b\, S_w}$$

where a and b are constants; the subscripts refer to non-wetting and wetting fluid, respectively; subscript $_i$ will indicate initial conditions. The above equation suggests that the relative permeability ratio, for any one given rock, is a function only of the relative saturation of the fluids present.

As mentioned above, a semilog plot of relative permeability ratios versus wetting-liquid saturation usually indicates a wide range of scattering of the individual curves. In order to reduce such scattering, it is suggested that all such curves be normalized by changing the wetting-liquid saturation axis into:

$$\frac{S_L - S_{wi}}{1 - S_{wi}}$$

By examining many different cores, it has been found that, if the pore size distribution does not vary too much, the above-mentioned plot usually reduces the scatter considerably. However, if the pore size distribution varies markedly, the above normalized plot often results in two or several distinctly different curves which should not be averaged any further, because the cores were probably obtained from different lithological beds.

If, after such a normalization, it is found that further averaging is still necessary, and providing that such a scatter is not caused by a change in lithology, i.e., the spread is caused only by random errors, Guerrero's least square approach *(The Oil & Gas Journal,* May 30, 1960, Vol. 58, No. 22) can be applied to obtain one ''average'' curve. However, it should be stressed again that the question of deciding just how well the curves actually represent the formation under consideration is of much more importance than the averaging methods to be applied. Also, because the suggested normalization is not actually an averaging method, but a means to bring all the available curves to a common ''denominator,'' that particular technique should always be applied first.

Capillary Pressures: The normalization of capillary pressure curves (P_c) is affected by changing the S_w-axis into:

$$\frac{S_w - S_{wi}}{1 - S_w}$$

By experience mainly, it has often been found that a better way is to plot the J-function against a normalized saturation:

$$J(S_w) = \frac{P_c}{\sigma}\sqrt{k/\Phi}$$

or:

$$J(S_w) = \frac{P_c}{\sigma \cos\theta}\sqrt{k/\Phi}$$

where:

P_c = capillary pressure (dynes/cm^2), σ = interfacial tension

343

(dynes/cm), k = permeability (cm^2), ϕ = fractional porosity, and Θ = contact angle.

Another method of averaging capillary pressure data is the statistical approach suggested by Guthrie *et al.* (22). This last method is described in some detail on page 158 of "Petroleum Reservoir Engineering" by Amyx, Bass and Whitting.

However, as in the case of averaging the relative permeability ratios, if the pore size distribution were markedly different among the various samples neither the above normalized methods nor the statistical approach will satisfactorily solve the scattering problem. In such a case, it may be mandatory to perform all the individual calculations for each different bed separately and then attempt to combine the final results numerically.

Method of Least Squares

A treatise on averaging geological and engineering data would not be complete without at least a brief discussion on one of the most misunderstood and misused methods of averaging procedures, namely the least squares method.

As this paper emphasizes and re-emphasizes, the results obtained from any geological-engineering study are no better than the quality of the basic data, averaging techniques and assumptions which are applied in arriving at the final solution of the problem. For these reasons, the nature and type of basic errors which are inherent in any solution must be known, carefully evaluated and thoroughly understood.

Nearly all measurements of physical quantities are afflicted with errors. From a physical-mathematical point of view, these may be classified into three categories:

Firstly, the *mistakes,* which may be small or large; however, for practical reasons one is concerned with large mistakes only.

Secondly, the *systematic errors;* i.e., the errors which could be predicted or evaluated logically.

Avoiding mistakes and eliminating or at least compensating for systematic errors, one is left with a third category of errors, which by definition defy logical prediction and thus are called *non-systematic errors.* There are several sources for these non-systematic errors: the physical equipment (instrumentation, meters, etc.), the methods of measurement, the environment, or the object of measurement itself (often the physical quantity to be measured is not absolutely defined, e.g., the porosity of some formation); and finally, last but not least, the human errors. Recognizing the limitations of the systematic-logical approach, one may go to the other extreme and, assuming (or, in fact, knowing) the probabilistic distribution of these non-systematic errors, one may call them *'random' errors* and then possibly attempt to find the 'true value' hidden by their "random deviation." Of course, we must always look for the means to avoid the after-effects of all these errors. There are available methods to guard against or to recognize mistakes (large ones at least); however, these procedures, each of which might be different, depending on the nature of the problem itself, are not within the scope of this discussion (for example—Quality Control Charts). A systematic error can be predicted logically—in which case, it hardly deserves the name error. Modifying either the mathematical-physical logic of the procedures of measurement, or compensating for the error can remedy the situation. This process of elimination leaves us finally with the 'random' errors only.

A method often practiced is to measure the same physical quantity repeatedly, either on the same object or on similar objects if they comprise physical quantities of the elements of some collective. Thus, there accrues a surplus of data which are afflicted with errors and which are most likely incompatible. Such a surplus of "erroneous" data necessitates some practical means of finding the best answer. To achieve that practical goal, Legendre suggested in 1806 the 'method of least squares,' in which he proposed to minimize the sum of the squares of such errors from the unknown "true" value. With this he established a mathematically sound standard procedure. C.F. Gauss has used the same method earlier (1795), and he managed to establish a logical foundation for the method using, and thus contributing to, the new concept of probability (Laplace, 1811). Indeed, he developed, using a few quite reasonable basic assumptions, the Gaussian error distribution function for the study of these random errors. Since those early attempts at studies of random errors, more powerful tools of probability theory have been applied successfully to the problem and better results have been obtained; for example, interval estimation instead of the point estimation, correlation theory, etc. When searching for numerical values of some *functional relationship* between one or more independent variables and some dependent variables from a surplus of 'observation' points, one may thus justifiably apply the method of least squares. However, if applied, we must always realize and keep in mind that this method does not and cannot, serve as a substitute for creative thinking, because it cannot in itself define, or even search for, the most logical type of functional relationship among the several variables and among the many available and possible solutions. However, having selected, based on applied physical logic, some type (class) of functional relationship, the method of least squares will obtain the 'most probable' one from all the classes (types) studied while it systematically considers all the given observation points.

Due to the internal logic of the actual evaluation process of the method of least squares, polynomials of the nth degree are most suitable for the variable classes of functions to be studied; this leads then to n-simultaneous equations, for which we have standard methods of solution. The classes of transcendental functions are usually less desirable, because simultaneous non-linear equations need to be solved (which is possible using digital computers, nevertheless). Having then decided on the method of least squares as a tool to be applied to smooth the randomness of our errors, we are somewhat committed to accept the sum of the squares of the errors as a measure of the 'non-fit' of the predetermined function. In the statistical interpretation, this leads actually to the probabilistic concept of variance (or some estimate of the variance). In fact, the method of least squares will select, from the types (classes) of functions studied, the one for which the estimate of that variance is a minimum. With all the above limitations and physical logic in mind, we have to agree that it is folly to attempt to use the same criterion which was used to study the randomness from some average to select a class (type) of function. Thus, while investigating the randomness, it is important to understand that random error pertains only to the operation of selecting a sample or making a measurement and that it does not judge at all the sample or the measurement itself.

Despite this limitation, there are available procedures which help in the selection of the best class (type) of polynomial to be used in a study of almost any problem on hand. For example, if the prime interest is to write some analytical form of functional relationship amongst several variables, then one might compare the estimates of the variances for several differnt functions and select the representation for which that estimate is the smallest. However, attention is called to a strongly growing branch of mathematics called applied statistics which determines more correctly the methods of solution in this general area, and any analyst working with randomly scattered data should be aware of it and should be at least partly conversant in the general aspects of its use. As a further limitation of the above-discussed general method of least squares, this technique presumes that only the independent variable is the variable afflicted with random errors; this is a technically unrealistic and often unfounded supposition. Methods are now available, however, which permit the fitting of selected classes (types) of functions when all variables are afflicted with random errors; that is, a least squares fit which minimizes the sum of perpendicular distances of the individual points from the 'best fit' curve; this is similar to 'eyeballing' data in order to fit the best curve.

From the above brief discussion on the application and use (or misuse) of the least squares method to average geological- engineering data, it is concluded that it is only a method which quantitatively investigates the randomness of the data, i.e., their 'spread' from some average. It should always be recognized that although random errors can be small, systematically the data and/or the solution might still be wrong. Thus, random describes only a method of obtaining and/or averaging data rather than some resulting property of the data discoverable after the observance of the sample.

RESUME ON STATIC DATA

The principal considerations in averaging static reservoir data are:

1.—The final usage, the purpose of interpretation and the averaging methods must be clearly defined, understood and kept in mind at all times.

2.—An understanding of the distribution of porosity and permeability within the reservoir must be obtained by independent means, if possible.

3.—All data should be considered to be statistical and subject to the laws of sampling. Unless the standard deviation or dispersion is at a low level (as might be the case with good fluid samples), at least thirty samples should be obtained if trends and quantitative results are desired.

4.—Wherever possible, corroboration should be obtained from other direct measurement tools such as:

a. geological-stratigraphic and lithological interpretations
b. log interpretations
c. material balance solutions
d. unsteady state tests
e. field production and pressure performance
f. pressure build-ups and draw-down analyses
g. decline curves

5.—Laboratory test conditions may not adequately duplicate the reservoir conditions *in-situ*.

6.—Only occasionally it will be unnecessary to make a decision between alternate selections (making the 'unknown correct' value of academic interest only), because the difference between the two has no signficant effect on the end result or because experimental or sampling errors are greater than the difference in resulting evaluations or because other information eliminates the necessity for developing the most probable value.

7.—The end result of any calculation should be reasonable, consistent with values on pools of a similar type and, as production-pressure performance becomes available, the prediction should always be reviewed in the light of the new information (23).

8.—The reservoir analyst should always be consulted before any geological-engineering data are to be collected. As a matter of fact, he should be the one who is, as a final authority on that subject, responsible for systematic and meaningful data collection.

9.—Unnecessary data collection should be considered a waste and should be eliminated.

10.—Least squares methods can be applied for eliminating random errors only. They might give a reasonable solution but still be systematically wrong.

BIBLIOGRAPHIC REFERENCES

Static Data

(1) Krumbein, W. C., *et al.:* "Permeability as a Function of the Size Parameters of Unconsolidated Sand," *Trans.* AIME, 1943.
Law, J.: "Discussion on Average Permeabilities of Heterogeneous Oil Sand," *Trans.* AIME, 1944.
Hatch, Rastal & Black: "The Petrology of the Sedimentary Rocks," *Thomas Murby & Co.*

(2) Kauffman, G. M.: "Statistical Decisions and Related Techniques in Oil and Gas Exploration," *Prentice-Hall.*

(3) Earlougher, R. C., *et al.:* "A Pore-Volume Adjustment Factor to Improve Calculations of Oil Recovery Based on Floodpot Data," API D&PP, 1962.

(4) Klinkenberg, L. J.: "The Permeability of Porous Media to Liquids & Gases," API D&PP, 1941.

(5) Katz, D. L., *et al.:* "Handbook of Natural Gas Engineering," *McGaw-Hill.*
Ramey, H. J., Jr.: "Non-Darcy Flow and Wellbore Storage Effects in Pressure Build-up and Drawdown of Gas Wells," *JPT,* Feb. 1965.
Rowan, G., *et al.:* "An Approximate Method for Non-Darcy Radial Gas Flow," *SPEJ,* June, 1964.
Dranchuk, P. M., *et al.:* "The Interpretation of Permeability Measurements," *The Journal of Canadian Petroleum*

Technology, Vol. 4, No. 3, 130-133, 1965.

(6) Muskat, M.: "Physical Principals of Oil Production," Pg. 138, Table 2, *McGraw-Hill,* 1949.

(7) Pirson, S. J.: "Elements of Oil Reservoir Engineering," First Edition, Pg. 147, *McGraw-Hill.*

(8) Collins, R. E.: "Determination of the Transverse Permeabilities of Large Core Samples from Petroleum Reservoirs," *Journal of Applied Physics,* June, 1952.

(9) Greenkorn, R. A., *et al.:* "Directional Permeability of Heterogeneous Anisotropic Porous Media," *SPEJ,* June, 1964.

(10) Bulnes, A. C.: "An Application of Statistical Methods to Core Analysis Data of Dolomite Limestone," *JPT,* May, 1946.

(11) Testerman, J. D.: "A Statistical Reservoir Zonation Technique," *Trans.* AIME, 1962.

(12) Douglas & Rachford: "On the Numerical Solution of Heat Conduction Problems in Two or Three Space Variables," *Trans.* Am. Math. Soc. (1965), 82, 421-439.
Irby, T. L., *et al.:* "Application of Computer Technology to Reservoir Studies," *The Journal of Canadian Petroleum Technology,* Vol. 3, No. 3, 130-135, 1964.

(13) McCarty, D. G., *et al.:* "The Use of High Speed Computers for Predicting Flood-Out Patterns," *Trans.* AIME, 1958.
Bechenbach, E. F.: "Modern Mathematics for Engineers," Second Series, *McGraw-Hill.*
Quon, D., *et al.:* "A Stable, Explicit, Computationally Efficient Method for Solving Two-Dimensional Mathematical Models of Petroleum Reservoirs," *The Journal of Canadian Petroleum Technology,* Vol. 4, No. 2, 53-58, 1965.

(14) Stiles, W. E.: "Use of Permeability Distribution in Waterflood Calculations," Petroleum Transactions Reprint Series #2, *Waterflooding Series.*
Dykstra, H., & Parsons, R. L.: "The Prediction of Oil Recovery by Waterflood," *Secondary Recovery of Oil in the United States,* Second Edition, API.
Suder, F. E., & Calhoun, J. C., Jr.: "Waterflooding Calculations," API D&PP, 1949.
Muskat, M.: "The Effect of Permeability Stratification in Complete Water-Drive Systems," *Trans.* AIME, 1950.
Muskat, M.: "The Effect of Permeability Stratification on Cycling Operations," *Trans.* AIME, 1949.
Warren, J. E., *et al.:* "Prediction of Waterflood Behaviour in a Stratified System," *SPEJ,* June, 1964.
Hiatt, W. N.: "Injected-Fluid Coverage of Multi-Well Reservoirs with Permeability Stratification," API D&PP, 1958.
Jacquard, P., & Jain, C.: "Permeability Distribution from Field Pressure Data," *SPE Journal,* Dec., 1965.

(15) Rapaport, L. A.: "Laboratory Studies of Five-Spot Waterflood Performance," *Trans.* AIME, 1958.
Prats, M.: "Prediction of Injection Rate and Production History for Multifluid Five-Spot Flood," *Trans.* AIME, 1959.
Aronofsky, J. S.: "Mobility Ratio—Its Influence on Injection or Production Histories in Five-Spot Waterflood," *Trans.* AIME, 1956.

(16) Warren, J. E., *et al.:* "Flow in Heterogeneous Porous Media," *Trans.* AIME, 1961.
Cardwell, W. T., Jr.: "Average Permeabilities of Heterogeneous Oil Sands," *Trans.* AIME, 1945.

(17) Muskat, M.: "Calculation of Initial Fluid Distributions in Oil Reservoirs," *Trans.* AIME, 1949.
Hassler, G. L., *et al.:* "The Role of Capillary Pressures in Small Core Samples," *Trans.* AIME, 1945.
Hassler, G. L., *et al.:* "The Role of Capillarity in Oil Production," *Trans.* AIME, 1945.
Leverett, M. C.: "Capillary Behaviour in Porous Solids," *Trans.,* AIME, 1941.
Hough, E. W.: "Interfacial Tensions at Reservoir Pressures and Temperatures," *Trans.* AIME, 1951.

(18) Slobod, R. L. *et al.:* "Method for Determining Wettability of Reservoir Rocks," *Trans.* AIME, 1952.

Bobek, J. E., *et al.:* "Reservoir Rock Wettability—Its Significance and Evaluation," *Trans.* AIME, 1958.

(19) Williams, M.: "Estimation of Interstitial Water from the Electric Log," *Trans.* AIME, 1950.

Calhoun, J. C., *et al.:* "The Intensive Analysis of Oil-Field Cores," API D&PP, 1953.

Edinger, W. M.: "Interpretation of Core-Analysis Results on Cores Taken with Oil or Oil-Base Mud" API D&PP, 1949.

Armstrong, F. E.: "A Study of Core Invasion by Water-Base Mud Filtrate Using Tracer Techniques," API D&PP, 1961.

(20) Thornton, O. F.: "Estimating Interstitial Water by the Capillary Pressure Method," *Trans.* AIME, 1946.

Law, J.: "A Statistical Approach to Interstitial Heterogeneity of Sand Reservoirs," *Trans.* AIME, 1944.

(21) Glanville, C. R.: "Large-Scale Log Interpretations of Water Saturation for Two Complex Devonian Reef Reservoirs in Western Canada," (unpublished paper presented to Logging Society in Dallas in Spring, 1965).

(22) Guthrie, R. K., *et al.:* "The Use of Multiple Correlation Analyses for Interpreting Petroleum Engineering Data," (unpublished), Spring API Meeting, New Orleans, La., March, 1955.

Buckles, R. S.: "Correlating and Averaging Connate Water Saturation Data," *The Journal of Canadian Petroleum Technology,* Vol. 4, No. 1, 42-52, 1965.

(23) Guerrero, E. T., & Earlougher, R. C.: "Analysis and Comparison of Five Methods to Predict Waterflood Reserves and Performance," API D&PP, 1961.

Arps, J. J.: "Analysis of Decline Curves," *JPT,* Sept., 1944.

Lefkovits, H. C.: "Application of Decline Curves to Gravity—Drainage Reservoirs in the Stripper Stage," *Petroleum Transactions,* Reprint Series No. 3, Oil & Gas Property Evaluation & Reserve Estimates.

Hovanessian, S. A.: "Waterflood Calculations for Multiple Sets of Producing Wells," *Trans.* AIME, 1960.

A few other additional useful references on the interpretation and use of Static Data are:

Law, J., *et al.:* "A Statistical Approach to Core-Analysis Interpretation," API D&PP, 1946.

Stoian, E.: "Fundamentals and Applications of the Monte Carlo Method," *The Journal of Canadian Petroleum Technology,* Vol. 4, No. 3, 120-129, 1965.

Warren, J. E.: "The Performance of Heterogeneous Reservoirs," SPE paper 964, presented at the 39th Annual Meeting of SPE in Houston, Texas, October, 1964.

Kruger, W. D.: "Determining Areal Permeability Distribution by Calculations," *JPT,* July, 1961.

Dupuy, M., & Pottier, J.: "Application of Statistical Methods to Detailed Studies of Reservoirs," *Revue de l'Institut Francais du Pétrole,* 1963.

ACKNOWLEDGEMENTS

Thanks are given to the management of the Hudson's Bay Oil and Gas Company Limited, specifically to the Exploration and to the Production Department, for their permission to spend time on this paper. Thanks are due also to Messrs. G. Meisner, H. D. Bagley, Dr. W. Nader and J. Rivette for their contribution and help in preparing certain portions of this presentation.

Dynamic Reservoir Data

By D. HAVLENA*

(19th Annual Technical Meeting, The Petroleum Society of CIM, Calgary, May, 1968)

(Part II—Interpretation, Averaging and Use of the Basic Geological-Engineering Data)

ABSTRACT AND INTRODUCTION

THIS PAPER IS THE SECOND HALF and an integral part of a previous presentation entitled: *"Interpretation, Averaging and Use of the Basic Geological-Engineering Data, STATIC RESERVOIR DATA (Part I),"* presented at the 17th Annual Technical Meeting of the Petroleum Society of CIM, in Edmonton, Alberta, in May 1966; published in JCPT, October-December, 1966 and republished by SPE of AIME as Paper SPE 1974 (1967).

The purpose of these two complemental papers is to emphasize that taking, assembling and averaging of the basic data for further use in an analysis of a reservoir requires a profound knowledge of many of the specialized disciplines of modern geology and petroleum engineering.

"DYNAMIC RESERVOIR DATA (Part II)," presented in this particular paper, investigates the level of energy, the potential and also the forces contained or induced into the reservoir. These aspects are specifically discussed under the following major subheadings:

A. *Pressures to be Used in Analysis of a Reservoir:*

(a) datum, (b) depth of measurements, (c) original reservoir pressure, (d) bubble point pressure, (e) pressures taken during the production life-averaging, (f) superposition theorem, (g) variable pressures-stepwise curve, and (h) incremental vs partial pressure drops.

B. *PVT Data:*

(a) sampling, (b) laboratory results, and (c) adjustment to different p_b.

C. *Resumé on Dynamic Data*

D. *Bibliographic References.*

DYNAMIC DATA

A. PRESSURES TO BE USED IN ANALYSIS OF A RESERVOIR:

In order to understand thoroughly the importance of representative BHP data for obtaining the optimum results with least expenditures and with the minimum of subjective interpretation, a review of the entire problem is presented.

(a) *Datum:*

The expansion of reservoir fluids is a function of the fluid pressure in any particular part of the reservoir. A high structural relief, with the resultant increase in pressure with depth, may cause the oil in the lower parts to be still undersaturated whereas in the higher parts of the accumulation the gas may already be liberated from the oil (1). To apply any quantitative engineering evaluation in such a case, the reservoir should be cut in horizontal slices and separate complete calculations should be made for each. An approximation, often not the best, can be made by using a different total (two-phase) expansion factor (B_t) for each

At time of writing—*Continental Oil Company, International Dept., New York, N.Y.*

Since August, 1968—*Technical Assistant to the President, Husky Oil Ltd., Calgary, Alta.*

slice and by determining their average by weighting them by their volume. This weighted total average expansion factor (B_{tavg}.) is then used in all the subsequent calculations.

In reservoirs of small structural relief (200 feet or less), it is usually sufficient to select a representative common datum level and convert all pressures to that depth. For the expansion factor (B_t), it is then assumed that, at any time, all the oil is at one common pressure, the average datum level pressure, and the B_t is a function of it. However, this simplification can cause serious inaccuracies, because the B_t curve has a parabolic shape and therefore forestalls any accurate averaging. If the above simplification is nevertheless adapted, as is nearly always done, the datum level should be that horizontal level which divides the prevailing oil or gas reservoir into equal volumes; as the remaining oil or gas reservoir volume changes, for instance by water encroachment and/or gas cap expansion, it may be necessary to change continuously this datum level so that at any time it divides the remaining hydrocarbon leg into equal volumes (2, 3).

To convert pressures measured in each well at the midpoint of the producing interval to pressures at field datum level, a static gradient g_ϱ is usually applied. An amount $g_\varrho\Delta h$ is the correction factor, Δh being the true vertical distance between the measurement at midpoint of the producing interval and the datum level.

The difficulty which arises in reservoirs of high structural relief is that the relation between composition and depth is not accurately known, and for this reason it cannot be taken for granted that the composition is the same throughout the accumulation. In accumulations with a primary gas cap, the p_b is equal to the reservoir pressure at the original oil-gas contact and, as the pressure increases with depth ($g_\varrho\Delta h$), the oil below this contact would be gradually more undersaturated. If the chemical-physical composition were uniform, the bubble point should then be the same at every depth except for a correction due to temperature and compression *(Figure 1)*. Two other possibilities could exist, however:

(a) The bubble-point pressure decreases with depth, indicating a decrease in the amount of lighter fractions in the fluid; this is accompanied by lower solution GOR, thus suggesting gravity segregation *(Figure 2a* and partly *Figure 2b;* see also Ref. 8).

(b) The p_b is everywhere more or less equal to the initial reservoir pressure corresponding to that depth, resulting from a larger amount of light components dissolved at greater depth and revealing itself by a corresponding higher solution GOR. The oil would be then initially saturated at all depths (as far as it is known, such a field case has never yet been conclusively reported in the literature—partly possibly *Figure 2b)*.

It is still problematical which of these above concepts, and under which conditions, is correct. Examples of reservoirs with changing composition are reported in References 1, and 4 to 8.

As mentioned previously, the reservoir pressures are usually converted to the datum by using the static gradient and then averaged arithmetically, areally or volumetrically (9). Usually, a more accurate method consists of vertically contouring the pressure measured in each well and arriving at the average pressure at the pertinent datum level graphically. This same vertical contouring can be used again to find the pressure at the original oil-water contact and eventually also at the gas-oil contact.

Figure 3 shows this procedure for a vertical cross section in one direction. The vertical solid lines represent the individual wells, and the circles bounding these lines are infinite extrapolated pressures ($p\infty$) of four closed-in wells, recorded at the midpoint of the producing interval in each of these wells. The crosses are the pressures for the field datum and for the original oil-water contact obtained from the curve drawn through the measured values. The same procedure could be repeated for other cross sections and results are then averaged by horizontal contouring, i.e. by isobars. The isobar method of determining average static reservoir pressure is suitable, but unfortunately it can be used only in water-driven reservoirs (see (e) below).

Furthermore, it should always be remembered that reservoirs with a primary (original) gas cap and a water table have three separate datums: that governing expansion of hydrocarbons within the oil zone, that within the gas cap and that at the original o/w contact (2, 3).

(b) Depth of Measurements:

Bottom-hole pressures are usually measured in the tubing some distance above the producing interval. To obtain the pressure at the midpoint of the producing interval the correction factor $g_\rho \Delta h$ is applied. This correction is found from pressure readings at different depths, usually taken after pressure build-up surveys. Unfortunately, this static gradient is normally determined above the point where the build-up survey was taken; i.e., above the run-depth of the pressure measuring instrument. The gradient in the column below, in which we are interested, is then found only by extrapolation. Appreciable errors can be caused by this method, particularly if water is produced with the oil, as the water tends to settle to the bottom when the well is closed-in and the proper gradient in such case is not known.

Such "settling" occurs also in sour gas, condensate or high-GOR wells, where the upper part is gaseous and the lower remains in the liquid phase. If tubing is hung high above the perforations, i.e. if there were a high open casing space below the bottom of the tubing, any pressure measurements in the tubing are virtually worthless. Similarly worthless data are obtained if the run-depth were far below the producing perforations.

During a pressure build-up survey, the water and/or condensate which settles to the bottom of the well may cause the build-up curve to show a downward trend, whereas the actual pressure in the reservoir is either increasing, constant or decreasing because of interference (12, 17). Furthermore, build-up surveys in a well taken at successive time intervals may show a pressure declining with time—in reality, the field pressure may still be more or less constant, but the increasing amount of water and the ever increasing error in extrapolated pressure gradients causes the pressure to appear to be declining. Similar discrepancies might occur in condensate or high-GOR wells.

A correct way of extrapolating pressure measurements is by using two pressure gradients for correcting the measured pressure to the reservoir datum. The extrapolated measured well gradient is employed to correct to the midpoint of the producing formation, and then a reservoir gradient is used to correct finally to the reservoir datum (Figure 3).

(c) Original—Initial—Reservoir Pressure ($p_i = p_o$):

This is the most important of all pressure measurements as its value influences, directly or indirectly, all subsequent engineering work. The initial pressure must be obtained from a detailed pressure build-up survey on one or more wells producing from the same reservoir. Such surveys have to be made at a time when each well is still producing as if from an infinite reservoir, that is without interference from other wells and before any depletion or water or gas movements within the reservoir take place. The data of the pressure build-up survey measured with a calibrated Amerada pressure bomb or other instruments of high accuracy, should be plotted versus log $[(t_D + \Delta t)/\Delta t]$, where t_D is the effective producing time (N_p/q_o), and Δt the shut-in time. The points, except for the initial ones, should fall on a straight line, which when extrapolated to infinite shut-in time, log $[(t_D + \Delta t)/\Delta t] = 1$ for $\Delta t = \infty$, gives, in a "new reservoir," the initial static reservoir pressure $p_i = p\infty$. This method is based on an equation expressing the pressure behaviour of a well which has been shut-in after being produced for time t_D with a constant

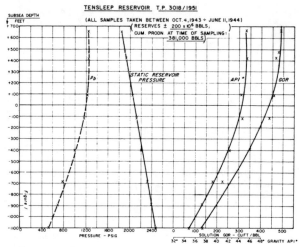

Figure 1.—Relation of Bubble-Point Pressure to Depth.

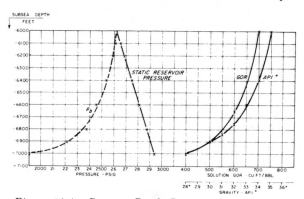

Figure 2(a).—Pressure-Depth Curves. (Ref. — Oil & Gas Journal, Vol. 75, No. 35, 1959).

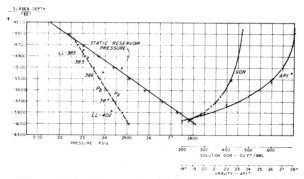

Figure 2(b).—Pressure-Depth Curves — LL-370 Area, Bolivar Coastal Field, T.P. 3653/1953.

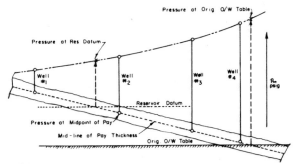

Figure 3.—Pressure Cross Section.

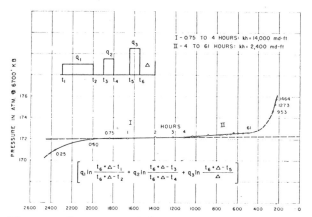

Figure 4.—Pressure Build-Up by Superposition.

$$q_1 \ln \frac{t_6 \cdot \Delta - t_1}{t_6 \cdot \Delta - t_2} + q_2 \ln \frac{t_6 \cdot \Delta - t_3}{t_6 \cdot \Delta - t_4} + q_3 \ln \frac{t_6 \cdot \Delta - t_5}{\Delta}$$

rate q_o (10) or a variable rate $q_o{}^*$ (14), from an infinite reservoir. The pressure at a distance r at time $(t_D + \Delta t)$ is then given by the equations of References 10, 13 and 14:

$$p_r = p_\infty + \frac{q_o\mu_o B_o}{4\pi kh} \left\{ ei\left[\frac{-\phi\mu_o c_o r^2}{4k(t_D + \Delta t)}\right] - ei\left[\frac{-\phi\mu_o c_o r^2}{4k\,\Delta t}\right]\right\} \quad (-1)$$

If this equation is applied to r_w, then, for Δt sufficiently large, the closed-in pressure can be represented by:

$$p_{\Delta t} = p_\infty - \frac{q_o\mu_o B_o}{4\pi kh} \ln \frac{t_D + \Delta t}{\Delta t} \quad (-2)$$

and finally using U.S. units and the Briggsian instead of Naperian logs we obtain:

$$p_{\Delta t} = p_\infty - \frac{162.6\, q_o\mu_o B_o}{kh} \log \frac{t_D + \Delta t}{\Delta t}. \quad (-3)$$

As mentioned previously, extrapolation of this straight line to $\log (t_D + \Delta t / \Delta t) = 1$ or what is the same but usually a preferable way, to $\log [\Delta t/(t_D + \Delta t)] = 1$ at $\Delta t = \infty$ will give the initial static reservoir pressure $p_i = p\infty$. If variations of rate were experienced prior to shutting a well in, either Horner's superposition approach (10) *(Figure 4)* or methods of Reference 14 must be used in evaluating such pressure build-ups. Procedures discussed above should be adapted throughout all subsequent pressure measurements during the producing life of the field. The practice of shutting-in a well for a certain period of time and taking one reading at the end of this period may cause such inaccuracies as to render the data completely worthless (15). Even after relatively very small cumulative production, such as a drill-stem test followed by a closed-in period of several hours or even days, the pressure read at the end of the closed-in period may differ considerably from the true static reservoir pressure. The size of the resulting error will depend on the cumulative producing time (t_D), the shut-in time (Δt) and the value of $q_o\mu_o B_o/kh$; at a high producing rate (q_o) and relatively low kh value, the error might be very appreciable.

To give an idea of the magnitude of possible errors resulting from accepting the instantaneous shut-in pressure spot reading (15) as representing the actual static reservoir pressure, the following examples are presented: Consider a well closed-in after a producing period of 2 hours (DST) and the same well after, let's

---

During the early life of a pool, there is usually a physical justification for straightline inter- and extrapolation of static reservoir pressures versus cumulative production; this applies to the entire reservoir as well as to the individual wells. However, there is no justification for such a straightline inter- and/or extrapolation if static reservoir pressures are plotted versus the real time.

say, four years of producing life. Instantaneous pressure readings were taken after the well had been shut-in for a certain period of time. The production rate just prior to shutting this well in was 80 B/D in each case; the permeability is 80 mds, pay thickness 12 feet, the oil has an underground viscosity of 0.6 centipoises and the shrinkage factor $(1/B_o)$ is 0.60. The value of $(162.6\, q_o\mu_o B_o/kh)$ is then 13.6; by applying equation (-3) above, we obtain:

1. *DST after 2 hours life and 24 hours shut-in.*

$$p_{24hrs} = p_{1\infty} - \left[13.6 \log\left(\frac{2+24}{24}\right)\right]$$
$$= p_{1\infty} - [13.6 \times 0.354] = p_{1\infty} - 0.48 \text{ psi}$$

2a. *After 4 years production and 24 hours shut-in:*

$$p_{24hrs} = p_{2\infty} - \left[13.6 \log\left(\frac{35040+24}{24}\right)\right]$$
$$= p_{2\infty} - [13.6 \times 3.167] = p_{2\infty} - 43 \text{ psi}$$

2b. *After 4 years production and 96 hours shut-in:*

$$p_{96hrs} = p_{2\infty} - \left[13.6 \log\left(\frac{35040+96}{96}\right)\right]$$
$$= p_{3\infty} - [13.6 \times 2.564] = p_{3\infty} - 35 \text{ psi}$$

As evident from this example, the errors inherent in the instantaneous pressure readings at later stages of the life are appreciable (from -0.48 psi to -43 psi after being shut-in for 24 hours; -35 psi after 96 hours); low kh value, high production rates and advancing life, i.e. greater t_D, will further increase their magnitude (15). The knowledge and some of the intricacies of correct interpretation of the pressure build-up curves for purposes of obtaining meaningful reservoir pressures and/or the inherent reservoir parameters are discussed comprehensively in References 16 and 17. Thorough understanding of the theories underlying these methods of interpretation is one of the fundamental requirements of all reservoir analysts (17). As the accurate determination of the original (initial) reservoir pressure is imperative for any reliable engineering studies, checks on it should be made by backward straight-line extrapolation of the static reservoir pressure versus cumulative production (but not versus time)* and by any other available means.

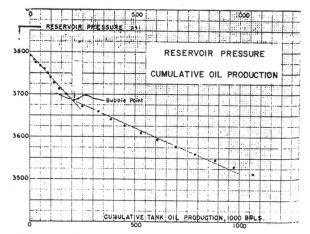

Figure 5.—Reservoir Pressure vs Cumulative Oil Production.

(d) *Determination of the Bubble Point Pressure (p_b):*

Next to the initial reservoir pressure (p_i), the original saturation of bubble-point pressure (p_b) is the most important. The bubble-point pressure is usually determined by the PVT analysis. However, as such determination requires that the analyzed PVT sample be representative of the reservoir fluids originally *in-situ,* its determination by the laboratory is influenced not only by the inaccuracy of the physical measurements in the laboratory, but mainly by the degree of the unknown representativeness of the reservoir fluids by the analyzed sample. For these two reasons, a laboratory determination of the p_b should never be taken for granted, and it is always imperative to check such laboratory value by studying and comparing it with the actual field production performance. Several methods are available:

I. A plot of reservoir pressure versus cumulative oil production often shows a definite break in slope if p_b was passed (*Figure 5*).

II. A plot of $N_p/\triangle p$ versus N_p or versus time, where N_p is the cumulative production of all fluids and $\triangle p$ is the cumulative pressure drop $(p_i - p_{Np})$, should have a sharp minimum at p_b [discontinuity is much more pronounced than in method I above (*Figures 6a* and *6b*)].

III. If the flowing bottom-hole pressure was higher than p_b and the flowing well-head pressure less than p_b, a plot of flowing pressure gradients versus depth and, therefore also versus pressure, should indicate a change of slope at p_b.

IV. The producing GOR of the undersaturated oil should remain more or less constant at fixed separator conditions. After p_b is passed, variations of GOR should be more pronounced (either less or more than solution GOR for fixed separator conditions).

V. Trial-and-error calculations using the equilibrium constants $(\Sigma n_i k_i = n)$ at proper convergence pressure should approximate not only the p_b but they also might indicate the relative size of gas-cap and oil, if both phases were originally present in the reservoir $(\Sigma\dfrac{n_i}{k_iG+L}=1)$.

VI. If the original gas-cap was tested and the reservoir is more or less homogeneous, the pressure at the original gas-oil interface should usually be close to p_b (as an exception see *Figure 1*).

VII. By successful solution of the material balance in a water-drive reservoir and finally by a plot of the resulting calculated cumulative water influx (W_e) versus $\triangle p/E_t$ quotient [cumulative pressure drop $(p_i - p_{Np})$ and corresponding composed expansion factor $E_t = (c_t\triangle p)$], a sharp break should be noticed if p_b was passed (*Figure 7*). If that happens, then it must be also possible to calculate the material balance separately for the time above p_b as well as below p_b — the resulting oil in place (N) should be the same in both cases. This last statement applies also to the MB solutions for a reservoir of the type illustrated by *Figure 5*.

VIII. If a number of PVT samples taken from the same reservoir were analyzed, the reported bubble points are plotted v e r s u s reported GOR's; if the initial, actual producing gas-oil ratio is known from the field data (after correction also for the usually not measured tank vapours), the corresponding p_b can be found from the above plot after accounting for the separator pressure and temperature.

IX. By correlation nomographs, such as in Standing's book: "Oil Field Hydrocarbon Systems."

If an appreciable difference is noticed between the p_b of the PVT sample and the p_b from the actual field performance data (viz. above), a careful analytical study should be made to find the most proper value of the p_b. The PVT data must be then always recalculated to this new p_b value determined from the actual field performance; for methods to be followed see Reference 18 and also Chapter B *(c)* of this write-up: "Adjustment of PVT Lab-Determined Data to a Different p_b."

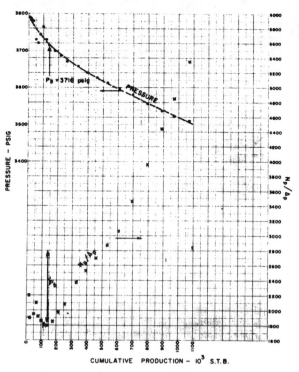

Figure 6(a).—Pressure and $N_p/\triangle p$ vs Cumulative Production.

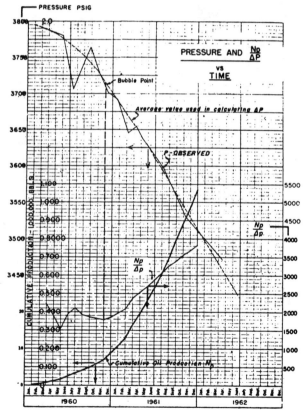

Figure 6(b).—Pressure and $N_p/\triangle p$ vs Time.

(e) *Pressures Taken During the Production Life—"Averaging"*

Because of physical differences in pressure distribution prevailing in the water-driven and in the depletion-type reservoirs

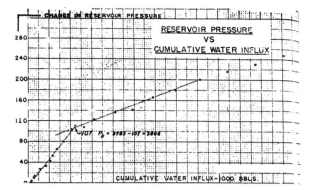

Figure 7.—Reservoir Pressure vs Cumulative Water Influx.

during the exploitation stages of their life, each of these two producing systems will require a different approach to the collection, averaging and use of BHP data.

In Water-Driven Reservoirs:

During the early life, the well or wells are producing as if from an infinite reservoir, without noticeable influence of any aquifer and/or boundaries. The wells, if all closed simultaneously, would indicate by extrapolation of the individual pressure build-up surveys the original reservoir pressure (at that particular subsea depth) at infinite closed-in time.

The correct pressure distribution, which is a function of the producing rates and the cumulative production of each well, theoretically should be determined without changing any of the current producing rates. The taking of pressure measurements in few strategically selected observation wells, which are never produced, is the ideal way of obtaining the true pressure distribution within that reservoir. The pressure in each of such observation wells is given below:

$$p = p_i - \overset{n}{\underset{1}{\Sigma}} \Delta p_j (t_{Dj})$$ (-4)

where: $[\Delta p_j (t_{Dj})]$ is the pressure drop at the observation well caused by the jth well, which has been producing for a time t_D and "n" is the total number of such interfering wells. By vertical contouring of pressures measured in the observation wells, the average pressure at datum levels (separate datum for the oil, the gas cap as well as for the original oil-water contact 2, 3) can be found. Unfortunately, in most cases not enough observation wells are available to make a reliable determination by this method. The exception in this respect might be Alberta, where the new proration system encourages the shut-down of unnecessary producers which could be thus used as pressure observation wells. Pressure measurements in producing wells (flowing pressure surveys) are useless for this particular purpose because the flowing pressure in each well is determined largely by the conditions in the immediate vicinity of the well (skin effect), and so it cannot yield information about the pressure distribution deep within the reservoir.

The next best method for a determination of the reservoir pressures in a water-driven reservoir is described below; it has its limitations, especially if the reservoir is drained by relatively few wells, but it gives the best approximation if a sufficient number of observation (permanently to be shut-in) wells is not available.

In a reservoir produced by "n" wells, the general equation for the flowing pressure in any one well is given by:

$$p_{wf} = p_i - \frac{q_o \mu_o B_o}{4\pi kh} \left[\ln \frac{kt}{\phi \mu c r_w^2} + 0.809 + 2S \right] - \overset{(n-1)}{\underset{1}{\Sigma}} \Delta p_j (t_{Dj})$$ (-5)

where the first part represents the flowing bottom-hole pressure of any well and the second part is the summation of the pressure

drops in that well caused by interference from all the others, i.e. (n-1) producing wells. If this particular well is closed-in, its shut-in pressure, after a certain time, is given again by:

$$p_{\Delta_t} = p_i - \frac{q_o \mu_o B_o}{4\pi kh} \ln \frac{t_D + \Delta t}{\Delta t} - \overset{(n-1)}{\underset{1}{\Sigma}} \Delta p_i (t_{Dj} + \Delta t_j) .$$ (-6)

The measured build-up times (Δt_j), are usually small compared to the producing time t_{Dj}, and equation (-6) for the straightline portion of the pressure build-up curve can be then simplified to:

$$p_{\Delta_t} = p_i - \frac{q_o \mu_o B_o}{4\pi kh} \ln \frac{t_D + \Delta t}{\Delta t} - \overset{(n-1)}{\underset{1}{\Sigma}} \Delta p_i (t_{Dj})$$ (-7)

where t_{Dj} replaces $(jt_{Dj} + \Delta t_{Dj})$. By plotting the pressure build-up survey data versus $\log [(t_D + \Delta t)/\Delta t]$ in a conventional way on semi-log paper, it follows from equation (-7) that a straight line results which, when extrapolated to $\log [(t_D + \Delta t)/\Delta t] = 1$ for $\Delta t = \infty$, yields an equation very similar to equation (-4) above:

$$p_\infty = p_i - \overset{(n-1)}{\underset{1}{\Sigma}} \Delta p_j (t_{Dj}) .$$ (-8)

Equation (-8) means that the extrapolated pressure $p\infty$ is the initial reservoir pressure (p_i) minus the pressure drop caused in the surveyed well by the interfering production of the (n-1) wells.

The same procedure is repeated in all other surveyed wells. We then obtain the true pressure distribution throughout the reservoir as required by equation (-4), except for the fact that the pressure drop at each location (well) is the sum of the pressure drops caused in it by all the other wells minus one. If the number of all wells is great, the error using this pressure distribution as a close approximation to the unknown true one is usually small as compared to the accuracy of the pressure measuring instruments. Besides, it can be more or less corrected by multiplying each $p\infty$ by:

$$\frac{n}{n-1} = \frac{\text{number of all wells}}{\text{number of all wells minus one}}$$

It is evident that this approach should be good if the oil in place (N) and the total number of wells are great. For small N and/or few producers, the errors can be considerable and the resulting pressures might be so doubtful as to make all the subsequent calculations based on such data completely worthless.

The above method, supplemented by isobaric contouring, is well suited for water-driven reservoirs as long as the aquifer can be considered infinite, which often is the case during the whole producing life of the reservoir. Even if this is not the case, the $p\infty$ may be used as a good approximation to describe the actual pressure distribution within the oil leg, as the errors of neglecting any eventual depletion will be usually insignificant. The above equations are applicable for the assumption of a uniform reservoir of oil and surrounding aquifer and presuppose more or less equal mobility ratios for oil and water. Similar equations for "non-unity" mobility ratios were also developed in the petroleum literature, but they are quite complicated and usually only of minor importance.

If the aquifer is very small the depletion of the oil-water reservoir begins at an early date. In this case of "semi-steady-state" pressure distribution, it is unimportant to know the pressure at the oil-water boundary—an average pressure in the whole reservoir system, partly oil volume and partly water volume, is then obtained according to the method suggested below for the depletion-type reservoirs.

In Depletion-Type Reservoirs (10, 11, 23):

For averaging pressures in depletion (volumetric) types of reservoirs, in which a completely different pressure distribution prevails than in the water-driven reservoirs discussed above, it is preferable to use, instead of isobars, a modified method described in very general terms by Horner in Reference 10 or, even better, by Matthews and Brons (Ref. 11) and, finally, specifically by Horner in Reference 23.

As long as transient conditions exist, the drainage areas of individual wells will not be constant; however, when all wells are produced at more or less constant, even if different, rates, after some time the pressure at any point within the system will drop at more or less the same rate (dp/dt = constant). When this condition of "semi-steady depletion" occurs, the drainage areas of the individual wells will not appreciably change any more. Then also, the change of the average pressure in each drainage area per unit of time will be just about the same as the change of the average pressure in the whole system per equal unit of time.

The way of finding a representative reservoir pressure for such depletion-type field was suggested in Ref. 10 and another method was described in detail in Ref. 11; finally, a field example is given in References 11 and 23. All these articles theoretically justify their particular approach and also present a numerical solution of specific examples. Because of the above detailed descriptions and lengthiness of the calculations, only the main physical phenomena which each of them describes are briefly mentioned in this paper.

Matthews and Brons (11) take into account the shape of the individual drainage areas and, because of that, their method does not give rise to inaccuracies in the case of an irregularly developed reservoir. The reservoir is divided into the drainage area of each individual producing well and the average pressure in each of these drainage areas is determined from the specific pressure build-up curve. The pressure gradient across the boundaries of the areas and at the limits of the reservoir is always considered to be zero. A pressure build-up curve should be available for each continuously producing well. The extrapolated pressure p* is corrected by an amount read from the available charts, in order to obtain the average pressure in the drainage volume. This correction is a function of production time, production rate, reservoir characteristics and size and shape of the drainage area. The mean reservoir pressure is obtained from these individual drainage volume pressures by means of a volumetric averaging process (9).

The method of Reference 11 is intended to apply in the transient as well as in the steady-state period. On simulated model studies it had been found that it gives a very good approximation to true behaviour in the transient period, while in the steady-state period it yields values very slightly higher than the true average reservoir pressure. Practical drawbacks of this theoretically very sound method are:

(a) Every continuously producing well in the reservoir should be pressure-surveyed at that particular time or at least previously in order that its pressure sink can be correctly evaluated and then taken into account.

(b) Pressure data obtained from continuously shut-in observation wells are not taken into consideration because their drainage areas are zero. These data are, however, the most accurate as they actually measure the reservoir pressure in that part of the reservoir. For that reason, corrections should be introduced to take such measuring points into account.

(c) The method is tedious and slow.

Horner's (10, 23) methods are not as accurate as the above, because they assume that the drainage area of each well has a circular shape. It is believed, however, if the reservoir is more or less regularly developed, that the error resulting from that simplification is relatively very small. The general method is based, similar to Park Jones' Reservoir Limit test, on a solution of a material balance equation (MBE) within the drainage volume of each particular well. An important advantage of the method is that the pressures of all the observation wells are fully taken into consideration. On the other hand, the edge wells present a special problem and, because of their eccentric position, these methods yield a low value for the average pressure within their respective drainage volume. From the physical standpoint, these

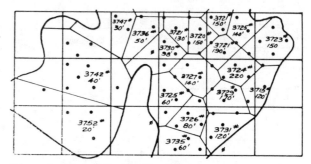

Figure 8.—Perpendicular Bisectors.

methods visualize a producing reservoir as a region of gradually varying pressure on which is superimposed a set of pressure sinks extending around each well. In the method of Ref. 10, the average reservoir pressure is again determined by volumetric averaging of all the individual average pressures. The method of Ref. 23 evaluates the average reservoir pressure "explicitly"; it thus yields not only the average static reservoir pressure within the drainage area of the well but also the active oil in place (N) to which such a pressure corresponds. The average reservoir pressure is again determined by volumetrically averaging (9) all of the individual average pressures.

Alternate Method (Water-drive and depletion type):

An alternate method of averaging static reservoir pressures, which is not as correct as the above-mentioned procedures and which has no such sound theoretical justification, but which, on the other hand, can be used in water-drive as well as in the depletion-type reservoirs, is the method of "perpendicular bisectors." This method assumes that the extrapolated static pressure of each pressure-surveyed well actually approximates the average static reservoir pressure within the outlined pressure area of such a well; because of this assumption, this method should be used only if the P.I.'s of all the individual wells are rather high and skin factors are relatively small. The area of a pressure-surveyed well is determined by bisecting perpendicularly the distance between any "two" pressure-surveyed wells and connecting together all such bisectors around it. *Figure 8* illustrates that procedure schematically. Such individual "pressure" areas are planimetered and are further weighted by the corresponding pay thickness to derive data for calculations of the average reservoir pressure. The advantage of this empirical method is its relative simplicity and, because of that, it could be used whenever the lack of data of high accuracy and confidence does not warrant the use of the other more exact procedures. Furthermore, the pressures of each individual "pressure" area could be approximately corrected by the method of Reference 23.

It should also be pointed out that volumetric weighting of pressures can be used only if such averages are used in "three-dimensional computations," as for example in the case of averaging reservoir pressure which controls the expansion of oil within the oil column. On the other hand, pressures used for calculations of the water influx (particularly in a bottom water drive reservoir) should be averaged areally across the original o/w table. Pressures controlling expansion of the gas cap should be weighted by volume of the gas, but if the gas cap is relatively small, they can be averaged areally at the gas-oil contact. In the case of an oil field with primary gas cap and suspected water influx, three completely different averaging (weighting) methods might be required (3):

(1) **volumetrically weighted reservoir pressure, which controls the expansion of liquids within the ever-changing remaining oil column (changing pressure datum) (2, 3);**

(2) areally weighted pressure at the original oil-water contact, which controls movements in or out of the original N (2); and

(3) volumetrically (areally) averaged pressure within the gas cap, as it exists at any particular time.

Depending on the configuration of the reservoir and on the exploitation practices, these three "average pressures" might be diverging or converging with time. Only in rare instances will they be parallel. Serious errors in results might occur by employing an incorrect method of averaging the pertinent reservoir pressures.

(f) Cumulative Water Influx (W_e) Caused by a Continuous Pressure Drop at the Original Oil-Water Boundary: Superposition Theorem:

This theorem is used to solve any problem when the underlying partial differential equation is linear (25). Two important rules help to solve such problems.

1. The result of any pressure drop is a multiple of the effect of the unit pressure drop.
2. The same results obtain irrespective of the time the drop is applied.

Hence, the theorem makes it possible to compute the cumulative production or the water influx caused by a variable pressure because the cumulative production and the influx per unit pressure drop (Q_{tD}) are known. It also permits computing the cumulative pressure drop caused by variable rates because the pressure drop caused by the unit rate (P_{tD}) is known. It should be noted, at this time, that the superposition theorem does not apply to functions which are solutions of non-linear differential equations (25), such as is the case if the exact analytical solution of the gas flow in gas reservoirs is studied.

The constant terminal pressure solution of the diffusity equation (19) applies only for a constant pressure drop Δp at the original oil-water table. In practice, however, this pressure drop Δp usually varies with time. The derived solutions can then still be used through application of the "superposition" theorem (15, 16):

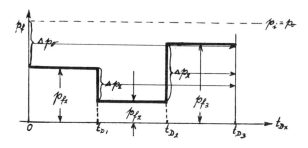

In order to calculate at any time t_{Dx} the cumulative production N_p, the actual pressure history may be regarded as composed of three parts in the above example:

Δp_o working (effective) from $t_D = 0$ to
$\qquad t_D = t_{Dx}$ causing a cumulative production N_{p1}

Δp_1 . . . . $t_D = t_{D1}$ to $t_D = t_{Dx}$ . . . .
$\qquad$ . . . . N_{p2}

Δp_2 . . . . $t_D = t_{D2}$ to $t_D = t_{Dx}$ . . . .
$\qquad$ . . . . N_{p3}.

Each of these individual cumulative productions (or water influx, which is the same) can be calculated with the constant terminal pressure case, and thus:

$N_p = N_{p1} + N_{p2} + N_{p3}$, or, working this out:

$N_{p1} = 2\pi\phi c_w r^2_b h \, \Delta p_o \, Q_{tD}$

$N_{p2} = 2\pi\phi c_w r^2_b h \, \Delta p_1 \, Q_{tD} - {}_{tD1}$

$N_{p3} = 2\pi\phi c_w r^2_b h \, \Delta p_2 \, Q_{tD} - {}_{tD2}$

For radial water influx calculations, instead of cumulative oil production (N_p), insert cum. water influx (W_e)

or finally: $N_p = W_e = 2\pi\phi c_w r^2_b h(\Delta p_o Q_{tD} + \Delta p_1 Q_{tD} - {}_{tD1} + \Delta p_2 Q_{tD} - {}_{tD2})$

or: $\qquad N_p = W_e = C\Sigma\Delta p_{tD}' Q'_{tD} - {}_{tD}{}^1 = C\Sigma\Delta p Q_{tD}$,

which can be extended indefinitely and which is the general solution of the constant terminal pressure case at the original oil-water boundary.

(g) Approximation of Variable Pressures by a Stepwise Curve:

Any continuous pressure drop Δp at the original oil-water boundary can therefore be represented approximately by a succession of instantaneous pressure drops $\Delta p'$ at regular (equal) time intervals, each such drop followed by a period of constant pressure. The required stepwise curve can be obtained from the smoothed curve of pressure by dividing the time scale into equal time intervals and considering the pressure drops at the beginning and end of each such equal time interval. This is illustrated in the schematic drawing below, where Δp, i.e. the cumulative pressure drop at any time from the initial pressure p_i, is plotted versus a dimensionless time t_D which is related to the real time t by a constant factor (for a radial case: $t_D = 0.00634 \, kt/\phi\mu cr^2$). The partial pressure drop $\Delta p'$ at the beginning of each interval is related to the total cumulative pressure drop

$$\Delta p = (p_i - p) \quad \text{by} \quad \Delta p'_n = \frac{\Delta p_{(n+1)} - \Delta p_{(n-1)}}{2}$$

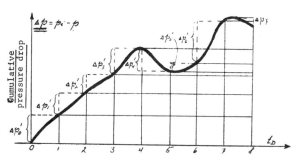

Consequently, referring to the above figure, the cumulative water influx* W_e at $t_D = 8$ is determined from the series of the individual partial (effective) pressure drops evaluated at equal time intervals from 0 to 8. At successive times $t_D = 0, 1, \ldots 7$, partial (effective) pressure drops $\Delta p_o'$, $\Delta p_1'$ . . . . $\Delta p_7'$ occur. At time $t_D = 8$, the partial pressure drop $\Delta p_o'$ will have been effective over period 8 (i.e. 8-0), $\Delta p_1'$ over period 7 (i.e. 8-1), etc. In general, at time t_D, $\Delta p_{tD}'$ will have been active over period $(t_D - t_D')$. The superposition theorem discussed above in general terms can now be written for this specific example:

$N_p = W_e = 2\pi\phi c_w r^2_b h(\Delta p'_o Q_8 + \Delta p_1' Q_7 + \ldots \Delta p'_7 Q_1) =$
$\qquad\qquad\qquad\qquad\qquad\qquad C\Sigma\Delta p'_t{}'_D \, Q_{tD} - {}_t{}'_D,$

which in practice is often incorrectly simplified to:
$N_p = W_e = C\Sigma\Delta p Q_{tD}.$

It is again emphasized that superposition theorem is only applicable because $W_e = f(\Delta p, t_D)$ is a solution of a linear differential equation. The superposition theorem does not apply to functions which are solutions of non-linear differential equations (25), as would be the case if the exact solutions for flow in gas reservoir were studied. Also, it should be understood that $\Delta p'$ is not the total pressure drop at t_D but rather the partial (effective) pressure drop from the previous pressure occurring at the time t_D'. The Table below illustrates a sample calculation of the summation term $\Sigma\Delta p'_{tD} Q_{tD-tD}'.$

(h) Incremental vs Partial (Effective) Pressure Drops:

As stressed before, the superposition theorem and all the above deductions are valid for cumulative water influx (W_e) as well as cumulative oil production (N_p).

353

Two simple methods of evaluating the effective pressure drops $\triangle p_n'$ to be used in the summation term of the water influx (or cumulative production) equation $\Sigma \triangle p'_{tD} Q_{tD} - t_D'$ are practical. Many engineers (20) evaluate the incremental pressure drop $\triangle p_n$ effective at time "n" as $\triangle p_n = p_n - p_{(n-1)}$. The method discussed in this review (see example calculation below) employs the partial pressure drops of Reference 21, defined as $\triangle p_n' = (\triangle p_{(n+1)} - \triangle p_{(n-1)})/2$, which appear to be superior to the incremental pressure drops $\triangle p_n$. To illustrate the difference in the summation terms resulting from these two different practically used approximations $\triangle p_n$ or $\triangle p_n'$, two example calculations have been prepared:

Example 1: evaluate the definite

integral: $\int_0^{4\pi} (\delta \triangle p / \triangle t_D) \, Q_{(t-tD)} \, \delta \triangle t_D$:

Exact value for the above limits
is: 8.000 = 100.000%
By incremental $\triangle p_n$ of Reference 20:
7.978 = −0.275%
By partial (effective) $\triangle p'_n$ of
Reference 21: 7.995 = −0.0625%

Example 2: evaluate the definite

integral: $\int_0^{10} (\delta \triangle p / \triangle t_D) \, Q_{(t-tD)} \delta \triangle t_D$ in

which: $\triangle p = (12 - t_D)^2 \, t_D$ and
$Q(t - t_D) = (t - t_D)^2$:
Exact value for the above limits
is: 18,000 = 100.00%
By incremental $\triangle p_n$ of Reference 20:
19,459 = +8.106%
By partial (effective) $\triangle p'_n$ of
Reference 21: 17,774 = − 1.256%

The two above numerical examples suggest that the partial $\triangle p_n'$ of Reference 21 is a much better approximation of the integral than the incremental $\triangle p_n$ of Reference 20.

B. PVT DATA

PVT data form the basic information for all reservoir calculations which analyze the past and predict the future performance of an oil reservoir. In the computations applied in reservoir engineering, the significant values are obtained as the differences of two numbers, so that a high degree of accuracy in the analytical data is required in order to obtain reliable results.

The equipment currently used by commercial laboratories in PVT analyses determines volumes with maximum errors of less than 0.01 per cent, pressures with errors of ± 5 to 10 psi and temperatures within ± 1 °F. Thus, considering the accuracy of other data, such as production and pressures and methods applied in averaging them, the physical accuracy of lab measurements of PVT data is more than satisfactory. Unfortunately, much less has been said about the importance of obtaining a truly representative sample of the original reservoir crude *in situ*. Therefore, any review of PVT's must first include a critical discussion of the sampling procedures.

(a) *Sampling:*

An excellent discussion of sampling methods and apparatus can be found in M.B. Standing's Chapter 2 of "Volumetric and Phase Behavior of Oil Field Hydrocarbon Systems." Because of that comprehensive review, only a very brief discussion of the methods follows:

1. *Bottom-hole sampling:* The samples are either of the open or vacuum type. In some cases, they are taken while the well is flowing at a reduced rate; in others, the well is closed-in for sampling. In any case, the selection of the well should be done very carefully in order to obtain a representative sample of the crude *in situ*. The well should have a large producing capacity (high P.I. and definitely a low skin factor) and it should produce without any water and/or free (gas-cap) gas.

2. *Separator Sampling:* The samples of oil and gas are obtained from the first separator—the gas usually near the orifice meter, the oil at an outlet point directly ahead of the oil dump valve. They are recombined in the lab in the ratio and conditions in which they were produced and separated. Accurate measurements of oil and gas volumes are the most important factors in this type of sampling.

3. *Special Devices:* can take samples before the well has produced any oil at all. However, because of very limited sampling volume, they are used only very seldom.

There has been, is, and will be a great deal of controversy as to the most appropriate sampling procedures; i.e., bottom-hole versus surface recombination samples. Because of the many uncertainties, one is seldom justified in directly applying the results of a single PVT determination to the reservoir fluid as a whole.

Rather, it should be recognized that the limitations imposed by sampling in the wellbore or at the surface usually require careful checking and eventually also a re-adjustment of the lab-reported PVT data to the original reservoir fluid conditions. The checking methods on representativeness were discussed in the chapter on bubble-point pressure; the adjusting of the lab-

TABULATION OF SAMPLE CALCULATION

t_D:	Q^*_{tD}:	p: psig	$\triangle p$: $(p_i - p)$	$\triangle p'_n$: $\left(\dfrac{\triangle p_{(n+1)} - \triangle p_{(n-1)}}{2} \right)$	$\Sigma \triangle p'_{t'D} \, Q_{tD-t'D}$:
0	0	pi = 3870	0	15	*0*
5	4.54	3840	30	25	(15 × 4.54) =
10	7.41	3820	50	15	(15 × 7.41) + (25 × 4.54) = *224.65*
15	9.95	3820	60	10	(15 × 9.95) + (25 × 7.41) + (15 × 4.54) =
20	12.32	3800	70	−5	(15 × 12.32) + (25 × 9.95) + + (15 × 7.41) + (10 × 4.54) = *590.10*
25	14.57	3820	50		(15 × 14.57) + (25 × 12.32) + + (15 × 9.95) + (10 × 7.41) + + (− 5 × 4.54) = *727.20*

*Q_{tD} taken from tables of Ref. 19.

reported PVT sample is outlined in the follow-up chapter of this paper.

Based on personal, and undoubtedly highly subjective, experience, the following comments are offered:

I. There is no sampling method known which will guarantee a truly representative sample of the reservoir crude.

II. The inability to obtain a representative sample is not caused by shortcomings in the sampling procedures but by flow processes occurring within the reservoir at the time of sampling.

III. The surface recombination method will yield more representative samples of the total fluids regardless of the presence of free gas in the flow-string.

IV. When free gas, for any reason, is present in the flow-string at the point of subsurface sampling, a representative homogeneous mixture of the total fluid will not be found. If a gas segregation appears within the flow-string either in the static or in the moving column of oil, the bottom-hole sample will be usually "undersaturated."

V. Because of III and IV above, the surface method of sampling should be considered as the more universal means of obtaining representative samples, except when the flowing bottom-hole pressure is higher than the p_b, in which case the results of both methods should be the same.

VI. Oil and gas samples should always be recombined in the ratio in which they were produced.

VII. In order to check the quality of the sample, duplicate samples should always be taken. If the reservoir contains greater number of wells and/or it has a high structural relief, such duplicate samples should be obtained on several wells (4 to 8).

VIII. Laboratory results on PVT samples must always be checked against the actual production-pressure performance of the reservoir.

IX. The necessity for an appreciable extrapolation, or interpolation, of PVT curves beyond the lab-determined p_b (adjustments greater than ± 10 per cent of the p_b value reported by the lab) should usually completely invalidate the representativeness of the sample.

X. In many of the flowing wells, it has been noted that the producing gas-oil ratio is a variable function of the well producing rate (for this often unrecognized phenomenon, see Reference 22 and *Figure 9*). If such were the case, no representative sample can be obtained from that well, irrespective of how carefully the sampling procedure is carried out (either subsurface or surface); because of this relatively common phenomenon, each well, prior to being PVT sampled, should be tested to see if such a relation does exist. It has been noted that quite a few wells in Western Canada exhibit such a behaviour; a detailed sampling and study of such a well in Venezuela proved that a representative sample, or even a duplicate sample at equal GOR, can never be obtained.

(b) *Laboratory Results:*

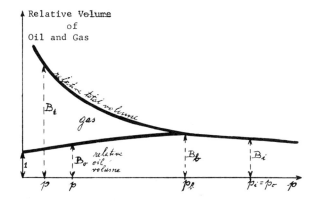

The above schematical diagram shows a pressure-volume relationship for a reservoir fluid at a reservoir temperature. At pressures higher than p_b, only one phase exists, namely the liquid, and the crude is undersaturated there. Below p_b, the gas is increasingly coming out of solution and, as well, the free phase expands, but the oil is shrinking in volume. By convention, the relative total volume is usually referred to p_b where it is unity, while the relative oil volume is unity at standard conditions of 0 psig and 60 °F; there is a small thermal shrinkage from 0 psig at reservoir temperatures to 0 psig at 60 °F.

Above p_b, i.e. in the undersaturated region, the relative total volume curve is expressed by:

$$V_t = V_i e^{-c_o(p_i - p)}$$

where: $V_t = \dfrac{\text{(Volume at a given pressure: p)}}{\text{(Volume at a saturation pressure: } p_b)}$

$V_i = $ (Volume at original pressure: p_i).

To calculate compressibility $(-c_o)$ from a given set of PVT data, the above equation, when plotted on semi-log paper, becomes:

$$\log V_t = \log V_i - \frac{c_o}{2.303} \Delta p \text{ where } (\Delta p = p_i - p)$$

from which c_o is evaluated by drawing the best straight line:

$$c_o = \frac{2.303}{\Delta p \ (1 \text{ cycle})}$$

Usually, over a wide range of pressures above the saturation pressure, the V_t suggests a curved trend which can be best expressed by:

$$\log V_t = \log V_i - \frac{c_o' \Delta p + \frac{\alpha}{2} c_o' \Delta p^2}{2.303}$$

from which c_o' and α can be calculated by the method of least squares. In the actual commercial PVT analysis, the compressibility (c_o) is usually reported as a constant over a certain pressure range:

$$c_o = \frac{2.303}{\Delta p \ (1 \text{ cycle})} \text{ between } p_1 \text{ and } p_2.$$

Below p_b, ever increasing volumes of gas are coming out of the solution and, furthermore, such free gas expands. The relationship between the V_t curve (partly oil and partly gas) can conveniently be expressed in that region with the help of a so-called "Y-curve." As the reliability of measurements of V_t just below p_b is usually rather poor, the Y-curve, calculated from data obtained at lower pressures and thus less sensitive to measurement errors, can be used to improve the accuracy of the V_t curve at, and also just below, the p_b. By definition, the Y-curve is:

$$Y = \frac{p_b - p}{p \ (V_t - 1)}$$

which is sometimes a straight-line function of pressure:

$$Y = a + bp.$$

This Y-pressure relationship can often be more exactly expressed as:

$$Y = a + bp + cp^2$$

from which a smoothed V_t is calculated from:

$$V_t = 1 + \frac{p_b - p}{ap + bp^2 + cp^3}$$

This empirical relationship is used to smooth the laboratory-measured V_t data as well as to define accurately the V_t close below the p_b, where the lab measurements are usually inaccurate. The constants a, b and eventually c are found by the least-square method, omitting the scattered Y-points just below p_b.

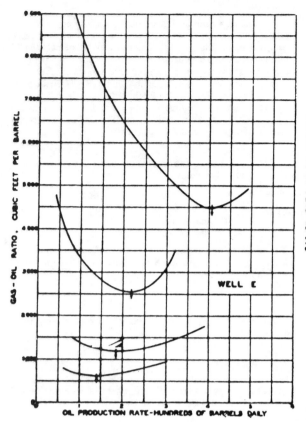

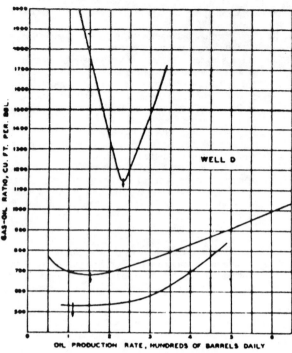

Figure 9.—Gas-Oil Ratio vs Production Rate — Well D and Well E.

This Y-relationship can be viewed as stating that the amount of gas released is proportional to the pressure drop p_b - p, and that its volume is inversely proportional to pressure p. These two proportionality factors combined give the a- and b-terms of the Y-function. In this simple expression, the effect of the released gas on the behaviour of the liquid volume of the mixture is not considered. Hence, the proportionality described by Y = a + bp will hold only for rather small pressure drops below the p_b or for only certain crude oils. The deviations from this simple form will give rise to the additional term c.

At the bubble point, although (p_b - p) and (V_t - 1) both approach zero, the Y-curve, expressed as:

$$ Y = \frac{1}{p \left(\dfrac{d V_t}{d p} \right)_b} , $$

retains its finite value and therefore the Y-curve calculated for the middle pressure range can always be extrapolated to calculate V_t accurately at, and also just below, the bubble-point pressure.

The Y-curve is also used for the recalculation of the lab-determined PVT analysis to another saturation (bubble-point) pressure, as will be shown in the next chapter. The use of a Y-curve makes unnecessary the reading of the V_t curve and it thus contributes to improve the over-all accuracy of the calculations.

In the lab, the reservoir fluid samples are analyzed by two methods of vaporization: the differential, which is usually the main subject of analysis, and the flash, reported usually as a separator test. In practice, the lab-reported differential process consists of a number of small step flash processes and, as such, the reported differential vaporization is actually already some kind of a combination process.

There are usually three sets of calculations which are performed on laboratory-reported fluid analyses before these can be used in the actual engineering calculations.

The *first step* depends upon the sample and the time when it was obtained within the life of the reservoir. As explained in the chapter on bubble-point pressure, the lab-determined p_b and

GOR must always be very carefully checked against the actual field performance information. If a discrepancy is discovered, this must be investigated and understood, and the lab-PVT must eventually be corrected to fit the actual field performance.

The *second step* involves the smoothing of the corrected data and is usually applied only to the relative total volume and to the differential oil volume data. However, an appropriate least-square smoothing procedure should be applied also to the B_g (gas expansion) data and eventually also to the oil viscosity data.

The *third step* involves computations of the combination formation volume factors and gas-oil ratios, accounting thus for the appropriate field separator conditions. This calculation is performed assuming that differential liberation occurs in the reservoir and that flash liberation occurs between the reservoir and the stock tank oil.

For further details on PVT analyses and their preparation for use in the actual engineering calculations, refer to Amyx, Bass and Whiting: "Petroleum Reservoir Engineering," pages 374 to 415.

(c) *Adjustment of PVT Lab-Determined Data to a Different p_b :*

Introduction

The data shown in Table I are the pertinent properties of a recombined surface sample from a field which was initially under-saturated. The sample was taken very early in the life of the field and the flowing bottom-hole pressure during the sampling process was some 750 psi above the saturation pressure. A laboratory saturation pressure 2,785 psia was obtained. Subsequent field performance indicates that the saturation pressure was actually 2,500 psia, suggesting that the recombined laboratory sample contained an excess of gas. The reference separator conditions for the field are 35 psig and 60 °F. The initial reservoir pressure was 3,240 psia and the reservoir temperature was 115 °F.

356

I	II	III	IV	V
Pressure Psia p	Flash Relative Total Volume V_t	Differential Relative Oil Volume V_o	Differential Gas Liberation Ratio D_F	Reference Shrinkage C_b
2785	1.000	1.000	0	0.681
2740	1.004	—	—	
2655	1.012	—	—	
2510	1.027	—	—	
2495	—	.974	90	
2395	1.041	—	—	
2282	1.058	—	—	
2210	—	.958	161	
2085	1.096	—	—	
1930	1.133	—	—	
1925	—	.925	257	
1710	1.208	—	—	
1660	—	.904	333	
1510	1.299	—	—	
1390	—	.882	411	
1385	1.375	—	—	
1280	1.452	—	—	
1110	—	.859	491	
1086	1.641	—	—	
961	1.831	—	—	
886	1.981	—	—	
828	—	.835	572	
801	2.155	—	—	
741	2.326	—	—	
687	2.501	—	—	
642	2.675	—	—	
568	3.026	—	—	
550	—	.810	652	
513	3.370	—	—	
468	3.703	—	—	
434	4.045	—	—	
310	—	.784	732	

TABLE I

PROPERTIES OF RECOMBINED OIL SAMPLE

SEPARATOR TESTS AT 60° F

Separator Pressure Psig	Gas-Oil Ratio Reference D_b	Reference Shrinkage C_b	API Gravity Tank Oil at 60° F.
0	1018	0.653	38.53
15	975	0.668	39.39
35	914	0.682	40.07
60	863	0.686	40.55
100	806	0.689	40.60

I. *Adjustment of Differential Vaporization Data* (1st Step):

A. Multiply each differential gas liberation ratio by the unadjusted bubble-point shrinkage, c_b, (Table I, Col. IV & V and Table II, Col. II & III) to convert the unadjusted differential gas liberation ratio to standard cubic feet per barrel of reservoir oil saturated at the unadjusted saturation pressure. Plot these ratios and the differential relative oil volumes, V_o, versus pressure *(Figure 10)* and interpolate (or extrapolate)* each curve to the new saturation pressure. The difference between the converted differential gas liberation ratio at the old saturation pressure and that at the new saturation pressure will be referred to as the corrective gas volume. The relative oil volume at the new saturation pressure, as read from the relative oil volume versus pressure curve, is the corrective relative oil volume.

B. Adjust the relative oil volume data (Table I, Col. III) by dividing each of the unadjusted values by the corrective relative oil volume read from the relative oil volume curve obtained in step I-A above (Table II, Col. IV & V).

Interpolate if pb (field) < pb (lab). Extrapolate if pb (field) > pb (lab).

C. Adjust the converted differential gas liberation ratios by subtracting (or adding) from each ratio the corrective gas volume obtained from the converted differential gas liberation ratio curve in Step I-A and by dividing each of these sums by the corrective relative oil volume (see Table II, Col. VI & VII). If the saturation pressure determined from field data is lower than the laboratory saturation pressure, the sample contained an excess of gas and the corrective gas volume is subtracted from each unadjusted differential gas liberation ratio. If the field saturation pressure is higher than the laboratory saturation pressure, the corrective gas volume is added to each unadjusted differential gas liberation ratio.

D. Plot the adjusted differential gas liberation ratios and relative oil values versus pressure to check the computations *(Figure 11)*.

E. An adjustment in the gravity of the residual oil is not required.

II. *Adjustment of Flash Vaporization Data* (2nd Step):

A. *Based on differential vaporization data* (preferred method).

1. Convert the unadjusted flash vaporization (separator tests) gas-oil ratios to ratios per barrel of saturated oil at the unadjusted saturation pressure by multiplying each ratio by the unadjusted flash vaporization shrinkage factor, c_b, at the indicated separator pressure (Table III, Col. II, III, IV).

2. Plot the converted gas-oil ratios obtained in Step II-A-I and the unadjusted shrinkage factors versus separator pressure in order to obtain the unadjusted gas-oil ratio (expressed on a saturated oil basis) and the unadjusted shrinkage factor at the desired separator pressure (the average separator pressure prevailing during the productive history of the reservoir) *(Figure 12)*.

3. From a plot of the adjusted differential vaporization data (Step I-D), determine a corrective gas volume and a corrective relative oil volume at the saturation pressure at which the flash data were obtained *(Figure 11)*.

4. Multiply the converted unadjusted gas-oil ratio obtained in Step II-A-2 by the corrective relative oil volume determined in Step II-A-3 (Table III, Col. V).

5. Subtract (if the PVT sample contains an excess of gas) from the gas-oil ratio determined in Step II-A-4 the corrective gas volume determined in Step II-A-3 to obtain the adjusted gas-oil ratio expressed relative to oil saturated at the new saturation pressure.

6. Adjust the shrinkage factor obtained in Step II-A-2 for the change in saturation pressure by multiplying it by the corrective relative volume obtained in Step II-A-3 (Table II, Col. VI).

7. Convert the adjusted separator gas-oil ratio (expressed relative to oil saturated at the same new saturation pressure) obtained in Step II-A-5 to a stock tank basis by dividing it by the adjusted shrinkage factor obtained in Step II-A-6 (Table III, Col. VIII).

8. An adjustment in the gravity of the residual oil is not necessary.

B. *Based on flash vaporization data* (alternate method—to be used only if A is not applicable for some reason) (See Table IV for illustration).

1. Convert the unadjusted flash vaporization gas-oil ratios (separator tests) to ratios per barrel of saturated oil at the unadjusted saturation pressure by multiplying each ratio by the corresponding unadjusted flash shrinkage factor at the indicated separator pressure.

2. Plot the converted separator gas-oil ratios obtained in Step II-B-1 versus separator pressure and extend the curve linearly from that ratio indicated at the highest separator pressure to zero cubic feet per barrel at the sample saturation pressure.

3. If necessary, extend the curve plotted in Step II-B-2 to its intersection with the new saturation pressure ordinate and read the corrective gas volume.

4. Plot the converted flash gas-oil ratios determined in Step II-B-1 versus separator pressure to obtain the unadjusted flash

gas-oil ratio at the desired separator pressure (the separator pressure prevailing during the productive history of the reservoir). A much shorter range of pressures is covered in this plot than in Step II-B-2, as the plot is not extended to the saturation pressure.

5. Plot the unadjusted shrinkage factors versus separator pressure to obtain the unadjusted shrinkage at the desired separator pressure.

6. Subtract the corrective gas volume obtained in Step II-B-3 from the unadjusted separator gas-oil ratio obtained in Step II-B-4 and divide this sum by the unadjusted shrinkage factor obtained in Step II-B-5 to determine the adjusted separator gas-oil ratio expressed relative to stock tank oil.

7. Determine the unadjusted change in volume per unit reservoir volume resulting from flash vaporizing the saturated oil at the highest separator pressure $(1 - C_b$ at the highest separator pressure).

8. From the "National Standard Petroleum Oil Tables" (U.S. Department of Commerce Circular C 410, U.S. Government Printing Office, Washington 25, D.C.), determine the volume that one barrel of oil at reservoir temperature occupies at 60°F.

9. Divide the unadjusted unit volume change at the highest separator pressure by the volume obtained in Step II-B-8.

10. Determine the change in unit volume caused by thermal shrinkage by subtracting the unadjusted unit volume change at the highest separator pressure from the value obtained in Step II-B-9.

11. Determine the unadjusted unit volume change due to gas liberation by subtracting the thermal shrinkage obtained in Step II-B-10 from the unadjusted unit volume change obtained in Step II-B-7.

12. Determine the shrinkage per cubic foot of gas liberated by dividing the unadjusted unit volume change due to gas liberation obtained in Step II-B-11 by the unadjusted gas-oil ratio at the highest separator pressure as obtained from Step II-B-1.

TABLE IV
BASED ON FLASH VAPORIZATION DATA (ALTERNATE METHOD)

The following items, which are numbered identical with those in the discussion, should illustrate the method.

(1)	Table III.
(2)	*Figure 13*
(3)	*Figure 13*
(4)	Not necessary in this case, as 35-psig separator test data available.
(5)	Same as (4).
(6)	$(623 - 58)/0.682 = 828$.
(7)	$(1 - 0.689) = 0.311$.
(8)	1 bbl. at 150°F = 0.9739 bbl. at 60°F
(9)	$0.311/0.9739 = 0.3193$.
(10)	$0.3193 - 0.3110 = 0.0083$.
(11)	$0.3110 - 0.0083 = 0.3027$.
(12)	$0.3027/555 = 0.0005454$.
(13)	$(0.0005454)(58) = 0.0316$.
(14)	$\dfrac{-0.0316 + (1.000 - 0.682)}{1 - 0.0316} = 0.2957.$
(15)	$1.0000 - 0.2957 = 0.7043.$

---

13. Determine the additional unit volume change due to liberation of the corrective gas volume by multiplying the value obtained in Step II-B-12 by the corrective gas volume as obtained in Step II-B-3.

14. Subtract the additional unit volume change obtained in Step II-B-13 from the unadjusted change in volume per unit volume (Step II-B-5) at the desired separator pressure and then divide this sum by one minus the additional unit volume change to determine the adjusted unit volume change.

15. Subtract the adjusted unit volume change determined in Step II-B-14 from one to obtain the adjusted shrinkage factor.

16. An adjustment in the gravity of the residual oil is not required.

III. *Adjustment of Total Volume Data* (3rd Step—See Table V):

A. Using the unadjusted total volume data, calculate the values of the Y-function at each of the pressures using the equation:

$$Y = \frac{p_b - p}{p\,(V_t - 1)}$$

where p_b is the unadjusted saturation pressure.

B. Using the method of least squares, determine the best values of the constants, a and b, in the equation: $Y = a + bp$ or a, b and c in the equation: $Y = a + bp + cp^2$ (the Y-curve is very often a curved line).

C. Calculate adjusted values for V_t using the equation:

$$V_t = 1 + \frac{p_b - p}{pY}$$

TABLE II
ADJUSTMENT OF DIFFERENTIAL VAPORIZATION DATA

I	II	III	IV	V	VI	VII
	Converted Gas Flashed Ratio, D_F Std.CuFt./ Sat.Res.Bbl.		*Adjusted Relative Oil Volume V_o*		*Adjusted Gas Flashed Ratio, D_F Std.Cu.Ft./ Sat.Res.Bbl.*	
Pressure Psia						
	@2785 psia				@ 2500 psia	
2495	90 x 0.681 = 61		0.974/0.976 = 0.998		(61-58)/0.976 = 3	
2210	161 x 0.681 = 110		0.958/0.976 = 0.982		(110-58)/0.976 = 53	
1925	257 x 0.681 = 175		0.925/0.976 = 0.948		(175-58)/0.976 = 120	
1660	333 x 0.681 = 227		0.904/0.976 = 0.926		(227-58)/0.976 = 173	
1390	411 x 0.681 = 280		0.882/0.976 = 0.904		(280-58)/0.976 = 227	
1110	491 x 0.681 = 334		0.859/0.976 = 0.880		(334-58)/0.976 = 283	
828	572 x 0.681 = 390		0.835/0.976 = 0.856		(390-58)/0.976 = 340	
550	652 x 0.681 = 444		0.810/0.976 = 0.830		(444-58)/0.976 = 395	
310	732 x 0.681 = 498		0.784/0.976 = 0.803		(498-58)/0.976 = 451	

TABLE III
ADJUSTMENT OF FLASH VAPORIZATION DATA

I	II	III	IV	V	VI	VII
Separator Pressure Psig	*Gas-Oil Ratio*	*Reference Shrinkage Factor, C_b*	*Converted Gas-Oil Ratio Cu. Ft./Sat.Res.Bbl.*	*Adjusted Gas-Oil Ratio Cu.Ft./Sat.Res.Bbl.*	*Adjusted Shrinkage C_b*	*Adjusted Gas-Oil Ratio Cu.Ft./STB*
0	1018	0.653	1018 x 0.653 = 665	(665 x 1.025) − 58 = 624	0.653 x 1.025 = 0.669	624/0.669 = 933
15	975	0.668	975 x 0.668 = 651	(651 x 1.025) − 58 = 609	0.668 x 1.025 = 0.685	609/0.685 = 889
35	914	0.682	914 x 0.682 = 623	(623 x 1.025) − 58 = 581	0.682 x 1.025 = 0.699	581/0.699 = 831
60	863	0.686	863 x 0.686 = 592	(592 x 1.025) − 58 = 551	0.686 x 1.025 = 0.703	551/0.703 = 784
100	806	0.689	806 x 0.689 = 555	(555 x 1.025) − 58 = 511	0.689 x 1.025 = 0.706	511/0.706 = 724

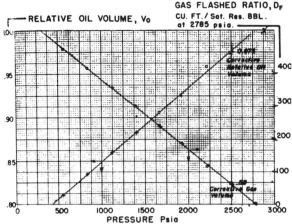

Figure 10.

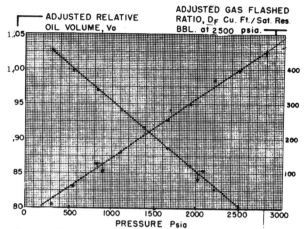

Figure 11.

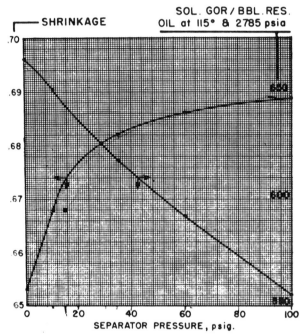

Figure 12.

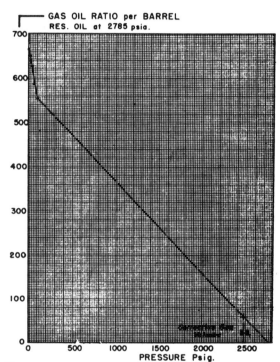

Figure 13.

where p_b is the adjusted saturation pressure and $Y = a + bp$ or $Y = a + bp + cp^2$, so that:

$$V_t = 1 + \frac{p_b - p}{ap + bp^2} \text{ or eventually}$$

$$V_t = 1 + \frac{p_b - p}{ap + bp^2 + cp^3} .$$

IV. Glossary:

Symbol: *Used Above*	*AIME([26])*	*Property*
c_b	$B_b = \dfrac{1}{c_b}$	shrinkage factor
v_o	$B_o = \dfrac{V_o}{c_b}$	relative oil volume
V_t	$B_t = \dfrac{V_t}{c_b}$	relative total volume.

C. RESUMÉ ON DYNAMIC DATA

This paper should make sufficiently clear the method of taking, determining, recording and using the correct reservoir pressures and PVT data. As each reservoir and even each individual well requires an intimate knowledge of its respective problems, including the prevailing reservoir mechanism, the planning, conducting, supervision and reporting of the dynamic data are the fundamental responsibilities of the reservoir engineers in the field and in the district offices. Because of the complex nature of the problem and the inherent difficulties of "tailoring," only a few very general and tentative suggestions can be put in writing:

1. The type of the prevailing driving mechanism in each particular reservoir requires a specific policy for the planning, taking and use of dynamic data. This report discusses, in sufficient detail, a few specific cases.

2. During the time of oversupply and resulting "shut-in" production potential and with the possibilities of transfer of allowables between wells, it might be possible to shut-in for a prolonged period (or "forever") a few strategically situated wells in each important producing reservoir. Such wells could be used as pressure observation wells, as explained in

(A) Pressure Psia	Relative Total Volume V_t	p_b-p (2785-p)	$p(V_t-1)$	Y
2785	1.000	0		
2740	1.004	45	10.960	4.10583
2655	1.012	130	31.860	4.08035
2510	1.027	275	69.025	3.98406
2395	1.041	390	98.195	3.97169
2282	1.058	503	132.356	3.80036
2085	1.096	700	200.160	3.49720
1930	1.133	855	256.690	3.33087
1710	1.208	1075	355.680	3.02238
1510	1.299	1275	451.490	2.82398
1385	1.375	1400	519.375	2.69555
1280	1.452	1505	578.560	2.60129
1086	1.641	1699	696.126	2.44065
961	1.831	1824	798.591	2.28402
886	1.981	1899	869.166	2.18485
801	2.155	1984	925.155	2.14451
741	2.326	2044	982.566	2.08027
687	2.501	2098	1031.187	2.03455
642	2.675	2143	1075.350	1.99284
568	3.026	2217	1150.768	1.92654
513	3.370	2272	1215.810	1.86871
468	3.703	2317	1265.004	1.83161
434	4.045	2351	1321.530	1.77900

TABLE V

ADJUSTMENT OF TOTAL VOLUME DATA

(B) $\Sigma Y = \Sigma B + \Sigma Ap$ $60.48111 = 22 B + 30,268 A$

$\Sigma pY = \Sigma pB + A \Sigma p^2$ $96,883.462 = 30269 B + 54,589,125 A$

$A = 0.001056139; \quad B = 1.296038$

(C) $Y = 1.296038 + 0.001056139 \, p$

$$V_t = 1 + \frac{p_b-p}{1.296038 \, p + 0.001056139 \, p^2}$$

the previous chapters: reliable dynamic data and good field coverage with relatively few, and instantaneous, surveys at small costs could be assured.

3. Quality and direct comparability between the results should be always preferred to a quantity of doubtful reliability.

4. A complete pressure build-up should be taken in each new well. It must be realized, however, that the time required for pressure stabilization continuously increases with the well cumulative production. Because of this fact, pressure build-ups on each well might have to be repeated after a few years in order to determine the new stabilization time. Instantaneous pressure readings should be discontinued as much as possible, and, if taken, they should be obtained only on the observation wells. Otherwise appropriate corrections must be applied (15).

5. If possible, pressures should be taken on wells which require a short stabilization period; the measuring depth should always be as close to the producing interval as is safely possible.

6. When build-ups are taken, the pressure bomb should be lowered to the measuring depth while the well is still producing. Flowing bottom-hole pressure should be taken in each case just prior to shutting-in the well for the build-up survey. Fluid gradients should always be recorded: flowing on the way down, static on the way up.

7. While the well is still flowing (viz. 6, above), and shortly prior to being shut-in, the pressure bomb should be pulled out by some 100 ft for a short period of time. That time should be accurately noted on the surface clock. After a few minutes at that (measuring − 100 ft) depth, the pressure bomb should be lowered back to the measuring depth and the corresponding time should be noted again. After a little while, the well can be shut-in for pressure build-up. The exact time is again to be noted. This pulling up and down (l00-ft jerk) should be repeated after the well is a few hours on build-up and once again shortly prior to pulling out of the hole. The exact time of starting, the midpoint and the

end of these "jerks" should be accurately clocked on the surface. With the help of these three "time-wise" exactly known jerks, the shut-in time on the bottom of the well can be determined more accurately than from the available surface data only. This might be important information, because the shut-in time on the surface need not always correspond to the shut-in time on the bottom. Such divergence might be particularly great in high-GOR wells, in the gas wells or in wells which are producing from fissures, fractures, etc. Interpretation of pressure build-ups (Δt) must always refer to the shut-in times experienced on the bottom of the well and not on the surface.

8. In fields jointly operated with other companies, unitized or not, it might be advantageous to plan and coordinate the surveys to obtain uniform data and better coverage at lower costs for all operators. All surveys should be taken within a short time interval.

9. In order to assure adequate pressure data for future studies, each reservoir should be surveyed about once every two years. On the basis of experience, it has been found that the maximum total number of pressure surveys taken in a reservoir during the two-year period does not normally need to exceed some 40 per cent of the average number of wells actually producing during that period.

To clarify this very general "rule of thumb," let's consider a reservoir with 100 active producers. The total number of pressure surveys taken during every two-year period does not need to exceed 40 measurements. As the number of wells increases, this "maximum" can become relatively smaller. Furthermore, if permanent pressure observation wells are utilized or good pressure build-ups are taken, the number of surveys can be sometimes smaller. In a pool with fewer producing wells, this suggested "maximum" might be a little higher (50 per cent).

10. In reporting the pressure data, it would be advantageous to distinguish clearly between the instantaneous pressure readings and the continuous pressure build-ups; a clarifying remark should be included in all the BHP forms and reports.

11. PVT samples should be obtained on each commercial reservoir. In reservoirs of relatively high relief and/or great areal extent, several PVT samples should be obtained and analyzed.

12. Before using any PVT, its representativeness must always be checked against the actual field performance at the proper separator conditions.

13. Static and dynamic reservoir data should supplement and complement each other at all times.

14. It should always be understood that, disregarding the physical and mathematical accuracy of the analytical methods, the answers will be only as good as the degree in which the basic data represent the actual reservoir conditions in-situ. If the basic data are poor, do not apply sophisticated analytical methods.

15. The quantity and quality of the derived information will depend on the characteristics of the reservoir, the type of drive, the quality and quantity of the basic data (static and dynamic) and, last but not least, on the experience, judiciousness and ingenuity of the analyst; every reservoir presents a challenge requiring knowledge, imagination, perseverance, accuracy, geological-engineering cooperation and, once again, profound theoretical knowledge.

BIBLIOGRAPHIC REFERENCES

(1) Grant, H. K.: "Material Balance Calculations," *The Oil and Gas Journal,* Aug. 24, 1959, Vol. 27, No. 35, p. 93.

(2) Havlena, D., and Paxman, D. S.: "Effects of Hydrodynamic Pressure Interference on Reservoir Performance, Buffalo Lake D-3," *Journ. Pet. Tech.,* 1966.

(3) McKibbon, J., Paxman, D. S., and Havlena, D.: "A Reservoir Study of the Sturgeon Lake South D-3 Pool," *Journ. Canad. Pet. Tech.,* 1963.

(4) Pym, L.A.: "The Measurement of Gas-Oil Ratios and Saturation Pressures and Their Interpretation," *Proceedings*

of World Pet. Congress London 1, p. 456 (1933).

(5) Comins, D.: "Gas Saturation Pressure of Crude Under Reservoir Conditions as Factor in the Efficient Operation of Oil Fields," *Proceedings of World Pet. Congress,* London 1, p. 459 (1933).

(6) Espach, R. H., and Fry, J.: "Variable Characteristics of the Oil in the Tensleep Sandstone Reservoir, Elk Basin Field, Wyoming and Montana," *Trans. AIME,* 192, p. 75 (1951).

(7) McCord, D. R.: "Performance Predictions Incorporating Gravity Drainage and Gas Gap Pressure Maintenance—LL-370 Area, Bolivar Coastal Field," *Trans. AIME,* 198, p. 231 (1933).

(8) Clinton, G. W., and Ruiz, J.: "Performance of Payoa Basal Eocene Block A Reservoir," *Journ. Pet. Tech.,* Jan. 1968, p. 31.

(9) Guerrero, E. T., and Steward, F. M.: "30. How to Find Average Reservoir Pressure," *The Oil and Gas Journal,* Dec. 26, 1960, pp. 157-159.

(10) Horner, D. R.: "Pressure Build-Up in Wells," *Proceedings of Third World Pet. Congress,* Section II, Example 4, pp. 517-519, (1951).

(11) Matthews, C. S. *et al.:* "A Method for Determining of Average Pressure in a Bounded Reservoir," *Trans. AIME,* T.P. 3876, 1954.

(12) Stegemeier, G. L., and Matthews, C. S.: "A Study of Anomalous Pressure Build-Up Behaviour," *Trans. AIME,* T.P. 8004, Vol. 213 (1958).

(13) Houpert, A.: "Eléments de Mécanique des Fluids dans des Milieux Poreux," Avril, Juin, Juillet, Août et Septembre, 1955, Mars et Avril, 1957. *Revue de l'Institut Francais du Pétrole.*

(14) Odeh, A. S.: "Pressure Build-Up Analysis, Variable-Rate

Case," *Journ. Pet. Tech.,* July, 1964, 790.

(15) Brons, F., and Miller, W.D.: "A Simple Method for Correcting Spot Readings" *Journ. Pet. Tech.,* August, 1961.

(16) Matthews, C. S.: "Analysis of Pressure Build-Up and Flow Test Data," *Journ. Pet. Tech.,* Sept. 1961, 862.

(17) Matthew, C. S., and Russell, D. G.: "Pressure Build-Up and Flow Tests in Wells," *Monograph Volume 1,* SPE of AIME, 1967.

(18) Clark, N. J.: "Adjusting Oil Sample Data for Reservoir Studies," *Journ. Pet. Tech.,* February, 1962, 143.

(19) van Everdingen, A. F., and Hurst, W.: "The Application of the Laplace Transformation to Flow Problems in Reservoirs," *Trans. AIME,* T.P. 2732 (1949).

(20) Chatas, A.: "A Practical Treatment of Nonsteady=State Flow Problems in Reservoir Systems," *Petroleum Engineer,* 1953, 25, No. 6, B-38.

(21) Craft & Hawkins: *Applied Petroleum Reservoir Engineering,* pp. 221-223.

(22) Sullivan, R. J.: "Gas-Oil Ratio in Flowing Wells," *Drilling and Production Practice,* 1957, pp. 103-117.

(23) Horner, D. R.: "Average Reservoir Pressure," *Proceedings, Fourth World Petroleum Congress,* Section II/C, Paper 3.

(24) Dietz, D. N.: "Determination of Average Reservoir Pressure from Build-Up Surveys," *Journ. Pet. Tech.,* August 1965, pp. 955-959.

(25) Odeh, A. S., and Jones, L. G.: "Pressure Drawdown Analysis, Variable Rate Case," *Journ. Pet. Tech.,* August 1965, pp. 960-964.

(26) SPE of AIME: "Standard Letter Symbols for Petroleum Reservoir Engineering, Natural Gas Engineering and Well Logging Qualities," *Journ. Pet. Tech.,* December, 1965, p. 1463.

Copyright 1944, Society of Petroleum Engineers of AIME.
Reprinted by permission. First published in *Petroleum Technology*, Transactions, V. 160, pp. 228-247.

AMERICAN INSTITUTE OF MINING AND METALLURIGAL ENGINEERS

Technical Publication No. 1758
(CLASS G. PETROLEUM DIVISION, NO. 221)

DISCUSSION OF THIS PAPER IS INVITED.

Discussion in writing (2 copies) may be sent to the Secretary, American Institute of Mining and Metallurgical Engineers, 29 West 39th Street, New York 18, N.Y. Unless special arrangement is made, discussion of this paper will close Dec. 30, 1944. Any discussion thereafter should preferably be in the form of a new paper.

Analysis of Decline Curves

By J. J. Arps,* Member A.I.M.E.

(Houston Meeting, May 1944)

ABSTRACT

Since production curtailment for other than engineering reasons is gradually disappearing, and more and more wells are now producing at capacity and showing declining production rates, it was considered timely to present a brief review of the development of decline-curve analysis during the past three or four decades.

Several of the commoner types of decline curves were discussed in detail and the mathematical relationships between production rate, time, cumulative production and decline percentage for each case were studied.

The well-known loss-ratio method was found to be an extremely valuable tool for statistical analysis and extrapolation of various types of curves. A tentative classification of decline curves, based on their loss ratios, was suggested. Some new graphical methods were introduced to facilitate estimation of the future life and the future production of producing properties where curves are plotted on semilogarithmic paper.

To facilitate graphical extrapolation of hyperbolic-type decline curves, a series of decline charts was proposed, which will make straight-line extrapolation of both rate-time and rate-cumulative curves possible.

INTRODUCTION

During the period of severe production curtailment, which is now behind us, production-decline curves lost most of their usefulness and popularity in prorated areas because the production rates of all wells, except those in the stripper class, were constant or almost constant.

While production-decline curves were thus losing in importance for estimating reserves, an increasing reservoir consciousness and a better understanding of reservoir performance developed among petroleum engineers. This fact, together with intelligent interpretation and use of electric logs, core-analysis data, bottom-hole pressure behavior and physical characteristics of reservoir fluids, eliminated a considerable part of the guesswork in previous volumetric methods and put reserve estimates, based on this method, on a sound scientific basis. At the same time, a number of ingenious substitutes were developed for the regular production-decline curve, which made it possible to obtain an independent check on volumetric estimates in appraisal work, even though the production rates were constant.

With the now steadily increasing demand for oil to supply the huge requirements of this global war, proration for reasons other than prevention of underground waste is gradually disappearing. More and more wells are, or will be, producing at capacity or at their optimum rates, as determined by sound engineering practice.

With this trend, the character of producing wells seems to regain, more or less, its "individuality," and the old and familiar decline curve appears to have had a comeback as a valuable tool in the hands of the petroleum engineer. It may be timely, therefore, to retrace the development of decline-curve analysis in the past by presenting a brief chronological review of bulletins and papers published during the past three or four decades, which have contributed to our present knowledge of this subject. Such a review will, at the same time, serve as a good basis for further analysis of the production-decline curve and its possibilities in this paper.

DEVELOPMENT OF DECLINE-CURVE ANALYSIS

The two basic problems in appraisal work are the determination of a well's most probable future life and the estimate of its future production. Sometimes one or both problems can be solved by volumetric calculations, but sufficient data are not always available to eliminate all guesswork. In those cases, the possibility of extrapolating the trend of some variable characteristic of such a producing well may be of considerable help. The simplest and most readily

---

*Manuscript received at the office of the Institute May 9, 1944.

**Chief Engineer, The British-American Oil Producing Co., Tulsa, Oklahoma.

PETROLEUM TECHNOLOGY. September 1944. Printed in U.S.A.

1

available variable characteristic of a producing well is its production rate, and the logical way to find an answer to the two problems mentioned above, by extrapolation, is to plot this variable production rate either against time or against cumulative production, extending the curves thus obtained to the economic limit. The point of intersection of the extrapolated curve with the economic limit then indicates the possible future life or the future oil recovery. The basis of such an estimate is the assumption that the future behavior of a well will be governed by whatever trend or mathematical relationship is apparent in its past performance. This assumption puts the extrapolation method on a strictly empirical basis and it must be realized that this may make the results sometimes inferior to the more exact volumetric methods.

The production rate of a capacity well, plotted against time on coordinate paper, generally shows a rapid drop in the beginning, which tends to decrease as time goes on. Changes in method of production, loss in efficiency of lifting equipment, shutdowns for work-over or pulling jobs, usually disrupt the continuity of a production decline curve, and for mathematical or statistical treatment some preliminary smoothing out is often necessary.

The first and most obvious mathematical approach to a declining production curve is to assume that the production rate at any time is a constant fraction of its rate at a preceding date or, in other words, that the production rates during equal time intervals form a geometric series. This also implies that the production drop over a given constant interval is a fixed fraction or percentage of the preceding production rate. The earliest reference in the literature of this type of decline was made by Arnold and Anderson[1] in 1908. This production drop, as a fraction, usually expressed in per cent per month, is called the decline. A considerable number of the decline curves encountered in appraisal work show this decline percentage to be approximately constant, at least over limited periods. A decline curve showing this characteristic is easy to extrapolate, since the rate-time curve will be a straight line on semilog paper and the rate-cumulative curve on coordinate paper.

The literature between 1915 and 1921 shows a considerable amount of research and study of production curves.[2-6] Much information from various sources was accumulated in the Manual for the Oil and Gas Industry.[7] J. O. Lewis and C. H. Beal, of the Bureau of Mines,[5] recommended the use of the percentage decline curve, which is an empirical rate-time curve, whereby the production rates during successive units of time are expressed as percentages of production during the first unit of time. This makes it possible to bring individual well or lease data to a comparable basis. The results can then be grouped together, either on regular coordinate or log-log paper. From such data on wells in the same area an empirical appraisal curve may be constructed to show the possible ultimate production as a function of the initial production rate.

W. W. Cutler,[8] in 1924, pointed out, after an intensive investigation of a large number of oil-field decline curves, that the assumption of constant percentage decline and a straight-line relationship on semilog paper generally gave results that were too conservative in the final stage. In his opinion, a better and more reliable straight-line relationship could be obtained on log-log paper, although some horizontal shifting usually was necessary. This implied that the decline curves showing such characteristics were of the hyperbolic rather than the exponential or geometric type. He also recommended the use of the family decline curve, either graphically constructed or statistically determined, which is a representative average decline curve for a given area based on a combination of the actual rate-time data from a number of wells in the area.

C. S. Larkey,[9] in 1925, showed how the method of least squares could be applied successfully to decline curves belonging to both the exponential and the hyperbolic types. He also demonstrated that the application of this well-known statistical method makes a strict mathematical extrapolation of a given decline trend possible.

H. M. Roeser,[10] in 1925, showed that equally reliable results can be obtained when, instead of the rigorous method of least squares, a somewhat simpler method of trial and error to determine the necessary constants is followed. He illustrated his method with examples of both the exponential and the hyperbolic types of decline curves. In his paper was also the first reference to the mathematical relationship between cumulative and time for hyperbolic type of decline.

C. E. Van Orstrand,[11] in 1925, investigated the empirical relationship of production curves representing the output of certain minerals by states or nations. Such a curve will rise from zero value at the time of first production to a maximum and then slowly decline, presumably to zero value. The possibilities of various mathematical relationships and different methods of curve fitting are described in this paper. The best results were obtained with a curve of the type:

$$P = At^m e^{-Bt}$$

R. H. Johnson and A. L. Bollens,[12] in 1927, introduced a novel statistical method for extrapolation of oil-well decline curves. With their so-called "loss-ratio method,"

[1]References are at the end of the paper.

2

the production rates are tabulated for equal time intervals, then the drop in production is listed in a second column and the ratio of the two, or "loss ratio," is listed in a third. A curve to be investigated with this method usually shows, after proper smoothing out, either a constant loss ratio or a constancy in the differences of successive loss ratios. Sometimes it may be necessary to take these differences two or three times before constancy is reached, and often additional smoothing out of the data is required. This procedure furnishes an easy and convenient method for extrapolation. It is only necessary to continue the column with the constant figures in the same manner and then work backward to the production-rate column.

H. N. Marsh,[13] in 1928, introduced the rate-cumulative curve plotted on coordinate paper and pointed out that this relationship generally appears to be or approaches a straight line. Although this is only mathematically exact for decline curves of the exponential type, as will be shown later, it was pointed out in his paper that the errors in estimating ultimate recovery with this method in most other cases were generally small or negligible. A distinct advantage of this type of curve is its simplicity in appraising the effect of different methods of production control on the same well.

R. E. Allen,[14] in 1931, mentioned four types of decline and classified them according to a simple mathematical relationship. The decline types were:

1. Arithmetic, or constant decrement decline.

2. Geometric, constant rate or exponential decline.

3. Harmonic, or isothermal decline.

4. Basic, or fractional power decline.

Type 1 is of little practical value for production-decline curves. Type 2 is the well-known straight-line relationship on semilog paper, and type 3 is the special case of hyperbolic decline where the decline is proportional to the production rate. It was not possible to reconcile the equation given for the type 4 decline, as the nominator and the denominator were of the same order, indicating a possible misprint.

S. J. Pirson,[15] in 1935, investigated the mathematical basis of the loss-ratio method and arrived at the rate-time relationships for production-decline curves having a constant loss ratio, constant first differences and constant second differences. Those of the first type appeared to be identical with the simple exponential or constant percentage decline curves, which straighten out on semilog paper; those of the second type were the hyperbolic type of decline curves, which can be straightened on log-log paper and those of the third type appeared to have such complicated mathematical equations as to be unsuitable for practical purposes.

During the period of production curtailment, interest centered upon suitable curves for reserve estimates that did not require the usually constant or almost constant actual rate of production.

H. E. Gross,[16] in 1938, showed the advantages of substituting oil percentage in gross fluid for the production rate in the Marsh rate-cumulative curve. This method, originated by A. F. van Everdingen in Houston, proved particularly valuable for prorated Gulf Coast water-drive production.

For depletion-type or gas-drive-type pools without water encroachment, however, a parameter other than oil or water percentage had to be found to replace the production rate.

W. W. Cutler and H. R. Johnson,[17] in 1940, showed how potential tests, taken periodically on prorated wells (or calculated from bottom-hole pressure and productivity-index data) can be used to reconstruct or calculate the production-decline curve, which the well would have followed if it had been permitted to produce at capacity.

H. C. Miller,[18] of the Bureau of Mines, introduced in 1942 the pressure-drop cumulative relationship on log-log paper and showed how changes in reservoir performance may be detected by abrupt changes in the slope of such a curve.

C. H. Rankin,[19] in 1943, showed how the bottom-hole pressure can sometimes be used to advantage as a substitute for the rate of production of the rate-cumulative curve on prorated leases. Apparently, this method applies only in pools where water drive is absent or negligible and where productivity indexes are constant.

In the Oklahoma City field, which is well known as a typical example of gravity drainage, a plot of fluid level against the cumulative production has been used successfully to estimate the reserves of wells with constant production rates.

P. J. Jones,[21] in 1942, suggested for wells declining at variable rates an approximation whereby the decline-time relationship follows a straight line on log-log paper. This corresponds to an equation:

$$\log D = \log D_0 - m \log t$$

in which D_0 designates the initial decline and m is a positive constant. Integration of this relationship will lead to a rate-time equation of the general form:

$$P = P_o e^{\frac{D_o t^{1-m}}{100(m-1)}}$$

It may be noted that this relationship will not straighten out on semilog or log-log paper, but shows the interesting characteristic of straightening out when the log-log of the production rate is plotted against the log of the time.

F. K. Beach,[20] in 1943, showed, with examples from the Turner Valley field, Canada, how cumulative-time curves sometimes can be extrapolated as straight lines in their last stage by plotting the antilog of the cumulative production against time. Such a straight-line relationship is mathematically correct only for the case of harmonic decline, where the decline itself is proportional to the production rate, as will be discussed later.

RESERVOIR CHARACTERISTICS AND DECLINE CURVES

In order to analyze what influence certain reservoir characteristics may have on the type of decline curves, it was first assumed that we are dealing with the idealized case of a reservoir, where water drive is absent and where the pressure is proportional to the amount of remaining oil. It was further assumed that the productivity indexes of the wells are constant throughout their life, so that the production rates are always proportional to the reservoir pressure.

In such a hypothetical case, the relationship between cumulative oil produced and pressure would have to be linear and, consequently, also the relationship between production rate and cumulative production.

This linear relationship between rate and cumulative is typical of exponential or semilog decline, as will be shown later (Eq. 4), and simple differentiation will lead to the basic equation for this type of decline in Eq. 1.

In most actual pools, however, the aforementioned idealized conditions do not occur. Pressures usually are not proportional to the remaining oil, but seem to decline at a gradually slower rate as the amount of remaining oil diminishes. At the same time the productivity indexes are generally not constant, but show a tendency to decline as the reservoir is being depleted and the gas-oil ratios increase. The combined result of these two tendencies is a rate-cumulative relationship, which, instead of being a straight line on coordinate paper, shows up as a gentle curve, convex toward the origin.

If the curvature is very pronounced, the curve can sometimes be represented by an exponential equation and the rate-cumulative relationship straightened out on semilog paper. This type is called harmonic decline, and its equation is identical with Eq. 14, derived on page 9. By differentiation, it can be shown that in this case the decline percentage is directly proportional to the production rate.

When the curvature of the rate-cumulative relationship is not pronounced enough to straighten out on semilog paper, it

can usually be represented as a straight line on log-log paper after some shifting. This identifies it as a hyperbola and it can be shown that it will fit Eq. 13 (p. 9) for the general case of hyperbolic or log-log decline.

From this general discussion, it is evident that the hyperbolic type of decline curve should be the most common and that harmonic decline is a special case, which occurs less frequently.

The exponential or semilog decline, however, although less accurate, is so much simpler to handle than the other two that it is still quite popular for quick appraisals and approximate estimates; particularly since a large number of decline curves actually show an apparent constant decline over limited intervals. The decline percentage in such calculations is then usually taken somewhat lower than the actually observed value in order to evaluate the possibility of a smaller decline in the final stage.

EXPONENTIAL DECLINE

Exponential decline, which is also called "geometric," "semilog" or "constant percentage" decline, is characterized by the fact that the drop in production rate per unit of time is proportional to the production rate.

Statistical Analysis and Extrapolation

The simplest method to recognize exponential decline by statistical means is the loss-ratio procedure.[12] With this method the production rates P at equal time intervals are tabulated in one column, the production drop per unit of time, ΔP in a second column and the ratio of the two (a = loss ratio) in a third. If this loss ratio is constant or nearly constant, the curve can be assumed to be of the exponential type. The mathematical basis for this will be discussed hereafter.

It will often be found, if time intervals of one month are used and when the decline percentage is small, that the general trend is disturbed considerably by irregularities in the monthly figures, and in such cases it is better to take the production rates further apart. As an example, Table 1 shows the data from a lease in the Cutbank field, Montana, where the monthly production rates are taken at six-month intervals. Since the loss ratio is defined as the production rate per unit of time divided by the first derivative of the rate-time curve, it is necessary in this case to introduce a factor 6 in the last column to correct the drop in production rate during the six months interval back to a monthly basis. The loss ratios in the fifth column of the table appear to be approximately constant. The average value over the period from July 1940 to January 1944 is 86.8 and this value was used to extrapolate the production rate to January 1947 in the lower half of the tabulation. The procedure followed in this extrapolation is self-explanatory; the same method that was used

TABLE 1.—*Loss Ratio on a Lease in the Cutbank Field, Montana*
(TYPICAL CASE OF EXPONENTIAL DECLINE)

Month	Year	Monthly Production Rate, P	Loss in Production Rate during 6 Months Interval, ΔP	Loss Ratio (on Monthly Basis), $a = 6\dfrac{P}{\Delta P}$
July......	1940	460		
January..	1941	431	−29	−89.2
July......	1941	403	−28	−86.4
January..	1942	377	−26	−87.0
July:.....	1942	352	−25	−84.5
January..	1943	330	−22	−90.0
July......	1943	309	−21	−88.3
January..	1944	288	−21	−82.3
July......	1944	269.4	−18.6	−86.8
January..	1945	252.0	−17.4	−86.8
July......	1945	235.7	−16.3	−86.8
January..	1946	220.4	−15.3	−86.8
July......	1946	206.1	−14.3	−86.8
January..	1947	192.7	−13.4	−86.8

Average loss ratio July 1940 to January 1944, 86.8

Decline percentage $\dfrac{100}{86.8} = 1.15$ per cent.

Extrapolation until January 1947 by means of average loss ratio, 86.8.

to arrive at the loss ratio from the known production rates in the upper half of the tabulation is used in reverse to find the unknown future production rates from the constant loss-ratio values.

Mathematical Analysis[15]

Rate-time Relationship.—The rate-time curve for the case of exponential decline has a constant loss ratio, as shown in the preceding section, which leads to the following differential equation (see list of symbols on page 15):

$$\frac{P}{dP/dt} = -a \qquad [1]$$

in which a is a positive constant. After integration of this equation, and after elimination of the integration-constant by setting $P = P_0$ for $t = 0$, the following rate-time relationship is obtained:

$$P = P_0 e^{-t/a} \qquad [2]$$

This expression obviously is of the exponential type and explains why such a rate-time curve can be represented as a straight line on semilog paper.

Rate-cumulative Relationship.—The expression for the rate-cumulative curve can be found by simple integration of the rate-time relationship, as follows:

$$C = \int P\,dt = \int P_0 e^{-t/a}\,dt \qquad [3]$$

which, after integration, and after elimination of the constant by setting $C = 0$ for $t = 0$, leads to:

$$C = a(P_0 - P) = 100\frac{(P_0 - P)}{D} \qquad [4]$$

This simple linear relationship indicates that the production rate plotted against the cumulative production should be a straight line on regular coordinate paper.[13]

Monthly Decline Percentage.—The monthly decline percentage as per definition can be represented by:

$$D = -100\frac{dP/dt}{P} \text{ per cent} \qquad [5]$$

or, with the use of Eqs. 1 and 4:

$$D = \frac{100}{a} = 100\frac{P_0 - P}{C} \text{ per cent} \qquad [6]$$

In other words, the decline percentage can be found directly from the loss-ratio tabulation ($100/86.8 = 1.15$ per cent in the example shown in Table 1) and also from the slope of the rate-cumulative curve.

Graphical Extrapolation and Practical Shortcuts

As pointed out before, the rate-time curve for exponential decline will show a straight-line relationship on semilog paper and can, therefore, be extrapolated by continuing the straight line.

The rate-cumulative curve shows a very simple linear equation (Eq. 4) and can, therefore, be represented by a straight-line relationship on regular coordinate paper.

In addition to these methods, some practical shortcuts have been developed recently, which were made possible by the fact that rate-time curves for exponential decline are usually plotted on semilog graph paper.

The gradient of the rate-time curve on semilog paper is constant and equal to

$$-\frac{1}{a}.$$

Since the decline percentage is a simple function of a (see Eq. 6), it is possible to make a calculator for standard semilog paper by plotting the constant drop in production rate per year for a given decline on a strip of paper or transparent film. This can be used, then, as a yardstick to read off the decline percentage immediately from the production drop over a one-year interval. By making the width of the calculator equal to one year on the horizontal time scale, the procedure can be simplified even more. Fig. 1 shows how such a calculator can be used for the purpose of determining the monthly decline percentage.

The relationship between cumulative production C and production drop $(P_0 - P)$ in Eq. 4 is a simple multiplication. Since we are already working on paper with a vertical logarithmic scale, it is easy to see that we can apply the slide-rule principle by using the paper on which the curve is plotted as one scale and a graduated strip with a similar logarithmic division as the other scale. By plotting the value of

366

5

$\dfrac{100}{D}$

on this strip for various values of the decline percentage D, it is possible to carry the multiplication out on the same graph paper used for the curves, and read the answer on its vertical log scale. Figs. 1 and 2 show how such a calculator, designed for determination of both decline percentage and future production, is used. The monthly decline percentage was read off from scale BC in Fig. 1 as 4 per cent and the constant, for 4 percent decline on scale AD was used to find the future production in Fig. 2. The economic limit was assumed to be 150 bbl. per month and the production drop from Janu-

ary 1944 until this economic limit will be reached is, therefore, $340 - 150 = 190$ bbl. per month. The constant for 4 per cent decline on scale AD is matched with this production rate of 190 bbl. per month and the future recovery is read off opposite arrow E as 4750 barrels.

HYPERBOLIC DECLINE

Statistical Analysis and Extrapolation

The hyperbolic or "log-log" type of decline, which occurs most frequently, can be recognized by the fact that the loss ratios show an arithmetic series and that, therefore, the first differences of the loss ratios

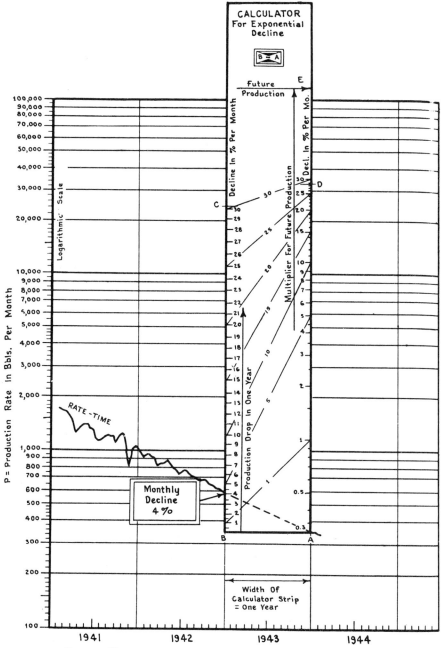

FIG. 1.—USE OF CALCULATOR TO DETERMINE DECLINE PERCENTAGE.

6

are constant or nearly constant.[12,15] As an example, Table 2 shows the loss ratio for production data from a lease producing from the Arbuckle lime in Kansas. This lease had been producing under conditions of capacity production since the completion of drilling and shows a rate-time curve on semi-log paper, curving steadily to the right (Fig. 3). To eliminate irregularities, it was necessary to smooth out the original data (see curve JB on Fig. 3). The production rates listed in Table 2 are identical with the circles on the curve in Fig. 3.

As in the case of exponential decline, the production rates were posted at six-month intervals to eliminate monthly fluctuations and to embrace the general trend of the curve without too much work. Since the loss ratio a is defined as the production rate divided by the first derivative of the rate-time curve, a factor 6 was introduced to find the proper values. The loss ratios thus obtained indicated a fairly uniform arithmetic series and consequently the differences between successive loss-ratio values b are reasonably constant. The average is 0.508.

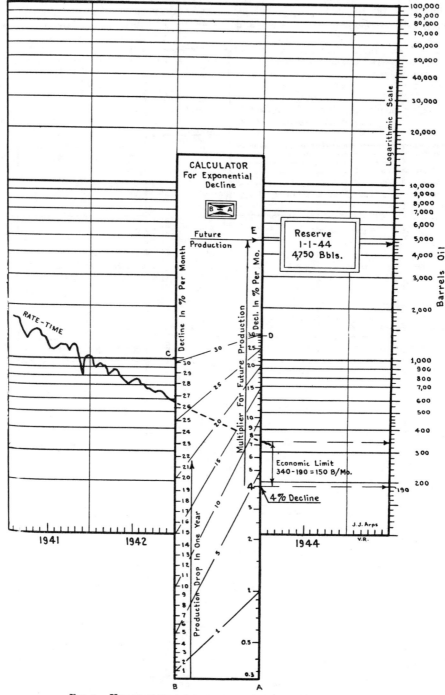

FIG. 2.—USE OF CALCULATOR TO DETERMINE FUTURE PRODUCTION.

368

These differences represent the derivatives of the loss ratios with respect to time, and since six-month intervals are used, a correction factor of 1/6 was introduced to find the proper values of b. The average value for b was used to extrapolate the curve to July 1948 by reversing the process used in the upper part of the tabulation. From these data, it is evident that the lease can be expected to reach its economic limit of 400 bbl. per month during the second half of 1947.

As will be shown later, the mathematical equations of the rate-time and rate-cumulative curves for hyperbolic decline are essentially of the same type and it is therefore also possible to use the loss-ratio method for extrapolation of rate-cumulative data. The only difference from the procedure in Table 2 is that the time column is replaced by cumulative production figures, and that the intervals therefore may not be constant. The loss ratio in that case is the production rate at a given point divided by the ratio of the drop in production rate to the total production during the preceding interval. In a similar way, the first derivative should be determined as the

TABLE 2.—*Loss Ratio for Lease Producing from Arbuckle Lime in Kansas*
(TYPICAL CASE OF HYPERBOLIC DECLINE)

Month	Year	Monthly Production Rate, P (Curve JB, Fig. 3)	Loss In Production Rate during 6 Months Interval, ΔP	Loss Ratio on Monthly Basis, $a = 6\dfrac{P}{\Delta P}$	First Derivative of Loss Ratio, $b = \dfrac{\Delta\left\{\dfrac{6P}{\Delta P}\right\}}{6}$
Jan...	1937	28,200			
July..	1937	15,680	$-12,520$	-7.52	
Jan...	1938	9,700	$-5,980$	-9.72	-0.37
July..	1938	6,635	$-3,065$	-12.97	-0.54
Jan...	1939	4,775	$-1,860$	-15.39	-0.40
July..	1939	3,628	$-1,147$	-18.96	-0.59
Jan...	1940	2,850	-778	-21.96	-0.50
July..	1940	2,300	-550	-25.08	-0.52
Jan...	1941	1,905	-395	-28.95	-0.64
July..	1941	1,610	-295	-32.76	-0.63
Jan...	1942	1,365	-245	-34.43	-0.28
July..	1942	1,177	-188	-36.97	-0.42
Jan...	1943	1,027	-150	-41.15	-0.70
July..	1943	904	-123	-44.20	-0.508
Jan...	1944	802	-102	-47.25	-0.508
July..	1944	717	-85	-50.30	-0.508
Jan...	1945	644	-73	-53.35	-0.508
July..	1945	582	-62	-56.40	-0.508
Jan...	1946	529	-53	-59.45	-0.508
July..	1946	483	-46	-62.50	-0.508
Jan...	1947	442	-41	-65.55	-0.508
July..	1947	406	-36	-68.60	-0.508
Jan...	1948	375	-31	-71.65	-0.508
July..	1948	347	-28	-74.70	-0.508

First derivative of loss ratios approximately constant; average $b = -0.508$.

Extrapolation until July 1948 by means of this average b value of -0.508.

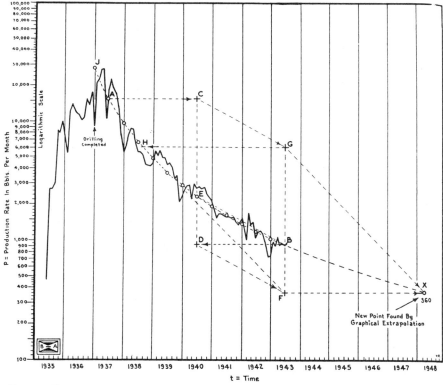

FIG. 3.—GRAPHICAL EXTRAPOLATION OF HYPERBOLIC RATE-TIME CURVE ON SEMILOG PAPER.

Type of curve: $P = P_o\left(1 + \dfrac{b}{a_o}t\right)^{-\frac{1}{b}}$

1. Smooth out the given curve AB.
2. Draw a vertical line CD midway between A and B.
3. Project A and B horizontally on this middle line and find points C and D.
4. Draw CG and DF parallel to EB.
5. Project G back horizontally on the curve and find point H.
6. Draw GX parallel to HF and find the unknown extrapolated point X at the intersection with the horizontal line through F.

369

8

increase in loss ratio over the given interval divided by the total production during the same interval. In hyperbolic decline, the first derivative should be approximately constant. To extrapolate the data and find the ultimate recovery for a given economic limit, the average first derivation can be used to extrapolate the tabulation in a manner similar to that of Table 2.

Mathematical Analysis

1. *Rate-time Relationship.*—When the first differences of the loss ratios are approximately constant, as in Table 2, the following differential equation can be set up:

$$\frac{d\left(\frac{P}{dP/dt}\right)}{dt} = -b \qquad [7]$$

in which b is a positive constant. Integration of Eq. 7 leads to:

$$\frac{P}{dP/dt} = -bt - a_0 \qquad [8]$$

in which a_o is a positive constant, representing the loss ratio for $t = 0$. Eq. 8 can be simplified to:

$$\frac{dP}{P} = -\frac{dt}{a_0 + bt} \qquad [9]$$

This second differential equation can be integrated and the constants eliminated by setting $P = P_o$ for $t = 0$, which results in the rate-time relationship for hyperbolic decline:

$$P = P_0\left(1 + \frac{bt}{a_0}\right)^{-1/b} \qquad [10]$$

This expression, which is obviously of the hyperbolic type, explains why such a curve can be straightened on log-log paper. It also shows that horizontal shifting to the right over a distance

$$\frac{a_0}{b}$$

is necessary for such straightening. The slope of the straight line on log-log paper thus obtained will be

$$-\frac{1}{b}.$$

Rate-cumulative Relationship.—To find the rate-cumulative relationship for this case, the above rate-time curve can be integrated as was done for the exponential decline curve:

$$C = \int P dt = \int P_0\left(1 + \frac{bt}{a_0}\right)^{-1/b} dt \qquad [11]$$

After carrying out the integration for the case where b is not equal to unity, and keeping in mind that the cumulative production $C = 0$ at time $t = 0$, the following relationship is obtained:

$$C = \frac{a_0 P_0}{b - 1}\left\{\left(1 + \frac{bt}{a_0}\right)^{1-1/b} - 1\right\} \qquad [12]$$

or after eliminating t with the rate-time relationship in Eq. 10:

$$C = \frac{a_0 P_0^b}{1 - b}\left(P_0^{1-b} - P^{1-b}\right) \qquad [13]$$

In the special case, where $b = 1$, the integration results in the expression for harmonic decline as can be easily verified:

$$C = a_0 P_0 (\log P_0 - \log P) \qquad [14]$$

The rate-cumulative relationship in Eq. 13 can apparently also be straightened on log-log paper after horizontal shifting on the cumulative scale, while the relationship in Eq. 14 can be represented by a straight line on semilog paper with the production rate plotted on the log scale.

Monthly Decline Percentage.—From Eq. 8, it can be found that the monthly decline for this case is:

$$D = -100\frac{dP/dt}{P} = \frac{100}{a_0 + bt} \text{ per cent} \qquad [15]$$

After elimination of t with Eq. 10, it is found that:

$$D = \frac{100}{a_0 P_0^b} P^b \text{ per cent} \qquad [16]$$

or, in other words, that in the case of hyperbolic decline, the decline percentage is proportional to the power b of the production rate. This is a very interesting result. It means that if a hyperbolic decline curve has a first difference in the loss ratio of say -0.5, the decline percentage is proportional to the square root of the production rate. This means that if such a well has a 10 per cent decline when the production rate was 10,000 bbl. per month, it will slow down to 1 per cent by the time the production rate has dropped to 100 bbl. per month.

Three-point Rule.—The hyperbolic decline curve shows another interesting feature, which can sometimes be used to advantage. It can be expressed as: "For any two points on a hyperbolic rate-time curve, of which the production rates are in a given ratio, the point midway between will have a production rate which is a fixed number of times the rate of either the first or last point, regardless of where the first two points are chosen."

In other words, if on a curve with an exponent $b = 0.5$, the first point has a production rate of $2A$ bbl. and the last point a rate of A bbl., the point midway between will have a value of $1.374A$ bbl., regardless of where the first set of points is selected on the curve and regardless of the time interval. The validity of this statement can be shown as follows:

According to Eq. 10, the production rates at time $t - v$, t and $t + v$ will be:

$$P_{t-v} = P_0 \left\{ 1 + \frac{b}{a_0} (t - v) \right\}^{-1/b} \qquad [17]$$

or

$$P_{t-v}{}^{-b} = P_0{}^{-b} \left\{ 1 + \frac{b}{a_0} (t - v) \right\}$$

$$P_t = P_0 \left(1 + \frac{b}{a_0} t \right)^{-1/b} \qquad [18]$$

or

$$P_t{}^{-b} = P_0{}^{-b} \left(1 + \frac{b}{a_0} t \right)$$

and

$$P_{t+v} = P_0 \left\{ 1 + \frac{b}{a_0} (t + v) \right\}^{-1/b} \qquad [19]$$

or

$$P_{t+v}{}^{-b} = P_0{}^{-b} \left\{ 1 + \frac{b}{a_0} (t + v) \right\}$$

By adding together the right sides of Eqs. 17 and 19, the time interval v is eliminated and an expression is obtained that is twice the value of the right side of Eq. 18. Therefore:

$$2P_t{}^{-b} = P_{t-v}{}^{-b} + P_{t+v}{}^{-b} \qquad [20]$$

If the rate at the first point is n times the rate at the last point, the value of the rate at the middle point (P_t) can be expressed as:

$$P_t = \left(\frac{n^{-b} + 1}{2} \right)^{-\frac{1}{b}} P_{t+v} \qquad [21]$$

This relationship was used advantageously for a simple graphical extrapolation construction for the hyperbolic-type decline curve on semilog paper as illustrated by Fig. 3 and discussed hereafter.

Graphical Extrapolation Methods

Log-log Paper.—As pointed out before, both the rate-time and rate-cumulative curves for hyperbolic decline can be represented and extrapolated as straight lines on log-log paper after some shifting. The rate-cumulative curve for the special case of harmonic decline where $b = 1$, however, can be straightened only on semilog paper.

Log-log paper extrapolation has the disadvantage of giving the least accuracy at the point where the answer is required; it is also somewhat laborious on account of the extra work involved in shifting until the best straight-line relationship is found.

Semilog Paper.—Although log-log paper is used to a large extent for production curves of the hyperbolic type, there are still some companies that continue to plot their production curves on semilog paper,

even though the decline may be of the hyperbolic type. The reason seems to be that this procedure allows a wide range in small space on the vertical log scale and at the same time has a simple linear horizontal time scale. The curvature in the rate-time relationship for this case, however, makes extrapolation difficult and uncertain.

With the help of the "three-point rule" for hyperbolic decline, it is now possible to extrapolate such a curved hyperbolic rate-time curve on semilog paper with a fair degree of accuracy by simple graphical construction. This procedure is shown on Fig. 3. Three points, *A, E* and *B,* are selected at equal time intervals on the smoothed-out curve *AB.* Then, according to the three-point rule the relative value of the middle point *E* is a simple function of the ratio of the first and third points *A* and *B,* regardless of the time interval or the location on the curve. Transfer of the value of these ratios is possible by drawing simple parallel lines, because the vertical scale is logarithmic. In the construction, the third point *B* is used as the middle point of a new set of three equidistant points whose ratios are identical with those originally selected. The third point of this new set of three is found by the construction shown on Fig. 3, which is self-explanatory, and it represents a new extrapolated point of the curve. The method can be used for both rate-time and rate-cumulative curves, provided they are of the hyperbolic type, and provided the construction is carried out on semilog paper.

Special Straight-line Charts.—It may be noted from Eq. 10 and 13 that the behavior of the hyperbolic-type decline curve is governed primarily by the value of the exponent *b,* the first differential of the loss ratio. When the value of b is zero, the decline curve is of the simple exponential or constant percentage type. Some mention is found in the literature of hyperbolic decline with a value of $b = 1$, which was called harmonic decline.

To find the practical range of this exponent b from actual production curves, the data assembled by W. W. Cutler[8] was used. He published the coordinates of a large number of hyperbolic field-decline curves. From his data the exponent b was calculated for each case. The results are shown in Table 3. According to this tabulation, the value of b in the majority of cases appears to be between 0.0 and 0.4. The b value equal to

TABLE 3.—*Value of b According to Cutler's Data*[8]

Exponent b Between	Number of Cases	Exponent b Between	Number of Cases
0.0 and 0.1......	19	0.4 and 0.5....	15
0.1 and 0.2......	41	0.5 and 0.6....	9
0.2 and 0.3......	27	0.6 and 0.7....	4
0.3 and 0.4......	34	Above 0.7.....	None

371

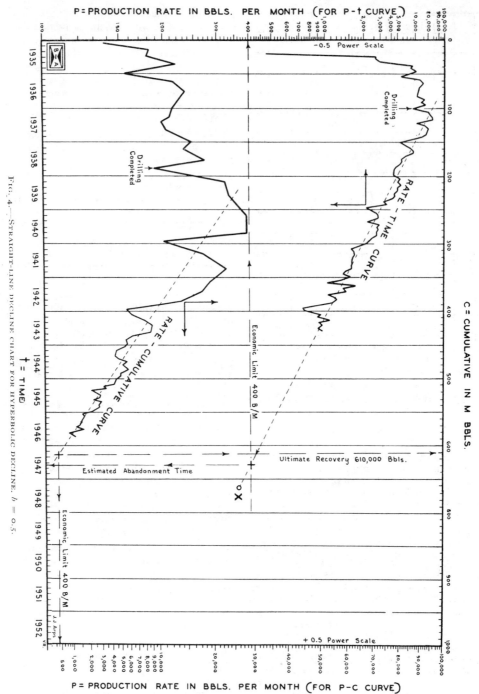

P = PRODUCTION RATE IN BBLS. PER MONTH (FOR P-t CURVE)

C = CUMULATIVE IN M BBLS.

P = PRODUCTION RATE IN BBLS. PER MONTH (FOR P-C CURVE)

Fig. 4.—Straight-line decline chart for hyperbolic decline, b = 0.5.

unity is, according to Cutler's data, very rare. In the writer's experience, however, this type decline does occur occasionally.

The rate-time and rate-cumulative relationship in Eqs. 10 and 13 can be rewritten as:

$$P^{-b} = P_0^{-b}\left(1 + \frac{b}{a_0}t\right) \qquad [22]$$

and

$$P^{1-b} = P_0^{1-b}\left(1 - \frac{1-b}{a_0 P_0}C\right) \qquad [23]$$

In both equations the right-hand side is linear in either time or cumulative while the left-hand side is an exponential function of the production rate P. The exponent in Eq. 22 is $-b$; in Eq. 23 it is $1 - b$. In other words, if a vertical scale could be arranged in such a manner that the ordinate for P would represent a distance P^{-b} for the rate-time curve and P^{1-b} for the rate-cumulative curve, a straight-line relationship should result for both. The horizontal scale could remain linear and no shifting would be necessary. At the same time, the accuracy of reading the extrapolated remaining life or the ultimate recovery on the linear scale would be better than with the log-log method.

Since most decline curves seem to be characterized by b values between 0 and 1, with

11

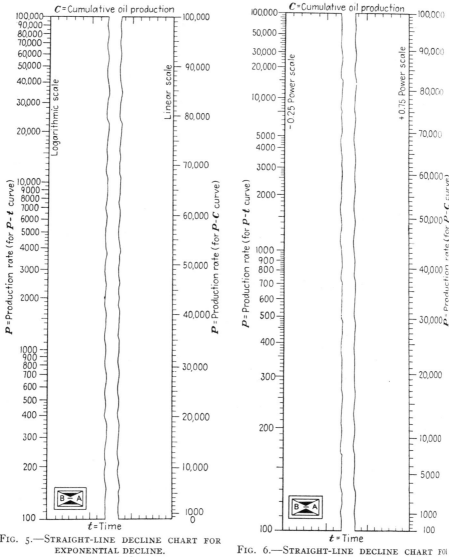

FIG. 5.—STRAIGHT-LINE DECLINE CHART FOR
EXPONENTIAL DECLINE.
a = constant; b = o.
(For curves with constant decline.)

FIG. 6.—STRAIGHT-LINE DECLINE CHART FOR
HYPERBOLIC DECLINE.
b = o.25.
(To be used if decline is proportional to the
¼ power of the production rate.)

the majority between 0 and 0.4, a set of so-called "straight-line decline charts" was prepared for successive values of b. The vertical scales were prepared simply by calculating and plotting a series of values for P^{-b} and P^{1-b}. It was found that a highly accurate determination of b is usually unnecessary for most practical purposes and that for ordinary appraisal work a set of charts for b values of 0, 0.25, 0.5 and 1.0 is sufficient.

The chart for $b = 0.5$ is shown in Fig. 4 and the data from Table 2 are plotted on this chart to show the straight-line extrapolation procedure. The scale on the right is designed to match the b value of the one on the left, so that it will fit the rate-cumulative relationship. The scale on the right should be used in conjunction with the linear cumulative scale on the top of the chart, while the scale on the left should be used in combination with the linear time scale on the bottom. Both curves

can then be plotted and extrapolated as straight lines, simultaneously. Vertical scales for similar charts, designed for b values of 0, 0.25 and 1.0 are shown in Figs. 5, 6 and 7, respectively.

To determine which chart should be used, the three-point rule can be used: two points are selected on the available curve in such a manner that the production rate of the first point is twice the rate at the last point. The production rate at the midway point is then read off and its ratio to the last point determined. If this ratio has a value between 1.414 and 1.396, the chart for $b = 0$ should be used; if it is between 1.396 and 1.383, the chart for $b = 0.25$ will be better; if it is between 1.383 and 1.352, the chart for $b = 0.50$ should be preferred, and if the ratio is between 1.352 and 1.333 the chart for harmonic decline ($b = 1$) will give the best results. If these ratios are too close together, other values can be calculated with the help of Eq. 21.

A simpler method is to plot the rate-time curve on semilog paper ($b = 0$) and if it shows a persistent curvature three representative points should be replotted on the chart for $b = 0.5$. If the three points do not lie on a straight line, but show curvature to the right, the chart for $b = 1$ should be selected; if the curvature is downward, the chart for $b = 0.25$ should give better results.

Another method is to set up a loss-ratio tabulation and actually determine the average value of the first differential b. The chart with the closest b value should then be chosen. This method was followed in Table 2, and since the b value obtained (0.508) was very close to 0.50, the chart for this latter value was used (Fig. 4).

OTHER EMPIRICAL DECLINE CURVES

In addition to the exponential type of decline, which is the simplest empirical relationship and has found widespread application for approximate estimates because of its simplicity, and the hyperbolic type of decline, which is more complicated, but also generally more accurate, there are several empirical equations that can sometimes be used to represent production-decline curves if the simpler types are inadequate. Three of the more important types are discussed in the following.

Loss Ratios Form a Geometric Series (Ratio Decline)

A curve of this type has the characteristic that the decline percentage-time relationship is similar to the rate-time relationship for exponential decline and can be plotted as a straight line on semilog paper. In other words, the decline fraction itself is declining at a constant percentage per month. The differential equation for the rate-time curve is:

$$\frac{P}{dP/dt} = a_0 r^t \quad [24]$$

in which r is the constant ratio of two successive values of the loss ratio a. After integration this leads to:

$$P = P_0 e^{\frac{(1 - r^{-t})}{a_0 \log r}} \quad [25]$$

The simplest way to recognize this type of decline, and to extrapolate it, is by means of the loss-ratio tabulation. The equation for the rate-cumulative curve, which can be found by integration of Eq. 25, is too complicated for practical use. As an example of the statistical treatment of production curves of this type, Table 4 shows a loss-ratio tabulation of the family decline curve from a Wilcox sand pool in Oklahoma. As before,

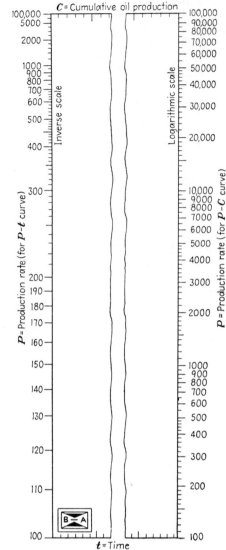

FIG. 7.—STRAIGHT-LINE DECLINE CHART FOR HARMONIC DECLINE.
$b = 1$.
(To be used if decline is proportional to the production rate.)

the per well production rates at equal time intervals are tabulated in column 3, the drop in production rate in column 4 and the loss ratio in column 5. In this case the loss ratios form approximately a geometric series. This is evidenced by the fact that the figures in column 6, which represent the ratios of successive loss-ratio values, are approximately constant. Their average value is 1.127 and this figure was used for the extrapolation of column 6 in the lower half of the table. The extrapolated values for the production rate were then found by reversing the process used in the upper part of the tabulation.

First Derivatives of Loss Ratios Form an Arithmetic Series

The first derivatives of the loss ratios form an arithmetic series and the second derivatives are constant. S. T. Pirson[15]

374

13

worked out the three possible mathematical solutions for the rate-time equations, and complete details may be found in his paper. It has been found that these equations are generally too complicated for practical use. The simplest way to extrapolate a curve, showing these characteristics, is by means of the loss-ratio method.

Straight-line Relationship between Decline Percentage and Time on Log-log Paper

This type of decline was discussed in a general way on pages 3 to 4, and for more details we refer to the original article by P. J. Jones.[21]

Aside from the fact that there is a straight-line relationship between decline and time on log-log paper, this type of curve can also be extrapolated as a straight line by plotting the log-log of the production rate against the log of the time. Statistical extrapolation by means of the loss-ratio method is possible but too complicated for practical use.

TENTATIVE CLASSIFICATION OF DECLINE CURVES, BASED ON LOSS RATIO

To summarize the discussions in this paper, a tabulation was prepared (Table 5) showing the mathematical interrelationship between the commoner types of decline curves. At the same time, it is shown how these decline curves can be classified according to the loss-ratio method.

If the loss ratio is constant, the decline curve must be of the exponential type. If the loss-ratio figures are not constant, but form an arithmetic series, the decline will be of the hyperbolic or harmonic type, depending on the value of the increment b. If the loss-ratio figures indicate a geometric series, the curve must be of the ratio-decline type.

On this table is shown also a summary of the graphical and other methods that can be used to extrapolate the different types of curves.

SUMMARY

Most production-decline curves can be classified into a few simple types, which can be recognized by graphical, statistical or mathematical means. There is a distinct interrelationship between these types and a detailed study revealed some new characteristics and possibilities for simplification of the extrapolation procedure. Among these, the most important are:

1. A decline calculator to be used for exponential decline curves plotted on semilog paper. This calculator, which is based on the slide-rule principle, makes it possible to read off the monthly decline percentage and the future reserve directly from the original curve.

2. The mathematical relationship between rate-time, rate-cumulative and rate-decline

TABLE 5.—*Tentative Classification of Decline Curves, Based on Loss Ratios*
Time, t; production rate, P; drop in production rate, P

	Loss Ratio, $a = \dfrac{P}{\Delta P}$	Differential of Loss Ratio, $b = \Delta a = \Delta\left(\dfrac{P}{\Delta P}\right)$		Ratio of Successive Loss Ratios, $r = \dfrac{a_n}{a_{n-1}}$
Loss ratios or a values	Constant	Arithmetic Series		Geometric Series
Type of decline	$a = \dfrac{P}{\Delta P}$ = Constant Then: Exponential or Constant Percentage Decline	$b = \Delta a$ = Constant ($0 < b < 1$) Then: Hyperbolic Decline	$b = \Delta a = 1$ Then: Harmonic Decline	$r = \dfrac{a_n}{a_{n-1}}$ = Constant Then: Ratio Decline
Rate-time relationship (P, t)	$P = P_0 e^{-t/a}$ (Straight line on semilog paper)	$P = P_0\left(1 + \dfrac{b}{a_0}t\right)^{-1/b}$ (Straight line on special decline charts. Straight line on log-log paper after shifting)	$P = \dfrac{P_0}{\left(1 + \dfrac{t}{a_0}\right)}$	$P = P_0 e^{a_0 \log r}^{(1-r^{-t})}$
Cumulative-rate relationship (C, P)	$C = a(P_0 - P)$ (Straight line on coordinate paper	$C = \dfrac{a_0 P_0^b}{1-b}(P_0^{1-b} - P^{1-b})$ (Straight line on log-log paper after shifting)	$C = a_0 P_0 (\log P_0 - \log P)$ (Straight line on special decline charts) (Straight line on semilog paper)	Too complicated
Decline percentage (D)	$D = \dfrac{100}{a}$ Decline constant	$D = \dfrac{100}{a_0 P_0^b} P^b$ (Straight line on log-log paper) Decline proportional to power b of production rate	$D = \dfrac{100}{a_0 P_0} P$ (Straight line on coordinate paper) Decline proportional to production rate	$D = \dfrac{100}{a} r^{-t}$ (Straight line on semilog paper) Decline diminishing with time at constant percentage (geometric series)
Graphical shortcuts on semilog paper	Special decline calculator	Graphical extrapolation construction based on "three-point rule"		

percentage for hyperbolic and harmonic decline.

3. A graphical construction method for extrapolation of hyperbolic-type decline curves, plotted on semilog paper. This method is based on the three-point rule, which is a mathematical connection between the production rates of three equi distant points on the curve.

4. The introduction of straight-line decline charts for hyperbolic decline. These charts have vertical scales arranged in such a manner as to make straight lines out of both rate-time and rate-cumulative curves, belonging to the hyperbolic type. Use of these charts facilitates extrapolation of this type of production curves considerably.

ACKNOWLEDGEMENT

The author wishes to express his appreciation to Mr. C. D. Miller, Vice-President in Charge of Production for The British-American Oil Producing Co., for permission to publish this paper.

REFERENCES

1. R. Arnold and R. Anderson: Preliminary Report on Coalinga Oil District. U.S. Geol. Survey *Bull.* 357 (1908) 79.

2. M. E. Lombardi: The Valuation of Oil Lands and Properties. *Western Engineering* 153, 6, (Oct. 1915).

3. R. W. Pack: The Estimation of Oil Reserves. *Trans.* A.I.M.E. (1917) 57, 968.

4. M. L. Requa: Methods of Valuing Oil Sands. *Trans.* A.I.M.E. (1918) 59, 526.

5. J. O. Lewis and C. H. Beal: Some New Methods for Estimating the Future Production of Oil Wells. *Trans.* A.I.M.E. (1918) 59, 492.

6. C. H. Beal: Decline and Ultimate Recovery of Oil Lands. U. S. Bur. Mines *Bull.* 173 (1919).

7. R. Arnold and Others: Manual for the Oil and Gas Industry. Bur. Internal Revenue, 1919.

8. W. W. Cutler, Jr.: Estimation of Underground Oil Reserves by Well Production Curves. U. S. Bur. Mines *Bull.* 275 (1924).

9. C. S. Larkey: Mathematical Determination of Production Decline Curves. *Trans.* A.I.M.E. (1925) 71, 1322.

10. H. M. Roeser: Determining the Constants of Oil-production Decline Curves. *Trans.* A.I.M.E. (1925) 71, 1315.

11. C. E. Van Orstrand: On the Mathematical Representation of Certain Production Curves. *Jnl.* Washington Acad. Sciences (Jan. 1925) 15, 19.

12. R. H. Johnson and A. L. Bollens: The Loss-ratio Method of Extrapolating Oil Well Decline Curves. *Trans.* A.I.M.E. (1927) 27, 771.

13. H. N. Marsh: Method of Appraising Results of Production Control of Oil Wells. Amer. Petr. Inst. and Prod. Eng. *Bull.* 202 (1928).

14. R. E. Allen: Control of California Oil Curtailment. *Trans.* A.I.M.E. (1931) 92, 47.

15. S. J. Pirson: Production Decline Curve of Oil Well May be Extrapolated by Loss-Ratio. *Oil and Gas Jnl.* (Nov. 14, 1935).

16. H. E. Gross: Decline Curve Analysis. *Oil and Gas Jnl.* (Sept. 15, 1938).

17. W. W. Cutler and H. R. Johnson: Estimating Recoverable Oil of Curtailed Wells. *Oil Weekly* (May 27, 1940).

18. H. C. Miller: Oil-Reservoir Behavior Based upon Pressure-Production Data. U.S. Bur. Mines *R.I.* 3634 (1942).

19. C. H. Rankin: Estimating Ultimate Recovery. *Petr. Eng.* (Nov. 1943).

20. F. K. Beach: Well Histories Aid in Estimating Oil Reserves. *Petr. Eng.* (Sept. 1943) 69.

21. P. J. Jones: Estimating Oil Reserves from Production-decline Rates. *Oil and Gas Jnl.* (Aug. 20, 1942) 43.

SYMBOLS

P, production rate, bbl. per month.

P_o, initial production rate, bbl. per month.

t, time elapsed since first production, months.

v, constant time interval.

C, cumulative production from completion until time t, bbl.

a, positive number, representing loss ratio on a monthly basis.

a_o, positive number, representing loss ratio during first month.

b, positive number, representing first derivative of loss ratio.

D, decline, per cent per month.

D_o, initial decline, per cent per month.

A, B, n, m, various constants.

r, ratio of successive loss ratios, log, natural logarithm.

e, base of natural logarithm.

Society of Petroleum Engineers

SPE 14271

A Simple Statistical Method for Calculating the Effective Vertical Permeability of a Reservoir Containing Discontinuous Shales

by S.H. Begg, *BP Research Centre*, and D.M. Chang* and H.H. Haldorsen,* *Sohio Petroleum Co.*

*SPE Members

ABSTRACT

This paper describes a quick, simple, easy to use method of calculating the effective vertical permeability of a reservoir region containing discontinuous shales. The method, which is derived in both two and three dimensions, can be applied to a layered medium in which the sand permeability anisotropies (in three mutually perpendicular directions) and the shale frequencies and dimensions can vary from layer to layer. The method is validated for specific cases by using numerical simulation and an analogue model of a real outcrop. The latter consists of conducting paper with slits cut in it to represent impermeable shales. Although strictly the method is only applicable to impermeable shales we show that it would appear to be generally valid when the real shale permeability is more than about two orders of magnitude lower than the effective permeability derived by assuming impermeable shales.

1. INTRODUCTION

The major gross inhomogeneity in many reservoirs may frequently be permeability barriers (on a scale less than the typical well spacing) embedded in the porous medium. Such permeability barriers are often referred to as stochastic shales[1,2]. We use the term "shales" since these are probably the most common type of barriers, however they could arise from other sources such as carbonate cementation. Whilst continuous correlatable shales can be handled with relative ease in

References and illustrations at end of paper

reservoir simulation models these stochastic shales historically have presented a difficult reservoir engineering problem, both in terms of their description and in accounting for their effects on fluid flow. Some important effects which they may have are given below:

1. The presence of shales above/below the perforations can provide protection from gas/water coning[3].

2. Shales may decrease viscous, capillary or gravitational cross-flow and increase dispersive cross-flow[4,5].

3. In reservoir simulators several physical parameters can be adjusted to yield the same response to observed pressure and production ratios. A good shale description could increase confidence in the uniqueness of the history match[6].

4. Any EOR process relying on the injection of a gas into the reservoir will be very sensitive to the presence of shales. In horizontal drives they may be advantageous by increasing the vertical sweep and retarding gas breakthrough at the producer. In vertical drives they may be a disadvantage due to decreased vertical sweep[7,8,9,10,11].

5. Shales may affect gravity drainage[10,12,13] by:

   - uneven advancement of gas-oil contact
   - hold-up of oil on top of shales
   - inhibiting upward segregation of liberated solution gas.

6. Shales may aid waterflooding either by reducing the slumping of water injected

near the top of reservoir and thus increasing the sweep efficiency or by retarding breakthrough when injecting into the aquifer[14,15].

In this paper we address the problem of calculating the reduction in the vertical permeability of a reservoir region due to the presence of stochastic shales. Haldorsen and Lake[1] provided a major step forward in this problem by describing a method of calculating an equivalent homogeneous effective vertical permeability for a reservoir simulation grid block containing both sand and shale. However, it was described only for the two-dimensional case and had the restrictions that the grid block length should be at least ten times its height and comparable to the mean shale length. It was also somewhat cumbersome to use and assumed that the sand permeability was homogeneous. Begg and King[2] then described four methods of calculating K_{VE}, in both two and three dimensions, which enabled most of these restrictions to be dropped. However, they still found it necessary to assume an isotropic, homogeneous sand permeability and impermeable shales. The present paper has four purposes:

- To further validate the statistical streamline method of Begg and King by comparing the results which it gives with those of an analogue model consisting of electro-conducting paper with slits cut in it to represent impermeable shales.

- To extend the statistical streamline method to the case of a layered medium with different anisotropic sand permeabilities and shale statistics in each layer.

- To investigate the importance of finite shale permeabilities and determine when the assumption of zero shale permeability is valid.

- To show that the statistical streamline method is a practical, quick, easy to use tool which the reservoir engineer can apply to real data to give a reasonably accurate estimate of K_{VE}.

2. METHOD

In this section we derive expressions in both two and three dimensions for the effective vertical permeability of a reservoir region containing shales (or other permeability barriers) which are small compared with the well spacing. The method can be used for problems with some or all of the following features:

- horizontally layered sand permeabilities

- anisotropic sand permeabilities, including:

  - different k_V/k_H ratios in different layers
  - different k_X/k_Y within a layer

- different shale dimensions and frequencies in different layers

- different shale statistics in the two horizontal dimensions

- use of either mean values or probability distributions to describe the shale dimensions.

The method has the following assumptions and limitations:

- Incompressibility

- Steady-state flow of a single phase

- The macroscopic flow is uni-directional and normal to layering

- The shales are impermeable with respect to the sands

- The shales are randomly distributed in space and their shape can be approximated by a rectangle.

Briefly, the method relies on the fact that the presence of shales increases the tortuosity of the streamlines in proportion to the length and frequency of the shales (see Figure 1). We will show that the effective vertical permeability is related to the streamline lengths and to any permeability variations which occur along them. The method of estimating these is an extension of that described by Begg and King[2] for a single layer with isotropic sand permeabilities. The data which are required, and their sources, are as follows:

- the number of shales which occur per metre in the vertical direction and the total fraction of shale present - from well cores and/or logs.

- sand permeabilities in the vertical and two horizontal directions - from core plugs.

- probability distributions, or mean values, for the shale lengths and widths - from measurements on outcrops of the same depositional facies or a geologist's interpretation of the known (well) data combined with his knowledge of likely depositional mechanisms and post-depositional events.

2.1 Calculation of K_{VE} for layered sands with anisotropic permeabilities

In order to describe the derivation as simply as possible we first take the two-dimensional case. Later we will show that the

378

three-dimensional derivation requires only minor changes. Consider a streamtube, such as shown in Figure 2, in which the permeability of the sand can vary along its length. The flux through this streamtube is given by Darcy's law:

$$Q = \frac{k_e}{\eta} W_e \frac{\Delta P}{S} \qquad (1)$$

where η is the viscosity, S the total length along the centre of the streamtube, W_e is the effective streamtube width and k_e is the effective permeability of the streamtube in the direction of flow. We now take an arbitrary sample of N_S streamtubes equally spaced across the region for which we wish to calculate an effective vertical permeability (see Figure 3a). Imposing the boundary conditions of constant pressures along the top and bottom faces and no flow through the sides gives the total flow out of the region to be the sum of the flows from each streamtube:

$$Q_T = \sum_i^{N_S} \left(\frac{k_{ei}}{\eta} W_{ei} \frac{\Delta P}{S_i} \right) \qquad (2)$$

We now suppose that the region can be modelled by an apparently homogeneous medium with an effective vertical permeability, K_{VE}, which gives the same total flow for the same boundary conditions (see Figure 3b). In this case:

$$Q_T = \frac{K_{VE}}{\eta} L \frac{\Delta P}{H} \qquad (3)$$

Equating the total flows gives the effective vertical permeability to be:

$$K_{VE} = \frac{H}{L} \sum_i^{N_S} \left(k_{ei} \frac{W_{ei}}{S_i} \right) \qquad (4)$$

The term W_{ei} can be eliminated from this equation in the following manner. Since ther is no flow between streamtubes, then given the above boundary conditions and the assumption of incompressibility we can further assume that the areas (in 2D) of the streamtubes are equal. That is, as a streamtube is lengthened by travelling around the shales it thins in order to keep the area which it encloses constant. The area of each streamtube is thus given by the total area of sand divided by the number of streamtubes.

$$\bar{W}_i S_i = \frac{(1 - F_S) H L}{N_S} \qquad (5)$$

where $\bar{W}_i$ is the arithmetic mean width of a streamtube and $1 - F_S$ is the fraction of sand, which is the net to gross ratio if the only non-pay areas are the shales.

Now, in equation (1) W_{ei} was used to denote the effective width of a streamtube. Ideally this should be calculated by taking the harmonic mean of the streamtube width measured at a number of points along its length. This gives the correct behaviour to equation (1) of the flow rapidly going to zero if the streamtube were to pinch out. However, in most cases the harmonic mean can be approximated by the arithmetic mean. The only situation where this could be a bad approximation is the case where a streamtube was severely constricted over only a short section of its length, which could arise from a sparse, non-statistically homogeneous shale distribution where only a small gap occurred between shales. This is somewhat unlikely since in most sparse shale distributions the shales will be well separated. Also, the problem does not occur for densely packed shales since the streamtubes will be constricted over most of their lengths and so there will not be a large difference between the harmonic and arithmetic means. Thus we can approximate W_{ei} by $\bar{W}_i$ and so equations (4) and (5) yield:

$$K_{VE} = \frac{(1 - F_S) H^2}{N_S} \sum_i^{N_S} \frac{k_{ei}}{S_i^2} \qquad (6)$$

The problem is now reduced to calculating the lengths and effective permeabilities of a sample of streamtubes, the k_{ei} terms arising from anisotropy and layering of the sand permeabilities. First we estimate the streamtube lengths.

Consider Figure 4a, b, c. Figure 4a shows the passage of a typical streamline around a distribution of shales. This can be approximated by horizontal and vertical straight line sections as shown in Figure 4b. From Figure 4c it can be seen that the length of the line is equivalent to the thickness of the box plus the sum of the perturbations due to the shales. In general the streamline is extended horizontally by a random fraction of the length of each shale encountered plus an extra bit at the top of the grid to allow for it returning to the same horizontal position as it started from - as it must do due to boundary conditions we have imposed. The streamline does not always pass round the end of the shale nearest to point where it intersects the shale (and thus be extended by, in general, a random fraction of half the shale length) since it will take the path which is shortest overall. Therefore it will usually keep as near as possible to its horizontal starting position which will entail it passing to the furthest end of a shale as often as to the closest end. Thus, for a problem with N_L layers with thicknesses h_i and different Cumulative Distribution Functions (CDF's) of shale lengths in each layer, the length of a streamline is given by:

$$S_i = \sum_j^{N_L} \left(h_j + \sum_k^{n_j} \left(r_{1ijk}\, l_{ijk} \right) \right) \qquad (7)$$

where i is the streamline subscript, j the layer subscript, k the shale subscript, and:

$$n_j = \quad h_j f_j \text{ for } j < N_L$$

$$h_j f_j + 1 \text{ for } j = N_L$$

$f_j = \quad$ number of shales per metre in layer j

$r_1 = \quad$ random number between 0 and 1

$l_j = \quad$ length of a shale chosen at random from the jth CDF

The inner summation in equation (7) is just the sum of the perturbations due to the presence of the shales and the outer summation the streamline lengths in each layer. We can now estimate the k_{ei} term in equation (6).

It was said earlier that k_{ei} was the harmonic mean of the permeability variations along a streamtube. Therefore, from Figure 4b, it can be seen that this can be approximated by the harmonic mean of the horizontal sand permeabilities for the sections which are parallel to the shales plus the vertical sand permeabilities for the sections normal to the shale alignment. Since we wish to model the case of a layered medium with different sand permeabilities and anisotropies in each layer the effective permeability of a streamtube is estimated by:

$$\frac{S_i}{k_{ei}} = \sum_j^{N_L} \left(\frac{S_{Vj}}{k_{Vj}} + \frac{S_{Hj}}{k_{Hj}} \right) \qquad (8)$$

where S_{Vj} and S_{Hj} are respectively the sums of lengths of the vertical and horizontal sections in the jth layer and k_{Vj} and k_{Hj} are the vertical and horizontal sand permeabilities. The S_{Vj} term is obviously just the thickness of layer j and the S_{Hj} term is given by the inner summation in equation (7). Finally, calling the S_i/k_{ei} term an effective length, S_{ei}, the effective vertical permeability for the whole reservoir region is:

$$K_{VE} = \frac{(1 - F_S)\, H^2}{N_S} \sum_i^{N_S} \frac{1}{S_i S_{ei}} \qquad (9)$$

where S_{ei} is now given by:

$$S_{ei} = \sum_j^{N_L} \left(\frac{h_j}{k_{Vj}} + \frac{1}{k_{Hj}} \left(\sum_k^{n_j} r_{1ijk}\, l_{ijk} \right) \right) \qquad (10)$$

and S_i is given by equation (7).

The above equations are for the general, two-dimensional case of a layered, anisotropic medium with different shale statistics in each layer if required. To derive the three-dimensional result we follow the same method but use an effective streamtube cross-sectional area instead of an effective width. Using the same arguments as before the arithmetic mean cross-sectional area will generally be close to the harmonic mean and by following through the algebra it is found that equation (9) also applies to the three-dimensional case. However, S_i and S_{ei} are different and are given as follows for the general case where the horizontal sand permeabilities, k_X and k_Y, are anisotropic and different CDF's are used for the shale lengths and widths.

In our original paper[2] we took the 3D equivalent to the $\sum r_1 l$ term in the streamline length estimation to be $\sum \min (r_1 l, r_2 w)$ where w is a shale width taken at random from a CDF (see Figure 5). That is, the streamline will generally travel the shorter of the two random lengths. This would also be the correct term to use here if the horizontal permeabilities were isotropic. However, since they are anisotropic the streamline will be extended by either, $r_1 l$ or $r_2 w$ whichever gives the minimum of:

$$\left(r_1 l\, \frac{r_1 l}{k_X} , \quad r_2 w\, \frac{r_2 w}{k_Y} \right)$$

Hence:

$$S_i = \sum_j^{N_L} \left(h_j + \sum_k^{n_j} d_{ijk} \right) \qquad (11)$$

and

$$S_{ei} = \sum_j^{N_L} \left(\frac{h_j}{k_{Vj}} + \sum_k^{n_j} d_{eijk} \right) \qquad (12)$$

where:

$$\left. \begin{array}{l} d_{ijk} = r_{1ijk}\, l_{ijk} \\[2mm] d_{eijk} = \dfrac{r_{1ijk}\, l_{ijk}}{k_{Xj}} \end{array} \right\} \text{if } \frac{(r_{1ijk}\, l_{ijk})^2}{k_{Xj}} < \frac{(r_{2ijk}\, w_{ijk})^2}{k_{Yj}}$$

$$\left. \begin{array}{l} d_{ijk} = r_{2ijk}\, w_{ijk} \\[2mm] d_{eijk} = \dfrac{r_{2ijk}\, w_{ijk}}{k_{Yj}} \end{array} \right\} \text{otherwise.} \qquad (13)$$

If the sand horizontal permeabilities are isotropic then:

$$d_{ijk} = \min(r_{1ijk} \, l_{ijk}, \; r_{2ijk} \, w_{ijk})$$

$$d_{eijh} = \frac{1}{k_{Hj}} \min(r_{1ijk} \, l_{ijk}, \; r_{2ijk} \, w_{ijk})$$

If the shale statistics are the same in each layer then the j subscript in the l_{ijk} and w_{ijk} terms is dropped.

2.2 Simplification of general formulae

The above formulae are derived for the general case of layered, anisotropic sands but they can be simplified in two ways. The first is by replacing the distributions of shale lengths and widths by mean values and the second is by choosing more simple models. In order to replace the shale dimension distributions we calculate the average horizontal extension to the streamtube length, $\bar{d}_j$ for each shale encountered and multiply this by the number of shales in that layer. In two-dimensions:

$$\bar{d}_j = \langle r_{1ijk} \, l_{ijk} \rangle = \frac{\bar{l}_j}{2} \qquad (15)$$

where $\bar{l}_j$ is the mean shale length in the jth layer. In Appendix I it is shown that the three-dimensional equivalent for isotropic, horizontal permeabilities is:

$$\bar{d}_j = \langle \min(r_{1ijk} \, l_{ijk}, \; r_{2ijk} \, w_{ijk}) \rangle = \frac{\bar{w}_j}{6\bar{l}_j}(3\bar{l}_j - \bar{w}_j) \qquad (16)$$

if $\bar{l}_j > \bar{w}_j$ with interchange of $\bar{l}_j$ and $\bar{w}_j$ for $\bar{w}_j > \bar{l}_j$. The number of shales in a layer is given by $h_j f_j$ plus one in the final layer. However, so long as $h_j f_j$ is large (greater than about 10), the final one can be dropped since the effective permeability converges to a value independent of the total thickness[2]. Further, each streamline will be approximately the same length for these conditions. Thus, using the above expressions, the reduced formulae for progressively less complex models are as given below:

1. Layered, anisotropic sands using mean shale lengths (and widths in 3D):

$$K_{VE} = \frac{(1 - F_S) \, H^2}{\sum\limits_{j}^{N_L}\left(h_j(1 + f_j \, \bar{d}_j)\right) \sum\limits_{j}^{N_L}\left(h_j\left(\frac{1}{k_{Vj}} + \frac{f_j}{k_{Hj}} \, \bar{d}_j\right)\right)} \qquad (17)$$

2. Isotropic, layered sands:

$$K_{VE} = \frac{(1 - F_S)}{\sum\limits_{j}^{N_L}\left(h_j(1 + f_j \, \bar{d}_j)\right) \sum\limits_{j}^{N_L}\left(\frac{h_j}{k_j}(1 + f_j \, \bar{d}_j)\right)} \qquad (18)$$

3. Anisotropic sand, single layer:

$$K_{VE} = \frac{(1 - F_S)}{(1 + f\bar{d})\left(\frac{1}{k_V} + \frac{f}{k_H} \, \bar{d}\right)} \qquad (19)$$

4. Isotropic sand, single layer ($S_{ei} = S_i$):

$$K_{VE} = \frac{k(1 - F_S)}{(1 + f\bar{d})^2} \qquad (20)$$

which is the same result as was previously[2] obtained.

It is possible to simplify only the models whilst retaining the distributions of shale dimensions. If this is done then equations (9), (11) and (12) reduce, for example, to those previously[2] derived for the case of a single layer with isotropic sand permeability.

Most of the above formulae require minimal input (net to gross, sand permeabilities, mean shale lengths and number of shales per metre) and can be easily executed to provide a quick estimate of K_{VE}. The more general formulae using distributions of shale dimensions provide equally quick estimates when coded into a reasonably short computer program.

One question still needs to be answered. What is the appropriate mean length that should be used to characterise a distribution of lengths? To answer this we calculated K_{VE} using the CDFs of triangular distributions of shale lengths (minimum, most likely and maximum) and compared the values with those obtained using the more simple formulae (Equation (20)) with the arithmetic means of the distributions. The results of this exercise are shown in Figure 6 for 11 distributions which had varying skewness and maximum lengths. The points are a reasonably good fit to a line of unit slope and so we conclude that the arithmetic mean is the correct one for triangular distributions.

Another reason for wanting to use a single length is so that sensitivities can be run about the value obtained by using a CDF. The technique here is to use a trial and error method to find a characteristic length which gives the same K_{VE} as using the CDF does. This length may well not be the arithmetic mean length for all

381

distributions. Having identified the characteristic length sensitivities can be run around it.

2.3 Method of Calculating K_{VE} for an explicit shale distribution

If the shales are not randomly arranged in space, or a particular distribution is known from an outcrop, the previously described method can be easily altered to treat an explicit distribution of shales. These are defined by the coordinates of their end points which may be obtained from the map of an outcrop or be produced by a shale generation scheme[1] (which need not be random). The following procedure can be carried out on a drawing or by a computer program.

First choose the number of streamlines to be calculated and hence work out the spacing between them which will give a uniform distribution of starting points along bottom of the reservoir region. Calculate the length of each streamline by tracing it upwards until it meets the first shale and then move to the end of the shale closest to the horizontal starting position. Repeat this process, each time moving to the end of the shale which gives the shortest effective path to the initial horizontal position from the current horizontal position, until the top of the reservoir is reached. As well as recording the total distance moved by each streamline it is necessary to record the sum of the distances moved in different permeabilities so that both its approximated real length and its effective length can be calculated. The effective vertical permeability is then calculated from equation (9).

The differences between this and the statistical method are:

i) the number of shales encountered by a streamline is not specified in advance of measuring its length,

ii) each shale has a specific length rather than one chosen at random,

iii) the streamline is extended by a specific fraction of the length of each shale encountered rather than by a random fraction.

With these exceptions the equations used are the same as those described in Section 2.1.

3. VALIDATION

In this section we validate the method by both numerical simulation and by applying it to the real case of a sandstone outcrop whose permeability anisotropy has been calculated from an analogue model. The numerical simulation is done in two-dimensions, one of the challenges for the future being to obtain the correct K_{VE} for some large three-dimensional problems with which the results obtained from the statistical method can be compared.

3.1 Analogue study

Our first validation is for the case of a sand with homogeneous, isotropic permeability. The data which we use come from a paper by Dupuy and LeFebvre[16] who analysed a precise mapping study[17,18] of the Assakao cliff in the Tassili region of the Central Sahara. This cliff is an outcrop of fluviatile sandstones with frequent fine silts which are virtually impermeable with respect to the sandstones. Figure 7 is a scaled map of a 145 x 100 metre section of the cliff showing the positions of the silts. Dupuy and LeFebvre determined K_{VE}/k values for this section by several independent means. In order to simulate single-phase vertical flow they reproduced Figure 7 on conducting paper by simply cutting a slit along the shales to represent the impermeable barriers. Then they measured the resistances to electrical flow in the vertical and horizontal directions by imposing equipotentials (conducting paint) on both the horizontal and vertical sides of the cross section (see Figure 8). From the measured resistances they derived the effective resistivities in the two directions of flow and thus found the coefficient of anisotropy to be 0.203. This model is exactly analogous to that outlined in Section 2.1 for a single layer with isotropic sand permeability.

To calculate K_{VE}/k by the statistical streamline method we first digitised Figure 7 in order to obtain the necessary input data. The barrier lengths were measured to give the PDF shown in Figure 9 and the number of shales per metre was calculated to be 0.33. A mean thickness of half a foot was chosen thus giving the fraction of shale to be 0.05. Equations (7), (9) and (10) were then used (with $N_L = 1$, $N_S = 10,000$ and $k_X = k_Y = k_V = k$) to calculate K_{VE}/k. The value obtained was 0.205 which is in very close agreement with that obtained by the conducting paper analogue.

We also calculated K_{VE}/k by the simple formula (equation 20) and the explicit streamline method described in this paper and the analytical method described by Begg and King[2]. The results are shown below:

Method	K_{VE}/k
Analogue	.203
Statistical streamline	
Using PDF of lengths	.205
Using mean length (6.83m)	.209
Explicit streamline	.216
Analytical	.202

It can be seen that all of the methods give similar results and that these are very close to that calculated from the analogue experiment.

We believe that this study, on a real outcrop, shows that the statistical streamline method is a quick, simple and practical tool for calculating the effective vertical permeability of a reservoir region containing stochastic shales. Further, it validates the method for at least the two-dimensional case with isotropic sand permeability and shows that a fairly accurate answer can be calculated from a knowledge of the shale density, fraction and mean length.

3.2 Simulation study

This method of validation uses the simulation technique described by Begg and King[2] in which the shales are explicitly defined on a fine-scale numerical grid which has the same boundary conditions as described in Section 2.1. The method gives an extremely accurate value of K_{VE} but obviously requires large amounts of CPU time if a sufficient number of grid blocks are used to give an acceptable resolution of the shale distribution.

To test the formulae that give K_{VE} for a layered, anisotropic medium we created a 2D hypothetical, three-layer model on a 200 x 100 grid in which each grid block represented a 5 x 1 metre^2 area. The shales were randomly generated using a triangular probability density function. The full parameters of the model are as follows.

Shale statistics:

Minimum length (m)	= 0.0
Most likely length (m)	= 25.0
Maximum length (m)	= 100.0
Thickness (m)	= 1.0
Number per metre	= 0.1

Sand permeabilities:

	Layer 1	Layer 2	Layer 3
Thickness (m)	30	40	30
k_H (md)	50	100	10
k_V (md)	20	50	2

The resulting shale distribution is shown in Figure 10.

The simulation method[2] yielded a K_{VE} of 1.816md for the above problem and required 5 hours CPU on a Vax 11/780. Equations (7), (9) and (10) required only a few seconds CPU time and gave a value of 1.243md. Again we also used the simple method (equation 17) and the explicit method. All of the results are given below:

Method	K_{VE}/k
Simulation	1.816
Statistical Streamline	
Using PDF of shale lengths	1.243
Using mean length (38.6m)	1.327
Explicit Streamline	1.700

Considering the assumptions made in Section 2.1 (the approximation of the streamlines by perpendicular sections with different permeabilities) we feel that these are still in good agreement. We would expect a better agreement for longer shales since the above approximation would be better.

Unfortunately we have not yet run a 3D simulation with a suitably large number of grid blocks. However Begg and King[2] in their original paper showed a good agreement between the explicit streamline method (see Section 2.3) and the simulation method for small 3D problems which had a single layer with isotropic sand permeabilities.

4. SENSITIVITY STUDIES

The major restriction of our method is that strictly it only applies to shales which are impermeable. In this section we investigate the effect of finite shale permeabilities in order to assess the likely values at which the shales can no longer be considered to be impermeable with respect to the sand. We also investigate the sensitivity of K_{VE} to the shale statistics and to the dimensionality of the model which is used.

4.1 Effect of finite shale permeability

To investigate the effect of finite shale permeability on K_{VE} we used the simulation method to calculate K_{VE} for models in which the shale permeability varied from zero to that of the sand. Two extreme examples of shale distributions were used. The first was a single shale which, with zero permeability, reduced K_{VE} to about two-thirds of the sand permeability. The second was a dense distribution of long shales which reduced K_{VE} to about 10^{-5} of the sand permeability. Figure 11 shows the percentage difference between the finite and zero permeability results for each example as a function of the ratio of the effective permeability to the shale permeability. This figure shows that the difference in K_{VE} caused by assuming a zero shale permeability is only a few percent (2% - 4%) so long as K_{VE} is greater than about 100 times the real shale permeability. If it is about a 1000 times greater the difference is almost zero. This is true for both the single shale and dense shale examples. Therefore, since most real shale distributions which we are interested in will lie somewhere between these two examples, it would seem plausible that a K_{VE} obtained by a method which assumes zero shale permeability will be applicable so long as the value of K_{VE} is greater than the real shale permeability by two or three orders of magnitude. Effectively, what we are saying is that so long as the contrast between the sand and shale permeabilities is sufficiently great (it depends on the particular shale distribution) to prevent a significant portion of the flow going through the shale, then the zero shale permeability assumption is valid. A much greater sand:shale permeability ratio is required for a distribution of frequent, long shales than for sparse, short ones. More detailed investigations of the relative proportions of transmitted and deflected flows are underway.

4.2 Sensitivity of K_{VE} to shale statistics

In this section we use the simplified formulae (equations (16) and (20)) to illustrate the effect of varying the shale dimensions and frequency on the 2D and 3D effective vertical permeabilities - see Table 1.

Cases 1 to 7 show the effect of increasing the size of square shales. It can be seen that there is a difference between the 2D and 3D results and that this difference increases as the size of the shales increases. Also, the effective permeability rapidly decreases as the shales increase. Note that the 2D results are for the plane in which the mean length lies.

Case 8 is for "rod-like" shales whose lengths are 100 times their widths. The 2D model gives a large reduction in K_{VE}/K whereas in 3D, as would be expected, they have little effect. Case 9 is for "ruler-like" shales whose lengths are about 1/10th of their width. Here the shales really can be considered to be two-dimensional objects with respect to the flow, this being borne out by closeness of the 2D and 3D results. However, the previous results show that a 2D model is generally inappropriate. Cases 8 and 9 indicate that the method is giving the sort of behaviour one would intuitively expect for such extreme examples.

The effect of keeping the shale area constant but varying the dimensions is shown by cases 9,10 and 11. These show that square shales reduce K_{VE} more than rectangular shales of the same area. Cases 12 and 13 just show that the effect of increasing the shale frequency by an arbitrary factor has the same effect as increasing the shale dimensions by the same factor.

Although not demonstrated here, Begg and King[2] also showed that if the total number of shales encountered in the vertical direction is small (less than about 5) then K_{VE} is a function of that number also - it increases as the number of shales decreases.

5. CONCLUSIONS

The main results of this paper are as follows:

(i) The statistical streamline method of Begg and King[2] for calculating the effective vertical permeability of reservoirs containing stochastic shales has been generalised to the case of a layered medium with different shale statistics and sand permeability anisotropies in each layer. The statistics and anisotropies can also differ in two perpendicular directions within each layer.

(ii) The method has been validated in 2D by both numerical simulation and an analogue model consisting of slits cut into conducting paper to represent shales in a homogeneous sand. The latter method was a scaled model of a real outcrop.

(iii) Strictly, the method only applies to impermeable shales. However, if the K_{VE} obtained by assuming impermeable shales is about two orders of magnitude greater than the real shale permeability then the assumption seems to give only a couple of percent error in K_{VE}. At three orders of magnitude the difference is negligible.

(iv) The method is sensitive to the shale statistics used and as the effective vertical permeability becomes increasingly small the difference between the 2D and 3D results becomes greater, the 3D value always being larger than the 2D.

We believe that these results demonstrate that we have developed a quick, accurate and practical method for calculating the effect of shales on vertical permeability for a wide range of models. The method shows that the effective vertical permeability is very sensitive to the shale statistics. Thus the dependence of shale continuity on depositional environment and the collection of adequate data by the geologist are of great importance.

Challenges for the future are to extend the method to the case of two-phase flow and to develop methods of calculating the effect of shales which cannot be considered to be impermeable with respect to the sand. The method could thus be generalised to other geological features which give rise to a permeability contrast and which can be described statistically.

6. NOMENCLATURE

CDF	=	Cumulative distribution function
$\bar{d}$	=	mean horizontal extension of streamtube
d	=	horizontal extension of streamtube
de	=	effective horizontal extension of streamtube
f	=	number of shales per metre
F_S	=	fraction of shale
h	=	layer thickness
k	=	isotropic sand permeability
k_H	=	horizontal sand permeability

k_V	=	vertical sand permeability
k_X, k_Y	=	anisotropic horizontal sand permeabilities
k_e	=	effective permeability of a streamtube
K_{VE}	=	overall effective vertical permeability
l	=	shale length taken from layer CDF
$\bar{l}$	=	mean shale length
n	=	number of shales
N_L	=	number of layers
N_S	=	number of streamlines
Q	=	flux
r_1, r_2	=	random numbers between 0 and 1
S	=	approximate, real streamline length
S_e	=	effective streamline length
w	=	shale width taken from layer CDF
$\bar{w}$	=	mean shale width
$\bar{W}$	=	arithmetic mean streamtube width
W_e	=	effective streamtube width

Subscripts:

i	=	streamline index
j	=	layer index
k	=	shale index

7. ACKNOWLEDGEMENTS

We would like to thank the managements of the British Petroleum Company plc and the Sohio Petroleum Company for permission to publish this paper. Special thanks are due to Dr. Peter King of BP for the derivation in the Appendix and for critical comment on the manuscript.

8. REFERENCES

1) Haldorsen, H.H. and Lake, L.W.: "A New Approach to Shale Management in Field-Scale Models", SPEJ August 1984, pp. 447-452.

2) Begg, S.H. and King, P.R. "Modelling the Effects of Shales on Reservoir Performance: Calculation of Effective Vertical Permeability", SPE Paper No. 13529 Presented at the 8th SPE Symposium on Reservoir Simulation held in Dallas, TX, February 10-13, 1985.

3) Addington, D.V.: "An Approach to Gas Coning Correlations for a Large Grid Cell Reservoir Simulation", J. Pet. Tech., November, 1981, pp. 2267-2274.

4) Zapata, V.J. and Lake, L.W.: "A Theoretical Analysis of Viscous Crossflow", SPE Paper No. 10111 Presented at the 56th Annual Technical Conference and Exhibition of the SPE in San Antonio, TX, October 5-7, 1981.

5) Yokoyama, Y. and Lake, L.W.: "The Effects of Capillary Pressure on Immiscible Displacements in Stratified Porous Media", SPE Paper No. 10109 Presented at the 56th Annual Technical Conference and Exhibition in San Antonio, Texas, October 5-7, 1981.

6) O'Brien, D.G., Brown, M.E. and Lederer, M.C.: "Use of Reservoir Performance Analysis in the Simulation of Prudhoe Bay", SPE Paper No. 13217, presented at the 59th Annual Technical Conference and Exhibition in Houston, TX, September 16-19, 1984.

7) Reitzel, G.A. and Callow, G.O.: "Pool Description and Performance Analysis Leads to Understanding Golden Spike's Miscible Flood", SPE Paper No. 6140 presented at the 51st Annual Technical Conference and Exhibition held in New Orleans, La., October 3-6, 1976.

8) Katz, M.L. and Stalkup, F.I.: "Oil Recovery by Miscible Displacement", 11th World Petroleum Congress Paper, London 1983 (RTD2).

9) Oglesby, K.D., Blevins, T.R., and Johnson, W.M.: "Status of the Ten-Pattern Steamflood, Kern River Field, California", SPE Paper No. 8833 Presented at the SPE 1980 Symposium on Enhanced Oil Recovery, Tulsa, OK, April 1980.

10) Ypma, J.G.J.: "Analytical and Numerical Modelling of Immiscible Gravity-Stable Gas Injection Into Stratified Reservoir", SPE Paper No. 12158 Presented at the 58th Annual Technical Conference and Exhibition in San Francisco, CA, October 5-8, 1983.

11) Bath, P.G.H., van der Burgh, and Ypma, J.G.J.: "Enhanced Oil Recovery in the North Sea", RTD2(2) Preprint of the 11th World Petroleum Congress, London, 1983.

12) Richardson, J.G., et al.: "The Effect of Small, Discontinuous Shales on Oil Recovery", J. Pet. Tech. (November 1978), pp. 1531-37.

13) Haldorsen, H.H., Rego, C.A., Chang, D.M., Mayson, H.J., and Creveling, D.M., Jr.: "An Evaluation of the Prudhoe Bay Gravity Drainage Mechanism by Complementary Techniques", SPE 13651, 1985 SPE California Regional Meeting.

14) Harpole, K.J.: "Improved Reservoir Characterization - A Key to Future Reservoir Management for The West Seminole San Andreas Unit", SPE Paper No. 8274 presented at the 54th Annual Technical Conference and Exhibition of the SPE in Las Vegas, Nevada, September 23-26, 1979.

15) Leonard, J.: "Big Waterflood to Boost Prudhoe Recovery", Oil & Gas Journal, April 18, 1982.

16) Dupuy, M., and LeFebvre Du Prey, E.: "l'anisotropie d'ecoulement en milieu poreux presentant des intercalations horizontales discontinues". Communication No. 34 Troisieme Colloque De L'Association De Recherche Sur les Techniques De Forage et De production, June 10-14, 1968, PAU, France.

17) Montadert, L.: "The sedimentology and detailed study of heterogeneities of a reservoir". ARTFP Colloquium, Rueil-Maison, June 10-14, 1963, Editions Technip, Paris, pp 2 1-257, 1963.

18) Montadert, L.: "The sandstones of the Cambrian-Ordovician (Tassilis mission). Structures in waterways and distribution of banks of silt". IFP Report (ARTFP), ref. No. 7647, July 1962.

APPENDIX

To calculate the mean horizontal extension to the streamline length for the three-dimensional model with rectangular shales, it is necessary to average:

$$\min (l,w)$$

where l is a random number taken uniformly from $[0,\bar{l}]$ and w from $[0, \bar{w}]$ where $\bar{l}$ and $\bar{w}$ are the length and width of the shales, all of which are assumed to be the same size.

Now suppose $\bar{w} < \bar{l}$ and perform the average over w first:

$$\langle\min\rangle = \frac{1}{\bar{w}} \int_0^l w\, dw + \frac{1}{\bar{w}} \int_1^{\bar{w}} 1\, dw \qquad 1 < \bar{w}$$

$$= \frac{1}{\bar{w}} \int_0^{\bar{w}} w\, dw \qquad\qquad 1 > \bar{w} \qquad (A1)$$

$$\langle\min\rangle = \frac{1}{2\bar{w}} \left[2l\bar{w} - 1^2 \right] \qquad 1 < \bar{w}$$

$$= \frac{\bar{w}}{2} \qquad\qquad\qquad 1 > \bar{w} \qquad (A2)$$

Now perform the 1 average:

$$\langle\min\rangle = \frac{1}{2\bar{w}\bar{l}} \int_0^{\bar{w}} (2l\bar{w} - 1^2)\, dl + \frac{\bar{w}}{2\bar{l}} \int_{\bar{w}}^{\bar{l}} dl$$

$$= \frac{\bar{w}}{6\bar{l}} \{3\bar{l} - \bar{w}\} \qquad\qquad (A3)$$

with interchange of $\bar{w}$ and $\bar{l}$ if $\bar{w} > \bar{l}$.

TABLE 1

Case	mean L m	mean W m	number /metre	Kve/K 2D	Kve/K 3D
1	10	10	.02	.824	.877
2	20	20	.02	.693	.777
3	50	50	.02	.443	.561
4	100	100	.02	.250	.359
5	200	200	.02	.111	.183
6	500	500	.02	.028	.053
7	1000	1000	.02	.008	.017
8	100	1	.02	.250	.978
9	100	1000	.02	.250	.258
10	200	500	.02	—	.134
11	316	316	.02	—	.103
12	100	100	.04	.111	.183
13	100	100	.10	.028	.053

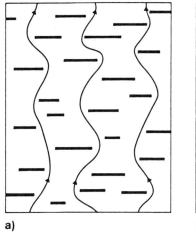

a) b)

Fig. 1—Effect of shale length on tortuosity: (a) short shales—low tortuosity and (b) long shales—high tortuosity.

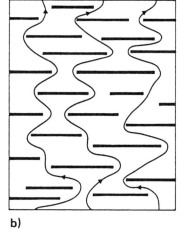

Fig. 2—Single stream-tube perturbed by shale.

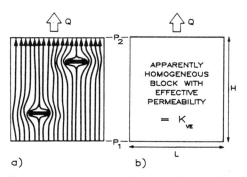

a) b)

Fig. 3—(a) Sample of N_s streamtubes in block of sand with shales and (b) equivalent block with homogeneous K_{VE}.

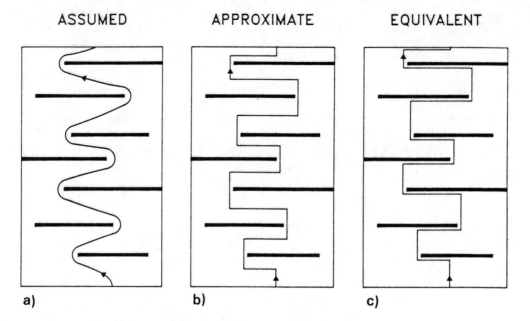

ASSUMED   APPROXIMATE   EQUIVALENT

a)   b)   c)

Fig. 4—Scheme for approximating streamline lengths.

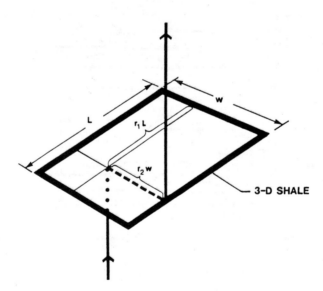

Fig. 5—Streamline perturbation by 3-D shale.

STATISTICAL STREAMLINE METHOD

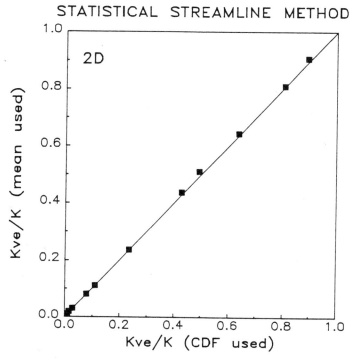

Fig. 6—K$_{VE}$ obtained by using mean shale length compared with K$_{VE}$ from a distribution of shale lengths.

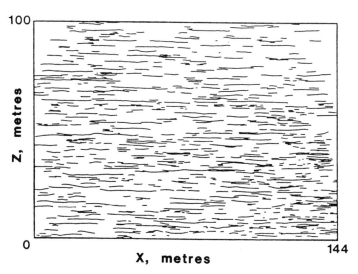

Fig. 7—Map of fine silts in the Assakao sandstone cliff.

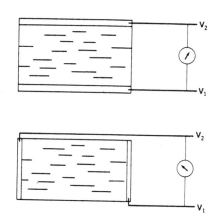

- ● VERTICAL RESISTANCE: 9,870 Ω:
- ● HORIZONTAL RESISTANCE: 4,150 Ω:
- ● HEIGHT: 50 cm:
- ● WIDTH: 72 cm:
- ● VERTICAL RESISTIVITY: 9,870 x 72/50 = 14,200 Ω
- ● HORIZONTAL RESISTIVITY: 4,150 x 50/72 = 2,880 Ω
- ▶ ANISOTROPY FACTOR: 0.203.

Fig. 8—Calculation of anisotropy factor from analog model of Assakao cliff.

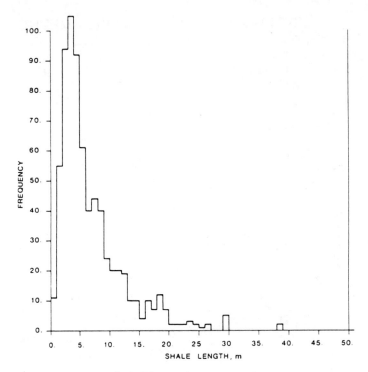

Fig. 9—PDF of silt lengths in Assakao sandstone.

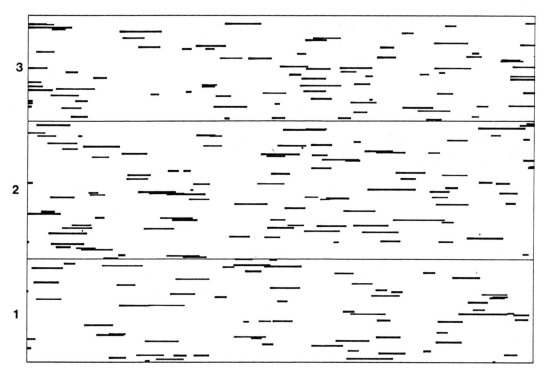

Fig. 10—Hypothetical shale distribution.

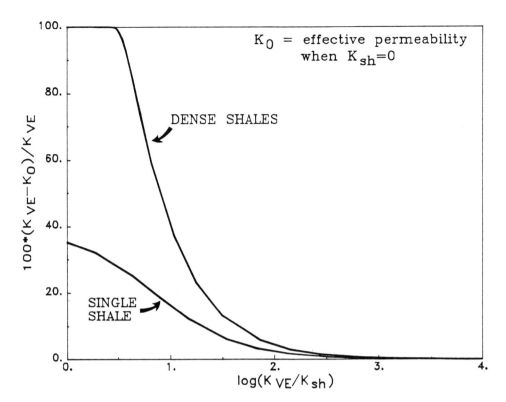

Fig. 11—Effect of finite shale permeability.

Copyright 1965, Society of Petroleum Engineers of AIME.
Reprinted by permission. First published in the *Journal of Petroleum Technology*, V. 17, No. 3 (March 1965), pp. 337-342.

PETROLEUM TRANSACTIONS

The Fry In Situ Combustion Test-Reservoir Characteristics

C. H. HEWLITT
J. T. MORGAN
— MEMBERS AIME —

MARATHON OIL CO.
LITTLETON, COLO.

ABSTRACT

The Fry cocurrent in situ combustion project was carried out in a 3.3-acre portion of a lenticular body of Robinson sandstone of Lower Pennsylvanian age. This particular sand body is about 12,000 ft long and 3,500 ft wide; it varies in thickness from 0 to 55 ft, trends northeast-southwest, and occurs at depths between 880 and 936 ft. The reservoir is made up of three distinct types of sandstone, each having characteristic reservoir properties.

Textural gradients, sedimentary structures, erosional basal contact, absence of marine fossils, and geometry of the sand body all indicate that the sandstone is of fluvial origin and part of an extensive complex of river-deposited sediments. Geographically oriented cores reveal that the direction of depositing currents was to the southwest, parallel with the sand-body trend. Directions of maximum permeability also parallel the northeast-southwest trend. The layered nature of the sandstone reservoir and its directional properties have controlled the areal distribution of produced gas, the ignition behavior, and the vertical profile and horizontal extent of the combustion zone.

INTRODUCTION

This paper is the first of a series of three papers that describe Marathon Oil Co.'s Fry in situ combustion test. This first paper contains the results of a detailed study of the geology and reservoir characteristics of the Robinson sandstone reservoir in which the Fry test was conducted. The second and third parts of this series report on the field operations[1] and performance.[2] The geologic work related to cores cut through the combustion zone, and to the effect of reservoir properties on the combustion process is included in the third paper.[2]

GEOLOGIC STUDY OF THE FRY RESERVOIR

The close spacing of continuous cores in the Fry combustion site provided an unusual opportunity to study the details of distribution of lithology and reservoir properties within a petroleum reservoir. The objectives of the study were:

1. To determine the distribution of rock types within the Robinson sandstone in the Fry area;

2. To determine the relationship of porosity and permeability to rock properties and establish the distribution and homogeneity of reservoir properties;

3. To determine the origin of the Robinson reservoirs so that attempts could be made to predict the size, shape, distribution, extent and continuity of Robinson reservoirs; and

4. To determine what effects, if any, the reservoir rocks might have on the combustion process, so that the process might be better evaluated.

AVAILABLE MATERIALS

Continuous cores were cut from 14 wells in the vicinity of the Fry site. Ten of the cores are from within the 3.3-acre pattern; six of the 10 were cut prior to combustion, and four were cut through the combustion zone. Cores from the Emma Fry No. 15, Wampler 0-2 and Fry B wells (Figs. 1, 2 and 9) were geographically oriented by the Eastman Oil Well Survey Co. Maps, electric logs, gamma ray logs, core descriptions, core analyses and production data were obtained from Marathon personnel in the Robinson and Terre Haute offices and utilized in this study.

METHODS OF STUDY

All cores were slabbed longitudinally on a diamond saw and examined for mineralogy, texture and sedimentary structures. Thin sections were prepared from several hundred samples and studied microscopically for details of mineralogy and texture. Additional mineralogical determinations were made by X-ray diffraction. Lewis Laboratories, in Robinson, Ill., made the usual determinations of porosity, permeability and oil and water saturations; Denver Research Center personnel made additional plug and whole-core analyses.

The data obtained from the core study were correlated with electric logs, so the distribution of the reservoir could be established where core control was not available.

GEOLOGY

Lower Pennsylvanian sandstones occurring between depths of 750 to 1,000 ft in Crawford County, Ill., have been termed the Robinson sands, after the city of Robinson. These sandstones are erratically distributed across the country and are oil productive on most structurally high features related to the La Salle anticline. The sandstones are commonly up to 50 ft thick, but are rarely continuous laterally for more than 1 or 2 miles.

The Robinson reservoir at the Fry site (Fig. 1) is near the crest of the La Salle anticline in the Robinson Main field. The top of

FIG. 1—INDEX MAP OF FRY COMBUSTION SITE IN CRAWFORD COUNTY, ILL.

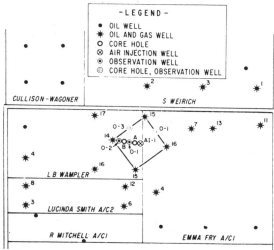

FIG. 2—MAP OF THE FRY COMBUSTION SITE SHOWING WELL
LOCATIONS.

the sandstone ranges in depth across the pattern from 882 to 894
ft and the thickness of the reservoir ranges from 42 to 56 ft
across the site.

ROCK CHARACTERISTICS

Properties of the sandstone are given in Table 1. The sandstone is generally fine- to medium-grained (on the Wentworth
scale), and is composed predominantly of sub- angular grains of
quartz, feldspar and muscovite and a matrix of the same three
minerals plus kaolinite and illite. Calcite, siderite and pyrite
occur as cementing and replacement minerals. Bedding surfaces
are commonly marked by coarse flakes of muscovite and/or carbonized fragments of plant material or clays.

The basal contact of the Robinson sandstone is sharp in all
wells. The base of the sandstone rests on a variety of siltstones
and shales within the pattern. In several of the wells, the base of
the sandstone is marked by a few pebbles of the underlying
lithology.

The upper contact of the Robinson is gradational. The fine-
grained sandstone that occurs at the top is marked by a few laminations of dark gray clay shale. Upward from the reservoir, the
shale laminations become thicker and more abundant as the
sandstone layers become finer grained and thinner. The transition from sandstone to shale occurs over a vertical interval of 2
to 6 ft.

Three distinctly different types of sandstone, each characterized by the abundance of a particular sedimentary structure,
occur in the Fry reservoir.

Each of these three types has been considered as a separate
zone in the following description. Although this zonation was
originally based solely on the distribution of rock types, it also
applies to the distribution of reservoir properties and is fundamental to an explanation and understanding of the air injection,
ignition, and combustion phases of the Fry test. The terminology that is used throughout these reports to designate these
zones is an upper A zone, a middle B zone, and a lower C zone.
Although the zones vary somewhat in thickness (Table 2), they
are present across the entire Fry pattern.

A ZONE

The distinguishing feature of the A zone is the presence of
small-scale, trough-shaped ripple structures (Fig. 3). These
troughs are concave upward, 1 to 3 in. across in plan view, and
commonly less than 1/2 in. thick.

The trough surfaces are marked by thin films of mica and clay
minerals that are somewhat darker colored than the adjacent
sand grains. The sandstone in this zone is very-fine-grained to
fine- grained.

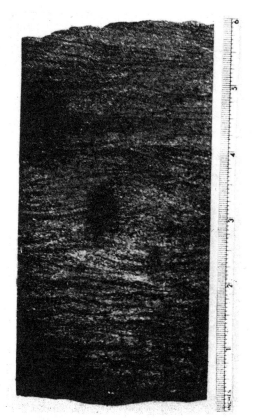

FIG. 3—PHOTOGRAPH OF SLABBED CORE SHOWING CHARACTERISTIC
TROUGH-SHAPED RIPPLES OF THE A ZONE.

B ZONE

The B zone is characterized by sets of cross stratification
(Figs. 4 and 5) that range in thickness from approximately 0.5 to
4 ft and average about 1 ft. The lower boundary of a set is erosional, truncates the set below it, and is horizontal or dips less
than 10°. The base of the set is marked by 0.1 to 0.5 ft of indis-
tinct ripples. Toward the top of the rippled zone, the ripple laminae become more continuous and planar. These planar surfaces
increase in dip upward through the set to a maximum of 30°.

A single cross lamina is less than 0.1 ft thick; its upper surface
is commonly marked by muscovite flakes, discontinuous clay
mineral accumulations, or carbonized wood fragments. Fragments of carbonized wood and pebbles of shale and siderite
occur isolated or in clusters along the cross laminae and also
commonly at the base of a set along the erosional surface.

At the base of the B zone is 0.5 to 1.5 ft of conglomerate. The
pebbles in the conglomerate are composed of gray clay shale that
has been partly or completely replaced by siderite ($FeCO_3$).
Medium-grained sand and fragments of carbonized wood sur-
round the pebbles. The conglomerate is tightly cemented with
siderite or a mixture of calcite, siderite and finely crystalline
pyrite. In two of the pattern wells, large pieces of carbonized
wood occur in this conglomerate.

The sandstone within the B zone is generally fine- to medium-
grained. Within a set, however, there is a subtle grain-size gradient. Finer sand occurs in the rippled zone and coarser sand in the
cross- stratified zone.

C ZONE

A distinctive, mixed lithology of medium-grained sandstone
and dark gray clay shale occurs in the C zone (Fig. 6). The sandstone layers are from 0.1 to 0.5 ft thick and the shale layers rarely
exceed 0.05 ft. The shale laminations mantle coarse ripples of
undulating form; they are commonly deformed, probably as a
result of compaction. Some of the shale laminations appear to

393

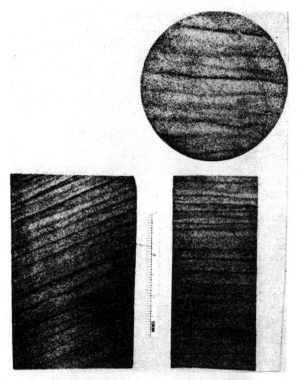

FIG. 4—SLABBED CORE SHOWING TOP, FRONT AND SIDE VIEWS OF CROSS STRATIFICATION IN THE B ZONE.

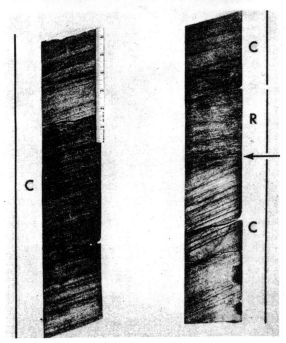

FIG. 5—TYPICAL SEQUENCE OF CROSS STRATIFICATION (C) AND RIPPLES (R) IN THE B ZONE. ARROW MARKS BOUNDARY BETWEEN SETS OF CROSS STRATIFICATION. (BOTTOM OF LEFT PHOTO FITS TOP OF RIGHT PHOTO.)

have been dessicated and cracked before the deposition of the overlying sandstone layer. Flakes and pebbles of the same shale, as well as fragments of carbonized wood, are fairly common in the sandstone layers. The shale is partly or completely replaced by gray-brown siderite. The replacement may be spotty or continuous throughout an entire shale unit across a core. A semiquantitative measure of the amount of shale in the C zone showed that in all of the pattern wells, the shale ranges from 13 to 16 per cent of the total C zone thickness. These data are some-

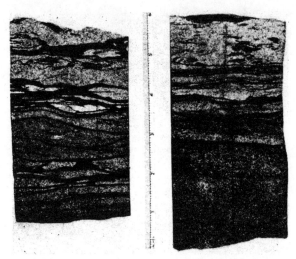

FIG. 6—TWO EXAMPLES OF LAMINATED SANDSTONE AND SHALE TYPICAL OF THE C ZONE.

what surprising in view of the heterogeneous appearance of the lithology (Fig. 6).

The sandstone units most commonly do not show any distinctive primary sedimentary structure. In a few places, trough-shaped ripples, approximately 3 to 4 in. across, can be observed. Less commonly, tabular sets of cross stratification that dip approximately 25° are apparent. The sandstone throughout the C zone is medium-grained, but it is coarser grained at the base and finer grained toward the top of the unit.

SEDIMENTARY STRUCTURES

Of the sedimentary structures described above, two are fairly well understood and give an indication of the depositional environment. The trough-shaped ripples of the A zone are known to be deposited under conditions of unidirectional flow. They have been observed in point-bar deposits of recent rivers.[3]

Sets of cross stratification similar to those in the B zone have also been observed in point-bar deposits and have formed under conditions of unidirectional flow. Where these structures have been observed in recent river deposits, they have a trough shape that is several feet across and approximately 1 ft. thick.[3]

DIRECTIONAL FEATURES

Both of the sedimentary structures described above indicate the direction and sense of flow of the depositing current. These directions were measured in three geographically oriented cores in the Fry area. In addition to the trough-shaped ripples and the cross stratification of the A and B zones, the long direction of shale and siderite pebbles was measured in the conglomerate at the base of the B zone and also a few pebble zones within the B zone.

All measurements for each well were calculated to a single vector resultant. The range for these measurements and their average direction are shown in Fig. 9. The directional features indicate that depositing currents of Robinson sandstone flowed toward the southwest and west at these three wells.

TEXTURAL FEATURES

The grain-size characteristics mentioned above are megascopically apparent in the slabbed cores and are consistent in all wells. Complete grain-size analyses were performed on 12 samples from the Fry No. 15 well so that these characteristics could be determined quantitatively. The grain-size analyses verified the visual observations (Fig. 7). The C zone is clearly coarser grained at the base and becomes finer grained toward the top. The B zone is quite consistent throughout, but it does become finer grained toward its top. Only one sample was analyzed from the A zone, and it is finer grained than any of the other samples.

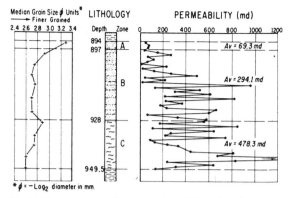

FIG. 7—CHARACTERISTIC RELATIONSHIP OF GRAIN SIZE, LITHOLOGY AND PERMEABILITY. (DATA FROM EMMA FRY A/C 1 NO. 15.)

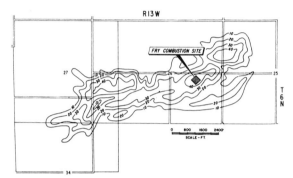

FIG. 8—ISOPACH MAP OF ROBINSON SANDSTONE, FRY AREA.

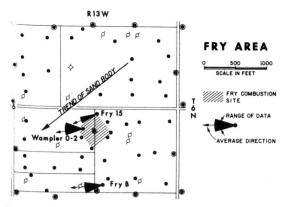

FIG. 9—SUMMARY OF DIRECTIONAL MEASUREMENTS FROM THREE GEOGRAPHICALLY ORIENTED CORES. (SEE TEXT FOR EXPLANATION OF MEASUREMENTS.)

Although there are grain-size gradients within individual zones, there is a consistent and progressive decrease in grain size upward through the reservoir. Such vertical grain-size gradients are often useful as one criterion indicating the depositional environment of a sandstone. Shoreline marine sandstones commonly become progressively coarser grained toward the top, whereas fluviatile (river) sandstones become finer grained toward the top. In conjunction with other criteria, this gradient suggests a fluviatile origin for the Robinson sandstone.

PALEONTOLOGY

The only fossil material megascopically observed during the detailed examination of the 14 cores was carbonized fragments of wood. No other fossils were found in the Robinson sandstone, or in any of the underlying or overlying lithologies. A suite of 25 samples representing the sandstone and shale in the reservoir as well as the shale above and below it was examined for microfossils. All samples were barren, except one which contained the remains of a megaspore.

MORPHOLOGY OF THE FRY SANDSTONE UNIT

After the characteristics of the Robinson sandstone had been determined from the cores at the Fry site, comparisons were made with electric logs from these wells so that the distribution of sandstone could be determined from electric logs outside the area of core control. Correlations were attempted throughout the area of the Fry waterflood unit by hanging the logs from a stratigraphic marker (Coal No. 2) that is consistent throughout most of the area of Robinson production.

The sandstone body in which the Fry test was conducted is approximately 12,000 ft long, 3,500 ft wide, 0 to 55 ft thick, and trends generally southwest (Fig. 8).

ORIGIN OF THE ROBINSON SANDSTONE, FRY SITE

The following characteristics of the Robinson sandstone are useful in determining its origin and depositional environment:

1. Sharp erosional basal contact with a variety of underlying rock types;
2. Vertical decrease in grain size from the base to the top of the reservoir (Fig. 7);
3. Presence of primary sedimentary structures (trough-shaped ripple laminations and cross stratification) that form under conditions of unidirectional flow (Figs. 3, 4 and 5);
4. Consistent directions of depositing currents parallel with the trend of the Fry sand unit (Fig. 9); and
5. Absence of marine megafossils or microfossils; presence of abundant carbonized wood.

Collectively, these characteristics suggest a non-marine, fluviatile environment of deposition for the Robinson sandstone at the Fry site. The Fry sandstone unit very likely was deposited along a series of migrating point bars in a meandering river that flowed from the northeast. The distribution of similar Robinson sandstone reservoirs along the crest of the La Salle anticline suggests a complex of fluviatile deposition by an extensive river that meandered across an alluvial valley, an alluvial plain, or perhaps a deltaic plain.

RESERVOIR PROPERTIES
POROSITY AND HORIZONTAL PERMEABILITY

A consistent pattern in the amount and distribution of porosity and horizontal permeability is apparent throughout the Fry pattern. In general, permeability and porosity decrease from base to top of the reservoir. A plot showing the relationship of porosity and permeability to the lithologic zones is shown in Fig. 10. The A zone shows a marked decrease upward from 800 md to less than 50 md. Porosity generally decreases from approximately 20 to 10 per cent. Within the B zone, the permeability is generally greater near the base, but considerable variation from 100 to 1,000 md is not uncommon throughout the zone; porosity varies between 15 and 25 per cent, but shows a general decrease toward the top of the zone. Extreme variation is the characteristic of the C zone. Part of the variation is a function of sampling for the core analysis. Some samples selected from the sandstone portions of the zone have permeability values in excess of 1 darcy, and samples containing shale laminations are as low as 100 md. The same type of variation is noted in the porosity of the C zone, between values of 15 and 25 per cent.

In addition to the general decrease in porosity and permeability from base to top, the reservoir displays a characteristic decrease in average values for each of the lithologic zones with decreasing depth. A summary of these average values for each zone and for the total reservoir for each of the wells appears in Table 2.

VERTICAL PERMEABILITY

The vertical permeability characteristics are also related to the three lithologic types of sandstone. Vertical core plug and whole-core analyses of permeability show that in the A zone the closely spaced clay films on the trough-shaped ripples effectively elimi-

TABLE 1—AVERAGE PROPERTIES OF SANDSTONE TYPES

Zone	Lithology	Average Thickness (ft)	Average Porosity (%)	Average Permeability (md)	Permeability Variation
Upper (A)	Ripple laminated sandstone	7	18	50	.79
Middle (B)	Cross stratified sandstone	24	19	300	.60
Lower (C)	Laminated sandstone (85%) and shale (15%)	19	20	425	.57
Total Reservoir		50	19.7	320	.68

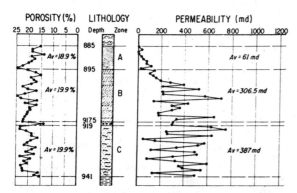

FIG. 10—CHARACTERISTIC RELATIONSHIP OF LITHOLOGY, POROSITY AND PERMEABILITY. (DATA FROM L. B. WAMPLER NO. 14.)

nate vertical permeability. However, within the B zone, the vertical permeabilities are commonly 75 to 95 per cent of the horizontal values. The shale and siderite laminations in the C zone also eliminate vertical flow within plugs or whole cores. Whereas vertical core plugs drilled completely within the sandstone portions of this zone have permeability values comparable to those in the B zone, plugs and whole cores cut across even the thinnest shale laminations have no vertical permeability. However, the large number of lateral terminations of shale laminations that are seen in the cores suggests that these laminations are not extensive, and that vertical permeability exists on a scale larger than the size of a core.

The conglomerate between the B and C zones presents an effective barrier to vertical flow. A pervasive cement of calcite and/or siderite seals all pore space. This conglomerate is not present across the entire pattern, but occurs in the Wampler AI-1, 0-2, No. 14 and No. 15.

Other barriers to vertical flow are the shale layers present in the transitional contact at the top of the reservoir. These layers are as thick as 0.5 ft and may well extend laterally for tens of feet.

DIRECTIONAL PERMEABILITY

Directional horizontal permeability was measured on a series of plugs drilled from the core at different angles and on whole cores as well. The size of the sedimentary structures imposes certain difficulties on directly measuring the directional permeability. Trough-shaped ripples are small enough so that portions of two or three trains of ripples are contained within a core. The anisotropy of this structural type can be measured directly. On the other hand, only a small part of a cross-stratified deposi-

tional unit is cut in a core. The anisotropy measured by core analysis is, therefore, local and probably caused by preferred grain orientation, but the anisotropy related to the geometry of the larger sedimentary unit cannot be measured.

Measurements in the A zone showed that the direction of maximum permeability parallels the axis of the small troughs. The minimum permeability is at right angles to the trough axis and varies 85 to 95 per cent of the maximum value. The dip direction of the cross-stratified sandstone is generally the direction of maximum permeability in the B zone, although several exceptions were noted. Minimum permeability is 90 to 95 per cent of maximum permeability. No consistent variations in horizontal permeability were apparent in the sandstone of the C zone.

In general, the direction of the maximum permeability tends to parallel the direction of the depositing currents. The data from the three geographically oriented cores (Fig. 9) show a correspondence between the long direction or trend of the sandstone body, the dip direction of the sedimentary structures, and the direction of sediment transport. If other conditions that control fluid flow are equal, the direction of maximum permeability would be expected in a northeast-southwest direction and the direction of minimum permeability in a northwest-southeast direction.

RESERVOIR HOMOGENEITY

The distribution of sandstone types across the Fry pattern indicates that the reservoir is made up of three layers each having distinctive characteristics of porosity, horizontal permeability, vertical permeability and directional permeability. The core analysis data suggest that the C zone is the best portion of the reservoir because of its consistently higher porosity and permeability. However, because the permeability value obtained in the C zone is so markedly controlled by the nature of the sample, these values are probably not representative of actual permeability in the reservoir. In addition, the abundance of shale layers and lenses presents discontinuous barriers to vertical flow that make tortuous flow paths on a scale larger than a core. If the values of porosity and permeability are considered along with the inherent sampling problem and the nature and scale of the sedimentary structures, the B zone is probably the most homogeneous and most conductive to fluid.

ROCK CONTROL OF RESERVOIR PROPERTIES

The increases of porosity and permeability with increasing depth that are discussed above appear to be closely related to the texture of the sandstone. No extensive post-depositional effects, such as development of authigenic kaolinite or abundant precipitation of chemical cements around or between grains, are found in the Fry reservoir. Uncommonly, calcite cement precipitated along planes of cross stratification in the B zone has locally reduced porosity and permeability. The increase in median grain size with increasing depth (Fig. 7) appears to be the major control on the vertical distribution of porosity and permeability.

SUMMARY

The reservoir at the Fry combustion site is a lenticular body of Robinson sandstone that is 12,000 ft long, 3,500 ft wide, 0 to 55 ft thick, and trends southwest. A detailed study of continuous cores from 14 wells indicates that the Fry reservoir is of fluviatile

TABLE 2—SUMMARY OF THICKNESS, POROSITY, AND PERMEABILITY FOR EIGHT WELLS IN THE FRY SITE
FRY COMBUSTION PROJECT

Well Location	Thickness (ft)				Porosity (per cent)				Permeability (md)			
	A	B	C	Total	A	B	C	Avg.	A	B	C	Avg.
Emma Fry a/c 1 No. 15	3	31	21.5	55.5	19.9	19.4	18.6	19.1	69.3	294.1	478.3	353.3
Emma Fry a/c 1 No. 16	2.5	26	17.5	46	16.1	20.3	20.8	20.2	42.7	329.8	367.4	328.5
*L. B. Wampler No. Al-1	11	19	21	51	17	19.8	21.7	20.0	92.0	427.8	615.7	432.7
L. B. Wampler No. 14	10	22.5	22	54.5	18.9	19.9	19.9	19.7	61	306.5	387	293.9
L. B. Wampler No. 15	7	22	13	42	18.8	19.4	20.5	19.6	30.3	333.2	323.3	279.6
L. B. Wampler No. 16	1.5	28	21	50.5	19.3	19.2	22.1	20.4	49	293.3	429.4	342.6
L. B. Wampler No. 17	14	21.5	20	55.5	16.6	19.0	19.2	18.5	9.2	247.9	359.8	228.0
L. B. Wampler No. 0-2	4.3	25.8	22.8	52.9	20.4	20.5	20.7	20.6	53.8	275.1	377.8	301.4
Average for 8 wells					17.9	19.7	20.4	19.7	48.4	309.9	423.1	319.8

*Partial analysis because of lost core.

origin and part of an extensive complex of river sediments. The reservoir is made up of three distinct lithologic types of sandstone, each having its own characteristic set of reservoir properties. Porosity and permeability are controlled by rock properties and can be related to primary sedimentary structures and textural variations.

The layered nature of the sandstone and its directional properties have controlled ignition behavior, areal distribution of produced gases, and vertical profile and horizontal extent of the combustion zone. Details of these controls are discussed in the third paper[2] of this series.

ACKNOWLEDGMENT

The authors wish to express appreciation to the management of the Marathon Oil Co. for permission to publish this paper. Thanks are due to the people in the Production and Research Depts. for their contributions to the success of this test.

REFERENCES

1. Clark, G. A., Jones, R. G., Kinney, W. L., Schilson, R. E., Surkalo, H., and Wilson, R. S.: ''The Fry In Situ Combustion Test-Field Operations'', *Jour. Pet. Tech.* (March, 1965) 343.
2. Clark, *et al:* ''The Fry In Situ Combustion Test-Performance'', *Jour. Pet. Tech.* (March, 1965) 348.
3. Harms, J. C., MacKenzie, D. B. and McCubbin, D. G.: ''Stratification in Modern Sands of the Red River, Louisiana'', *Jour. of Geology* (Sept., 1963) **71,** 566.
4. Potter, P. E. and Pettijohn, F. J.: *Paleocurrents and Basin Analysis,* Springer-Verlag, Berlin (1963).

Original manuscript received in Society of Petroleum Engineers office July 20, 1964. Revised manuscript received Feb. 1, 1965. Paper presented at SPE 39th Annual Fall Meeting, held in Houston, Oct. 11-14, 1964.
[1]References given at end of paper.
Discussion of this and all following technical papers is invited. Discussion in writing (three copies) may be sent to the office of the *Journal of Petroleum Technology.* Any discussion offered after Dec. 31, 1965, should be in the form of a new paper. No discussion should exceed 10 per cent of the manuscript being discussed.

Reprinted by permission. First published in the *Journal of Petroleum Technology*, V. 35, No. 7 (July 1983), pp. 1355-1365.

Nugget Formation Reservoir Characteristics Affecting Production in the Overthrust Belt of Southwestern Wyoming

Sandra J. Lindquist, Amoco Production Co.

Summary

The Jurassic/Triassic Age Nugget sandstone of the southwestern Wyoming overthrust belt is a texturally heterogeneous reservoir with anisotropic properties that have been inherited primarily from the depositional environment but also have been modified by diagenesis and overprinted by tectonism.

Predominantly eolian processes deposited crossbedded, low-angle to horizontally bedded and rippled, very-fine- to coarse-grained sand in dunes, interdune areas, and associated environments. Original reservoir quality has been somewhat modified by compaction, cementation, dissolution, clay mineralization, and the precipitation or emplacement of hydrocarbon asphaltenes or residues. Low-permeability gouge- and carbonate-filled fractures potentially restrict hydrocarbon distribution and negatively affect producibility, whereas discontinuous open fractures enhance permeability in some intervals. Contrast in air permeability between dune and interdune deposits ranges over four to five orders of magnitude. Dune and interdune intervals are correlatable locally with the aid of core log, conventional log, and stratigraphic dipmeter data. Stratigraphic correlations then can be utilized to model the lateral and vertical extent of directional properties in the reservoir.

Introduction

A large proportion of the sizable hydrocarbon reserves in the Utah/Wyoming overthrust belt are within the Jurassic/Triassic Age Nugget formation. Upper Nugget data were obtained from three fields in Uinta County,

southwest Wyoming—Clear Creek, East Painter, and Glasscock Hollow—all of which are closed anticlinal structures on the upper plate of the Absaroka thrust (Fig. 1). Nugget pay fluids are probable retrograde gas condensates and volatile oils.

In the producing trend, the Nugget sandstone is primarily a clastic eolian deposit ranging (north to south) from 800 to $> 1,000$ ft (250 to > 300 m) in thickness. It probably is conformable with the underlying Triassic Ankareh formation and certainly unconformable with the overlying Jurassic Twin Creek formation. Lithology is predominantly sandstone, with some siltstone. Siltier lithologies are more abundant toward the base of the formation, concurrent with probable increased influence of moisture, water tables, and/or water processes in the depositional environment. Many eolian sandstone reservoirs mistakenly have been considered homogeneous from a reservoir performance standpoint because the importance of subtle, facies-related controls on reservoir properties within these thick sandstone sequences has been overlooked. Textural heterogeneity results from bedding variations in grain size and sorting, which are directly related to the geometry of eolian facies. Within the Nugget formation, these primary depositional heterogeneities are amplified by diagenesis and further modified by tectonic stresses that folded and faulted the region.

Reservoir Characteristics Inherited From the Depositional Environment

The upper 500 ft (150 m) of Nugget formation represented in the cores studied comprises stacked dune

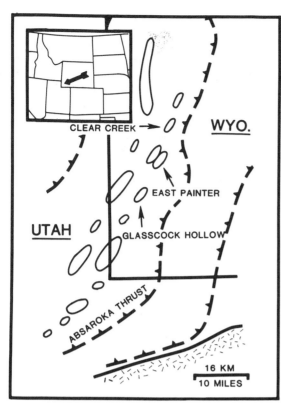

Fig. 1—Location map of producing fields on the Absaroka thrust plate of the Utah-Wyoming overthrust belt.

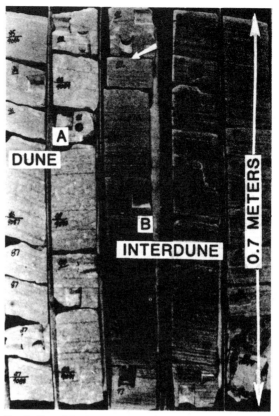

Fig. 2—Dune/interdune sequence from a well in Clear Creek field.

and mostly dry-to-damp interdune deposits. Because of varied amounts of erosion associated with the post-Nugget unconformity, this equivalent porous and permeable stratigraphic interval is not located at the top of the formation in all producing fields, especially to the south. Facies distinction within Nugget depositional environments is based on textural (grain size and sorting) and reservoir-quality criteria rather than on major lithological variation (such as the presence of interbedded shales, carbonates, or evaporites).

Dune Deposits

Nugget dune deposits are crossbedded, very-fine- to coarse-grained sandstones containing laminations of the coarsest material available in the depositional environment. The crossbeds are accretionary laminae that record each successive location of the dune slipface as sand migrated in a downwind direction. Deposition on the preserved dune downwind slope occurred from grainfall and avalanche processes as described by McKee et al.[1] and by Hunter.[2]

Grainfall occurs as sand grains saltating and creeping up the backside of the dune reach the crest and drop into the zone of flow separation and lower wind velocity on the slipface. The resultant deposit consists of thin (millimeter scale), tabular laminae of segregated grain size. Continued grainfall deposition concentrated near

the dune crest can result in oversteepening of the upper slipface beyond the angle of repose. Avalanche processes (slumping or sandflow caused by slipface oversteepening or by other physical phenomena in the environment) commonly redistribute grains from oversteepened portions to a position farther down the dip slope. Resultant avalanche deposits are generally thicker (centimeter scale) than grainfall deposits and may be virtually structureless or subtly graded. Sand on the dune slipface and especially at the toe of the slipface is in many cases reworked and more tightly packed by deflected winds to form ripples and translatent (climbing ripple) strata. The predominant bedding type within a preserved dune deposit depends on the original size and shape of the dune, the amount of slipface preserved, and other climatic and geographic factors in the depositional environment.

Dune crossbed porosity and permeability have been described as being best in thickly laminated avalanche deposits, poorer in thinly laminated grainfall deposits, and poorest in ripples at the dune toe.[2,3] Directional attributes of permeability also vary considerably with respect to bedding type. For example, the larger the grain size contrast between laminae, the greater the along-bedding component of permeability is, compared with that across-bedding. The height, orientation, and length of the dune slipface therefore determine the lateral and vertical extent of preferred directional permeability within the crossbedded sand.

Interdune Deposits

According to the classification of Ahlbrandt and Fryberger,[4] interdune areas are either depositional or erosional environments further categorized by association with water (i.e., predominantly dry, wet, or evaporitic). Each type of interdune area can leave a characteristically distinct rock record with unique reservoir properties in comparison with those of dune deposits. In contrast to the dune sequences, these upper-Nugget interdune deposits consist of a variable but overall finer grain size and contain thin, low-angle to horizontal laminae, ripple forms and translatent strata, and irregular bedding. Laminae as thin as one or several grain diameters are not uncommon. Most of these interdune areas probably were dry to ephemerally dampened or wetted. Carbonate and evaporite lithologies and obvious water-laid deposits characteristic of wetter interdune environments are not common in the upper-Nugget cores examined for this paper. But carbonate-cemented laminae, centimeter-scale trough stratification, wavy bedding, bioturbated and brecciated zones, and distorted bedding are present within the Nugget formation elsewhere in the overthrust area and, especially, lower in the stratigraphic section (not cored in these wells).

With respect to reservoir quality, thickness and extent of Nugget interdune deposits are variable, ranging from discontinuous individual units less than 1 ft (30 cm) thick to extensive "packages" of small dune and interdune sands more than 10 ft (3 m) thick. Ahlbrandt and Fryberger[4] related depositional interdune variability to complexity of dune form and wind regime. For example, transverse (straight-crested) dunes are likely to be associated with more laterally extensive interdune deposits than barchan (crescentic) dunes, and interdunes in a complex wind regime are likely to be thicker than those in a unidirectional wind system.

Texture and Reservoir Quality of Dune and Interdune Deposits

Figs. 2 through 4 compare texture and reservoir quality in thicker-bedded dune facies with that in thinly laminated interdune facies. Fig. 2 shows approximately 13 ft (4 m) of core from a well in Clear Creek field. The upper part of the core (above the arrow) is the lower portion of a dune slipface. It is thickly crossbedded, light in color, and contains white blotches of carbonate cement, which are common in most permeable Nugget facies. Interdune deposits below the arrow are darker in color (in this case, darker red), flatter-bedded, more thinly laminated and rippled, and lacking in nodular carbonate cement.

A thin-section photomicrograph of Sample A from the dune interval is shown in Fig. 3. Grain size is homogeneous on a microscopic scale, and both horizontal and vertical air permeability exceed 1 darcy measured at surface conditions. The interdune Sample B at the same magnification (Fig. 4) illustrates rippled microlaminae of various grain sizes that have contrasting pore and pore-throat sizes. The horizontal (along-bedding) air permeability of 50 md and vertical (across-

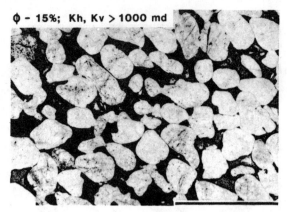

Fig. 3—Thin-section photomicrograph: Sample A, dune crossbedded sandstone. Bar scale is 1.0 mm.

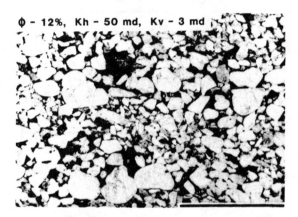

Fig. 4—Thin-section photomicrograph: Sample B, interdune ripple-laminated sandstone. Bar scale is 1.0 mm.

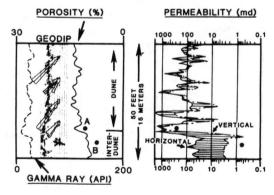

Fig. 5—Relationship of porosity and directional permeability to dune/interdune facies shown in Figs. 2 through 4. Disparity between vertical and horizontal permeability is shaded to highlight directional heterogeneity within the interdune sequence. Facies are easily distinguished with porosity, permeability, and stratigraphic dipmeter data; the gamma ray is less consistently diagnostic.

400

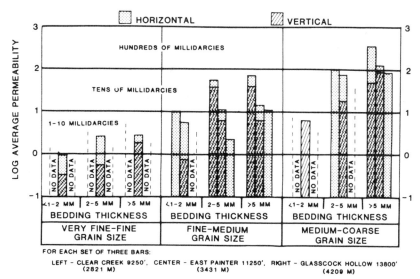

FOR EACH SET OF THREE BARS:
LEFT - CLEAR CREEK 9250', CENTER - EAST PAINTER 11250', RIGHT - GLASSCOCK HOLLOW 13800'
(2821 M)   (3431 M)   (4209 M)

Fig. 6—Effects of grain size, bedding thickness, and depth on permeability.

bedding) air permeability of 3 md reflect an average of alternating laminae that probably have permeabilities ranging from hundredths to hundreds of millidarcies on a microscopic scale.

Fig. 5 illustrates the locations of these two samples from the base of a compound set of dune crossbeds and the underlying ("dry") interdune sequence. The figure contains a plot of the log gamma ray and core porosity on the left side, with horizontal and vertical air permeability displayed on a logarithmic scale to the right. A display of stratigraphic bedding dip also is shown on the left (the structural dip component has been removed from the data). Of the four sets of data displayed, the gamma ray is least consistent as a facies indicator. Permeability ranges over several orders of magnitude both directionally *within* the interdune deposit and *between* the dune and interdune sequences (i.e., from hundreds of millidarcies to tenths of millidarcies). Disparity between vertical and horizontal permeability (shaded separation between the two permeability curves) is most pronounced in the thin, horizontally laminated and rippled interdune interval. Thus, in addition to being generally poorer in overall reservoir quality, Nugget interdune deposits are potential vertical permeability impediments to fluid flow.

Effects of grain size, bedding thickness, and depth on permeability were examined (Fig. 6) for the three fields that have present depths of burial ranging from about 9,200 to 13,800 ft (2800 to 4200 m). Core plugs were separated into three visually estimated, *average* grain size categories (very-fine- to fine-grained, fine- to medium-grained, and medium- to coarse-grained), each of which was in turn subdivided into three groups based on bedding thickness (increasing from left to right within each grain-size category). These bedding-thickness groups are approximately correlative to a gradual and overlapping transition from thinly laminated and rippled interdune deposits to thin dune-grainfall laminae to thickest dune-avalanche bedding. Each bedding thickness/grain-size subdivision contains three adjoining

bars, one for each field with its unique present burial depth. Each set of three bars is arranged shallowest to deepest in a left-to-right direction [Clear Creek on the left at 9,250 ft (2820 m), East Painter in the center at 11,250 ft (3429 m), and Glasscock Hollow on the right at 13,800 ft (4206 m)]. Superimposed on each bar are both the average along-bedding horizontal (dot pattern) and average across-bedding vertical (diagonal slashes) plug permeabilities on a logarithmic scale. Absence of one pattern on a bar indicates that no plug data were available for that category. The plugs used did *not* contain significant macroscopically visible volumes of diagenetic cements that would notably affect the reservoir quality associated with these primary textural characteristics (see Effect of Diagenesis on Reservoir Properties). Approximately 875 plug samples are represented in the graph. Conclusions from the bar graph are as follows.

1. For constant average grain size, vertical and horizontal permeability generally show more disparity in samples with thin bedding; therefore, directionality is more important in thinly laminated grainfall and rippled deposits than in thicker avalanche bedding.

2. For constant average grain size, a comparison of permeability with bedding thickness illustrates that one-half to one and one-half orders-of-magnitude range in overall permeability is explained by bedding-thickness variability alone (subtle laminar-scale grain size and sorting differences, which are related to the genesis of the bedform).

3. For constant bed thickness, a comparison of permeability to grain size suggests that one-half to one and one-half orders-of-magnitude range in permeability is explained by grain-size variability alone (a function of sediment availability and processes of deposition).

4. For constant grain size *and* bedding thickness, a comparison of permeability by field shows that the present total range in depth of burial between fields [about 4,500 ft (1372 m)] can account for perhaps one-half to a

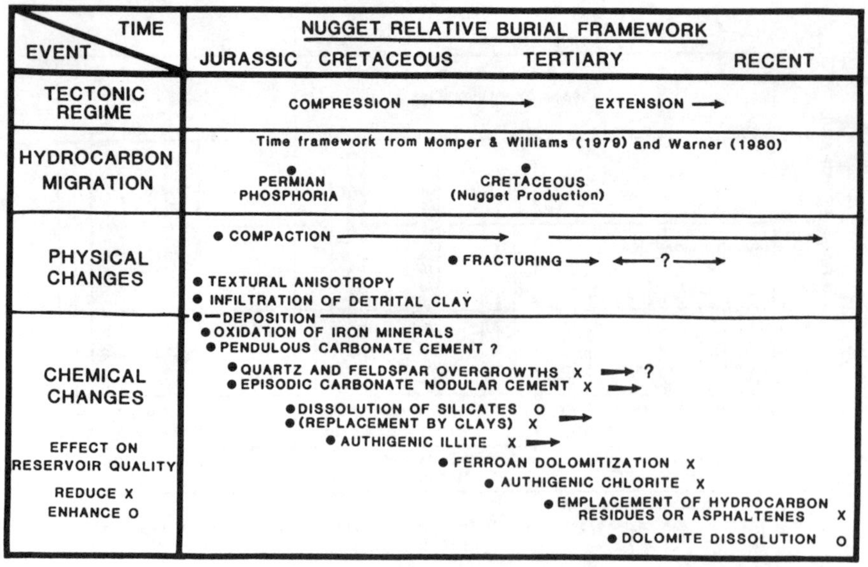

TIME / EVENT	NUGGET RELATIVE BURIAL FRAMEWORK
	JURASSIC CRETACEOUS TERTIARY RECENT

Fig. 7—Burial history of Nugget formation based on data from three overthrust fields. Location of dot indicates relative time of first occurrence for each event. Arrows indicate continued or episodic events.

full order-of-magnitude difference in average reservoir permeability. (If loss of permeability with increased depth of burial is caused primarily by compaction from overburden in these fields, this permeability/depth relationship indicates that present burial depths in the three fields are comparatively proportionate to their maximum burial depths before erosion of overlying Cretaceous and Tertiary rocks.)

Effect of Diagenesis on Reservoir Properties

Physical and chemical changes occurred within the sediment in the depositional environment and during burial. A simplified sequence of burial-history events and their qualitative effect on reservoir quality is summarized in Fig. 7. Relative ordering of events is based primarily on textural evidence from thin-section and scanning electron microscope (SEM). (Somewhat similar Nugget diagenetic sequences have been described previously for surface and subsurface sections.[5-8,*] The overall effect of diagenesis on these upper-Nugget reservoirs has been to magnify the reservoir quality previously inherited from the depositional environment (grain size, bedding, and sorting characteristics). Somewhat different sequences of diagenetic events with more variable effects on reservoir quality may be characteristic of the Nugget formation either lower in the stratigraphic section or elsewhere in the overthrust area.

Events that probably began early in depositional and burial history include the oxidation of iron minerals and incipient nodular carbonate cementation, mostly within permeable laminae (Fig. 8a). Locally, nodules have become sufficiently abundant to coalesce and to cement laminae completely, but in most cases overall reservoir quality is not seriously affected. Some thin pendulous rims of carbonate cement may have nucleated on framework grains from evaporating surface waters (rain and dew[9]) that percolated downward along most-permeable bedding surfaces (as opposed to early cementation of *less*-permeable laminae by fluid capillary rise from a water table where wetter interdunes are present). Cathodoluminescence delineates rare examples of this pendulous asymmetry to thin (volumetrically insignificant) zones of carbonate cement partially surrounding framework grains within nodules. More commonly observed with cathodoluminescence are examples of episodic stages of continued nodule growth (zoning) that probably occurred later in the diagenetic sequence and with greater burial depth. Stained thin-sections reveal that some nodules are calcite, some are ferroan dolomite, and others are a combination of both carbonate minerals.

Fox[7] documented episodic carbonate (calcite and dolomite) cementation with cathodoluminescence and microprobe data from Nugget cores in nearby Painter Reservoir field. Her geochemical data suggest varying but generally reducing (subsurface) conditions for precipitation of most of that carbonate, both as nodules similar to those described here and as interstitial dolomite preferentially distributed in finer-grained laminae lower in the Nugget section.

Quartz and feldspar overgrowths impair reservoir quality locally in these samples. In rare cases, hydrocarbon inclusions are present at the contact between the detrital grain and its overgrowth (as revealed by fluorescence observation**), indicating either additional earlier stage(s) of hydrocarbon migration[10] or later episodic stages of overgrowth formaion with continued burial.

Alteration and dissolution of silicate minerals, as well as authigenic clay mineralization (mostly illite) probably began relatively early in burial history. Dissolution of rock fragments and feldspars produced at most several percent secondary intragranular porosity in some of the

*Thomas, J.B.: personal communication, Amoco Production Co., Denver (1978)

**Prezbindowski, D: personal communication, Amoco Research Co., Denver (1982).

Fig. 8a—Thin-section photomicrograph. Nodule of carbonate cement restricted to porous and permeable lamina on right half of photograph. Porosity is darkest gray. Bar scale is 1.0 mm.

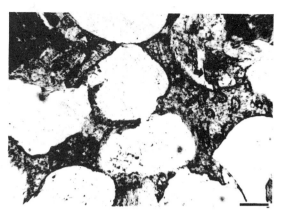

Fig. 8d—Thin-section photomicrograph. Ferroan dolomite as intragranular cement within dissolution porosity (arrow) and as replacement of calcite nodule (remnants of original calcite within the dashed lines were identified by staining techniques). Bar scale is 0.1 mm.

Fig. 8b—Thin-section photomicrograph. Dissolution porosity in feldspar grain (top center) and in rock fragment (left center). Bar scale is 0.1 mm.

Fig. 8e—SEM photograph. Chlorite (arrow) preferentially distributed within open pore spaces. Euhedral crystal faces are quartz overgrowths.

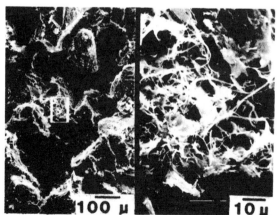

Fig. 8c—SEM photograph. Fibrous authigenic illite, which is highly microporous and commonly bridges pore throats.

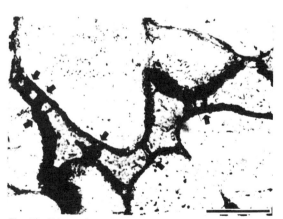

Fig. 8f—Thin-section photomicrograph. Black hydrocarbon residue or asphaltene surrounding framework grains and bridging pore throats (at arrows). Locally, this material can completely occlude porosity. Bar scale is 0.1 mm.

Fig. 8—Diagenetic events in the Nugget formation.

403

most permeable zones (Fig. 8b), presumably because these zones experienced a greater volume of fluid movement and more water/rock interaction through time. Illitization had a more negative effect on permeability than on porosity because illite has a microporous, fibrous morphology and is preferentially distributed in the pore throats of the rock (Fig. 8c).

Ferroan dolomitization is related in time to a stage of tectonism and fracturing, and probably overlaps the period of migration and entrapment of Cretaceous-source hydrocarbons[11] that are currently produced from Nugget reservoirs. Ferroan dolomite occurs as fracture (cavity) fill, as inter- and intragranular cement, and as replacement of calcite nodules (Fig. 8d). Fluorescent hydrocarbon inclusions were also observed in some dolomite fracture fills.*

Most authigenic chlorite postdates a stage of fracturing and ferroan dolomitization. Chlorite is most common as an open-fracture lining and as a pore fill (contrasted to the pore *throat* preference of illite, Fig. 8e). Its iron content and distribution in permeable pathways in the rock make formation damage a possibility in the presence of unsequestered acid.

Precipitation of chlorite was followed by the locally heavy precipitation or emplacement of black hydrocarbon asphaltenes or residues in pores and pore throats, decreasing permeability more and porosity to a lesser (but locally significant) extent (Fig. 8f).

A late (postresidue) stage of dolomite dissolution is of relatively minor volumetric importance to overall Nugget reservoir quality. Earlier and multiple stages of carbonate dissolution and/or replacement are possible but not readily documentable from these data.

Effect of Fracturing on Reservoir Properties

Structural overprinting is strongly evident in Nugget reservoirs. Its major effect is to reduce overall reservoir quality and perhaps compartmentalize the formation. It is likely that Nugget fractures record changing stress fields through time, especially as structural fold hinges migrated with respect to rock position. None of the cores for this study were from highly deformed, steeply dipping fold limbs or present hinge zones. Many observed fractures most closely resemble extension-conjugate shear sets associated with folds (an extension fracture bisecting the acute angle between two intersecting conjugate shear fractures[12]). Although Nugget fractures assume a wide range of spatial orientations, predominant occurrence in the subsurface is at high angles approaching verticality.

Nugget fractures range in type from gouge-filled microfaults and larger shear zones to carbonate-filled varieties to completely open fractures. Gouge is created by the shearing and pulverizing of silicate matrix material along the plane of movement. Most common widths for individual gouge-filled or open fractures are tenths of millimeters, but gouge-filled shear zones and deformation bands associated with larger displacements commonly extend over several feet of core. The type, extent, and width of individual fractures change over short (millimeter or less) distances. Fracture spacing is

Fig. 9a—Thin-section photomicrograph. Cross-cutting, low-permeability gouge-filled fractures in high-porosity (darkest gray) matrix rock. Bar scale is 1.0 mm.

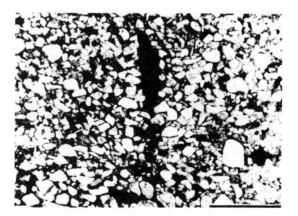

Fig. 9b—Thin-section photomicrograph. Carbonate-filled (ferroan dolomite) fracture within thinly laminated rock of low porosity. Bar scale is 1.0 mm.

Fig. 9c—Thin-section photomicrograph. Discontinuous, open fracture in finely laminated rock. Fracture and adjacent underlying matrix are lined with black hydrocarbon. Bar scale is 1.0 mm.

Fig. 9—Fracture types in the Nugget formation.

*Prezbindowski, D.: personal communication, Amoco Research Co., Denver (1982).

also variable, but in many cases, "parallel" sets repeat over intervals less than a foot apart.

Gouge-filled fractures indicate ductile deformation and are most common in, but not restricted to, Nugget sandstone with maximum porosity and permeability (Fig. 9a). Gouge material is in many cases slickensided and, in most cases, contains porosity on a submicron scale observable with SEM. Gouge-filled fractures are of sufficiently low permeability to restrict potentially the distribution of hydrocarbons in the reservoir and to affect their producibility negatively, particularly the heavier liquid phases. Pittman[13] described similar reservoir compartmentalization and reservoir quality reduction by gouge-filled fractures in Ordovician-age Simpson Group sandstones from Oklahoma.

Carbonate-filled (ferroan dolomite) fractures (Fig. 9b) are more common in less porous and permeable, thinly laminated sequences. In contrast to gouge-filled fractures, matrix material adjacent to carbonate-filled fractures commonly is not pulverized, indicating a more brittle mode of deformation (or perhaps different conditions or time of deformation; see Fig. 7). All gradations between gouge- and carbonate-filled fractures exist. Jamison[14] discussed brittle vs. ductile deformation behavior and similar effects of texture and lithology on deformation for the crossbedded Wingate sandstone of western Colorado.

Open fractures (Fig. 9c) are most commonly irregular, short in length (millimeters or less), and lacking associated gouge in adjacent matrix material, similar to the carbonate-filled varieties. Presence of hydrocarbon asphaltenes or residues as coatings along open fractures is an indication that these discontinuous, open fractures do transmit fluids in the subsurface but in most cases probably in direct conjunction with matrix permeability. In zones of greater deformation than cored in these wells, open fractures may be more continuous and may have greater effectiveness as fluid-transfer mechanisms.

The type, width, extent, variability, and distribution of tectonic fractures depend on complex relationships between the strength (texture and mineralogy) of the rock, the thickness of the textural (strength) unit within the vertical sequence, and the specific stress/strain history (magnitude, direction, pressures, temperature, and rate) that the formation has undergone. It is possible that there were several distinct episodes of fracturing that occurred during the time of Nugget deformation (see Fig. 7) and that some fields along the producing trend might have undergone stages of fracturing at slightly different times in their burial history relative to other diagenetic events. Consistent quantitative relationships could not be made between fracture-type/density and matrix textural parameters for these data, probably because of the varying influence of other parameters just described.

Fig. 10 is a bar graph, similar to Fig. 6, comparing horizontal (along-bedding) matrix permeability not affected by the presence of fractures with horizontal permeability measurements made across a visible gouge-filled fracture in matrix rock of approximately the same average grain size and bedding thickness as the corresponding unfractured samples. Fig. 10 represents 275 samples from East Painter field. It was impossible to tell by inspection whether the fracture visible on the outside of the plug was continuous throughout the center. Thus,

although all fractures were gouge-filled on the perimeter of the plug, fractures may have thickened, thinned, become mineralized, or completely opened within the interior of the plug. No classification distinctions were made for fracture width because of the common high variability in fracture width around the circumference of the plugs. Therefore, this graph shows only an average or overall effect of fractures on the reservoir. The population probably includes some discontinuous open fractures as well as gouge- and carbonate-filled varieties. Conclusions drawn from the graph are as follows.

1. In the best reservoir rock, the overall effect of this (mostly gouge-filled) fracture population was to reduce horizontal permeability by an average of nearly an order of magnitude; that is, from permeability of hundreds of millidarcies to permeability of tens of millidarcies.

2. In poorer reservoir rock, the effect of fractures was less important because more unobserved open fractures exist and/or because the reservoir properties across many fracture planes are comparable to those of this poorer matrix rock.

The most porous and permeable facies apparently behaved predominantly in a ductile manner during deformation and, therefore, were most detrimentally affected by the formation of gouge-filled fractures. Facies with poorer primary reservoir properties are generally characterized by more brittle deformation—i.e., discontinuous open or mineralized fractures whose reservoir properties (measured at surface conditions) can be comparable to, or in some cases better than, those of the matrix rock. Similar observations on orientation and reservoir quality of fractures in other Nugget reservoirs have been made previously.*

Stratigraphic Modeling of Reservoir Properties

As shown in Fig. 5, Nugget crossbed dip azimuths from structurally corrected stratigraphic dipmeters are predominantly southwesterly. Because lithified eolian crossbeds are the preserved lower portions of dune slipfaces, a commonly observed high degree of crossbed unidirectionality indicates that many Nugget dunes were consistent in orientation and morphology—i.e., relatively straight-crested, compound transverse to sinuous barchanoid ridges (Fig. 11) in which the slipfaces dip predominantly in a consistent downwind direction.[15] Also common are crossbed sets with wider azimuth scatter, perhaps recording the presence of morphologies having less consistent lateral (strike) continuity of slipfaces such as crescentic barchan dunes and more complex or compound dune bedforms. Because best permeability is likely to be along bedding planes, preferred permeability directions in a dune sequence of crossbed sets (see Fig. 11) are (1) up and down the individual crossbeds (limited by thickness of crossbed set) and (2) along the trend (strike) of the crossbeds (limited by lateral extent of dune and perpendicular to wind direction for transverse and barchanoid dune systems).

Fig. 12 summarizes several hypotheses for facies preservation (net deposition) in an eolian environment.[16,17] Note that lateral downwind correlation of potentially continuous interdune deposits differs in angle

*Nelson, R.A.: personal communication, Amoco Production Co., Denver (1981)

for each case. The middle situation best explains net deposition that occurred within the Nugget formation—i.e., an unknown thickness of upper dune sediment was truncated by a time-transgressive, downwind-climbing interdune deposit that is largely preserved in the stratigraphic section. This hypothesis can serve as a model for *local* stratigraphic correlations within the Nugget formation. Kocurek[16] reported that in eolian Entrada sandstone (Jurassic) outcrops of northeastern Utah, angle of interdune climb is generally less than 1°, with tenths and hundredths of a degree common. [Over a distance of 2 miles (3.2 km), a 0.5° angle of climb results in a vertical elevation change of about 100 ft (30 m), which can be significant for correct stratigraphic correlation.]

For Nugget wells, when the effects of structural dip and deviated boreholes are removed to correct logs to approximate compacted, true stratigraphic thickness, major dune and interdune packages can be correlated *locally* within some fields, assuming a very small angle of downwind climb necessary to preserve sediment in an eolian system with downwind-climbing depositional units. Within the Nugget sandstone of Clear Creek field, major interdune packages can be documented to climb in a southwesterly downwind direction at angles similar to those described for Entrada dunes.[16] Because interdune deposits are the zones of poorest Nugget reservoir quality, their accurate correlation aids in the delineation of major, *potentially* fieldwide permeability impediments or layers. Depending on the fluid phases present and on the in-situ relative permeabilities of each phase, these layers may not necessarily obstruct fluid flow, but instead affect *rates* and preferred *paths* of fluid movement. Each genetic dune package between interdune barriers then can be evaluated quantitatively for continuity and directional changes in permeability.

Fig. 13 illustrates the type of correlations that can be used in reservoir models. Correlation lines denote erosional surfaces at which interdune deposits truncate underlying dune sediments in the core. Dipmeter and reservoir quality data also aid in the identification of these facies and surfaces. Notice the compound, smaller-scale sequences and overall poorer reservoir quality of correlative zones in Well 11-3B when compared with those in Well 23-9B. If these zones are indeed correlative migrated dune systems, the downwind (northeast to southwest) variability of sequence suggests that Nugget dune size and/or morphology varied with time and over small geographic areas. Also, these correlations cannot be extrapolated so believably through every well in this field.

Conclusions

The eolian Nugget sandstone of the southwest Wyoming overthrust belt is heterogeneous in its reservoir properties. The wide range in porosity and permeability is primarily a result of textural properties (grain size and sorting) inherited from the depositional environment. Diagenesis generally has slightly reduced reservoir quality by intergranular cementation and clay mineralization; however, small quantities of dissolution porosity are present in the most permeable units. Tectonism has primarily further degraded and compartmentalized the reservoir by producing low-permeability, gouge-filled, and mineralized fractures, although discon-

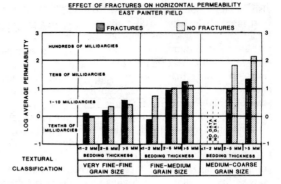

Fig. 10—Overall effect of fractures on average horizontal permeability.

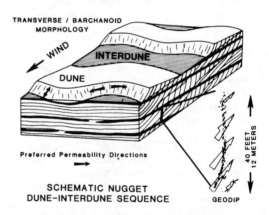

Fig. 11—Schematic correlation of Geodip (structurally corrected stratigraphic dipmeter data) to a transverse/barchanoid dune morphology. Many Nugget dunes may have been slightly more complex versions of this dune type. Preferred permeability directions along the dune dip slopes are marked by arrows. Note the predominant southwesterly dip for Nugget crossbeds.

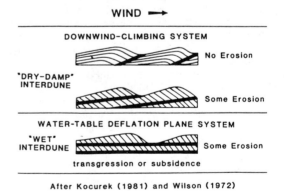

Fig. 12—Hypotheses for net preservation of eolian sand. Interdune deposits are black. Note the varied angles of downwind interdune correlation (exaggerated) for each case. Refer to Kocurek[16] for further discussion.

406

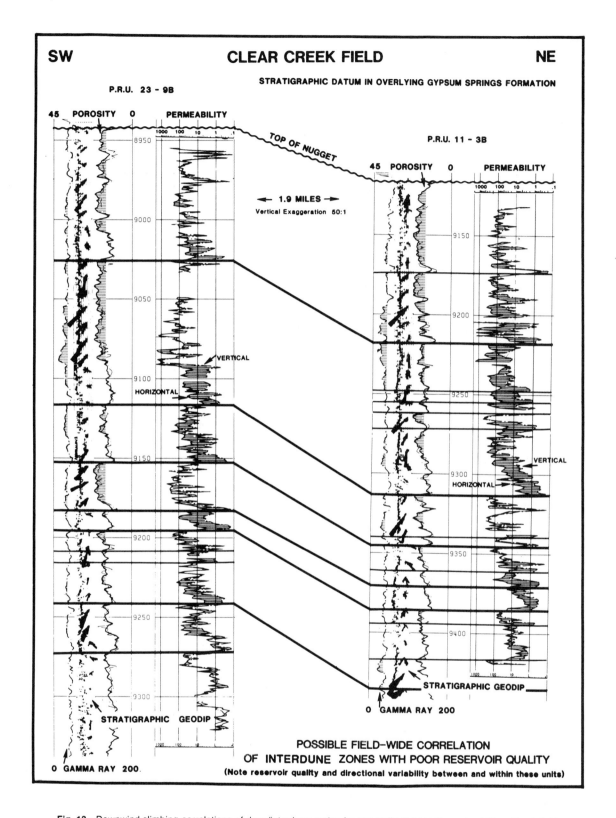

Fig. 13—Downwind-climbing correlations of dune/interdune cycles for two wells at opposite ends of Clear Creek field.

tinuous open fractures enhance matrix permeability locally. Stratigraphic correlation and reservoir modeling of directional properties are possible with the combined use of conventional log, core, and stratigraphic dipmeter data.

Nomenclature

k_h = horizontal air permeability, md
k_v = vertical air permeability, md
ϕ = porosity, %

Acknowledgments

I thank Amoco Production Co. for permission to publish these data and other individuals who shared their expertise and experience, especially Sarah Andrews, Ron Nelson, Dennis Prezbindowski, and Jack Thomas.

References

1. McKee, E.D., Douglas, J.R., and Rittenhouse, S.: "Deformation of Lee-Side Laminae in Eolian Dunes," *Bull.*, GSA (1971) **82**, 359–78.
2. Hunter, R.E.: "Basic Types of Stratification in Small Eolian Dunes," *Sedimentology* (1977) **24**, 361–87.
3. Schenk, C.J.: "Porosity and Textural Characteristics of Eolian Stratification," *Bull.*, AAPG (1981) **65**, 986.
4. Ahlbrandt, T.S. and Fryberger, S.G.: "Sedimentary Features and Significance of Interdune Deposits," Special Publication 31, Soc. of Economic Paleontologists and Mineralogists (1981) 293–314.
5. Pacht, J.A.: "Diagenesis of the Nugget Sandstone: Western Wyoming and North-Central Utah," *29th Field Conference Guidebook*, WY Geological Assn. (1977) 207–19.
6. Picard, M.D.: "Facies, Petrography and Petroleum Potential of Nugget Sandstone (Jurassic), Southwestern Wyoming and Northeastern Utah," *Proc.*, Symposium on Deep Drilling Frontiers, Rocky Mt. Assn. of Geologists (1975) 109–27.
7. Fox, L.: "Porosity and Permeability Reduction in the Nugget Sandstone, Southwestern Wyoming," Master's thesis, U. of Missouri, Columbia (1979).
8. Podsen, D.W.: "Diagenesis of the Nugget Sandstone in the Overthrust Belt of Southwestern Wyoming," Master's thesis, U. of Colorado, Boulder (1981).
9. McKee, E.D.: "Sedimentary Structures in Dunes of the Namib Desert, Southwest Africa," Special Paper 188, GSA (1982) 1–64.
10. Momper, J.A. and Williams, J.A.: "Geochemical Exploration in the Powder River Basin," *Oil and Gas J.* (1979) **77**, 129–34.
11. Warner, M.A.: "Source and Time of Generation of Hydrocarbons in Fossil Basin, Western Wyoming Thrust Belt," *Bull.*, AAPG (1980) **64**, 800.
12. Stearns, D.W.: "Certain Aspects of Fracture in Naturally Deformed Rocks," *NSF Advanced Science Seminar in Rock Mechanics*, R.E. Rieker (ed.), Special Report AD 6693751, Air Force Cambridge Research Laboratories, Bedford, MA (1968) 97–118
13. Pittman, E.D.: "Effect of Fault-Related Granulation on Porosity and Permeability of Quartz Sandstones, Simpson Group (Ordovician), Oklahoma," *Bull.*, AAPG (1981) **65**, 2381–87.
14. Jamison, W.R.: "Laramide Deformation of the Wingate Sandstone, Colorado National Monument: A Study of Cataclastic Flow," PhD dissertation, Texas A&M U., College Station (1979).
15. Ahlbrandt, T.S. and Fryberger, S.G.: "Eolian Deposits in the Nebraska Sand Hills," Geologic and Paleoecologic Studies of the Nebraska Sand Hills Series, USGS, Professional Paper 1120-A (1980).
16. Kocurek, G.: "Significance of Interdune Deposits and Bounding Surfaces in Aeolian Dune Sands," *Sedimentology* (1981) **28**, 753–80.
17. Wilson, I.G.: "Aeolian Bedforms—Their Development and Origins," *Sedimentology* (1972) **19**, 173–210.

SI Metric Conversion Factors

ft $\times$ 3.048*	E−01	= m
in. $\times$ 2.54*	E−02	= mm
mile $\times$ 1.609 344*	E+00	= km

*Conversion factor is exact. **JPT**

Original manuscript received in Society of Petroleum Engineers office July 18, 1982. Paper accepted for publication Feb. 1, 1983. Revised manuscript received April 27, 1983. Paper (SPE 10993) first presented at the 1982 SPE Annual Technical Conference and Exhibition held in New Orleans, Sept. 26–29.

THE AMERICAN ASSOCIATION OF PETROLEUM GEOLOGISTS BULLETIN
V. 51, NO. 10 (OCTOBER, 1967), P. 2122-2132, 9 FIGS., 1 TABLE

INFLUENCE OF GEOLOGIC HETEROGENEITIES ON SECONDARY RECOVERY FROM PERMIAN PHOSPHORIA RESERVOIR, COTTONWOOD CREEK FIELD, WYOMING[1]

JAMES A. McCALEB[2] AND ROBERT W. WILLINGHAM[2]

Denver, Colorado

ABSTRACT

Cottonwood Creek field is on the east side of the Big Horn basin in northwestern Wyoming. It is on a west-southwest-dipping monoclinal surface on the southwest flank of the Hidden Dome anticline, with about 5,500 ft of structural relief through the productive interval. This field is an excellent example of a stratigraphic trap resulting from an updip facies change. The field was discovered in 1953 and has produced more than 22 million bbl of oil from 14,200 productive acres. Most of the field development was complete by 1958.

The clastic reservoir zones interfinger with a red-shale–anhydrite facies on the north, east, and southeast, which formed a stratigraphic trap for hydrocarbons where the impervious strata of the redbed sequence grades into the carbonate facies. Within the productive area outlined by the Cottonwood Creek field, shoaling and bioclastic thickening in the upper Phosphoria occurred together, with a concomitant porosity increase within what generally is a fine-grained dolomite facies. The producing interval contains oölites, lithoclasts, pellets, and residual fossil bioclasts. Vugs are numerous; some appear to be fossil molds.

In 1958, a gas-injection program was begun to maintain reservoir pressure and increase ultimate oil recovery. This program resulted in rapid movement of gas to producing wells and decline in oil-production rates. Water-injection programs were begun in 1959 and 1962, and resulted in the channeling of injection water to producing wells and rapid decrease in oil production. To account for the reservoir performance, all available geological, engineering, and production data were reviewed. This study indicated that reservoir performance is dependent on both the primary (matrix) rock characteristics and a superimposed fracture system. The fracture system was the primary reason for poor injection performance.

The geologic concept of the reservoir was found to correlate well with field performance and resulted in a rational explanation of the poor secondary-recovery performance. The results obtained from this study are being used to change field-operating practices extensively to improve reservoir performance. Initial results have been quite favorable.

INTRODUCTION

The Cottonwood Creek field is on the east side of the Big Horn basin in Washakie County, Wyoming, about 15 mi east of the town of Worland, along State Highway 16, in T. 47 N., R. 90–91 W. (Fig. 1).

Since discovery in 1953, the Cottonwood Creek oil field has produced more than 22 million bbl of oil from the Permian Phosphoria dolomite. As noted by Pedry (1957, p. 823), Cottonwood Creek was the first major discovery in Wyoming which was drilled specifically as a stratigraphic-trap prospect. The field is near the Phosphoria shale to carbonate facies change (Thomas, 1965,

p. 1872, Fig. 13) and the west-southwest flank of the Hidden Dome anticline. The field includes about 5,500 ft of structural relief in the producing oil column. Most of the field development was complete by 1958 after 14,200 acres were proved productive. During June 1958, an up-structure gas-injection program was started to maintain pressure and increase ultimate oil recovery. Gas injections resulted in rapid movement of gas to nearby producing wells, and many oil-producing rates dropped. An up-structure water-injection program was started in 1959 and was expanded to mid-structure during 1962. This program first appeared promising but continued operations resulted in rapid water breakthrough to producing wells.

It became apparent that the reservoir was not responding to secondary recovery in the desired manner. To explain reservoir performance, all available geological, engineering, and production data were reviewed thoroughly. Data indicated that reservoir behavior resulted from both primary (matrix)-rock characteristics and a superim-

[1] Read before the Rocky Mountain Section of the Association, October 23, 1966, Denver, Colo.; the Rocky Mountain Section, American Institute of Mining, Metallurgical, and Petroleum Engineers, May 22, 1967, Casper, Wyo.; and the Rocky Mountain Association of Geologists, June 9, 1967, Denver, Colo. Published by permission of Pan American Petroleum Corp. Manuscript received and accepted, June 5, 1967.

[2] Pan American Petroleum Corp.

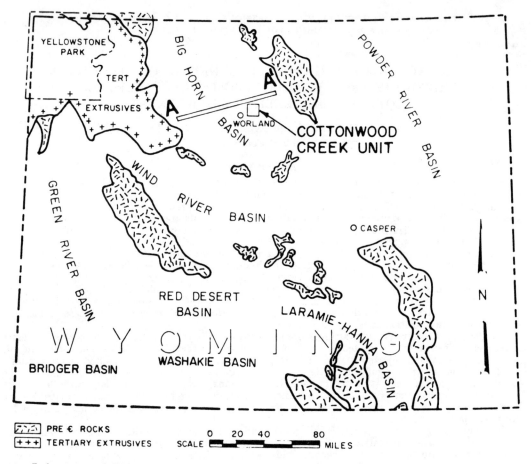

Fig. 1.—Index map of Wyoming showing location of Cottonwood Creek field. Section A-A' is Figure 2.

posed fracture system, as had been noted in the North Dakota Sanish pool by Murray (1965). Poor secondary recovery performance is due primarily to the fracture system.

The geological concept of the reservoir correlates well with field performance and has resulted in a rational explanation of field performance under primary- and secondary-recovery programs. Various methods were used to determine areal geological variations in the reservoir. The fracture system was delineated by core studies, injected-fluid movement, bottom-hole pressure data, well performance, and a structural residual map. Variations of matrix-rock properties were determined by lithologic studies, log analyses, and production data.

The reservoir analysis indicated that extensive changes in reservoir operations would be needed to effect optimum reservoir performance. Major recommendations included relocating water injections from highly fractured areas to less frac-

tured areas, and converting former water-injection wells to producing status. As a result, field-producing rates increased from 2,100 to 3,500 b/d of oil during the first year of revised operating practices.

DEVELOPMENT HISTORY

Cottonwood Creek field was discovered in August 1953, when Pan American Petroleum Corporation completed Unit well No. 1, Sec. 2, T. 47 N., R. 91 W., for an initial potential of 200 b/d of oil. During August 1954 the field-confirmation well was completed for 400 b/d of oil. Development proceeded rapidly and by January 1959 most of the field was defined. A maximum field-producing rate of 14,300 b/d was obtained in May 1958 (Fig. 8). A total of 80 producing wells has been drilled on 14,200 productive acres. Well depths range from 5,000 ft on the eastern side of the field to more than 10,000 ft on the western edge. The Permian Phosphoria dolomite

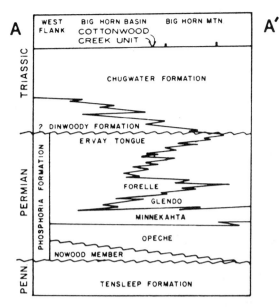

FIG. 2.—West-east correlation section A-A' of Phosphoria Formation (after Pedry, 1957). Location on Figure 1.

reservoir has been developed using 160-acre spacing. The Cottonwood Creek unit was formed prior to drilling the discovery well. Pan American Petroleum Corporation is the unit operator.

GEOLOGY

A thorough review of available geologic information was made after the reservoir failed to respond to fluid-injection programs in the desired manner. Rapid breakthrough of injected fluid to injection wells and lack of significant production response indicated that Phosphoria fracture permeability could be a significant factor contributing to poor fluid-injection performance. In addition, many unexplained producing anomalies indicated that large variations in lithology existed within the reservoir envelope. Examination of geologic data in conjunction with engineering and production data appeared to be the best approach to improve reservoir performance.

Stratigraphy.—Detailed discussions of stratigraphy, depositional environments, and character of the producing zones in the Cottonwood Creek field and surrounding area have been presented, primarily by Pedry (1957; see Fig. 2, this paper), and also by McCue (1953), Harris (1955), Sheldon (1957, 1963), Boyd (1958), and Thomas (1965).

Isopachous-lithofacies studies indicate that during Permian deposition the western half of Wyoming was the site of a large arcuate marine embayment, having a westward connection with the Cordilleran miogeosyncline. Throughout the time of Phosphoria deposition, the boundary between the carbonate facies and red shale-anhydrite facies fluctuated along a north-south trend in the eastern part of the Big Horn basin. The Cottonwood Creek field is on the boundary of the two facies (Fig. 3). The reservoir zones interfinger toward the north, east, and southeast with the red-shale–anhydrite facies. The stratigraphic trap for hydrocarbons is the change from carbonate to redbed facies. In the area outlined by the Cottonwood Creek field, there were shoaling and bioclastic thickening in the upper Phosphoria with a concomitant porosity increase within what is generally a very fine- to fine-grained dolomite.

Within the Cottonwood Creek field limits there is a complete lateral transition, west to east (Fig. 4), from dark, very fine-grained, carbonaceous dolomite, to litho-bioclastic particulate dolomite, to a red-bed–anhydrite sequence. This lateral transition from source beds, to reservoir rocks, to sealing beds is ideal for petroleum accumulations in a stratigraphic trap. The sealing beds are updip facies of the reservoir beds which are an updip facies of the source beds. The transgressive and regressive relations resulted in an intertonguing of source, reservoir, and trap beds. The fine-grained dolomite in the down-structure part of the field is poor reservoir rock. The clastic dolomite in the mid-structure part of the field has more porosity and permeability. Up-structure, the sealing redbeds are impervious and totally nonproductive.

The upper Phosphoria, Ervay Tongue, ranges in thickness from 30 to 100 ft, and contains net-pay intervals from a few feet to 70 ft thick. Porosity through this interval averages 10 percent (Fig. 5) and core permeability values average 16 md. The increased porosity of the producing interval is due to the particulate carbonate material, and is interrupted by discontinuous, irregular zones of dense, very fine-grained dolomite. The producing interval contains oölites, lithoclasts, and pellets, with residual fossil bioclasts of algae, coelenterates, gastropods, brachiopods, bryozoans, and echinoderms. Vugs are evident and some appear to be fossil molds. In general, the rock seems to be genetically related to

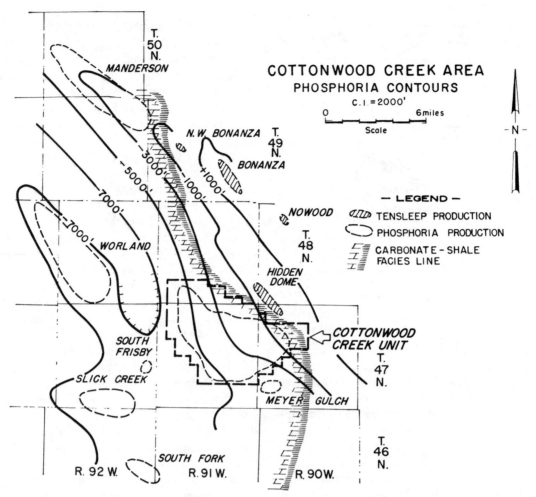

Fig. 3.—Structural map, Cottonwood Creek area. Structural datum is top of Phosphoria. Phosphoria carbonate-shale facies indicated.

some form of organic accumulation, probably more biostromal than biohermal. The organic accumulation is associated with fairly high-energy wave action and intermittent periods during which restricted hypersaline environments probably developed.

The Phosphoria Formation has been assigned to the latest Early Permian, basal Word Formation, by Glenister and Furnish (1961, p. 683), and Nassichuk *et al.* (1965, p. 4, Table I). This correlation is based on the occurrence of a distinctive ammonoid fauna from the Meade Peak Phosphatic Shale Member in the lower part of the Phosphoria Formation along the Idaho-Wyoming border.

Structure.—The Cottonwood Creek field is on the east side of the Big Horn basin in an area of northwest-southeast-trending folds (Fig. 3). Lo-

cally, the field is on the southwest flank of the Hidden Dome anticline. Structural relief across the productive area is about 5,500 ft. A high-angle reverse fault is in the northeast part of the field (Fig. 4). Electric logs indicate upward fault displacement of the Phosphoria Formation on the west side. The displacement ranges from 110 to 350 ft, with maximum amount being at the south end of the fault. This fault displacement is sufficient to juxtapose the 50–70-ft Phosphoria producing interval against the red siltstone of the Chugwater Formation. However, injection and production data indicate that the fault does not form a barrier to fluids throughout its length (Fig. 6).

Fractures.—Core data show that much of the Phosphoria dolomite is extensively fractured, predominantly in vertical planes. Primary well per-

412

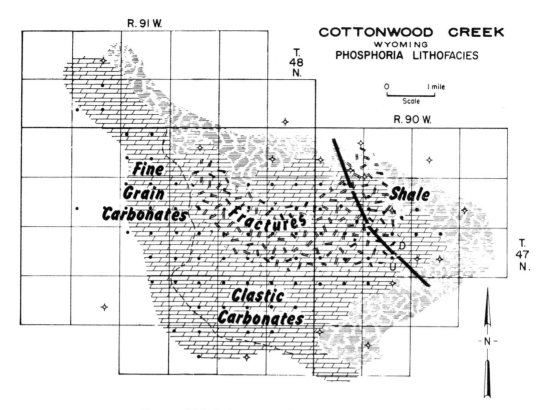

Fig. 4.—Lithofacies map of Cottonwood Creek field.

formance, fluid-injection performance, pressure-buildup data, structural mechanics, and core data indicate that at least one extensive fracture system is in the field. Probable orientation of the fracture system is shown in Figure 4.

Gas- and water-injection performance and core data verify the existence of a fracture system. Up-structure-gas injection was started in June 1958 with almost immediate gas breakthrough to offset wells. The movement of the injected-gas front along the fracture system is additional support for an oriented fracture system. In 1959, water injections were begun at up-structure wells, and in mid-structure wells in 1962. The subsequent movement of injection water is shown in Figure 6. Injected water rapidly followed the same path previously indicated for the injected gas (Fig. 6).

A comparison of core permeability values with inter-well permeability values determined from bottom-hole pressure buildup data also indicates the existence of fractures in the area shown on Figure 4. Where available, the ratio of bottom-hole pressure permeability and core permeability

has been determined. This ratio should be relatively large where fractures are contributing significantly to well capacity. The largest ratios correlated with the areas of indicated maximum fracturing. A comparison of ratios may be seen in the ratio of $603/16 = 38$ in the fractured area and $14/14 = 1$ in the unfractured area.

Therefore, where developed, the fracture systems probably offer the primary means available for fluid movement in the reservoir.

RESERVOIR DATA

Original oil in place in the Phosphoria at Cottonwood Creek is 182,000,000 BSTO (Table I). Produced crude is a black 30° API oil with an original solution gas-oil ratio ranging from 313 to 452 CFPB. Laboratory tests indicate that the reservoir is oil-wet. Primary producing mechanism was a solution-gas drive. Field-productive limits consist of a carbonate to shale facies change toward the northeast and lack of porosity development toward the northwest, southwest, and southeast (Figs. 4, 5). No water-oil contact has been found in the field.

413

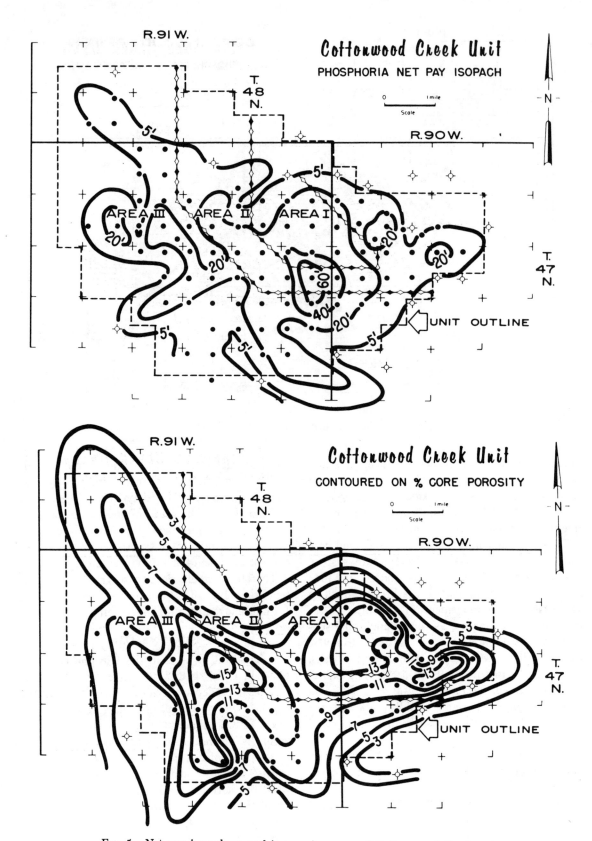

Fig. 5.—Net-pay isopachous and isoporosity maps of Cottonwood Creek unit.

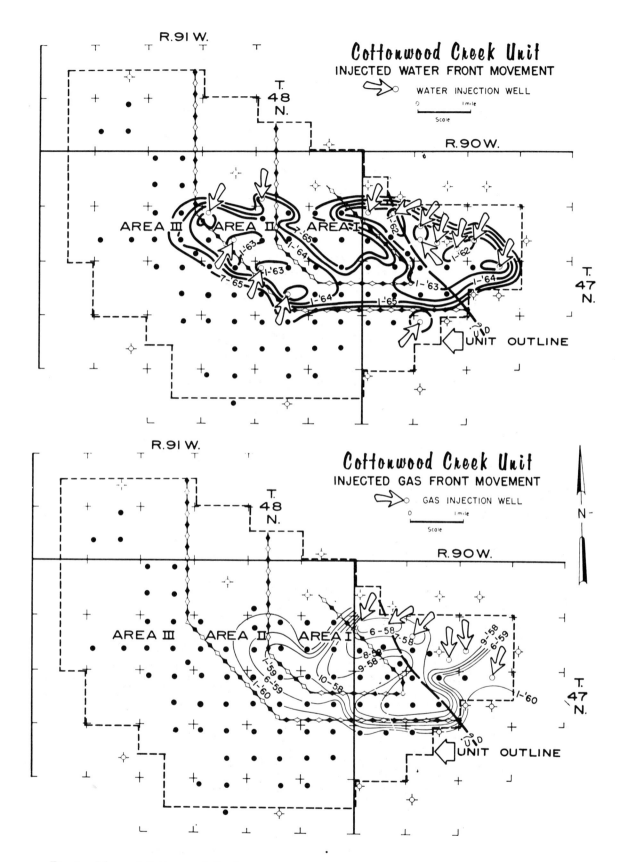

FIG. 6.—Maps of Cottonwood Creek unit showing gas-front and water-front movements with time. Gas-front contours on monthly basis; water-front contours on yearly basis.

In June 1958 a gas-injection program was started to return produced gas to the reservoir. This program was started early in the life of the field to achieve maximum benefit from pressure maintenance. At first, gas injections were confined to the up-structure area. In certain areas, injected gas quickly broke through to surrounding wells, usually with oil-production decreases caused by increased gas-oil ratios. Figure 6 shows the movement of the injected-gas front with time. Gas movement generally followed the area of extensive fractures shown on Figure 4. Field fluid-injection histories are shown on Figure 8.

A pilot water-injection program was started in the up-structure area during October 1959 to determine if waterflood would be a more efficient means of depleting the reservoir. Excellent production response was obtained in the pilot area and up-structure water injections were expanded in 1960 and 1961. Gas injections were moved to the mid-structure area as up-structure water injections were increased. Figure 6 shows the location of up-structure and mid-structure wells used as gas and water injectors.

After increasing reservoir water injection to more than 20,000 bbls per day of water, the producing gas-oil ratios began declining (Fig. 8).

TABLE I. PERTINENT DATA SHEET, PHOSPHORIA
RESERVOIR, COTTONWOOD CREEK UNIT,
WASHAKIE COUNTY, WYOMING

GENERAL

Date of discovery	Aug., 1953
Type structure	Monocline
Trapping mechanism	Stratigraphy
Primary producing mechanism	Soln. gas drive

CRUDE CHARACTERISTICS

Oil gravity, °API	30
Bubble point pressure, PSIA	1,126 to 1,810
Original solution gas-oil ratio—CFPB	313 to 452
Oil viscosity @ reservoir temperature, Cp.	2.75

AREA AND WELLS

Proved productive acres	14,200
Spacing pattern, acres/well	160
Wells	80

FORMATION CHARACTERISTICS

Aver. depth, top of pay, ft.	7,900
Aver. net pay thickness, ft.	20.5
Aver. por., %	10.4
Aver. perm. (core), md.	16

OIL IN PLACE AND PRODUCTION

Orig. oil in place, BSTO	182,000,000
Orig. oil in place, BSTO/AF	625
Cum. prod., August, 1966, BSTO	22,600,000

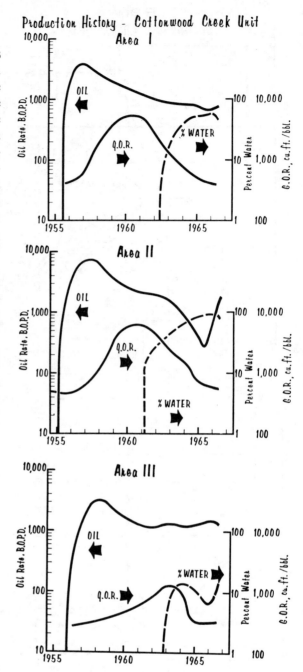

FIG. 7.—Production history, Cottonwood Creek unit Areas I, II, and III.

This resulted in less produced gas for injections and, by late 1964, gas injections were stopped entirely. By that time, water was being injected at a rate of 22,000 bbls per day of water into both the up-structure and mid-structure areas.

The declining gas-oil ratio trend, started in late 1962, indicated that reservoir pressure was in-

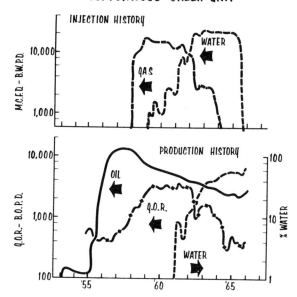

FIG. 8.—Combined injection-production history, Cottonwood Creek unit.

permeability values determined from cores. This indicates that fracture permeability exists. Most Phosphoria cores from *Area 2* wells also are highly fractured. *Area 2* wells have been the most prolific in the field because of the high-capacity fractures. *Area 2* initial-well potentials were the

creasing. This normally results in increased rates of oil production. However, total field oil rates continued to decline (Fig. 8). To explain this behavior, it was necessary to divide the field into three areas as shown on Figures 5 and 6. Geological characteristics and reservoir behavior of each area were determined as follows.

Area 1 has good pay thicknesses and good porosity (Fig. 5) primarily because of deposition of clastic sediments (Fig. 4). Gas movements shown in Figure 6 indicate the possibility that fracturing may be present. However, *Area 1* producing history has shown a definite leveling of the production decline trend since 1962. Good oil recovery by waterflooding is expected in this area and water injections were started at one well in the area during June 1966.

Area 2 reservoir rocks have matrix properties similar to *Area 1, i.e.,* clastic carbonate with good porosity and thick pay sections (Fig. 5). In addition, the area contains the extensive fracture system shown in Figure 4. *Area 2* production characteristics reflect a reservoir in which extensive interconnected fractures have been superimposed on a matrix system with properties as shown in Figure 5. For example, *Area 2* inter-well permeability values determined from bottom-hole pressure data are several times greater than matrix-rock

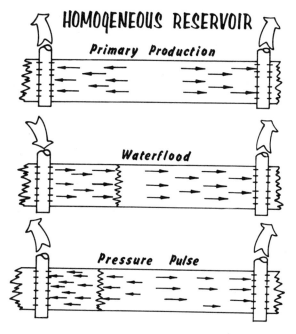

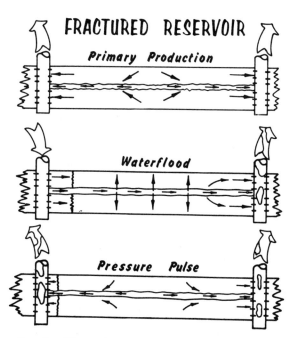

FIG. 9.—Theoretical schematic diagram of production history and characteristics of homogeneous reservoir *versus* heterogeneous fractured reservoir.

417

highest in the field, ranging from 200 to 1,100 bbls per day of oil.

A maximum producing rate of more than 7,000 b/d was reached in early 1958 (Fig. 7). After starting gas injections, several *Area 2* wells had to be shut-in due to the high gas-oil ratios. This has been the primary area in which injected fluids have channeled rapidly to producing wells.

Figure 7 shows that producing water cuts increased rapidly with a concurrent decrease in oil rates after full-scale water injections were started in up-structure areas in late 1961.

Well performance indicated that water injections in *Area 2* were primarily flooding the fractures and that very little oil was being displaced from the matrix part of the reservoir rock. This is indicated on Figure 6 by the rapid migration of injected-fluid fronts through *Area 2*. This performance suggested that recoverable oil could be trapped and bypassed in the rock matrix if *Area 2* water injections were continued. Recent laboratory work by Owens and Archer (1966) indicates that additional oil recovery can be obtained from a heterogeneous rock system similar to *Area 2* by waterflood pressure pulsing. This recovery method alternates injection cycles with producing cycles to use the compressibility of the rock-fluid system to recover oil from areas unswept by injected fluids. This process uses water to increase pressure in the reservoir system. After repressuring, injections are stopped and fluids are withdrawn to depressure the reservoir. Expansion of gas and oil in the rock matrix drives oil to the fracture systems and on to producing wells for recovery. The cycle is repeated after pressure becomes too low to permit desired producing rates. Figure 9 is a series of theoretical idealized schematic diagrams illustrating the expected results of waterflood pressure pulsing in both a homogeneous and a fractured reservoir.

Because all evidence suggested that *Area 2* is highly fractured and that severe bypassing of oil was occurring, it was decided to stop injections to start the depressuring cycle of the pressure-pulse recovery program.

All *Area 2* water injections were stopped during January 1966. Injection wells were backflowed to hasten reservoir depressuring and to determine if water injections had bypassed oil close to injection wells. During January 1966, four up-structure injection wells were produced by flowing. At

the beginning, all wells flowed 100 per cent water. However, within 2–3 weeks, oil cuts began to increase. This was encouraging and, by September 1966, eight former water-injection wells had increased production to total rates of 1,050 b/d of oil and 2,900 b/d of water (73 per cent water cut).

Total water injection into the eight wells was 11.9 million bbls of water. Total injections into individual wells ranged from 400,000 bbls to 1.8 million bbls of water, with an average of 1.5 million bbls of water per well.

Area 2 response due to producing water-injection wells is shown on both the total field curve (Fig. 8) and on the *Area 2* production curve (Fig. 7).

Pressure-pulse production response has been encouraging and increasing oil rates are expected as water cuts decrease and other injection wells are returned to production.

Area 3 is the largest of the three areas. Much of the area is composed of relatively unfractured, fine-grained dolomite with poor reservoir characteristics. Performance of *Area 3* is shown on Figure 7. The area has responded well to mid-structure water injections. The rapid injected-water breakthrough to producing wells which took place in *Area 2* has not occurred in *Area 3* (Fig. 7). Production response and lack of water breakthrough show that *Area 3* is flooding in a conventional manner. This indicates that additional water injections will be beneficial to recovery and producing rates. Additional wells are being converted for this purpose.

CONCLUSIONS

Phosphoria heterogeneity caused by lithologic variations with a superimposed fracture system has resulted in complex reservoir performance in the Cottonwood Creek field. By applyng geological and engineering technology, a rational explanation of response to the secondary-recovery program can be explained.

An understanding of geological conditions has led to major changes in operating practices to improve performance. One major operating change involved starting a pressure-pulsing progam to improve reservoir performance. Based on initial results, it appears that pressure pulsing will increase oil rates and improve ultimate oil recovery.

REFERENCES

Boyd, D. W., 1958, Observations on the Phosphoria reservoir rock, Cottonwood Creek field, Wyoming: Denver, Petroleum Information, Geol. Record, p. 45–53.

Glenister, B. F., and W. M. Furnish, 1961, The Permian ammonoids of Australia: Jour. Paleontology, v. 35, no. 4, p. 673–736.

Harris, L. E., 1955, The Manderson field: Denver, Petroleum Information, Geol. Record, p. 67–73.

McCue, J. J., 1953, Facies changes within the Phosphoria Formation in the southeast portion of the Big Horn basin, Wyoming: Univ. Wyoming, unpubl. M.A. thesis.

Murray, G. H., Jr., 1965, Quantitative fracture study—Sanish pool, McKenzie County, North Dakota (abs.): Am. Assoc. Petroleum Geologists Bull., v. 49, no. 9, p. 1570.

Nassichuk, W. W., W. M. Furnish, and B. F. Glenister, 1965, The Permian ammonoids of Arctic Canada: Canada Geol. Survey Bull. 131, 56 p.

Owens, W. W., and D. L. Archer, 1966, Waterflood pressure pulsing for fractured reservoirs: Jour. Petroleum Technology, v. 17, no. 6, p. 745–752.

Pedry, J. J., 1957, Cottonwood Creek field, Washakie County, Wyoming, carbonate stratigraphic trap: Am. Assoc. Petroleum Geologists Bull., v. 41, no. 5, p. 823–838.

Sheldon, R. P., 1957, Physical stratigraphy of the Phosphoria Formation in northwestern Wyoming: U. S. Geol. Survey Bull. 1042-E, p. 105–185.

——— 1963, Physical stratigraphy and mineral resources of Permian rocks in western Wyoming, U. S. Geol. Survey Prof. Paper 313-B, p. 269.

Thomas, L. E., 1965, Sedimentation and structural development of Big Horn basin: Am. Assoc. Petroleum Geologists Bull., v. 49, no. 11, p. 1867–1877.